ASTRONOMICAL DATA ANALYSIS SOFTWARE AND SYSTEMS X

COVER ILLUSTRATION:

SAO Image ds9 image display tool shows an X-ray image of supernova remnant Cassiopeia A, as presented by Stephen S. Murray in these proceedings (p. 13). SAO Image ds9 is an astronomical imaging and data visualization application that supports FITS images and binary tables, multiple frame buffers, region manipulation, and many scale algorithms and color maps. It provides easy communication with external analysis tasks and is highly configurable and extensible.

A SERIES OF BOOKS ON RECENT DEVELOPMENTS IN
ASTRONOMY AND ASTROPHYSICS

Publisher

THE ASTRONOMICAL SOCIETY OF THE PACIFIC
390 Ashton Avenue, San Francisco, California, USA 94112-1722
Phone: (415) 337-1100   Fax: (415) 337-5205
E-Mail: catalog@aspsky.org   Web Site: www.aspsky.org

ASP CONFERENCE SERIES - EDITORIAL STAFF
Managing Editor: D. H. McNamara   LaTeX-Computer Consultant: T. J. Mahoney
Associate Managing Editor: J. W. Moody   Production Manager: Enid L. Livingston

PO Box 24453, 211 KMB, Brigham Young University, Provo, Utah, 84602-4463
Phone: (801) 378-2111   Fax: (801) 378-4049   E-Mail: pasp@byu.edu

ASP CONFERENCE SERIES PUBLICATION COMMITTEE:
Alexei V. Filippenko   Geoffrey Marcy
Ray Norris   Donald Terndrup
Frank X. Timmes   C. Megan Urry

A listing of all other ASP Conference Series Volumes and IAU Volumes
published by the ASP is cited at the back of this volume

ASTRONOMICAL SOCIETY OF THE PACIFIC
CONFERENCE SERIES

Volume 238

ASTRONOMICAL DATA ANALYSIS
SOFTWARE AND SYSTEMS X

Proceedings of a meeting held at
Boston, Massachusetts, U.S.A.
12- 15 November 2000

Edited by

F. R. Harnden, Jr.
Smithsonian Astrophysical Observatory, Cambridge, Massachusetts, USA

Francis A. Primini
Smithsonian Astrophysical Observatory, Cambridge, Massachusetts, USA

and

Harry E. Payne
Space Telescope Science Institute, Baltimore, Maryland, USA

© 2001 by Astronomical Society of the Pacific. All Rights Reserved

No part of the material protected by this copyright notice may be reproduced or utilized in any form or by any means – graphic, electronic, or mechanical including photocopying, taping, recording or by any information storage and retrieval system, without written permission from the publisher.

Library of Congress Cataloging in Publication Data
Main entry under title

ISSN: 1080-7926
ISBN: 1-58381-075-7

ASP Conference Series - First Edition

Printed in United States of America by Sheridan Books, Chelsea, Michigan

Contents

Preface xv
Participant List xix
Conference Photograph xxxiii

Part 1. Enabling Technologies for Astronomy

The National Virtual Observatory 3
    A. S. Szalay

Chandra X-Ray Observatory - First Year Science Highlights 13
    S. S. Murray

Chandra X-ray Observatory Operations 22
    R. J. Brissenden

The SIRTF Science Center Is Alive And Well And Living In Pasadena . . 32
    W. B. Green

Advanced Architecture for the Infrared Science Archive 36
    G. B. Berriman, N.-M. Chiu, J. Good, T. Handley, A. Johnson, M. Kong,
    S. Monkewitz, S. W. Norton, and A. Zhang

A Parallel, Distributed Archive Template for the VO 40
    A. R. Thakar, P. Z. Kunszt, and A. S. Szalay

Tools for Coordinating Planning Between Observatories 44
    J. Jones, L. Maks, M. Fishman, V. Grella, U. Kerbel, D. Misra, and
    V. Pell

The Scientist's Expert Assistant Simulation Facility 48
    K. R. Wolf, C. Li, J. Jones, D. Matusow, S. Grosvenor, and A. Koratkar

Data Information Services at the Infrared Science Archive 52
    J. Good, M. Kong, G. B. Berriman, N.-M. Chiu, T. Handley, A. Johnson,
    W.-P. Lee, and J. Ma

The CDS Information Hub 56
    F. Genova, F. Bonnarel, P. Dubois, D. Egret, P. Fernique, S. Lesteven,
    F. Ochsenbein, M. Wenger, and P. F. Ortiz

The PyRAF Graphics System 59
    M. D. De La Peña, R. L. White, and P. Greenfield

AstroRoute 63
    P. Fernique and D. Durand

Hera: Building a Web-Based Analysis Environment at the HEASARC . . 67
    C. W. Heikkila, W. D. Pence, T. A. McGlynn, P. Wilson, J. Peachey,
    N. Gan, E. Sabol, and K. A. Arnaud

MAISON: A Web Service of Creating Composite Images On-the-fly for
    Pointing and Survey Observational Images 70
    M. Watanabe, K. Aoki, A. Miura, N. Yasuda, and S. Uno

Contents

The ALADIN Survey Integrator 74
    F. Bonnarel, P. Fernique, F. Genova, O. Bienaymé, and D. Egret

Visual Exploration of Astronomical Documents 78
    S. Lesteven, P. Poinçot, and F. Murtagh

The CARMA Data Viewer 82
    M. W. Pound, R. Hobbs, and S. Scott

The Design of Solar Web, a Web Tool for Searching in Heterogeneous Web-based Solar Databases 86
    I. Scholl

Quick-look Applet Spectrum 2 90
    M. Dolensky, B. Pirenne, S. Binegar, M. Brandt, N. Gaffney, C. Arviset, and J. Hernandez

The New JCMT Holographic Surface Mapping System - Implementation . 93
    I. A. Smith, F. Baas, F. Olivera, N. P. Rees, Y. Dabrowski, R. Hills, J. Richer, H. Smith, and B. Ellison

The 4-m International Liquid Mirror Telescope Project (ILMT) 97
    J. Poels, E. Borra, J. F. Claeskens, C. Jean, J. Manfroid, F. Montfort, O. Moreau, Th. Nakos, J. Surdej, J. P. Swings, E. van Dessel, and B. Vangeyte

Part 2. Education & Public Outreach

Hands-On Universe: A Global Program for Education and Public Outreach in Astronomy 103
    M. Boër, C. Thiébaut, H. Pack, C. Pennypaker, M. Isaac, A.-L. Melchior, S. Faye, and T. Ebisuzaki

Space Science Education Resource Directory 107
    C. A. Christian and K. Scollick

Hands-On TAROT: Intercontinental Use of the TAROT for Education and Public Outreach 111
    M. Boër, C. Thiébaut, A. Klotz, G. Buchholtz, A.-L. Melchior, C. Pennypaker, M. Isaac, and T. Ebisuzaki

A Japan-U.S. Educational Collaboration: Using the Telescopes in Education (TIE) Program via Intelsat 115
    P. L. Shopbell, E. Hsu, N. Kadowaki, and G. Clark

Using the Net for Education and Outreach in Astronomy 119
    L. Benacchio

Part 3. Sky Surveys

The Uccle Direct Astronomical Plate Archive Centre UDAPAC—A New
International Facility for Inherited Observations 125
    J.-P. De Cuyper, E. Elst, H. Hensberge, P. Lampens, T. Pauwels,
    E. van Dessel, N. Brosch, R. Hudec, P. Kroll, and M. Tsvetkov

Providing Improved WCS Facilities Through the Starlink AST and NDF
Libraries 129
    D. S. Berry

A System for Low-Cost Access to Very Large Catalogs 133
    K. Kalpakis, M. Riggs, M. Pasad, V. Puttagunta, and J. Behnke

An Observer's View of the ORAC System at UKIRT 137
    G. S. Wright, A. B. Bridger, D. A. Pickup, M. Tan, M. Folger, F. Economou,
    A. J. Adamson, M. J. Currie, N. P. Rees, M. Purves, and R. D. Kackley

ASTROVIRTEL: Accessing Astronomical Archives as Virtual Telescopes   141
    F. Pierfederici, P. Benvenuti, A. Micol, B. Pirenne, and A. Wicenec

PICsIM – the INTEGRAL/IBIS/PICsIT Observation Simulation Tool for
Prototype Software Evaluation 144
    J. B. Stephen and L. Foschini

An Enhanced Data Flow Scheme to Boost Observatory Mine-ability and
Archive Interoperability 148
    A. Micol and P. Amico

INES Version 3.0: Functionalities and Contents 152
    E. Solano, R. González-Riestra, A. Talavera, F. Rodríguez, A. J. de la
    Fuente, I. Skillen, J. D. Ponz, and W. Wamsteker

INES: The Next Generation Astronomical Data Distribution System . . . 156
    R. González-Riestra, E. Solano, A. Talavera, F. Rodríguez, J. García,
    J. Martínez, B. Montesinos, L. Sanz, A. J. de la Fuente, I. Skillen,
    J. D. Ponz, and W. Wamsteker

The OaPd System for Web Access to Large Astronomical Catalogues . . 160
    L. Benfante, A. Volpato, A. Baruffolo, and L. Benacchio

The Submillimeter Array Data-Handling System 164
    J.-H. Zhao and T. Tsutsumi

See Spike Run... Run, Spike, Run 168
    L. Zimmerman

CIA V5.0—the Legacy Package for ISOCAM Interactive Analysis 170
    S. Ott, R. Gastaud, B. Ali, M. Delaney, M.-A. Miville-Deschênes,
    K. Okumura, M. Sauvage, and S. Guest

The Multimission Archive at the Space Telescope Science Institute 174
    P. Padovani, D. Christian, M. Donahue, C. Imhoff, T. Kimball, K. Levay,
    M. Postman, M. Smith, and R. Thompson

STPOA—The New Pipeline Package for the HST Post-Operational
Archive 178
    A. Alexov, P. Bristow, F. Kerber, and M. Rosa

The SuperCOSMOS Sky Surveys 182
    M. A. Read and N. C. Hambly

Part 4. Software Development Methodologies & Technologies

SOFIA's CORBA Experiences: Instances of Software Development 189
    J. Graybeal, R. Krzaczek, and J. Milburn

Software Engineering Practices for the ESO VLT Programme 199
    G. Filippi, P. Sivera, and F. Carbognani

Exploiting VSIPL and OpenMP for Parallel Image Processing 209
    J. Kepner

CORBA as an Interoperability Tool for Astronomy 213
    M. Wenger and L. Frisée

Specifics on a XML Data Format for Scientific Data 217
    E. Shaya, B. Thomas, and C. Cheung

Declarative Metadata Processing with XML and Java 221
    D. R. Guillaume and R. Plante

Funtools: An Experiment with Minimal Buy-in Software 225
    E. Mandel, S. S. Murray, and J. B. Roll

Embedded Astrophysics Query Support Using Informix Datablades 229
    A. Zhang and T. Handley

msg: A Message-Passing Library 233
    J. B. Roll and J. Mandel

DASH—Distributed Analysis System Hierarchy 237
    M. Yagi, Y. Mizumoto, M. Yoshida, G. Kosugi, T. Takata, R. Ogasawara,
    Y. Ishihara, Y. Morita, H. Nakamoto, and N. Watanabe

ADASS Web Database XML Project 241
    M. I. Barg, E. B. Stobie, A. J. Ferro, and E. J. O'Neil

DISCoS—Detector-Independent Software for On-Ground Testing and Calibration of Scientific Payloads Using the ESA Packet Telemetry and
Telecommand Standards 245
    F. Gianotti and M. Trifoglio

CAOS Simulation Package 3.0: an IDL-based Tool for Adaptive Optics
Systems Design and Simulations 249
    M. Carbillet, L. Fini, B. Femenía, A. Riccardi, S. Esposito, É. Viard,
    F. Delplancke, and N. Hubin

The CAOS Application Builder 253
  L. Fini, M. Carbillet, and A. Riccardi

Software Fault Tolerance for Low-to-Moderate Radiation Environments . 257
  R. Sengupta, J. D. Offenberg, D. J. Fixsen, D. S. Katz, P. L. Springer,
  H. S. Stockman, M. A. Nieto-Santisteban, R. J. Hanisch, and J. C. Mather

Extending the XImtool Display Server Using Components 261
  M. Fitzpatrick and D. Tody

Linux BoF 265
  P. N. Daly

Part 5. Science Data Pipelines

The SDSS Imaging Pipelines 269
  R. H. Lupton, J. E. Gunn, Z. Ivezić, G. R. Knapp, and S. Kent

The Networked Telescope: Progress Toward a Grid Architecture for
    Pipeline Processing 279
  R. Plante, D. Mehringer, D. R. Guillaume, and R. Crutcher

The ESO Imaging Survey Project: Status and Pipeline Software 283
  R. N. Hook, S. Arnouts, C. Benoist, L. da Costa, R. Mignani, C. Rité,
  M. Schirmer, R. Slijkhuis, B. Vandame, and A. Wicenec

GIGAWULF: Powering the Isaac Newton Group's Data Pipeline 287
  R. Greimel, N. A. Walton, D. C. Abrams, M. J. Irwin, and J. R. Lewis

Using OPUS to Perform HST On-The-Fly Re-Processing (OTFR) 291
  M. S. Swam, E. Hopkins, and D. A. Swade

HST Data Flow with On-The-Fly Reprocessing 295
  D. A. Swade, E. Hopkins, and M. S. Swam

Automated Reduction and Calibration of SCUBA Archive Data
    Using ORAC-DR 299
  T. Jenness, J. A. Stevens, E. N. Archibald, F. Economou, N. Jessop,
  E. I. Robson, R. P. J. Tilanus, and W. S. Holland

The Chandra Automatic Processing/Archive Interface 303
  S. Subramanian and D. A. Plummer

A High Throughput Photometric Pipeline 306
  M. L. Reid, D. J. Sullivan, and R. J. Dodd

Interactive Analysis and Scripting in CIAO 2.0 310
  S. Doe, M. Noble, and R. Smith

Infrared Spectroscopy Data Reduction with ORAC-DR 314
  F. Economou, T. Jenness, B. Cavanagh, G. S. Wright, A. B. Bridger,
  T. H. Kerr, P. Hirst, and A. J. Adamson

Infrared Imaging Data Reduction Software and Techniques 317
   C. N. Sabbey, R. G. McMahon, J. R. Lewis, and M. J. Irwin

ADS's Dexter Data Extraction Applet 321
   M. Demleitner, A. Accomazzi, G. Eichhorn, C. S. Grant, M. J. Kurtz,
   and S. S. Murray

The OPUS CORBA Blackboards and the New OPUS Java Managers . . 325
   W. W. Miller III and J. F. Rose

An Extragalactic Point Source Simulator for SIRTF Pipeline Testing . . . 329
   S. Kolhatkar, F. Fang, and M. M. Moshir

Interfacing Real-time Linux and LabVIEW 333
   P. N. Daly

The Basic Calibrated Data Processing Pipeline for SIRTF IRS 337
   F. Fang, R. Narron, C. Waterson, J. Li, and M. M. Moshir

Part 6. Software History

Astronomical Software—A Review 343
   K. Shortridge

Reflections on a Decade of ADASS 349
   R. A. Shaw, E. B. Stobie, and M. I. Barg

The Evolution of GIPSY—or the Survival of an Image Processing System 358
   M. G. R. Vogelaar and J. P. Terlouw

Part 7. Software Applications

The LWS Interactive Analysis (LIA) Package 365
   S. J. Chan, S. D. Sidher, B. M. Swinyard, M. G. Hutchinson, S. Lord,
   S. Molinari, S. J. Unger, and S. J. Leeks

Prototypical Operations Support Tools for NASA Interferometer Missions:
   Applications to Studies of Binary Stars Using the Palomar Testbed
   Interferometer 369
   R. J. Bambery and C. Backus

Correction of Systematic Errors in Differential Photometry 373
   J. Manfroid, P. Royer, G. Rauw, and E Gosset

Projective Transform Techniques to Reconstruct the 3-D Structure and
   the Temporal Evolution of Solar Polar Plumes 377
   A. Llebaria, A. Thernisien, and P. Lamy

Pile-up on X-ray CCD Instruments 381
   J. Ballet

ATV: An Image-Display Tool for IDL 385
   A. J. Barth

A New Field-Matching Method for Astronomical Images 388
  C. Thiébaut and M. Boër

A System for Web-based Access to the HSOS Database 392
  G. Lin

Uniform Data Sampling: Noise Reduction & Cosmic Rays 396
  J. D. Offenberg, D. J. Fixsen, M. A. Nieto-Santisteban, R. Sengupta,
  J. C. Mather, and H. S. Stockman

New Tools for the Analysis of ISOPHOT P32 Mapping Data in PIA . . . 400
  C. Gabriel and R. J. Tuffs

AIRY: Astronomical Image Restoration in interferometrY 404
  S. Correia, M. Carbillet, L. Fini, P. Boccacci, M. Bertero, A. Vallenari,
  A. Richichi, and M. Barbati

Parallelization of Widefield Imaging in AIPS++ 408
  K. Golap, A. Kemball, T. Cornwell, and W. Young

Ionospheric Corrections in AIPS++ 411
  O. Smirnov

Redesign and Reimplementation of XSPEC 415
  B. Dorman and K. A. Arnaud

A Flexible Object Oriented Design for Page Formating 419
  N. R. Adams-Wolk

The Stellar Spectra Acquisition, Reduction, and Archiving Systems at the
  Ondřejov Observatory 2-meter Telescope 423
  P Škoda, J. Honsa, and M. Šlechta

HST NICMOS Residual Bias Removal Techniques 427
  H. Bushouse

Deconvolution of "Drizzled" and Rotated HST/WFPC2 Images: Faint-
  object Photometry in Crowded Fields 431
  R. F. Butler, A. Golden, and A. Shearer

The Chandra X-ray Observatory PSF Library 435
  M. Karovska, S. J. Beikman, M. S. Elvis, J. M. Flanagan, T. Gaetz,
  K. J. Glotfelty, D. Jerius, J. C. McDowell, and A. H. Rots

On the Fly Bad Pixel Detection for the Chandra X-ray Observatory's
  Aspect Camera 439
  M. Cresitello-Dittmar, T. L. Aldcroft, and D. Morris

The Sliding-Cell Detection Program for Chandra X-ray Data 443
  T. Calderwood, A. Dobrzycki, H. Jessop, and D. E. Harris

SLIM: A Program to Simulate the ACS Spectroscopic Modes 447
  N. Pirzkal, A. Pasquali, R. N. Hook, J. R. Walsh, R. A. E. Fosbury,
  W. Freudling, and R. Albrecht

Automated Spectral Extraction for High Multiplexing MOS and IFU
    Observations 451
    M. Scodeggio, A. Zanichelli, B. Garilli, O. Le Fèvre, and G. Vettolani

The VIMOS Mask Preparation Software 455
    D. Bottini, B. Garilli, and L. Tresse

IFU Data Products and Reduction Software 459
    A. Allan, J. Allington-Smith, J. Turner, R. Johnson, B. Miller, and
    F. G. Valdes

The ST5000: An Attitude Determination System with Low-Bandwidth
    Digital Imaging 463
    J. W. Percival and K. H. Nordsieck

An Object Oriented Design for Monitoring the Chandra Science Instru-
    ment X-ray Background 467
    J. G. Petreshock, S. J. Wolk, M. Cresitello-Dittmar, and T. Isobe

Kalman Filtering in Chandra Aspect Determination 471
    R. Hain, T. L. Aldcroft, R. A. Cameron, M. Cresitello-Dittmar, and
    M. Karovska

The Chandra Automatic Data Processing Infrastructure 475
    D. A. Plummer and S. Subramanian

The FITS Embedded Function Format 479
    A. H. Rots, J. C. McDowell, M. Wise, X. H. He, and P. E. Freeman

New Elements of Sherpa, CIAO's Modeling and Fitting Tool 483
    P. E. Freeman, S. Doe, and A. Siemiginowska

Converting FITS into XML: Methods and Advantages 487
    B. Thomas, E. Shaya, and C. Cheung

SVDFIT: An IRAF Task for Eigenvector Sky Subtraction 491
    D. J. Mink and M. J. Kurtz

BIMA Xfiles: Empowering the Observer with Tcl/Tk Applications .. 495
    T. Yu

Immersive 4-D Interactive Visualization of Large-Scale Simulations .. 499
    P. J. Teuben, P. Hut, S. Levy, J. Makino, S. McMillan,
    S. Portegies Zwart, M. Shara, and C. Emmart

An Integrated Procedure for Tree N-body Simulations: FLY and
    AstroMD 503
    U. Becciani, V. Antonuccio-Delogu, F. Buonomo, and C. Gheller

ACE: Astronomical Cataloging Environment 507
    F. G. Valdes

Automated Photometric Calibration Software For IRAF 511
    L. E. Davis

CCD Charge Shuffling Improvements for ICE 515
    R. L. Seaman

Enhancements of MKRMF 518
    X. H. He, M. Wise, and K. J. Glotfelty

Development of Radio Astronomical Data Reduction Software
    NEWSTAR 522
    M. Ikeda, K. Nishiyama, M. Ohishi, and K. Tatematsu

ESO C Library for an Image Processing Software Environment (eclipse) . 525
    N. Devillard

Developing a Wavelet CLEAN Algorithm for Radio-Interferometer
    Imaging 529
    S. Horiuchi, S. Kameno, and M. Ohishi

Author Index 533
Subject Index 538

Preface

This volume of the ASP Conference Series contains papers presented at the tenth annual conference for Astronomical Data Analysis Software and Systems, known as ADASS X or ADASS 2000. The ADASS Conference Series provides a forum for scientists and computer specialists concerned with development of software for enabling technologies for Astronomy, sky surveys, science data pipelines, software development methodologies and technologies, education and public outreach, and software history.

ADASS X was held at the Swissôtel in Boston, Massachusetts, from November 12 to 15, 2000. The conference was hosted by the Smithsonian Astrophysical Observatory (SAO). There were 287 registered participants at the meeting, topping last year's attendance by one and again setting a record. Those attending included 94 people representing 18 countries outside the United States.

1. Conference Overview

The Conference Reception convened on the evening of Sunday, November 12, 2000, and over the next three days, attendees participated in 22 BoFs and demos, listened to 35 (16 invited, 19 contributed) oral presentations distributed in nine sessions, and viewed 115 posters. The manuscript submission rate of 72% was about the same as that last year.

The banquet, on Tuesday evening, was held at the New England Aquarium on Boston Harbor and featured traditional New England fare served not only by caterers to the ADASS participants, but also by Aquarium divers to the fish in the three-story tall central tank!

2. Proceedings Overview

Approximately 120 papers are presented in these Proceedings. The editors have re-arranged them (from the order in which they were presented during the conference) into seven subject-driven sections. In each section, invited oral presentations are followed by contributed talks, demos, posters, and BoF summaries.

Sixteen invited talks were presented at the conference; unfortunately, not all of these speakers were able to submit manuscripts. The theme of "enabling technologies" was addressed by A. Szalay (reviewing the genesis, challenges, and opportunities of the virtual observatory), J. Spyromilio (commissioning the VLT), and J. Tarter and D. Werthimer (the science and technical challenges of the SETI project). S. Murray and R. Brissenden described the early scientific successes and operational systems for the Chandra X-ray Observatory, and B. Glendenning gave an overview of the software engineering challenges facing the ALMA telescope project. Key papers on software development technologies included J. Graybeal (CORBA), G. Filippi (software development methodologies employed at ESO), C. Kesselman (computational grids), J. Manuel Filgueira (distributed objects), and V. Yodaiken (real-time Linux). In the area of science data processing pipelines, R. Lupton described the image reduction pipeline for the Sloan Digital Sky Survey, and R. Cutri described the 2MASS project under the topic of sky surveys. K. Shortridge gave an insightful and amusing historical perspective of astronomical software development, and D. Shaw gave an historical retrospective on ten years of ADASS conferences. The oral program also included a number of excellent contributed papers, and daily poster sessions (with over 120 posters presented in total) included presentations on all areas of software development and systems. Fourteen groups demonstrated their latest software systems and astronomical information services.

3. The People Behind the Conference

The ADASS 2000 Program Organizing Committee consisted of Rudi Albrecht (ST-ECF/ESO), Dick Crutcher (UIUC/NCSA), Daniel Durand (NRC/HIA), Brian Glendenning (NRAO), Tom Handley (IPAC) F. Rick Harnden (SAO), Richard Hook (ESO), Gareth Hunt (NRAO), Glenn Miller (STScI), Jan Noordam (NFRA), Dick Shaw (STScI), Betty Stobie (Steward Obs.) Doug Tody (NOAO), Christian Veillet (CFHT), and Patrick Wallace (RAL). In conjunction with the conference, Dick Shaw (STScI) assumed the responsibilities of Chair from F. Rick Harnden (SAO), who had served in that capacity since February, 1997.

The Local Organizing Committee consisted of Frank Primini (Chair), Patricia Buckley, Paul Grant, Rick Harnden, Michelle Henson, Susan Tuttle, and Wendy Roberts. This core team was assisted by numerous other employees and resources made available from the hosting institution, SAO. Among these are: Steve Beikman, Christopher Dingle, Elizabeth Bohlen, Kathleen Campbell, Guenther Eichhorn, Thomas Fine, Lisa Paton, Dan Rehner, and James Simms.

Finally, we gratefully acknowledge the technical and financial support of the conference sponsors: Akibia, Inc., Canada France Hawaii Telescope (CFHT), Infrared Processing and Analysis Center/SIRTF Science Center (IPAC/SSC). National Aeronautics and Space Administration (NASA), National Optical Astronomy Observatory (NOAO), National Radio Astronomy Observatory (NRAO), National Science Foundation (NSF), Smithsonian Astrophysical Observatory (SAO), and Space Telescope Science Institute (STScI).

4. ADASS Information

Details about ADASS 2000 are available on the Web at:

http://hea-www.harvard.edu/adass .

Past conferences are listed in the ADASS webpage:

http://www.adass.org.

ADASS 2001 will be hosted by the Canadian Astronomy Data Centre, operated by the Herzberg Institute of Astrophysics for the National Research

Council of Canada, in Victoria, BC, Canada at the Victoria Conference Centre, from September 30 through October 3, 2000. For further information, visit their website (http://cadcwww.hia.nrc.ca/adass_2001/) or send mail to adass@hia.nrc.ca.

The ADASS X Editors

F. Rick Harnden, Jr., Frank A. Primini
Smithsonian Astrophysical Observatory

Harry E. Payne
Space Telescope Science Institute

May 2001

Participant List

Mark Abernathy, Space Telescope Science Institute, 3700 San Martin Drive, Baltimore, MD, 21218, U.S.A. (abernathy@stsci.edu)

Alberto Accomazzi, SAO, 60 Garden Street, Cambridge, MA, 02138, U.S.A. (aaccomazzi@cfa.harvard.edu)

Nancy Adams-Wolk, SAO, Mail Stop 81, 60 Garden Street, Cambridge, MA, 02138, U.S.A. (nadams@head-cfa.harvard.edu)

Rudolf Albrecht, ST-ECF, Karl-Schwarzschild-Str. 2, Garching, D-85748, Germany (ralbrech@eso.org)

Anastasia Alexov, European Southern Observatory, Karl-Schwarzschild-Str. 2, Muenche Bavaria, D-85748, Germany (aalexov@eso.org)

Alasdair Allan, Starlink (Keele), Dept. of Physics, Keele University, Keele, Staffordshire, ST5 5BG, U.K. (aa@astro.keele.ac.uk)

Troy Ames, NASA/GSFC, Code 588, Greenbelt, MD, 20771, U.S.A. (Troy.Ames@gsfc.nasa.gov)

Hajime Baba, Dept. of Astronomy, Kyoto University, Kitashirakawa-Oiwake-cho, Sakyo-ku, Kyoto, Kyoto, 606-8502, Japan (baba@kusastro.kyoto-u.ac.jp)

Jean Ballet, DSM/DAPNIA/SAp, CEA Saclay, Bt 709, Gif/Yvette Cedex, F-91191, France (jballet@cea.fr)

Raymond Bambery, JPL/Caltech, 4800 Oak Grove Drive, MS: 168-414, Pasadena, CA, 91109, U.S.A. (Raymond.J.Bambery@jpl.nasa.gov)

Paul Barrett, STScI, 3700 San Martin Drive, Baltimore, MD, 21218, U.S.A. (barrett@stsci.edu)

Aaron Barth, SAO, 60 Garden St., Mail Stop 20, Cambridge, MA, 02138, U.S.A. (abarth@cfa.harvard.edu)

Ugo Becciani, Osservatorio Astrofisico Catania, via S. Sofia, 78, Catania, I-95125, Italy (ube@sunct.ct.astro.it)

Stephen Beikman, SAO, 60 Garden Street, MS-70, Cambridge, MA, 02138, U.S.A. (sbeikman@cfa.harvard.edu)

Stephane Beland, CASA/UC - Boulder, 1255 38th Street, Campus Box 593, Boulder, CO, 80309, U.S.A. (sbeland@colorado.edu)

David Bell, NOAO, 950 N. Cherry Ave., P.O. Box 26732, Tucson, AZ, 85726, U.S.A. (dbell@noao.edu)

Lucio Benfante, Osservatorio Astronomico - Padova, via Monteriondo, 14, Monteforte d'Alpone, VR, 37032, Italy (benfante@dei.unipd.it)

Kevin Bennett, ESA, SCI-SA ESTEC, Keplerlaan 1, Noordwijk, 2200 AG, Netherlands (kbennett@astro.estec.esa.nl)

G. Bruce Berriman, IPAC, Caltech, Caltech Mail Stop 100-22, 770 South Wilson Avenue, Pasadena, CA, 91125, U.S.A. (gbb@ipac.caltech.edu)

David Berry, Starlink, Centre for Astrophysics, University of Central Lancashire, Preston, Lancs, PR1 2HE, U.K. (dsb@sa1.star.uclan.ac.uk)

Martin Bly, Starlink, RAL, UK, Rutherford Appleton Laboratory, Chilton, DIDCOT, Oxfordshire, OX11 0QX, U.K. (bly@star.rl.ac.uk)

Michel Boër, CESR/CNRS, 9, ave. du Colonel Roche, BP 4346, Toulouse Cedex, 31028, French Guiana (Michel.Boer@cesr.fr)

Participant List

Stefan Bogun, European Southern Observatory, Karl-Schwarzschild-Str. 2, Garching, 85748, Germany (Stefan.Bogun@eso.org)

Kirk Borne, Raytheon ITSS, GSFC, Code 630, Goddard Space Flight Center, Greenbelt, MD, 20771, U.S.A. (Kirk.Borne@gsfc.nasa.gov)

Todd Boroson, NOAO, P.O. Box 26732, Tucson, AZ, 85726-6732, U.S.A. (tboroson@noao.edu)

Dario Bottini, Istituto di Fisica Cosmica G.Occhialini, Via Bassini,15, Milano, I-20133, Italy (bottini@ifctr.mi.cnr.it)

Alan Bridger, UK Astronomy Technology Centre, Royal Observatory, Blackford Hill, Edinburgh, Midlothian, EH9 3HJ, U.K. (ab@roe.ac.uk)

John Bright, SAO, 60 Garden St., MS 21, Cambridge, MA, 02138-1516, U.S.A. (jbright@head-cfa.harvard.edu)

Roger Brissenden, SAO, 60 Garden Street, MS-6, Cambridge, MA, 02138, U.S.A. (rbrissenden@cfa.harvard.edu)

Paul Bristow, ST-ECF, Karl-Schwarzschild-Straße 2, Garching bei Muenchen, Munich Bavaria, D-85748, Germany (bristowp@eso.org)

Patricia Buckley, SAO, 60 Garden Street, MS 2, Cambridge, MA, 02138, U.S.A. (tricia@cfa.harvard.edu)

Howard Bushouse, Space Telescope Science Institute, 3700 San Martin Dr, Baltimore, MD, 21218, U.S.A. (bushouse@stsci.edu)

Ivo Busko, STScI, 3700 San Martin Drive, Baltimore, MD, 21218, U.S.A. (busko@stsci.edu)

Raymond Butler, National University of Ireland, Galway, University Road, Galway,, Republic Of Ireland (ray@itc.nuigalway.ie)

Tom Calderwood, SAO, 60 Garden St, cambridge, MA, 02138, U.S.A. (tcalderw@head-cfa.harvard.edu)

Kathy Campbell, SAO, 60 Garden Street, MS 70, Cambridge, MA, 02138, U.S.A. (kathyc@head-cfa.harvard.edu)

Marcel Carbillet, Osservatorio Astrofisico di Arcetri, Largo Enrico Fermi, 5, Firenze, 50125, U.S.A. (marcel@arcetri.astro.it)

Josephine Chan, Rutherford Appleton Laboratory, Space Science Department, Rutherford Appleton Lab, Chilton,Didcot, Oxfordshire, OX11 0QX, U.K. (s.j.chan@rl.ac.uk)

Alberto Maurizio Chavan, European Southern Observatory, Karl-Schwarzschild-Str. 2, Garching, 85748, Germany (amchavan@eso.org)

Bing Chen, Johns Hopkins University, 3400 North Charles Street, Baltimore, MD, 21218, U.S.A. (bchen@pha.jhu.edu)

Judy Chen, SAO, 60 Garden Street, Cambridge, MA, 02138, U.S.A. (jchen@head-cfa.harvard.edu)

Cynthia Y. Cheung, NASA/GSFC, Code 631, Greenbelt, MD, 20771, U.S.A. (cynthia.y.cheung.1@gsfc.nasa.gov)

Carol Christian, STScI, 3700 San Martin Drive, Baltimore, MD, 21218, U.S.A. (carolc@stsci.edu)

Maureen Conroy, SAO, 60 Garden St., Cambridge, Mass., 02138, U.S.A. (mo@cfa.harvard.edu)

Tim Cornwell, NRAO, PO Box 0, Socorro, NM, 87801, U.S.A. (tcornwel@nrao.edu)

Serge Correia, Osservatorio Astrofisico di Arcetri, Largo E.Fermi, 5, Firenze, 50125, Italy (correia@arcetri.astro.it)

Participant List

Mark Cresitello-Dittmar, SAO, 60 Garden St.; MS 81, Cambridge, MA, 02138, U.S.A. (mdittmar@cfa.harvard.edu)

Richard Crutcher, NCSA/University of Illinois, 1002 W. Green Street, Urbana, IL, 61801, U.S.A. (crutcher@uiuc.edu)

Malcolm Currie, Joint Astronomy Centre, 660, N. Aohoku Place,, Hilo, Hawaii, 96720, U.S.A. (mjc@jach.hawaii.edu)

Roc Cutri, IPAC/Caltech, MS 100-22, Pasadena, CA, 91125, U.S.A. (roc@ipac.caltech.edu)

Philip Daly, NOAO, 950 N. Cherry Avenue, P O Box 26732, Tucson, AZ, 85726-6732, U.S.A. (pnd@noao.edu)

Dennis Davidson, Ctr for Cartographic Design, 104 8th Ave. 5fs, New York, New York, 10011, U.S.A. (davidson3d@earthlink.net)

Lindsey Davis, National Optical Astronomy Observatories, P.O. Box 26732, Tucson, AZ, 85726-6732, U.S.A. (davis@noao.edu)

Jean-Pierre De Cuyper, Royal Observatory of Belgium, Ringlaan 3, Ukkel, B1180, Belgium (Jean-Pierre.DeCuyper@oma.be)

Michele De La Pena, STScI, 3700 San Martin Drive, Baltimore, MD, 21218, U.S.A. (delapena@stsci.edu)

Edward DeLuca, SAO, MS 58, 60 Garden St., Cambridge, MA, 02138, U.S.A. (edeluca@cfa.harvard.edu)

Markus Demleitner, SAO, 60 Garden Street, Cambridge, MA, 02138, U.S.A. (msdemlei@tucana.harvard.edu)

Janet DePonte, SAO, 60 Garden Street, MS 81, Cambridge, MA, 02138, U.S.A. (janet@cfa.harvard.edu)

Nicolas Devillard, ESO-European Southern Observatory, Karl-Schwarzschild str. 2, Garching, D-85748, Germany (ndevilla@eso.org)

Christopher Dingle, SAO, 60 Garden Street, MS 70, Cambridge, MA, 02138, U.S.A. (cmd@head-cfa.harvard.edu)

Adam Dobrzycki, SAO, Center for Astrophysics, 60 Garden Street, Cambridge, MA, 02138, U.S.A. (adobrzycki@cfa.harvard.edu)

Stephen Doe, SAO, 60 Garden Street, MS 81, Cambridge, MA, 02138, U.S.A. (sdoe@head-cfa.harvard.edu)

Markus Dolensky, ST-ECF, Karl-Schwarzschild-Str. 2, Garching, D-85748, Germany (mdolensk@eso.org)

Benjamin Dorman, Raytheon ITSS - NASA/GSFC, Code 664, NASA/GSFC, Greenbelt, MD, 20771, U.S.A. (Ben.Dorman@gsfc.nasa.gov)

Anil Dosaj, SAO, 60 Garden St. MS 21, Cambridge, MA, 02138, U.S.A. (anil@head-cfa.harvard.edu)

Daniel Durand, National Research Council Canada / CADC, 5071 W. Saanich Rd, Victoria, BC, V9E 2E7, Canada (Daniel.Durand@nrc.ca)

Frossie Economou, Joint Astronomy Centre, 660 N. A'ohoku Pl., Hilo, HI, 96720, U.S.A. (frossie@jach.hawaii.edu)

Guenther Eichhorn, SAO, 60 Garden Street, MS-83, Cambridge, MA, 02138, U.S.A. (gei@cfa.harvard.edu)

Alesha Estes, SAO, 60 Garden Street, Cambridge, MA, 02474-8730, U.S.A. (aestes@head-cfa.harvard.edu)

Ian Evans, SAO, 60 Garden Street, MS-27, Cambridge, MA, 02138, U.S.A. (ievans@cfa.harvard.edu)

Participant List

Hajime Ezawa, Nobeyama Radio Observatory, Nobeyama, Minamimaki, Minamisaku, Nagano, 384-1305, Japan (ezawa@nro.nao.ac.jp)

Giuseppina Fabbiano, SAO, 60 Garden St,, Cambridge, MA, 02138, U.S.A. (pepi@cfa.harvard.edu)

Fan Fang, Caltech/IPAC, 770 S. Wilson Ave., Pasadena, CA, 91125, U.S.A. (fan@ipac.caltech.edu)

Tony Farrell, Anglo Australian Observatory, 167 Vimiera Road, Eastwood, N.S.W., 2122, Australia (tjf@aaoepp.aao.gov.au)

Pierre Fernique, C.D.S., 16 rue de la gare, Strasbourg, 67000, France (fernique@astro.u-strasbg.fr)

Anthony Ferro, University of Arizona, Steward Obs./NICMOS, 933 N. Cherry Ave., Tucson, AZ, 85721-0065, U.S.A. (tferro@as.arizona.edu)

Jose M. Filgueira, GTC Project, Avda. Via Lactea, s/n, La Laguna, Canary Islands, 38200, Spain (jmfilgue@ll.iac.es)

Giorgio Filippi, E.S.O., Karl-Schwarzschildstr.2, Garching, D-85748, Germany (gfilippi@eso.org)

Luca Fini, Osservatorio di Arcetri, L.go E. Fermi, 5, Firenze, 50125, Italy (lfini@arcetri.astro.it)

Mark Fishman, AppNet, Inc., 1100 West Street, Laurel, MD, 20707, U.S.A. (mfishman@appnet.com)

Michael Fitzpatrick, NOAO/IRAF Group, 950 N Cherry Ave, PO Box 26732, Tucson, AZ, 85719, U.S.A. (fitz@noao.edu)

James Fowler, Hobby•Eberly/McDonald Obs, P.O. Box 1337-MCD, Fort Davis, TX, 79734, U.S.A. (jrf@astro.as.utexas.edu)

Peter Freeman, SAO, SAO MS-81, 60 Garden St., Cambridge, MA, 02138, U.S.A. (pfreeman@head-cfa.harvard.edu)

Marian Frueh, McDonald Observatory, P.O. Box 1337, Fort Davis, TX, 79734, U.S.A. (frueh@astro.as.utexas.edu)

Carlos Gabriel, ESA, VILSPA Satellite Tracking Station, Madrid, 28080, Spain (cgabriel@iso.vilspa.esa.es)

Niall Gaffney, STScI, 3700 San Martin Drive, Baltimore, MD, 21218, U.S.A. (gaffney@stsci.edu)

Séverin Gaudet, National Research Council Canada, 5071 West Saanich Rd., Victoria, BC, V9E 2E7, Canada (severin.gaudet@nrc.ca)

Gregg Germain, SAO, 60 Garden St. MS-34, Cambridge, Ma., 02138, U.S.A. (gregg@head-cfa.harvard.edu)

David Giaretta, Starlink,RAL,UK, Rutherford Appleton Laboratory, Chilton, Didcot, Oxon, OX11 0QX, U.K. (d.giaretta@rl.ac.uk)

Idan Ginsburg, SAO, 60 Garden St MS-34, Cambridge, MA, 02138, U.S.A. (iginsburg@cfa.harvard.edu)

Mark Giuliano, STScI, 3700 San Martin Drive, Baltimore, MD, 21218, U.S.A. (giuliano@stsci.edu)

Brian Glendenning, NRAO, PO Box O, Socorro, NM, 87801, U.S.A. (bglenden@nrao.edu)

Kenny Glotfelty, SAO, 60 Garden St, MS 81, Cambridge, MA, 02138, U.S.A. (kjg@head-cfa.harvard.edu)

Kumar Golap, NRAO, P.O. Box O, Socorro, NM, 87801, U.S.A. (kgolap@aoc.nrao.edu)

Participant List

Anthony Gonzalez, SAO, 60 Garden Street, Cambridge, MA, 02138, U.S.A. (anthony@head-cfa.harvard.edu)

Rosario González-Riestra, LAEFF, P.O. Box 50727, Madrid, Madrid, 28080, Spain (ch@laeff.esa.es)

John Good, IPAC / Caltech, Caltech, MS 100-22, Pasadena, CA, 91125, U.S.A. (jcg@ipac.caltech.edu)

Paul Grant, SAO, 60 Garden Street, MS 70, Cambridge, MA, 02138, U.S.A. (paul@head-cfa.harvard.edu)

John Graybeal, SOFIA, NASA Ames Research Center, M/S 207-1, Moffett Field, CA, 94035, U.S.A. (jgraybeal@mail.arc.nasa.gov)

William Green, Caltech/IPAC/SSC, 1800 Oak Grove Drive, Pasadena, CA, 91109, U.S.A. (bgreen@ipac.caltech.edu)

Gretchen Greene, Space Telescope Science Institute, 3700 San Martin Drive, Baltimore, MD, 21784, U.S.A. (greene@stsci.edu)

Perry Greenfield, STScI, 3700 San Martin Drive, Baltimore, MD, 21218, U.S.A. (perry@stsci.edu)

Robert Greimel, Isaac Newton Group, Apartado de Correos, 321, Santa Cruz deLaPalma, Canary Is., 38700, Spain (greimel@ing.iac.es)

Frank Gribbin, Isaac Newton Group of Telescopes, Apartado de correos 321, S. Cruz de La Palma, Tenerife, E-38700, Spain (fjg@ing.iac.es)

Sandy Grosvenor, Booz-Allen / Goddard SFC, Goddard Space Flight Center, Mail Stop 450.F, Greenbelt, MD, 20771, U.S.A. (sandy.grosvenor@gsfc.nasa.gov)

Rainer Gruber, MPE, Giesenbachstr.1, Garching b/Muenchen, Bavaria, 85740, Germany (gru@mpe.mpg.de)

David Grumm, SAO, 60 Garden St., M.S. 81, Cambridge, MA, 02138, U.S.A. (dmg@head-cfa.harvard.edu)

Damien Guillaume, UIUC / NCSA, 218 Astronomy Bldg, 1002 W. Green St., Urbana, IL, 61801, U.S.A. (dguillau@astro.uiuc.edu)

Roger Hain, SAO, 60 Garden St., Cambridge, MA, 02139, U.S.A. (rhain@head-cfa.harvard.edu)

Robert Hanisch, Space Telescope Science Institute, 3700 San Martin Drive, Baltimore, MD, 21218, U.S.A. (hanisch@stsci.edu)

Peter Harbo, SAO, 60 Garden Street, MS 81, Cambridge, MA, 02138, U.S.A. (pnh@head-cfa.harvard.edu)

F. R. Harnden, Jr., SAO, 60 Garden St. MS-2, Cambridge, MA, 02138, U.S.A. (frh@head-cfa.harvard.edu)

Helen He, SAO, 60 Garden Street, MS 81, Cambridge, MA, 02138, U.S.A. (hhe@cfa.harvard.edu)

Andre Heck, Strasbourg Observatory, 11, rue de L'Université, Strasbourg, F-67000, France (heck@astro.u-strasbg.fr)

Martin Heemskerk, Astronomical Institute 'Anton Pannekoek,' Kruislaan 403, Amsterdam, 1098 SJ, Netherlands (martin@astro.uva.nl)

Dietmar Heger, European Space Agency, Robert-Boschstr. 5, Darmstadt, D-64293, Germany (dheger@esa.int)

Christina Heikkila, NASA GSFC / Raytheon ITSS, NASA Goddard Space Flight Center, Mailstop 664.0, Bldg T2 rm 77, Greenbelt, MD, 20771, U.S.A. (cwh@lheamail.gsfc.nasa.gov)

Michelle Henson, SAO, 60 Garden Street, MS 3, Cambridge, MA, 02138, U.S.A. (michelle@cfa.harvard.edu)
Norman Hill, NRC/Canadian Astronomy Data Center, 5071 W. Saanich Rd., Victoria, B.C., V9E 2E7, Canada (Norman.Hill@hia.nrc.ca)
Philip Hodge, Space Telescope Science Inst., 3700 San Martin Drive, Baltimore, MD, 21218, U.S.A. (hodge@stsci.edu)
Guo Hongfeng, Beijing Stronomical Observatory, 20A Datun Rd., Chaoyang District, Beijing, n/a, 100012, China (ghf@bao.ac.cn)
Richard Hook, ST-ECF, Karl-Schwarzschild-Str. 2, Garching, D-85748, Germany (rhook@eso.org)
Toshihiro Horaguchi, National Science Museum, Tokyo, 3-23-1 Hyakunin-cho, Shinjuku, Tokyo, 169-0073, Japan (horaguti@kahaku.go.jp)
Shinji Horiuchi, NAO Japan, 2-21-1, Ohsawa, Mitaka, Tokyo, 181-8588, Japan (horiuchi@nao.ac.jp)
Allan Hornstrup, Danish Space Research Institute, Juliane Maries Vej 30, Copenhagen, 2100, Denmark (allan@dsri.dk)
Wolfgang Hovest, Max-Planck-Institut fuer Astropysik (MPA), Karl-Schwarzschild-Str. 1, Garching, 85741, Germany (hovest@mpa-garching.mpg.de)
Jinchung Hsu, STScI, 3700 San Martin Drive, Baltimore, MD, 21218, U.S.A. (hsu@stsci.edu)
Edwin Huizinga, STScI, 3700 San Martin Drive, Baltimore, MD, 21218, U.S.A. (huizinga@stsci.edu)
Gareth Hunt, NRAO - Green Bank, Route 28/92, Green Bank, WV, 24944, U.S.A. (ghunt@nrao.edu)
Miho Ikeda, National Astoronomical Observatory, Osawa 2-21-1, Mitaka, Tokyo, 181-8588, Japan (miho.ikeda@nao.ac.jp)
Yasuhide Ishihara, Fujitsu Limited, mihama-ku 1-9-3, Chiba, Chiba, 261-8588, Japan (ishi@ssd.se.fujitsu.co.jp)
George Jacoby, WIYN, P.O. Box 26732, Tucson, AZ, 85726, U.S.A. (jacoby@noao.edu)
Tim Jenness, Joint Astronomy Centre, 660 N. Aohoku Pl, Hilo, HI, 96720, U.S.A. (t.jenness@jach.hawaii.edu)
Jeremy Jones, NASA/GSFC, Code 588, Greenbelt, MD, 20771, U.S.A. (Jeremy.E.Jones@gsfc.nasa.gov)
William Joye, SAO, 60 Garden Street, Cambridge, MA, 02138, U.S.A. (wjoye@cfa.harvard.edu)
Marcus Juette, Astronomisches Inst. der Ruhr-Universitaet, Universitaetsstrasse 150, Bochum, 44780, Germany (juette@astro.ruhr-uni-bochum.de)
Margarita Karovska, SAO, 60 Garden Street, Cambridge, MA, 02138, U.S.A. (karovska@cfa.harvard.edu)
Athol Kemball, NRAO, 1003 Lopezville Road, Socorro, NM, 87801, U.S.A. (akemball@nrao.edu)
Jeremy Kepner, MIT Lincoln Lab, 244 Wood St, Lexington, MA, 02420, U.S.A. (kepner@ll.mit.edu)
Carl Kesselman, USC Information Sciences Institute, 4676 Admiralty Way Suite 1001, Marina Del Rey, CA, 90292, U.S.A. (carl@isi.edu)

DongChan Kim, ASIAA, Taiwan, Academia Sinica, Inst. of Asrtronomy & Astroph., Nankang, Taipei, 115, Taiwan (kim@asiaa.sinica.edu.tw)
David L. King, NRAO, P.O. Box O, Socorro, NM, 87801, U.S.A. (dking@aoc.nrao.edu)
Sonali Kolhatkar, SIRTF Science Center, Caltech, California Institute of Technology, Mail Code 314-6, Pasadena, CA, 91125, U.S.A. (sonali@ipac.caltech.edu)
Mih-seh Kong, IPAC / Caltech, Caltech, MS 100-22, Pasadena, CA, 91125, U.S.A. (mihseh@ipac.caltech.edu)
Anuradha Koratkar, Space Telescope Science Institute, 3700 San Martin Drive, Baltimore, MD, 21218, U.S.A. (koratkar@stsci.edu)
Peter Kunszt, Johns Hopkins University, Center for Astrophysical Sciences, 3701 San Martin Drive, Baltimore, MD, 21218, U.S.A. (kunszt@pha.jhu.edu)
Wayne Landsman, Raytheon ITSSS, Code 681, NASA/GSFC, Greenbelt, MD, 20771, U.S.A. (landsman@mpb.gsfc.nasa.gov)
Youngung Lee, KAO/TRAO, Korea Astronomy Observatory, Whaam-dong 61-1, Yusong, Taejon, 305-348, U.S.A. (yulee@trao.re.kr)
Soizick Lesteven, CDS - Strasbourg Observatory, 11, rue de l'Université, Strasbourg, 67000, France (lesteven@astro.u-strasbg.fr)
James Lewis, Institute of Astronomy, Madingley Road, Cambridge, Cambridgeshire, CB3 0HA, U.K. (jrl@ast.cam.ac.uk)
Ganghua Lin, Beijing Astronomical Observatory, Datun Rd. #20A, Chaoyang District, Beijing, 100012, China (lgh@sun10.bao.ac.cn)
Don Lindler, ACC Inc., 11518 Gainsborough Road, Potomac, Maryland, 20854, U.S.A. (lindler@rockit.gsfc.nasa.gov)
Antoine Llebaria, Lab. Astroph. Marseille (CNRS), 13376 Marseille Cedex 12, Traverse du Siphon, Marseille, BDR, 13376, France (antoine.llebaria@astrsp-mrs.fr)
Stephen Lubow, STScI, 3700 San Martin Drive, Baltimore, MD, 21218, U.S.A. (lubow@stsci.edu)
Robert Lucas, IRAM, 300 rue de La Piscine, Ste Martin d'Heres, F-38406, France (lucas@iram.fr)
Robert Lupton, Princeton University, Ivy Lane, Princeton, NJ, 08544, U.S.A. (rhl@astro.princeton.edu)
Daniella Malin, University of Massachusetts, LGRT104, Umass, Amherst, MA, 01003, U.S.A. (daniella@lmtgtm.org)
Eric Mandel, SAO, 60 Garden Street, Cambridge, MA, 02138, U.S.A. (eric@cfa.harvard.edu)
Jean Manfroid, Univ. of Liege, Avenue de Cointe 5, B-4000 Liege,, Belgium (manfroid@astro.ulg.ac.be)
Joseph Masters, SAO, 60 Garden St, MS-81, Cambridge, MA, 02138, U.S.A. (jmasters@head-cfa.harvard.edu)
David Matusow, NASA/GSFC, Code 588, Bldg. 23, Greenbelt, MD, 20771, U.S.A. (dmmatuso@pop500.gsfc.nasa.gov)
Steve McDonald, Silicon Spaceships, 74 Bay State Ave, Somerville, MA, 02144, U.S.A. (steve@SiliconSpaceships.com)
Brian McLean, STScI, 3700 San Martin Drive, Baltimore, MD, 21218, U.S.A. (mclean@stsci.edu)

Participant List

Alberto Micol, ST-ECF, Karl-Schwarzschild-Str. 2, Garching, D-85748, Germany (amicol@eso.org)
Joseph Miller, SAO, 60 Garden Street MS-34, Cambridge, ma, 02138, U.S.A. (gregg@head-cfa.harvard.edu)
Dave Mills, NOAO, 950 N Cherry Ave, Tucson, AZ, 85719, U.S.A. (dmills@noao.edu)
Doug Mink, SAO, 60 Garden St., Cambridge, MA, 02138, U.S.A. (dmink@cfa.harvard.edu)
Anthony Minter, NRAO, PO Box 2, Green Bank, WV, 24944, U.S.A. (tminter@nrao.edu)
John Moran, SAO, 60 Garden Street, Cambridge, MA, 02138, U.S.A. (jfmoran@head-cfa.harvard.edu)
Patrick Murphy, NRAO, 520 Edgemont Road, Charlottesvill, VA, 22903-2575, U.S.A. (pmurphy+ad2k@nrao.edu)
Stephen S. Murray, SAO, 60 Garden Street, MS 2, Cambridge, MA, 02138, U.S.A. (ssm@cfa.harvard.edu)
Robert Narron, Caltech/IPAC/SSC, M/S 314-6, 1200 E. California Blvd., Pasadena, CA, 91125, U.S.A. (bob@ipac.caltech.edu)
Joy Nichols, SAO, 60 Garden St., Mail Code 34, Cambridge, MA, 01940, U.S.A. (jnichols@head-cfa.harvard.edu)
Michael Noble, SAO, 60 Garden Street, M/S 81, Cambridge, MA, 02138, U.S.A. (mnoble@cfa.harvard.edu)
Jan Noordam, ASTRON, P.O.Box 2, Dwingeloo, Drenthe, 7990AA, Netherlands (noordam@nfra.nl)
Earl O'Neil, University of Arizona/Steward Observatory, 933 N. Cherry Avenue, Tucson, Arizona, 85721, U.S.A. (eoneil@as.arizona.edu)
Joel Offenberg, Raytheon ITSS, Code 681, NASA's Goddard Space Flight Center, Greenbelt, MD, 20771, U.S.A. (Joel.D.Offenberg@gsfc.nasa.gov)
Ryusuke Ogasawara, Subaru Telescope, NAOJ, 650 North Aohoku Place, Hilo, HI, 96720, U.S.A. (ryu@subarutelescope.org)
Stephan Ott, ESA, VilSpa, P.O. Box 50727, Madrid, Madrid, E 28080, Spain (sott@iso.vilspa.esa.es)
Ryan Overbeck, SAO, 202 Whitwell St., Quincy, MA, 02169, U.S.A. (r_overbeck@yahoo.com)
James Overly, SAO, Smithsonian, 60 Garden St. MS 81, Cambridge, MA, 02138, U.S.A. (joverly@head-cfa.harvard.edu)
Hughes Pack, Northfield Mount Hermon School, 206 Main St., Box 4706, Northfield, MA, 01360-1050, U.S.A. (hughes_pack@nmhschool.org)
Paolo Padovani, STScI, 3700 San Martin Drive, Baltimore, MD, 21218, U.S.A. (padovani@stsci.edu)
Clive Page, University of Leicester, Physics and Astronomy Department, University Road, Leicester, LE1 7RH, U.K. (cgp@star.le.ac.uk)
Stephane Paltani, Harvard-Smithsonian CFA, 60, Garden St. MS 81, Cambridge, MA, 02138, U.S.A. (spaltani@cfa.harvard.edu)
Lisa Paton, SAO, 60 Garden St., Cambridge, MA, 02138, U.S.A. (lpaton@cfa.harvard.edu)
Harry Payne, STScI, 3700 San Martin Drive, Baltimore, MD, 21218, U.S.A. (payne@stsci.edu)

Nestor Peccia, ESA / ESOC, Robert Bosch str. 5, Darmstadt, 64293, Germany (npeccia@esoc.esa.de)
William Pence, NASA/GSFC, Code 662, Greenbelt, MD, 20771, U.S.A. (pence@tetra.gsfc.nasa.gov)
Steven Penton, Center for Astrophysics & Space Astronomy, 1255 38th Street, Campus Box 593, Boulder, CO, 80303, U.S.A. (spenton@casa.colorado.edu)
Jeffrey Percival, Space Astronomy Laboratory, 1150 University Ave, 6295 Chamberlin Hall, Madison, WI, 53706, U.S.A. (jwp@sal.wisc.edu)
James Petreshock, SAO, Smithsonian Astrophysical Observatory, 60 Garden Street MS-81, Cambridge, MA, 02138, U.S.A. (jpetres@head-cfa.harvard.edu)
Francesco Pierfederici, ST-ECF, Karl-Schwarzschild-Str. 2, Garching, D-85748, Germany (fpierfed@eso.org)
Benoît Pirenne, ST-ECF, Karl-Schwarzschild-Str. 2, Garching, D-85748, Germany (bpirenne@eso.org)
Norbert Pirzkal, ST-ECF, Karl-Schwarzschild-Str. 2, Garching, D-85748, Germany (npirzkal@eso.org)
Raymond Plante, NCSA, 103 Astronomy, 1002 W. Green St., Urbana, IL, 61801, U.S.A. (rplante@ncsa.uiuc.edu)
David Plummer, SAO, 60 Garden St., MS-81, Cambridge, MA, 02138, U.S.A. (dplummer@cfa.harvard.edu)
Joel Poels, Astrophysics Institute, University of Lige, Avenue de Cointe 5, Liège, Liège, B-4000, Belgium (joel.poels@ulg.ac.be)
Joseph Pollizzi, STScI, 3700 San Martin Drive, Baltimore, MD, 21218, U.S.A. (pollizzi@stsci.edu)
Marc Pound, University of Maryland, Astronomy Department, College Park, Md, 20742, U.S.A. (mpound@astro.umd.edu)
Francis Primini, SAO, 60 Garden Street, Mail Stop 3, Cambridge, MA, 02138, U.S.A. (fap@head-cfa.harvard.edu)
Vasundhara Puttagunta, University of Maryland, 4714 Aldgate Green, Baltimore, MD, 21227, U.S.A. (vputta1@csee.umbc.edu)
Michael Raley, SAO, 219 Harvard St., Apt. 19, Brookline, MA, 02446, U.S.A. (mraley@head-cfa.harvard.edu)
Michael Read, IfA (Edinburgh), IfA, Royal Observatory, Blackford Hill, Edinburgh, Scotland, EH9 3HJ, U.K. (mar@roe.ac.uk)
Michael Reid, Victoria Univeristy of Wellington, SCPS, PO Box 600, Wellington, Wellington, N/A,, New Zealand (Michael.Reid@vuw.ac.nz)
Gastaud Rene, CEA, CEA SACLAY/DAPNIA SEI, Gif Sur Yvette, n/a, 9119L, France (rgastaud@cea.fr)
Michael Riggs, UMBC, 337 Ida ave, Baltimore, MD, 21221, U.S.A. (mriggs1@umbc.edu)
Wendy Roberts, SAO, 60 Garden St., Cambridge, MA, 02138, U.S.A. (wroberts@cfa.harvard.edu)
John Roll, SAO, 60 Garden Street, Cambridge, MA, 02138, U.S.A. (john@cfa.harvard.edu)
James Rose, STScI/CSC, Space Telescope Science Institute, 3700 San Martin Drive, Baltimore, MD, 21218, U.S.A. (rose@stsci.edu)

Arnold Rots, SAO, 60 Garden Street MS 81, Cambridge, MA, 02138, U.S.A. (arots@head-cfa.harvard.edu)
Chris Sabbey, IoA, U. of Cambridge, Madingley Road, Cambridge,, U.K. (sabbey@ast.cam.ac.uk)
David Schade, Canadian Astronomy Data Centre, Herzberg Institute of Astrophysics, 5071 West Saanich Rd., Victoria, BC, V9E 2E7, Canada (David.Schade@nrc.ca)
Skip Schaller, Las Campanas Observatory, Casilla 601, La Serena,, Chile (skip@lco.cl)
Eric Schlegel, SAO, 60 Garden Street, Cambridge, MA, 02138, U.S.A. (eschlegel@cfa.harvard.edu)
Dennis Schmidt, SAO, Harvard-Smithsonian CfA, 60 Garden St., MS-34, Cambridge, MA, 02138, U.S.A. (dennis@head-cfa.harvard.edu)
Isabelle Scholl, IAS-CNRS, Batiment 121, Université Paris XI, ORSAY, F-91405, France (scholl@medoc-ias.u-psud.fr)
German Schumacher, NOAO/SOAR Telescope, 950 N. Cherry Avenue, Tucson, AZ, 85719, U.S.A. (gschumacher@noao.edu)
Joseph Schwarz, ESO, Karl-Schwarzschild-Str. 2, Garching, 85748, Germany (jschwarz@eso.org)
Marco Scodeggio, Istituto di Fisica Cosmica, via Bassini 15, Milano, I-20133, Italy (marcos@ifctr.mi.cnr.it)
Stephen Scott, Caltech/OVRO, POB 968, Big Pine, CA, 93513, U.S.A. (scott@ovro.caltech.edu)
Rob Seaman, NOAO/IRAF, 950 N. Cherry Ave., Tucson, AZ, 85719, U.S.A. (seaman@noao.edu)
Ratna Sengupta, Raytheon ITSS, Bldg. 3, Room 211, Code 582, GSFC, NASA, Greenbelt, MD, 20770, U.S.A. (rsengupta@hst.nasa.gov)
Richard Shaw, STScI, 3700 San Martin Drive, Baltimore, MD, 21218, U.S.A. (shaw@stsci.edu)
Ed Shaya, NASA/ADC/ITSS, GSFC/Code 631, Greenbelt, MD, 20771, U.S.A. (Edward.J.Shaya.1@gsfc.nasa.gov)
Patrick Shopbell, Caltech, Dept. of Astronomy, MC 105-24, Caltech, Pasadena, CA, 91125, U.S.A. (pls@astro.caltech.edu)
Keith Shortridge, Anglo-Australian Observatory, P.O.Box 296, Epping, NSW, 2121, Australia (ks@aaoepp.aao.gov.au)
Petr Škoda, Astronomical Inst., Academy of Sciences, Stellar Dept., Fricova 1, Ondrejov, 25165, Czech Republic (skoda@sunstel.asu.cas.cz)
Oleg Smirnov, ASTRON, P.O. Box 2, Dwingeloo, 7990 AA, Netherlands (smirnov@astron.nl)
Ian Smith, Joint Astronomy Centre, 660 N. A'ohoku Place, University Park, hilo, hi, 96720, U.S.A. (ismith@jach.hawaii.edu)
Enrique Solano Marquez, LAEFF, ESA-VILSPA Campus, P.O. Box 50727, Madrid, 28080, Spain (esm@vilspa.esa.es)
Kamal Souccar, UMASS, LGRT 815, LMT Project, Amherst, MA, 01003, U.S.A. (souccar@astro.umass.edu)
Jason Spyromilio, ESO, Karl-Schwarzschild-Str 2, Garching, D-85748, Germany (jspyromi@eso.org)
Scott Stallcup, STScI, 3700 San Martin Drvie, Baltimore, MD, 21218, U.S.A. (stallcup@stsci.edu)

Participant List

John Stephen, Istituto TeSRE/CNR, Via P. Gobetti 101, Bologna, BO, 40129, Italy (stephen@tesre.bo.cnr.it)
David Stern, Research Systems, 4990 Pearl East Cr, Boulder, CO, 80301, U.S.A. (dave@rsinc.com)
Elizabeth Stobie, University of Arizona/Steward Observatory, 933 N. Cherry Avenue, Tucson, Arizona, 85721, U.S.A. (bstobie@as.arizona.edu)
Sreelatha Subramanian, SAO, 60 Garden St. MS-81, Cambridge, MA, 02138, U.S.A. (latha@head-cfa.harvard.edu)
Daryl Swade, STScI, 3700 San Martin Drive, Baltimore, MD, 21218, U.S.A. (swade@stsci.edu)
Michael Swam, STScI, 3700 San Martin Drive, Baltimore, MD, 21218, U.S.A. (mswam@stsci.edu)
Raymond Swartz, Jet Propulsion Laboratory, 4800 Oak Grove Drive, M/S 301-486, Pasadena, CA, 91109, U.S.A. (Raymond.Swartz@jpl.nasa.gov)
Alexander Szalay, Johns Hopkins, Dept. Physics and Astronomy, The Johns Hopkins University, Baltimore, MD, 21218, U.S.A. (szalay@jhu.edu)
Arpad Szomoru, Joint Institute for VLBI in Europe, PO Box 2, Dwingeloo, 7990 AA, Netherlands (szomoru@jive.nl)
Jill Tarter, SETI Institute, 2035 Landings Drive, Mountain View, CA, 94043, U.S.A. (tarter@seti.org)
Johannes P. Terlouw, Kapteyn Astronomical Institute, P.O. Box 800, Groningen, 9700 AV, Netherlands (terlouw@astro.rug.nl)
Peter Teuben, Astronomy Department, University of Maryland, College Park, MD, 20742, U.S.A. (teuben@astro.umd.edu)
Aniruddha Thakar, The Johns Hopkins University, Ctr for Astrophyical Sciences, 3701 San Martin Drive, Baltimore, MD, 21218-2695, U.S.A. (thakar@pha.jhu.edu)
Carole Thiebaut, CESR-CNRS, 38 rue de l'Etoile, Toulouse, 31000, France (thiebaut@cesr.fr)
Brian Thomas, GSFC-NASA/RITSS, Code 631, Goddard Space Flight Center, Greenbelt, MD, 20771, U.S.A. (thomas@adc.gsfc.nasa.gov)
Doug Tody, National Optical Astronomy Observatories, 950 N. Cherry Ave, P.O. Box 26732, Tucson, AZ, 85726, U.S.A. (tody@noao.edu)
Susan Tokarz, SAO, 60 Garden Street, Cambridge, MA, 02138, U.S.A. (stokarz@cfa.harvard.edu)
Ralph Tremmel, MPI-Astronomie, Koenigstuhl 17, Heidelberg, 69117, Germany (tremmel@mpia-hd.mpg.de)
Massimo Trifoglio, CNR/ITESRE, Via Gobetti, 101, Bologna, 40129, Italy (trifoglio@tesre.bo.cnr.it)
Euan Troup, Jodrell Bank Observatory, Jodrell Bank Observatory, Macclesfield, Cheshire, SK11 9DL, U.K. (etroup@jb.man.ac.uk)
Takahiro Tsutsumi, SAO, 60 Garden St. MS78, Cambridge, MA, 02138, U.S.A. (ttsutsumi@cfa.harvard.edu)
Susan J. Tuttle, SAO, 60 Garden Street, MS 2, Cambridge, MA, 02138, U.S.A. (logan@cfa.harvard.edu)
Francisco Valdes, NOAO, P.O. Box 26732, Tucson, AZ, 85726, U.S.A. (fvaldes@noao.edu)
Robert Vallance, Birmingham University, Edgbaston park Rd, Edgbastob, Birmingham, W Midlands, B15 2TT, U.K. (rjv@star.sr.bham.ac.uk)

Kjeld van der Schaaf, ASTRON, Postbus 2, Dwingeloo, 7990 AA, Netherlands
   (schaaf@astron.nl)
Ger van Diepen, ASTRON, P.O.Box 2, Dwingeloo, 7990 AA, Netherlands
   (diepen@nfra.nl)
Gustaaf van Moorsel, NRAO, P.O. Box 0, Socorro, NM, 87801, U.S.A.
   (gvanmoor@nrao.edu)
Christian Veillet, CFHT, P.O. Box 1597, Kamuela, Hawaii, 96743, U.S.A.
   (veillet@cfht.hawaii.edu)
Andy Vick, UK Astronomy Technology Centre, ROE, Blackford Hill,
   Edinburgh, Midlothian, EH9 3HJ, U.K. (A.Vick@roe.ac.uk)
Vladimir Vitkovskij, SAO RAS, h. 1, fl.11, Niznij Arkhys,
   Karachaevo-Cherkessia, 357147, Russia (vvv@sao.ru)
Martin Vogelaar, Kapteyn Astronomical Institute, Postbus 800, Groningen,
   groningen, 9700 AV, Netherlands (vogelaar@astro.rug.nl)
Heather Volatile, SAO, 60 Garden Street, MS-34, Cambridge, MA, 02138,
   U.S.A. (hvolatile@cfa.harvard.edu)
Claudio Vuerli, OATs, Via G.B. Tiepolo 11, Trieste, Trieste, I-34131, Italy
   (vuerli@ts.astro.it)
Gary Wallace, University of Massachusetts, LGRT 104, Amherst, MA, 01003,
   U.S.A. (wallace@astro.umass.edu)
Patrick Wallace, Rutherford Appleton Laboratory, CLRC, Chilton, Didcot,
   Oxfordshire, OX11 0QX, U.K. (ptw@star.rl.ac.uk)
Rein H. Warmels, ESO, Karl-Schwarzschild-Straße 2, Garching, 85748,
   Germany (rwarmels@eso.org)
Masaru Watanabe, Japan Science and Technology Corporation, 3-1-1
   Yoshinodai, Sagamihara, Kanagawa, 229-8510, Japan
   (mwatanab@db01.mtk.nao.ac.jp)
Donald C. Wells, NRAO, 520 Edgemont Road, Charlottesville, VA,
   22903-2475, U.S.A. (dwells@nrao.edu)
Marc Wenger, CDS, M, Rue De L'Université, Strasbourg, 67000, France
   (wenger@astro.u-strasbg.fr)
Dan Werthimer, University of California, Berkeley, Space Sciences Lab,
   Berkeley, CA, 94720-7450, U.S.A. (danw@ssl.berkeley.edu)
Richard L. White, STScI, 3700 San Martin Dr, Baltimore, MD, 21218, U.S.A.
   (rlw@stsci.edu)
Andreas Wicenec, ESO, Karl-Schwarzschild-Str. 2, Garching, 81737, Germany
   (awicenec@eso.org)
Sherry Winkelman, SAO, 60 Garden Street, Cambridge, MA,, U.S.A.
   (swinkelman@cfa.harvard.edu)
Karl Wolf, AppNet, Inc., 1100 West Street, Laurel, MD, 20707, U.S.A.
   (karl.wolf@appnet.com)
Scott Wolk, SAO, 60 Garden St., Cambridge, MA, 02138, U.S.A.
   (swolk@cfa.harvard.edu)
Shui Kwan Wong, SAO, Smithsonian Astrophysical Observatory, 60 Garden
   Street, MS 81, Cambridge, MA, 02138, U.S.A.
   (swong@head-cfa.harvard.edu)
William Wyatt, SAO, 60 Garden St., m.s. 20, Cambridge, MA, 02138, U.S.A.
   (wyatt@cfa.harvard.edu)

Masafumi Yagi, NAOJ, 2-21-1 Osawa, Mitaka, Tokyo, 181-8588, U.S.A. (yagi@optik.mtk.nao.ac.jp)
Victor Yodaiken, FSMLabs, P.O. Box 1822, Socorro, NM, 87801, U.S.A. (yodaiken@fsmlabs.com)
Nelson Zarate, NOAO, 950 N Cherry Av, Tucson, AZ, 85719, U.S.A. (zarate@noao.edu)
Jun-Hui Zhao, SAO, 60 Garden St MS 78, Cambridge, MA, 02138, U.S.A. (jzhao@cfa.harvard.edu)
Leslie Zimmerman, STScI, 3700 San Martin Drive, Baltimore, MD, 21218, U.S.A. (zimmerman@stsci.edu)
Martin Zombeck, SAO, 60 Garden Street, Cambridge, MA, 02138, U.S.A. (mzombeck@cfa.harvard.edu)

Conference Photograph

Part 1. Enabling Technologies for Astronomy

The National Virtual Observatory

Alexander S. Szalay

Department of Physics and Astronomy, The Johns Hopkins University, Baltimore, MD 21218

Abstract. Technological advances in telescope and instrument design during the last ten years, coupled with the exponential increase in computer and communications capability, have caused a dramatic and irreversible change in the character of astronomical research. Large-scale surveys of the sky from space and the ground are being initiated at wavelengths from radio to X-ray, thereby generating vast amounts of high quality, irreplaceable data. The potential for scientific discovery afforded by these new surveys is enormous. Entirely new and unexpected scientific results of major significance will emerge from the combined use of the resulting datasets, science that would not be possible from such sets used singly. However, their large size and complexity require tools and structures to discover the complex phenomena encoded within them.

1. Introduction: Scope and Vision of the NVO

Astronomy faces a data avalanche. Breakthroughs in telescope, detector, and computer technology allow astronomical surveys to produce terabytes of images and catalogs. These datasets will cover the sky in different wavebands, from γ and X-rays, optical, infrared, to radio. In a few years it will be easier to "dial-up" a part of the sky than wait many months to access a telescope. With the advent of inexpensive storage technologies and the availability of high-speed networks, the concept of multi-terabyte on-line databases interoperating seamlessly is no longer outlandish.[2,3] More and more catalogs will be interlinked, query engines will become more and more sophisticated, and the research results from on-line data will be just as rich as that from "real" telescopes. The planned Large Synoptic Survey Telescope will produce over 10 petabytes/year by 2008! These technological developments will fundamentally change the way astronomy is done. These changes will have dramatic effects on the sociology of astronomy.

---

[1] This paper represents the current views of the NVO Collaboration, a group of over 30 astronomers, computer scientists, statisticians, and physicists, who have been working on developing the concept of the Virtual Observatory. Over the last year the group has written a 'White Paper' about our ideas, and has submitted a large proposal to the National Science Foundation. This paper presents a short synopsis of these.

[2] http://www.astro.caltech.edu/nvoconf/presentations.html

[3] http://www.astro.caltech.edu/nvoconf/white_paper.pdf

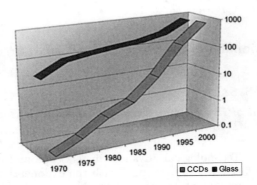

Figure 1. The total area of astronomical telescopes in m^2, and CCDs measured in Gigapixels, over the last 25 years. The number of pixels and the data double every year.

On-line astronomy demands new IT approaches that will yield tools and methodologies for data access, analysis, and discovery that are scalable to this regime. New needs lead to opportunities in IT research for data mining, for sophisticated pattern recognition, for large-scale statistical cross-correlations, and for the discovery of rare objects and sudden temporal variations. With a billion objects, statistical algorithms requiring N^3 steps would take billions of processor-years–even $N \log N$ algorithms will take a long time, creating challenges in their own right! Moreover, there is a growing awareness, both in the US and abroad, that the acquisition, organization, analysis, and dissemination of scientific data are essential elements to a continuing robust growth of science and technology. These factors demand efficient and effective synthesis of these capabilities both for astronomy and for the broader scientific community.

Recognizing these trends and opportunities, the National Academy of Sciences Astronomy and Astrophysics Survey Committee, in its decadal survey[4] recommends, as a first priority, the establishment of a National Virtual Observatory. The NVO would be a "Rosetta Stone:" linking the archival data sets of space- and ground-based observatories, the catalogs of multi-wavelength surveys, and the computational resources necessary to support comparison and cross-correlation among these resources. The NVO will benefit the entire astronomical community. It will democratize astronomical research: the same data and tools will be available to students and researchers, irrespective of geographical location or institutional affiliation. The NVO will also have far-reaching education potential. Astronomy occupies a very special place in the public eye: new discoveries fascinate both the large number of amateur astronomers and the general public alike. The NVO will be an enormous asset for teaching astronomy, information technology, and the method of scientific discovery. Outreach and education will be key elements: the NVO will deliver rich content via the Internet to a wide range of educational projects from K-12 through college and to the public.

---

[4]http://www.nap.edu/books/0309070317/html/

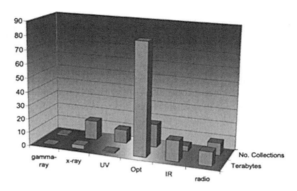

Figure 2. The chart shows the sizes and numbers of the collections expected to participate in the NVO from the beginning.

This paper describes a short and incomplete overview of the NVO's scientific opportunities and Information Technology challenges. It presents an implementation strategy and management plan to create an initial federation, and a foundation for further tools and applications. The NVO will challenge the astronomical community with new opportunities for scientific discovery and it will challenge the information technology community with a visionary but achievable goal of distributed access and analysis of voluminous data collections. The scope of the effort is international.

By its very nature the NVO brings together groups with different talents. We recognize that the ability to fully exploit the federation of all major astronomy archives as an integrated resource requires cooperation between data providers from space missions, ground-based telescopes, and special surveys, from leading astronomical institutions throughout the country. The currently participating groups have committed to federate more than 100 Terabytes of astronomical data, consisting of over 50 collections in the NVO.

2. The NVO Science and IT Paradigm

2.1. Enabling New Science – The New Astronomy

In order to be an engine of discovery for astronomy, and enable qualitatively new advances, the NVO must be driven by science goals. We think of the NVO as a genuine observatory that astronomers will use from their desks. It must supply digital archives, metadata management tools, data discovery, access services, programming interfaces, and computational services. Astronomers may develop their own custom programs to answer specific questions, sifting through the vast digital sky to identify rare objects, compare data with numerical models, and make discoveries through advanced visualizations and special statistical analyses. In addition, students, teachers, under-represented groups, and the general public have equal access to this cutting-edge resource from tailored portals. Before discussing the IT challenges, we first select a few cutting edge astronomy topics and discuss the NVO's scientific impact on them:

Comparing the Local and the Distant Universe: Combining IR and optical observations has opened a new window on the distant universe. Having a broad range of colors for distant galaxies enables us to estimate not only photometric redshifts, but also spectral type, and study detailed star formation history. The NVO will let us create rest-frame selected samples from combined UV, optical, and IR datasets. Comparing local and distant samples, we can study the evolution of physical properties, such as the infrared-radio correlation in star-forming galaxies. Statistical analyses of samples defined via multi-wavelength queries in the NVO will measure spatial clustering patterns as a function of redshift, revealing density fluctuations in the early universe and constraining values of the fundamental cosmological parameters. Massive cluster surveys will enable us to trace the complex evolution of large-scale structure, from its origins in the cosmic microwave background to the amazing diversity we see today; MAP and PLANCK promise views of the infant universe at high resolution. X-ray missions (Chandra, XMM, ROSAT), and deep ground based observations show structure to $z \sim 4$, and we will later probe to $z \sim 30$ with NGST. SIRTF and other IR surveys will fill in the gap at intermediate redshifts. Distortions in optical galaxies around these clusters provide direct measures of dark matter.

Digital Milky Way: We know surprisingly little about the origin and evolution of our own galaxy. Federating all existing information on the Milky Way will enable systematic mining of multi-wavelength catalogs and surveys, both existing and yet to be created. The Milky Way galaxy is a complex entity consisting of multiple stellar components (bulge, disk, halo), each with its own mass function, metallicity, age, and kinematics. The interstellar medium is equally complex. Current surveys covering 100 square degrees already reveal halo kinematic substructures in the form of star streams and accretion remnants. Current surveys capture all halo giants out to 100 kpc and all but the faintest dwarfs to 15 kpc; deeper surveys will allow for the first time a definitive study on the origins of the Galactic halo and thick disk. An important challenge, for example, is to understand the role of fast encounters with other galaxies (typically smaller than the Milky Way) in the evolution of our Galaxy, which can create new stellar components, impact the evolution of existing components, and can directly induce star formation. The planned Large Synoptic Survey Telescope will provide high precision proper motions from co-added images that will help address these crucial problems.

Rare and Exotic Objects: Large surveys detect significant numbers of outliers in the statistical distributions of derived parameters. These anomalies are often not immediately obvious from a single observation; and are not followed up scientifically, since the discoverers may not be able to devise a compelling model for the phenomena using the limited data at their disposal. The NVO will enable astronomers to find objects that can only be identified by being statistically unusual when multiple-wavelength catalogs are compared. Early multicolor surveys (SDSS, 2MASS, DPOSS2) have led to the discovery of distant quasars and brown dwarfs. The search for new classes of objects in far larger volumes of parameter space will remain untouched until the NVO is created. The search for rare objects in the temporal domain could yield some of the most exciting new results from the NVO: rapid identification of transient objects by comparing new observations with prior epochs in real-time may reveal distant supernovae and gravitational microlensing.

Census of Active Galactic Nuclei (AGN): There has been a major shift in our understanding of the role of supermassive black holes in galactic formation and evolution. These black holes have masses in the range of 10^6–10^{10} M$_\odot$ and we now believe that most galaxies harbor such black holes at their centers. AGN are "beacons" which signal the presence of a black hole in a galaxy, shining by converting the energy from accreting matter into radiation. Much of the radiative output of the early universe may have been emitted by these accreting supermassive black holes. However, despite decades of effort, a census of AGN and their place in the galactic evolutionary scheme still eludes us. The chief problem is dust obscuration: hidden AGN may outnumber their unobscured counterparts by an order of magnitude. Glimpses of this population are now being seen in X-rays. The NVO will include the deepest X-ray data ever obtained (Chandra); the most extensive redshift catalogs (SDSS); the highest resolution optical observations (Hubble); the largest infrared archives (2MASS, then SIRTF); the high-resolution spectra from the VLT and Gemini; radio data from the Very Large Array; and a vast panoply of planned future surveys at various wavelengths. It will enable a panchromatic census of AGNs, allow us to probe the connection between galaxies and supermassive black holes, and reveal the cosmic history of energy production from both nucleosynthesis (star formation activity) and accretion (AGNs).

Search for Extra-Solar Planets: The search for extra-Solar planets is a major goal of twenty-first century astronomy; it carries with it the scientist's hunger for new understanding, the philosopher's inquiry into the meaning of life, and the public's desire to know the answer to a simple question: "are we alone?" The spectacular recent progress with the radial velocity technique has already overturned the standard models for planet formation, but also revealed its frustrating limitations. Other techniques are being pursued: the planet-transit technique readily scales to surveys around very large numbers of stars, and would benefit enormously from the federation of data from multiple future surveys. In particular, the data taken by the Large Synoptic Survey Telescope, coupled with infrared surveys (2MASS and its successors) and astrometric surveys (FAME), will enable a survey for planetary transits around billions of stars in the Milky Way and in several nearby galaxies.

Theoretical Astrophysics: The NVO will make possible, for the first time, truly significant interactions between large datasets and the equally large-scale theoretical simulations of astrophysical systems that are just now becoming available. In a few years, use of massively parallel terascale computing systems will allow: a) the calculation of the orbits and evolution of every star in a globular cluster, including stellar collisions, the formation of tight binary systems, core evolution, and the effect of all these on the stellar evolutionary tracks; b) the details of galaxy encounters and mergers, including both the fate of the stars and the interstellar medium; and c) the evolution of the large scale structure of the Universe, including the formation of galaxies, clusters of galaxies, and clusters of clusters. All of these and similar theoretical calculations will produce datasets comparable in size with those of the large scale observational surveys, and it will be possible to mine these datasets just like observational data. Definitive comparisons will be made between complex theoretical calculations and observational datasets large enough to be statistically significant in all parameters. These studies, which will be carried out within the framework of the NVO, will lead to solutions of some of the most outstanding and significant astrophysical problems of our time.

2.2. The NVO as a Semantic Web

The primary focus of the NVO is data federation, fusion, and exploration. In order to achieve these goals, the NVO will break new ground as a large-scale prototype of a semantic web (Berners-Lee, Hendler, & Lassila 2001). The NVO will federate a large number of heterogeneous data sets distributed around the world. Some of these data sets are small, some are tens of terabytes in size, some are under database management, and others are not. The data include catalogs of objects with attributes, image data of varying resolution and wavelength, spectral data, temporal data, and ancillary reference literature. An ambitious goal of the NVO is to federate the data and information of an entire scientific discipline. Computational grid technologies will provide access to distributed computing resources that enable the creation of the terascale analysis pipelines that will be required for some investigations.

The NVO will require a close collaboration between computer scientists and application scientists, and the experience gained from development of the NVO can be expected to significantly impact Information Technology research in the future. Astronomy provides an ideal environment for developing a large scale data grid prototype: the community of astronomers is moderate in size (a few thousand) but not too large; the data sets are heterogeneous, but not exceedingly so, providing interesting, tractable challenges for metadata standards and protocols; the data repositories are widely distributed, but typically already electronically accessible; security is not usually emphasized in astronomy; and finally, astronomy is of widespread interest to the public, providing an interesting proving ground for a knowledge network that engages many levels of society.

2.3. The NVO as Facility for Data Publication

A long-standing problem in astronomy and other sciences has been how to publish data in addition to a scientific paper. The advent of the Web has shown us how easy it is to "publish" a web page, and we intend to make data publishing just as easy for astronomers, yet in a semantically-rich fashion that allows readers to find it, read it, assess its provenance, and compute with it—in other words, to make full use of it.

The NVO must recognize the three different roles of author, publisher/curator, and reader. Data authors will work with publishers to "publish" their data, i.e., to provide full digital access to information which has traditionally only been available in graphs or tables in printed papers. When this process is complete, the data moves to the archive along with the metadata and documentation. In addition, a "standard" form of the data will be generated: we need to translate measurements into standard units wherever possible, translate data into the standard representations supported by the archive (data models), and perhaps define a few new measurements and representations. Publishing astronomical data must become far less difficult than it is today. It is the task of the NVO to develop the tools, templates, and standards that make it easy to document and publish data, and to make it cost-effective for the archivists to manage and curate published data. In the NVO era an astronomer wishing to publish data will be able to characterize it through an NVO "publication portal" which captures metadata describing the content in NVO compliant terms, identifies the access mechanisms, and registers the resource with the NVO. In this process the NVO publishing standards do not obscure the raw data; users will always be able to "drill down" to data in the original format. The standardization is used only for locating and federating disparate data sources.

3. The Main Challenges

Meeting the NVO's unique IT challenges will both enable new science and advance our IT technologies into a petascale data grid, soon to be a frontier for US business as well as science. Knowledge extraction from billion-object catalogs requires new indexing and summarization techniques. Petascale pixel image analysis from multiple distributed archives will require integration of digital library and grid middleware. As new classifications are discovered, understood, and archived, the NVO catalog will have to evolve. A data management system will provide a uniform data access layer for data pipelining, archiving, and retrieval of terabytes of distributed astronomical images. An information management system will support inserting, querying, and evolving billions of objects, each with thousands of attributes. A knowledge support system will provide software tools for correlation, visualization, and statistical comparisons of both cataloged data and original image pixel data.

The NVO architecture will be based on middleware that integrates federated, distributed, autonomous archives. It will connect users to analysis services and data services. The analysis-oriented service will support massive data analysis of catalog and pixel image data. The key functional requirements for the middleware are:

Handling distributed collections and resources. The NVO will federate existing astronomical data into a cooperating system. As such, interoperability and autonomy are key attributes of any design. Each participant must expose a uniform data access layer, but the internals of each member of the federation will quite likely remain unique. Collection integration will require mediation across the diverse semantic conventions used to describe all wavelengths and both ground- and space-based sensors.

Providing a uniform astronomy information infrastructure. The NVO will maintain international links to similar efforts in other countries, and in other disciplines. We will work to promote international standards for data and information access. This will lead to direct enhancement of international collaborations in astronomy.

Enabling tera/petascale data analysis. The NVO will provide integrated access to terascale computing and data facilities in a way useful to astronomers. The emphasis is on seamless scalability from desktop to supercomputer, so what is learned at one stage need not be unlearned later.

Capitalizing on advancing technology. We will incorporate commercially available technology where possible, while developing the needed large-scale statistical analysis and image analysis capabilities that are not currently available. We will focus on interoperability and open standards, and plan to adapt as new technologies emerge.

3.1. Data Federation and Fusion

Data federation and fusion is a prime focus of NVO. Combining existing datasets can create new knowledge; knowledge that does not require a telescope or a rocket launch. A prerequisite for data federation is interoperable metadata standards, but for large data, there are additional requirements. Caching and replication services can save the results of complex joins of multiple databases for later reuse. An efficient proximity join of a billion sources requires that the data are clustered so that nearby objects in the sky are nearby in the stream: we plan to use the HTM indexing developed at Johns Hopkins to achieve this (Kunszt, Szalay, & Thakar 2001).

Users have the most control and interactivity with their desktop workstations, so small datasets will probably be brought to the desktop for visualization and for experimentation. However, the initial stages of the pipeline may involve huge data volumes distributed over a wide area. Algorithms must be moved to the data. Thus agent-code portability is as important as the portability of the data format. While many operations can be controlled by menus and numerical parameters sophisticated users will want to write compiled code that can execute near the data.

3.2. Metadata Standards and Protocols

The NVO framework will work toward widely accepted astronomical metadata standards and protocols. These must be extensible into the far future. XML will be our fabric for structured information, including interoperation between Astronomical XML, Astronomical Markup Language,[5] Astrores,[6] Extensible Scientific Interchange Language,[7] and other astronomical data representations. FITS is a standard for structured datathat predates XML, and a first milestone of this proposal is a software toolbox for FITS/XML interoperation.

The NVO will help extend the "profiles" defined by NASA's Space Science Data System, and its prototype implementation in the Astrobrowse system. In addition to the interfaces among web services, we must extend the metadata semantics so that programs can be written against a "metadata API." Other, non-astronomical objects must also be described for a successful NVO, and we hope to borrow from other projects and from the commercial world for adequate semantic and syntactic descriptions.

3.3. Scalability

Scalability is a major challenge: The NVO must handle billions of objects in high-dimensional petascale astronomical catalogs. It must also scale to an international federation of hundreds of institutions: some huge and some tiny. The computational grid will help by providing massive online storage, parallel computing, resource management, and high-speed network access, but the NVO must solve the problems of data organization, data access, data analysis, and

---

[5] http://monet.astro.uiuc.edu/dguillau/these/

[6] http://vizier.u-strasbg.fr/doc/astrores.htx

[7] http://www.cacr.caltech.edu/XSIL/

data visualization. We will attack these problems by a combination of cunning and brute force: where possible the NVO will build sophisticated indices, precompute popular aggregates, and cluster the data for efficient access. But, in the end, the curse of dimensionality, or the ad hoc nature of some queries will force bulk data scans. In these cases, the NVO will use brute force: providing very high speed sequential access to the data, and compact replicas of the data (bitmaps or tag objects) that minimize the amount of data that must be scanned to answer a query. Of course, all these operations will be done in parallel using both parallel processing and parallel IO. This parallelism should come from the data management tools, but if required, the NVO will implement these mechanisms.

In the end, the system will be judged by how quickly users can pose questions, and understand the answers. This means that the system will be responsible for translating high-level, non-procedural queries into efficient execution plans, executing the plans so as to minimize data movement, and then delivering the results to the visualization tool as quickly as possible, perhaps allowing the user to steer the computation as it progresses.

The diverse user communities will access the NVO through networks of varying speeds; the tools and human interfaces must be usable for both low and high-speed connections. Some analyses will require very-large-memory machines and computing speeds, while others will fit well on Beowulf class systems. The computational grid and the NVO federation will include and support both computation styles.

NVO will serve a large user community. The number of astronomers worldwide is only a few thousand; however, the education and public access functions will have user communities numbered in the millions. While each such user will impose a modest load, the aggregate will be substantial. The NVO framework must be designed to handle large numbers of small users as well as a few very large users.

3.4. Data Understanding

Things are changing in the way that data analysis takes place in astronomy: intensive use of algorithms on data is replacing the personal astronomer/data relationship of the previous generation. The NVO will accelerate this process, so that thousands of astronomers can benefit from the data without drowning in it. There will be many astronomers wishing to analyze the data in many different ways. We will provide tools and algorithms that support statistical and data mining queries needed by astrophysicists, and interface these to the framework. These include spatial access methods such as kd-trees, R-trees, metric trees and newer generations thereof, and also condensed representations such as sparse datacubes and binned grids. These structures support queries regarding n-point correlations and non-parametric density estimates (Connolly et al. 2001). We will provide example implementations that will serve as well-documented tutorials on interfacing one's own analysis algorithm with the NVO and will also provide directly useful tools. Examples of such analyses involve the identification of rare objects, or automated shape finders searching for atypical objects, like the gravitationally lensed arcs.

4. Summary

We plan to build the NVO framework both through coordinating diverse efforts already in existence and providing a focus for the development of capabilities that do not yet exist. The NVO we envisage will act as an enabling and coordinating entity to foster the development of further tools, protocols, and collaborations necessary to realize the full scientific potential of large astronomical datasets in the coming decade. The components of the NVO include not only the archives, but also metadata standards, a standardized data access layer, query and computing services, and data mining applications. This involves significant technical and managerial challenges. No single group can build the NVO–the effort must include the existing data archiving efforts in astronomy even as it develops new capabilities and structures. The NVO must be able to change and respond to the rapidly evolving world of IT technology. In spite of its underlying complex software, the NVO should be no harder to use for the average astronomer than today's brick-and-mortar observatories and telescopes. Development of these capabilities will require close interaction and collaboration with the information technology community and other disciplines facing similar challenges. We need to ensure that the tools that we need exist or are built; we do not duplicate efforts, but rather rely on relevant experience of others.

The new capabilities of the NVO will be essential to realize the full value of the tera/petabyte datasets that are in hand or soon to be created. Rapid querying of multiple large-scale catalogs, establishment of statistical correlations, discovery of new data patterns and temporal variations, and confrontation with sophisticated numerical simulations are all avenues for new science that will be made possible through the NVO. Future surveys will have well defined templates available that will enable them to publish their data much more easily. The NVO, through its rich content and special portals designed for students, teachers and the public, will have major impact on a wide range of science education and public outreach projects. Computer science will be able to capitalize on the public appeal of astronomy.

Acknowledgments. I would like to acknowledge the enormous contributions from my friends and colleagues of the NVO Collaboration. It has been an extremely stimulating and enjoyable process working with such a knowledgable and highly motivated group, and getting a step closer towards our ambitious common goals.

References

Berners-Lee, T., Hendler, J., & Lassila, O. 2001, Scientific American, 284, 5 (May 2001), 34

Connolly, A. J., Genovese, C., Moore, A. W., Nichol, R. C., Schneider, J., Wasserman, L. 2000, AJ, in press

Kunszt, P. Z., Szalay, A. S., & Thakar, A. R. 2001, Proc. Mining the Sky, A. Banday, ed., Kluwer, 2001, in press

Astronomical Data Analysis Software and Systems X
ASP Conference Series, Vol. 238, 2001
F. R. Harnden Jr., F. A. Primini, and H. E. Payne, eds.

Chandra X-Ray Observatory - First Year Science Highlights

Stephen S. Murray

Smithsonian-Harvard Center for Astrophysics, Cambridge, MA 02138

Abstract. The *Chandra* X-ray Observatory achieved orbit on July 23, 1999, and began science observing in October 1999. Now over a year since the first light image of Cas-A, *Chandra* has observed several hundred science targets and many calibration objects. A few highlights from these data are presented - including a discussion of the spatially resolved spectrum of supernova ejecta in Cas-A and X-ray emission from the central regions of M31, the nearby AGN Cen-A, and the quasar 3C273.

1. Introduction

The launch, on July 23, 1999 of the *Chandra* X-ray Observatory heralded the second golden age of X-ray astronomy. *Chandra* brings to the field sub-arcsecond imaging that is comparable to ground based optical astronomy. With the Advanced CCD Imaging Spectrometer (ACIS; Garmire 97) and the High Resolution Camera (HRC; Murray 98) in the focal plane, and two Transmission Grating Spectrometers (LETGS, Brinkman 2000; HETGS, Markert 1994), *Chandra* gives astronomers a new view of the X-ray Universe that is sharper and deeper than ever before. This science highlight presentation will illustrate the power and potential of *Chandra* by showing some of the early results that have been obtained.

2. Cas-A

This supernova remnant (SNR) has long been studied. It was selected as the official first light image for *Chandra* because of its complex spatial extent and its high X-ray brightness. As seen in Figure 1 the X-ray emission is complex - it is easy to note the nearly circular shape of the SNR and the shell-like appearance that was already known from previous observations. However, in viewing the central part of the SNR *Chandra* for the first time revealed a point-like source that lies very near the center of expansion of the source (Tananbaum 1999). This compact central object was one of the first surprises from *Chandra*, and if truly associated with Cas-A, it is likely to be the end product of the core collapse of the progenitor star that lead to the supernova explosion of some 320 years ago.

The X-ray data (taken with the ACIS) also provide the X-ray energy distribution at each part of the image (spatially resolved spectroscopy); this is illustrated in Figure 1 by the various colors assigned to the image pixels. The lowest energy photons are colored red, intermediate energies are green, and the highest energies are blue. This representation shows how the spectrum of X-ray

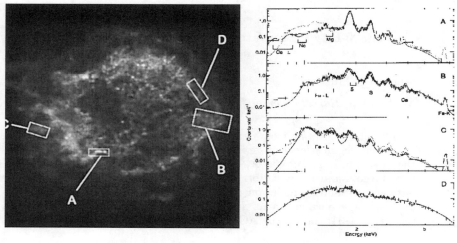

Hughes, Rakowski, Burrows, & Slane (1999)

Figure 1. Cas-A spatially resolved spectra, from Hughes et al. (2000).

emission changes over the remnant. This effect can be seen in more detail in the work done by Hughes et al. (2000). As shown in Figure 1 several regions were selected from the image, and the X-ray spectrum was plotted versus an expected spectrum for a particular model of SNR ejecta. As discussed by Hughes et al. the selected spectra vary considerably, and represent material thrown out from different layers of the core of the collapsing star during the supernova explosion. Interestingly, the material from the inner-most layers of the core has traveled out to the edge of the remnant as indicated by the presence of iron in the X-ray spectrum. This effect is probably due to differences in the ejection velocity of the core layers as the collapse evolves. It provides important observational data for the theorists to match in their detailed numerical calculations for these events.

Figure 2 shows a 50 ksec observation of Cas-A that was taken with the HRC. In this image, the central compact object is clearly seen. There are about 4,000,000 events from Cas-A in this image, and only about 1500 of them are from the central source. The excellent quality of the *Chandra* mirrors, particularly their high angular resolution and low scattering, is essential for separating the point source from the surrounding nebula. The purpose of this long HRC observation is to search for a periodic signal from the compact object. This signal, if found, would confirm the association of the point source with a neutron star that should have been formed during the supernova explosion. At the time of this writing, the data are still undergoing analysis, and results are not yet available.

These observations demonstrate the importance of coupling image processing and display systems with data analysis software. Defining regions of interest in an image to select data of interest from a non-spatially sorted data set (as for example the photon event list from *Chandra*) is an important tool needed for science analysis. Similarly the tools needed for detailed timing analysis present

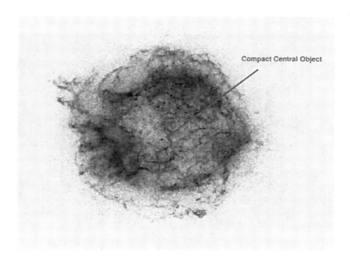

Figure 2.   HRC image of Cas-A

a challenge to the *Chandra* data system. Not only must the proper events be easily selected (as for the spectral analysis case), but there are systematic time corrections, such as accounting for orbital motion, that need to be accurately applied to the data before any temporal analysis can be performed. Searching for periods is known to be a compute intensive task. Developing efficient algorithms to implement the search (e.g., FFT, period folds, and other statistical tools) is an important need of the scientist.

3. M31

The *Chandra* monitoring program for M31 (our nearest neighbor galaxy, Andromeda) has provided several interesting results. Figure 3 shows a comparison of ROSAT and *Chandra* images of the central region of M31. These images are on the same spatial scale, and illustrate that the sharpness of the *Chandra* telescope achieved in just 15 ksec what took much longer with ROSAT.

Figure 4 shows the central region of the galaxy. Here the circle represents a $5''$ radius about the galaxy center. Where ROSAT detected a single source, *Chandra* resolves five individual point sources. One of these is likely to be associated with the actual center of M31, where there is a $3 \times 10^7 M_\odot$ black hole (BH). The two *Chandra* sources closest to the BH location are just $0.5''$ apart.

It will take some additional work to determine which of these sources is most likely the central source. One has an unusually soft X-ray spectrum, a characteristic associated with a class of objects known as super-soft sources (SSS's). The other appears to be typical of most galactic X-ray sources. The SSS is quite variable in intensity (about a factor of 5-10 within a few months). However,

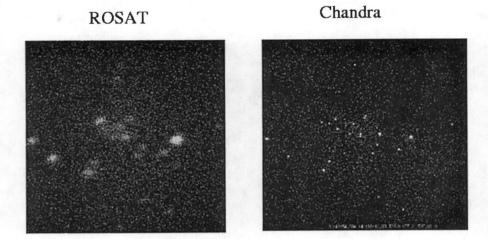

Figure 3. Comparison of ROSAT and *Chandra* images of the central region of M31 (Andromeda), our nearest neighbor spiral galaxy

none of these properties is sufficient to confidently determine an association of the black hole with the center of M31. Monitoring of the region is continuing to improve the absolute locations. HST images of the center of M31 provide matches between *Chandra* and HST sources so that the coordinates can be accurately aligned. If successful, this work will eventually lead to 0.2″ precision in locating the center of M31 with respect to the X-ray sources.

The role of astronomical software in helping to process, analyze, and understand the data from M31 is critical to progress. In the case of M31, the entire galaxy is too large to be observed in a single detector field of view. Mosaic images are needed to give a complete view. This process requires algorithms for translating images onto a larger frame, retaining all of the information and taking into account edge effects from the detectors and telescope. Exposure maps and corrections are needed, particularly in areas of overlap, so that proper source intensities and light curves can be constructed. Stacking repeated images, as in the central region of M31, is another challenge for software developers. Easy methods for co-aligning images and calculating combined exposures are required for such observations. Matching data from *Chandra* and HST involves careful calculations of source centroids, transformations of coordinates, and accounting for detailed differences in the astrometry from each mission.

4. Cen-A

The *Chandra* Observatory makes it possible to study in great detail the X-ray properties of the nearest active galactic nucleus (AGN) source Cen-A (or NGC 5128). This galaxy is about 3 Mpc distant, implying that 1′ corresponds to about 1 kpc in linear dimension. Thus with 1″ image quality, it is possible to observe

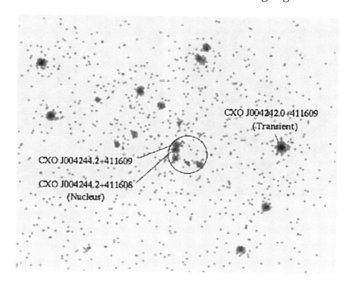

Figure 4.   The nucleus of M31 as seen in X-rays by *Chandra*

X-ray phenomena on the linear scale of about 15 pc (45 ly). This is the scale of distances between our sun and neighboring stars.

Figure 5 shows X-ray images obtained with the HRC and ACIS. In both images it is easy to see the strong central nucleus that is being powered by a super massive black hole (about $3 \times 10^7 M_\odot$) and a long one sided jet of emission directed to the northeast. The jet consists of many bright knots embedded in what appears to be a well collimated diffuse emission region. The jet extends at least 6′ from the nucleus.

In Figure 6, the X-ray emission is compared with radio data taken at 13 and 6 cm wavelengths. The 13 cm radio data emphasizes the large scale structure from Cen-A and shows what are called the inner radio lobes that extend to the northeast and southwest, as well as the jet. There is excellent correspondence between the X-ray and radio data when viewed on a large scale. Of special interest is the bright X-ray ridge of emission that corresponds to the edge of the southwest radio lobe. This ridge was barely detected in previous X-ray observations (e.g., *Einstein* and ROSAT) but is now seen clearly. It is very closely aligned with the edge of the radio emission. The southwest radio lobe is actually filled with diffuse X-ray emission, with the region of peak radio emission inside the lobe corresponding to the minimum X-ray emission region. Understanding the details of the physical conditions at Cen-A that give rise to these correspondences is the goal of this investigation. The ACIS can extract spatially resolved X-ray spectra at different locations in the image for studying the spectral changes as a function of location as an indicator of changes in the properties of the emitting material.

The second panel in Figure 6 shows the radio emission at 6 cm plotted over the X-ray image along the jet. Again there is good general correspondence between the radio and X-ray image. However, at the detailed level, there are some

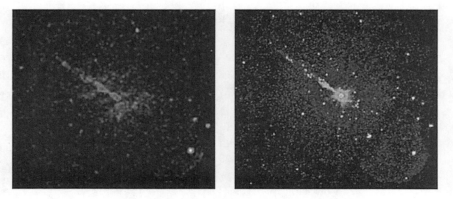

Figure 5. *Chandra* images of Cen-A HRC and ACIS

striking differences. While peaks in the X-ray and radio emission correspond for the innermost knot in the jet, farther out from the nucleus the correspondence becomes less precise, and the peaks in X-ray emission lie closer to the nucleus than the corresponding radio peaks. This change in the relative locations of peak X-ray and radio emission indicates that there is likely to be a lot of shock formation, particle re-acceleration, and radiation going on along the path of the jet. The knots may well be sites of shocks from which accelerated particles propagate outward, losing energy to radiation. The higher energy particles lose their energy faster so that the spectrum softens farther from a shock. The detailed processes along the jet are much more complex at the scale of *Chandra* resolution than was evident from lower resolution observations.

The astronomical software impact for studying these data follows many of the points already discussed. A new process used for looking at complex images is seen in the adaptively smoothed image of Figure 6. This is an example of image processing that helps to highlight features on differing spatial scales making visual inspection easier. Various algorithms for this class of image processing are possible, and typically there are many parameters that can be set for these algorithms. Having well constructed software that can run efficiently and reliably to implement these processes is critical to their use. Being able to quickly run many cases of an adaptive smooth, for example, varying parameters to bring out desired features or suppress noise, allows a researcher to develop a sense of what is important and what is real in an image. Similarly, detecting sources in complex images such as Cen-A is an intriguing problem. Using wavelet decomposition, percolation techniques, or standard sliding box detect algorithms, each gives a particular advantage and disadvantage in the task of extracting information from an image. In practice all of these methods (as well as others not discussed here) should be tried and compared in order to obtain the most reliable results. The ease of use of these techniques, as well as their running time, often dictates what is actually done. Astronomical software developers need to work with scientists to make this task as easy and complete as possible.

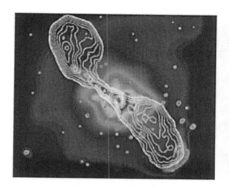 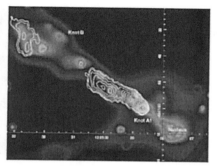

Figure 6.   Radio and X-ray comparison for Cen-A

5. 3C273

A final example of recent *Chandra* results, shown in Figure 7, is the image obtained using the HRC to observe the quasar 3C273, a well known bright QSO. Associated with the source is a jet that extends over hundreds of kpc. Thanks to the resolution and low mirror scatter of *Chandra*, this jet is easily detected even though there are just 1300 photons in that part of the image, as compared with over 225,000 photons from the central point source. *Chandra* image quality allows X-ray emission from the jet to be resolved into at least three components. The first bright knot along the jet may itself be slightly extended. The other two knots are too faint to detect any such extent.

Figure 8 shows a composite of images of the jet of 3C273 taken in the radio, optical and X-ray bands. It is apparent that the jet behaves differently at these three wavelengths: the X-ray image becomes fainter farther from the nucleus, the radio image gets brighter, and the optical remains largely constant - although fragmented into many knots. As for Cen-A, this comparison suggests that the physical processes responsible for emission are varying along the jet. The closest knot is well fit to a synchrotron model, while this model fits less well for the jet region farther along and fits rather poorly at the jet's end. Physical conditions along the jet are changing – not surprising since the jet is the size of our Milky Way Galaxy. More complex models must be considered if radio, optical, and X-ray emission is to be jointly understood and interpreted.

The observation of 3C273 illustrates the need to be able to simultaneously consider the multi-dimensional problem of spatially resolved spectroscopy across broad wavelength bands. The development of data-cubes, for representing the information, and then processes that can act on these large data items, appears to be the direction for the future. *Chandra* is only the first of several next generation observatories that will require such data types and processes.

6.   Conclusion

The launch and operation of *Chandra* has brought a flood of exciting new data into the hands of observers. The tremendous detail and richness of *Chandra*

Figure 7. HRC Image of 3C273

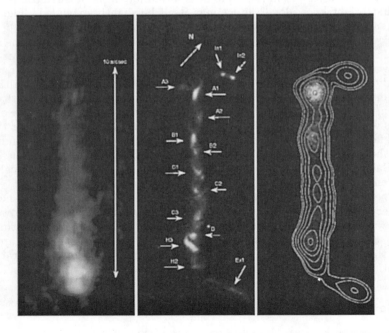

Figure 8. Jet of 3C273 shown in radio, optical and X-ray bands

data, especially when combined with observations in other bands (radio, IR, optical, UV, etc) places new demands on astronomical software. The volume of data is growing along with its complexity. This growth leads to needs for faster, more capable computer systems to deal with the data, but more importantly, it places greater demands on the types of analysis and presentation software. The modern astronomer needs interactive image display, data selection tools, filters and algorithms that work in an integrated fashion to allow for coherent, rapid data analysis. The science productivity of a mission such as *Chandra* depends as heavily on the capabilities of astronomical software as it does on the technological innovations of the observatory. From these few examples of what *Chandra* data look like and how their analysis and interpretation depend on the data systems available, it is clear that much progress toward the required software capabilities for processing *Chandra* data has already been made (if this weren't true, the above images could not have been presented!). Further progress is almost certainly needed in the area of science analysis tools that realize the full potential for interpretation and understanding of the information in images such as the ones presented here. It is these latter needs that the ADASS community must address. One can hope that the new data systems software to emerge in the near future will produce many satisfied customers.

Acknowledgments. This work was supported by NASA Contract NAS8-38248. I would like to thank the many members of the HRC Team for their efforts in building the HRC, particular thanks to Mr. Gerald Austin the Project Engineer for his tireless efforts. I also want to thank the many scientists at SAO who have worked with me on the HRC both in the development of the instrument and in the subsequent science program. Particular thanks for this presentation go to Mike Garcia, Ralph Kraft, and Dan Schwartz for permission to discuss their work.

References

Brinkman, B. C. et al. 2000, Proc. SPIE, 4012, 81

Garmire, G. G. 1997, BAAS, 190, 34.04

Hughes, John P., Rakowski, Cara E., Burrows, David N., & Slane, Patrick O. 2000, ApJ, 528, L109

Markert, T. H., et al. 1994, Proc SPIE, 2280, 168

Murray, S. S., et al. 1998, Proc SPIE, 3356, 974

Tananbaum, H. D. 1999, IAU Circ. No. 7246

Astronomical Data Analysis Software and Systems X
ASP Conference Series, Vol. 238, 2001
F. R. Harnden Jr., F. A. Primini, and H. E. Payne, eds.

Chandra X-ray Observatory Operations

Roger J. Brissenden

Harvard-Smithsonian Center for Astrophysics

Abstract. The Chandra X-ray Observatory was launched from the Space Shuttle on 23 July 1999 and has completed the first year of operations with outstanding results. We present a description of the Chandra Observatory, the Chandra mission operations concept, ground system architecture, selected operations metrics and an example where operational processes have required modification due to on-orbit events and experience.

1. Introduction

The Chandra X-ray Observatory (CXO) (formerly the Advanced X-ray Astrophysics Facility, or AXAF) is the third of NASA's Great Observatory missions; it follows the Hubble Space Telescope (HST; 1990–) and the Compton Gamma-Ray Observatory (CGRO; 1991–2000) and precedes the Space Infrared Telescope (SIRTF; 2002). Chandra is a space-based Observatory containing a high resolution (0$''$5) X-ray telescope responsive to the energy range 0.1–10 keV and a complementary set of imaging and spectroscopic instruments. The mission was designed with a minimum 5-year lifetime and a goal of 10+ years, and provides an order-of-magnitude advance in spatial and spectral resolution over previous X-ray telescopes.

The Chandra Program is managed by NASA's Marshall Space Flight Center. Science and mission operations for the Program are carried out at the Chandra X-ray Center (CXC) and Operations Control Center (OCC) located in Cambridge, MA, using facilities of the Smithsonian Astrophysical Observatory (SAO) and the Massachusetts Institute of Technology (MIT). Observing time is awarded through an annual peer review. Selected targets are scheduled in weekly segments and command loads to implement the mission schedule are uplinked to the spacecraft from the OCC via NASA's Deep Space Network. Telemetry and data are downlinked approximately every 8 hours, monitored for state of health at the OCC, and passed to the CXC for science processing, archiving, and distribution to the Observer. In addition, CXC also provides an Education and Outreach program, and administers the Chandra Grants and Fellowship programs.

The Chandra Observatory was launched on the Space Shuttle Columbia (STS-93) on 23 July 1999. Following launch and orbital insertion, Chandra underwent a 2 month Orbital Activation and Checkout phase before starting 2 months of Guaranteed Time Observations in September 1999. The first cycle

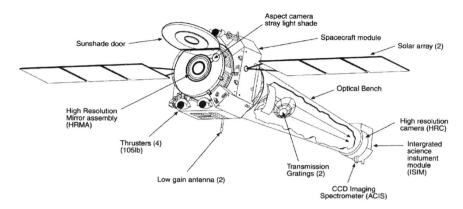

Figure 1.   Chandra X-ray Observatory showing selected components.

of General Observer observations began in November 1999, the second cycle in November 2000, and the third is scheduled for November 2001.

In this paper, we provide a brief overview of the Chandra Observatory and discuss the launch, orbital insertion, and activation (§2), describe the science-phase mission operations concept (§3), and give a summary of selected operational metrics used to monitor mission progress and efficiency (§4). In §5 we provide an example of how an on-orbit anomaly can impact all of these systems and processes, and discuss the importance of a system-wide approach (for both the space and ground segments) to problem response. For a discussion of the scientific aspects of the mission, see Murray (2001).

2. Mission and Observatory Description

2.1. Observatory Description

Figure 1 shows Chandra in its deployed configuration with selected components labeled. The Observatory consists of the telescope system, the science instruments, the Command, Control, and Data Management system (CCDM), the Pointing, Control, and Attitude Determination system (PCAD), the Electrical Power System (EPS), thermal control, and propulsion systems. Chandra is a physically large spacecraft with a wing-span of 19.5 m, length with sun-shade door open of 11.8 m and an on-orbit mass of 4800 kg.

The principal components of the telescope system are the High Resolution Mirror Assembly (HRMA), which consists of four pairs of grazing incidence Wolter Type I mirrors with focal length 10 m, and an optical bench assembly that connects the mirror assembly to the Integrated Science Instrument Module (ISIM) that houses the focal plane science instruments. The mirror assembly has an effective area of 800, 400, and 100 cm^2 at 0.25, 5.0, and 8.0 keV, respectively. It provides a ghost-free field of view of 30$'$ diameter with a plate scale of 48.8 μm arcsec^{-1}, and a point spread function whose full width at half maximum including detector effects is 0$''$.5.

X-rays are focused onto one of two selectable focal plane instruments, the Advanced CCD Imaging Spectrometer (ACIS) and the High Resolution Camera (HRC), which are optimized for the higher and lower portions of the energy range, respectively. The ACIS focal plane is populated by ten CCDs each 1024×1024 format with 24 μm pixel size, configured as a 2×2 imaging array and a 1×6 spectroscopic array. Two of the ACIS CCDs are back illuminated and eight are front illuminated. Image data from the ACIS CCD chips are acquired and processed on-board by the ACIS flight software. The software is hosted by a Mongoose controller based on a MIPS R3000 chip set. By means of ground specified parameter blocks, it supports configurable science runs with a variety of observing modes including timed exposure, continuous clocking, spectroscopy, calibration, and diagnostic.

The HRC is a microchannel plate instrument with two detector regions configured as imaging and spectroscopy detectors. The HRC provides time resolution up to 16 μs. ACIS and HRC are complemented by two movable gratings, the High Energy Transmission Grating (HETG) and the Low Energy Transmission Grating (LETG), that can be inserted into the optical path to disperse a spectrum across one of the spectroscopic detectors. The HETG contains two sets of gratings, each with a different period, mounted on the same structure. The outer set disperses X-rays from the outer two mirror pairs, and the inner set disperses from the inner two mirrors pairs. The spectra form an X shaped pattern centered on the undispersed zeroth order. The HETG is matched with the ACIS to provide high resolution spectroscopy with $E/\Delta E$ up to 1000 over 0.4–10 keV. The LETG is matched with the HRC to provide high resolution spectroscopy with $E/\Delta E \geq 1000$ over 0.07–7.29 keV.

The CCDM system provides the command, telemetry, and data management functions for the spacecraft. The major components include two low gain antennas, two S-band transponders, two 1.8 Gb (giga-bit) solid state recorders, redundant 16-bit CDI 1750A On-Board Computers (OBC), redundant LSI 16-bit 1750A computers for control during safe mode, and redundant Interface Units, Command and Telemetry Units, and Remote Command and Telemetry Units for communications with the spacecraft systems. The principal telemetry rate is 32 kbps (kilo-bits per second), with 24 kbps allocated to the science instruments and 8 kbps to spacecraft engineering data. The spacecraft supports a command uplink rate of 2 kbps and a downlink range of 32–1024 kbps. The OBC flight program consists of an executive responsible for state control, interrupts, and task control, and a set of functional software systems for the control of PCAD, CCDM, EPS, Telescope (thermal and gratings), radiation monitor and health, and status. The CCDM function handles the Science Instrument telemetry.

The PCAD system provides the sensors and control hardware used to point the observatory, slew to new targets, and perform solar array positioning and momentum unloading. Chandra's pointing requirements are modest ($30''$) compared with the Hubble Space Telescope (for example) due to the photon counting nature of the detectors. The pointing direction is obtained from the gyroscopes and aspect camera as it tracks 5–8 optical stars pre-selected for each target; during ground processing the position of each photon is transformed from detector to sky coordinates using the aspect camera star data ("image reconstruction"). The performance of the aspect system on-orbit yields an absolute celestial pointing of 3″, an image reconstruction of 0.″3 and celestial location of 0.″76. Other components of the PCAD system include six reaction wheels for attitude control, and coarse and fine sun sensors used for pointing control modes that do not use the aspect camera.

The Propulsion System consists of the Integral Propulsion System (IPS) used during the transfer orbit and now deactivated, and the Momentum Unloading Propulsion System (MUPS) used to unload momentum from the reaction wheels. Momentum build-up occurs due to the gravity gradient of the Earth as the spacecraft passes through perigee, and from solar pressure. The IPS system includes both 105 lbf liquid apogee engines and 20 lbf Reaction Control Thrusters (RCS), while the MUPS system contains 0.2 lbf thrusters. Both the RCS and MUPS systems use liquid fuel hydrazine. The fuel is the only mission expendable. Projections based on usage during the first year of operations indicate sufficient fuel for more than 20 years of operation.

The Electrical Power System generates, stores, and distributes electrical power to the spacecraft. The major components are the solar arrays (two 3-panel wings), which provide 2112 watts of power, and three NiH_2 batteries with 120 A-hr capacity. The batteries provide power during earth eclipses and occasional lunar eclipses. Chandra's orbit was chosen to ensure a battery depth of discharge no more than 80% for any eclipse. This requirements results in eclipses generally shorter than 2 hours.

The Thermal Control System contains passive elements such as multi-layer insulation blankets and a range of radiator materials, and active elements including temperature sensors, thermostats, and heaters. The on-board computer controls the active heaters to maintain the mirror assembly temperature at $70 \pm 2.5°$ F, a key requirement for maintaining the image quality.

2.2. Chandra Launch, Orbital Insertion, and Activation

Following Chandra's 23 July 1999 launch on the Space Shuttle Columbia, the spacecraft was deployed together with its attached two-stage Inertial Upper Stage (IUS) rocket motor. The IUS took Chandra to its transfer orbit before separating from the satellite approximately 11 hours after launch. Chandra traveled to its final orbit via a series of five firings of its own onboard propulsion system (IPS). The final orbit of $\sim 140,000 \times 10,000$ km was reached on 7 August 1999 with Keplerian orbital parameters as shown in Table 1. For comparison,

Parameter		7 Aug 1999	1 Jan 2001
Semi major axis	a	80798.5 km	80790.1 km
Eccentricity	e	0.802	0.756
Inclination	i	28.5°	35.5°
Right Ascension of Ascending Node	Ω	194.1°	148.9°
Argument of Perigee	ω	271.1°	305.1°
True Anomaly	ν	180.1°	172.0°
Period	P	63.491 hrs	63.481 hrs

Table 1. Chandra Keplerian Orbital Element Comparison.

the parameters are shown as of 1 Jan 2001 and are consistent with a predicted orbit circularization trend through 2005. The orbit provides approximately 80% viewing time above the radiation belts that extend to ~ 60,000 km altitude around the earth. The orbital parameters resulting from the fourth burn were propagated over a 10 year mission using multiple computers running copies of the same code. The calculation was time critical since the fifth (trim) burn parameters were required within 24 hours in order to be able to uplink and execute the burn. The final orbit parameters were adjusted by the fifth burn to ensure an optimal eclipse duration mission profile. The propagation calculation yielded a clear best choice resulting in a final orbit with no earth or lunar eclipse exceeding a battery depth of discharge of 80%.

Once final orbit was attained, the spacecraft systems were activated over several weeks. Following the opening of the sunshade door on 8/12/99, the pointing system was activated, and the official first light image of Casseopia A was taken on 8/19/99. Other observations taken during the activation phase included aspect camera field distortion measures, and science instrument calibrations to measure the focus, determine bore-sight and optical axis, characterize the on-axis and off-axis response, effective area and plate scales, and measure standard candles.

3. Chandra Operations Concept

3.1. Principal Operations Thread

The principal operational thread for the CXC is shown in Figure 2 and shows the end-to-end flow for a Chandra proposal submitted by an observer from top-to-bottom in the center of the figure. Rectangular boxes map to functional components of the CXC Data System (CXCDS).

In response to an annual NASA Research Announcement, an observer prepares and submits a proposal through Proposal Support (Figure 2) using observation modeling software and remote proposal submission tools. Following the annual peer review, the targets from the accepted proposals (Proposal Data) populate the Observing Catalog (OCAT) and form the input for the long term schedule that is generated by the Science Mission Planning Team using the *SPIKE* scheduling software. A weekly oversubscribed target list is submitted to the OCC as an Observation Request (OR) list. The OR list includes calibration targets, which account for an average of 8% of the observing time.

The Flight Mission Planning team incorporates any required engineering activities and develops an optimized operations schedule using the Off-Line System (OFLS) software. The OFLS includes software for mission planning and scheduling, command load generation, attitude determination, sensor calibration, ephemeris and orbit events generation, and engineering analysis. The schedule of spacecraft slews, mechanism motions, momentum dumps, and other actions is generated to ensure that none of over 800 pre-defined constraints are violated. Command loads are generated using the OFLS, and a series of additional constraint and timing checks are performed by the flight and science team prior to uplink to the spacecraft. Command loads typically run for 2 days, and three loads at a time are placed on-board for sequential execution. As the OBC executes the sequence of maneuvers, instrument movements, and configurations, telemetry is recorded at 32 kbps on the spacecraft solid state recorder.

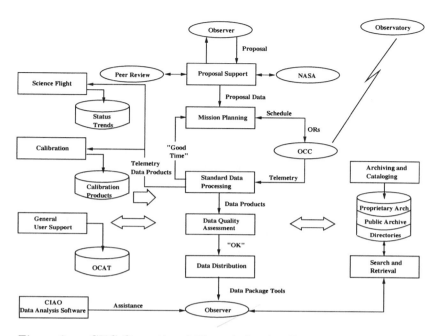

Figure 2. CXC Operational Thread showing Proposal Path.

The data are downlinked during a real-time pass, usually at 512 kbps. Chandra is operated with three passes of 1–2 hours per day through the DSN. Two types of telemetry reach the OCC: "real-time" telemetry of science and engineering data that are generated on the spacecraft while it is in contact with the OCC; and "dump" telemetry of data that were stored on the spacecraft solid state recorder between contacts.

Real-time commanding and monitoring are performed using the On-line System (ONLS) and GRETA[1] software systems. The ONLS includes software to uplink commands and data in real-time to the spacecraft and to monitor downlink telemetry. GRETA provides software for real-time monitoring and long-term trending of engineering data. Engineers check the state of the spacecraft against the predicted state at each pass; $\sim$ 1100 real-time telemetry points are checked against a limit data base with automatic out-of-limit notification. Dump data are transferred from the DSN to the OCC within a few hours of each pass and then to the CXC for science processing.

Chandra dump data are processed by the CXCDS Standard Data Processing function through a series of levels using an Automated Processing (AP) system. Level 0 decommutates telemetry and processes ancillary data, Level 1 performs event processing, standard filtering, and aspect reconstruction, Level 2 performs source detection and generates derived source properties, and Level 3 will create catalogs spanning multiple objects and observations. A determina-

---

[1]Heritage code from the CGRO mission, modified for use with the Chandra Program.

tion of the useful science time on a target ("good time") is made during AP and fed back to the mission planning function to trigger additional observations if needed. An example of "bad time" is a period of unacceptable aspect solution. See Plummer & Subramanian (2001) for a detailed discussion of the CXCDS Automated Processing system.

The Data Quality Assessment function includes both automated and manual checks of the data products generated by AP, including quality of aspect solution, background rates, observation and instrument configuration parameters. Following a successful data quality check (called Verification and Validation, or V&V), L0–L2 data products are made available to the observer through the archive and on distribution media. Data have a 1 year proprietary period and are available through the archive only to the observer during that period. Non-proprietary data (e.g., calibration observations) are publically available once V&V is complete. Data analysis by observers is supported by the Data Analysis Software function with the Chandra Interactive Analysis of Observations (*CIAO*) software system (Doe, Noble, & Smith 2001). *CIAO* provides a package of data analysis tools that include an interactive fitting and plotting environment (*sherpa*), an interactive filter composition (*filterwindow*), quick interaction with observation files (*firstlook*), a plotting and imaging package (*chips*), a file browser (*prism*), and tools for high resolution spectral analysis (*guide*).

In addition to carrying out the primary proposal thread shown as the central path through Figure 2, the CXC performs other functions. These include science monitoring and trending of the science instruments and spacecraft subsystems (Science Flight), maintaining the evolving calibration of the optics and science instruments (Calibration), providing software in support of the annual peer review and other Chandra programs, and providing Education and public Outreach (General User Support).

4. Mission Metrics

A set of mission metrics were defined prior to launch and have proven valuable during the first year and a half of operations in detecting process bottle-necks, taking corrective action, and monitoring their resolution. The specific metrics, shown in Table 2, were chosen to cover mission expendables, end-to-end data throughput, observing efficiency, and user interaction.

A breakout of the key metric Observing Efficiency, which is defined as time collecting science photons as a fraction of total orbit time, is shown in Table 3 for Cycle 1. Chandra cannot observe in the radiation belts or during the Charge Transfer Inefficiency (CTI) measurements (see §5), so the actual on-target observing efficiency of 66% is out of a possible maximum of 75%. The present efficiency is within 10% of optimal; however effort is on-going to increase the efficiency by optimizing the target acquisition time and reducing the CTI measurement time through refined modeling of the radiation belts.

Momentum Unloading Fuel	Fuel remaining by month vs. prediction. Projection supports > 20 year mission.
Observing Efficiency	On-target time divided by total orbit time per month. Cycle 1 average 66%.
Data Delivery Effectiveness	Time from observation to data accessibility to observer. Average reduced from 60 days at mission start to 13 days.
Cumulative Observing Time	Cumulative observing time (ks) by month for calibration and non-calibration targets with linear projection.
Scheduled Observing Time	Absolute observing time (ks) by month for calibration and non-calibration targets. Average 1600 ks non-cal, 140 ks cal.
Help Desk Statistics	Help tickets open and closed per month, numbers active and deferred. Average open and closed per month 100 tickets with peaks around NRA.
Data Loss Statistics	Table of data loss from spacecraft to DSN, DSN to CXC. Average loss < 0.25%.
Grant Award Time	Days from data distribution to grant award. Average of 30 days at cycle 1 start reduced to current average of 12 days.
Data Archive Growth Rate	Archived 650 GB (giga-bytes) in 14 months with 2:1 compression, growth 500 GB/yr, average 25 GB/month retrieval.

Table 2. Chandra Mission Metrics.

5. Operational Process Example: Reacting to Instrument Damage from the Radiation Belts

The Problem. Shortly after the start of science observations, a sudden and unexpected degradation was detected in the energy resolution of the front-side illuminated CCD chips of the ACIS instrument. The energy resolution as characterized by the Charge Transfer Inefficiency (CTI) was seen to have become a function of row number, being nearer the pre-launch value in rows close to the frame store portion of the array and substantially degraded in the farthest row.

The Cause. Analysis of on-orbit data and operational conditions, and ground tests of similar CCDs pointed to the likely cause as damage due to low energy protons encountered during the radiation belt passages reflecting off the X-ray mirrors and onto the focal plane. The back-side illuminated chips were not damaged, as would be expected for the soft proton scenario given the depth of their buried channels.

The Operational Response. Immediately following the identification of the cause, operational procedures were modified to ensure that ACIS was moved out of the focal plane during radiation belt passage. In addition, measurements of the internal ACIS calibration source were scheduled during a period before

On Target Observing	66%
GTO/GO/DDT[a] Targets	58%
Calibration Targets	8%
CTI Measurements (radiation zone margins)	10%
Maneuver and SIM motions	5%
Star Acquisitions	2%
Idle (includes safe modes and large solar flares)	2%
Radiation Zone	15%

[a] Guaranteed Time Observer/General Observer/Director's Discretionary Time

Table 3. Chandra Cycle 1 Observing Efficiency.

and after each belt entry, to ensure that no further damage was sustained and to monitor degradation levels. These calibration measurements provide a measure of the CTI each orbit and are included in Table 3. Since this operational change, no further degradation has been detected and observations with ACIS have continued effectively through use of both the undamaged back-side chips and the front-side chips.

Mitigation Approach. A series of mitigation approaches are being investigated to improve the energy resolution of the front-side chips, including running the ACIS focal plane at the lowest possible temperature (implemented) and clocking the chips using a new mode that distributes particle induced charge to fill the radiation-induced charge traps.

The Operational Impacts. A systems approach was taken in responding to the CTI problem that involved all groups within the Chandra Program. The full range of impacts was unexpected and illustrates the importance of taking a systems approach when dealing with operational issues. The following steps were taken:

- The mission scheduling approach was modified to ensure that ACIS was removed from the focal plane before, during, and immediately after radiation belt passage.
- ACIS CTI calibration measurements were defined and added before and after belt passages. Analysis and monitoring software were developed to perform on-going analysis and trends.
- Spacecraft software was changed to protect ACIS in the event of a safing action. The original software was designed to leave ACIS at the mid-point between the focal point and the newly determined safe position. The flight software modifications now ensure that ACIS is always moved fully to the safe position for safing actions.
- ACIS processing, monitoring, and analysis software were modified to accommodate CTI degradation.
- A ground and in-orbit test program was undertaken to develop mitigation techniques.
- The Cycle 1 observing program was re-organized to accommodate the new instrument performance. All observers were contacted and new chip con-

figurations were determined, observations were rescheduled and documentation was changed.
- A new calibration program was developed, planned, and implemented to characterize ACIS at a new operating temperature, and to fully characterize the chip response. Further calibration will be required in the event of implementation of other mitigation approaches.

The Chandra team was well organized to respond to this anomaly effectively. The ACIS instrument team reproduced the problem on the ground and developed possible mitigation approaches; CXC and MSFC science and data system staff modeled the radiation belts and provided operational parameters for safe operation; the spacecraft team rapidly and safely modified and verified complex safing flight software; the mission planning staff and system responded with essentially no loss in observing efficiency; and a set of existing management processes ensured that the multiple impacts were assessed and responded to on a system-wide basis. The latter was accomplished through the use of a configuration management control board with membership from all the teams (e.g., spacecraft engineering, flight operations, science, and ground data system teams), and ensured that changes were made in a controlled and coordinated manner throughout the system.

6. Conclusion

We have described the Chandra mission, the Observatory, the operations concept, mission metrics, and an example of a single anomaly with system-wide impacts that the team has responded to effectively. Where possible, future missions should design processes prior to launch that assume system level responses will be required.

Acknowledgments. The success of the Chandra mission is due to the hard work and creativity of a large team of dedicated people at many institutions including MIT, NASA, PSU, SAO, SRON, TRW, and their contractors. This paper summarizes their outstanding work.

This work is supported by the Chandra X-ray Center under NASA contract NAS8-39073.

References

Doe, S., Noble, M., & Smith, R. 2001, this volume, 310
Murray, S. 2001, this volume, 13
Plummer, D. & Subramanian, S. 2001, this volume, 475
Chandra Proposers' Observatory Guide, Version 3.0, 28 December 2000: http://asc.harvard.edu/udocs/docs/POG/MPOG/index.html
CIAO web location: http://asc.harvard.edu/ciao/
Chandra X-ray Center web site: http://asc.harvard.edu/

The SIRTF Science Center Is Alive And Well And Living In Pasadena

William B. Green

IPAC/SIRTF Science Center, California Institute of Technology, Pasadena, CA 91344

Abstract. The Space Infrared Telescope Facility (SIRTF) will perform an extended series of science observations at wavelengths ranging from 20 to 160 microns for five years or more. The science payload consists of three instruments delivered by instrument Principal Investigators located at University of Arizona, Cornell, and Harvard Smithsonian Astrophysical Observatory. The California Institute of Technology has been selected as the home for the SIRTF Science Center (SSC)[1]. The SSC is responsible for evaluating and selecting observation proposals, providing technical support to the science community, performing mission planning and science observation scheduling activities, instrument calibration during operations and instrument health monitoring, production of archival quality data products, and management of science research grants. The SSC is currently engaged in development of a major data processing and archive system. The first components of that system have now been operational for over a year, and have been used by Guaranteed Time Observers to submit over 3000 observation requests. In addition, the solicitation for the Legacy Science Proposals has been completed, and proposal teams utilized tools and systems developed by the SSC to submit their proposals electronically for evaluation and selection.

1. Introduction

The Space Infrared Telescope Facility (SIRTF) is now scheduled for launch in July, 2002. The SIRTF Science Center (SSC) on the Caltech campus in Pasadena, CA, will have responsibility for scientific operations during the mission. The responsibilities include solicitation of proposals from the science community, development of integrated science observation planning and scheduling products, transfer of integrated observation schedules to the Jet Propulsion Laboratory for execution by the Observatory, pipeline processing of data from the three science instruments, and development and maintenance of the science data archive for community access. The SSC is responsible for design, development, and operation of the Science Operations System (SOS) which will support the functions assigned to the SSC by NASA.

---

[1]http://sirtf.caltech.edu

In the past year, the SOS has made the transition from a design on paper to an operational system that is currently supporting proposal submission, observation planning by the Guaranteed Time Observers and the general science community, mission planning and scheduling of the first set of post-launch observations, and pipeline design and processing of preflight instrument data. Innovative web-based tools for planning observations with SIRTF have been successfully downloaded to hundreds of user sites, and the observation database at the SSC currently contains thousands of individual observation requests. The proposal submission system has been used to support the call for proposals for the SIRTF Legacy Science program, and proposals will be in evaluation by the time ADASS 2000 takes place.

The SIRTF spacecraft, mission profile, and science instrument design have undergone almost ten years of refinement. SIRTF development and operations activities are highly cost constrained. The cost constraints have impacted the design of the SOS in several ways. The Science Operations System has been designed to incorporate a set of highly efficient, easy to use tools which will make it possible for scientists to propose observation sequences in a rapid and automated manner. The use of highly automated tools for requesting observations will simplify the long range observatory scheduling process, and the short term scheduling of science observations. Pipeline data processing will be highly automated and data-driven, utilizing a variety of tools developed at JPL, the instrument development teams, and Space Telescope Science Institute to automate processing. An incremental ground data system development approach has been adopted, featuring periodic deliveries that are validated with the flight hardware throughout the various phases of system level development and testing. This approach minimizes development time and decreases operations risk.

2. System Design

The system architecture is modular and functional. The most mature elements of the system include the system elements used to plan and schedule science observations.

The Science User Tools consist of a variety of Java-based software elements that are downloaded to the user's PC or workstation. These tools support observation planning in a variety of ways. Users can maintain a catalog of observation targets on their home machines. They can select from one of seven observing modes across the three payload instruments, and input observation parameters to generate an AOR (Astronomical Observation Request). Users can build a library of desired observations on their home systems, and then download the final set of observation requests to the SSC when completed.

The AOR/IER Expansion Editor is used to expand each individual AOR (or Instrument Engineering Request-the IER's) into spacecraft and instrument commands. The user is provided with the resource estimate for each observation following expansion of the AOR or IER into a command sequence. This feedback enables users to provide observation programs to the SSC that fit within their time allocations.

The Science Observations Data Base (SODB) is used at the SSC to hold all information submitted by the community, and will eventually encompass the project archival data bases as well. Command expansions are transferred to the flight operations systems at JPL for execution via transfer from the SODB to JPL operational data systems. Downlink data is transferred to the SODB by the JPL flight operations system when data are acquired by the Observatory and transferred to the JPL ground data support systems from the Deep Space Network.

The SSC pipeline processing is an automated process that is initiated when data requested by the uplink planning system are delivered to the SODB. The SSC is currently designing and implementing the pipeline systems in conjunction with the instrument Principal Investigators and their support teams.

3. Results to Date

Over three thousand individual AORs were submitted to the SSC prior to the call for Legacy Science Proposals by the GTO community. The SSC processed the GTO requests and produced a Reserved Object Catalog, so the Legacy proposal teams were aware of targets that had been reserved by the GTOs. The database submitted by the GTOs represents a substantial resource that will be used in future testing of the ground data system with the Observatory. It also represents a significant percentage of the first year's observations, and will be used extensively in developing the mission plan for the first year of operations. The Legacy Proposal Teams will add their detailed AORs to the SODB following their selection, so that the majority of first year observations will be available well before launch for mission planning purposes. The GTO database has been analyzed by SSC staff, and the analysis indicates that the design estimate of an average of 30 minutes per observation was a good choice. Analysis also indicates an excellent sampling across the three instruments, the seven available instrument operations modes, and sampling of various areas of the sky.

4. Future Work

Future work includes completion of the expansion logic for three of the seven observing modes that have not yet been completed, and increasing focus on development of the pipeline processing software system. The three instruments were integrated into the cryogenic assembly in December 2000, and have passed their functional tests at operating temperatures. Extensive analysis of available detector data from the three instruments is well along, and algorithms for processing and analysis of the instrument data are in various stages of development and testing. SIRTF project plans include a "week in the life" test of the full Observatory and flight qualified ground data system prior to launch, at which time a typical week-long sequence of science and engineering activities will be executed with the flight hardware.

Acknowledgments. It is a pleasure to represent the efforts of the outstanding staff of the SIRTF Science Center in designing, developing, implementing, testing and operating the systems that will be used to support SIRTF operations, and their efforts in supporting the external scientific community. The cooperation received from the three instrument Principal Investigators, Dr. Giovanni

Fazio, Dr. James Houck, and Dr. George Rieke, is gratefully acknowledged. The SSC Director, Dr. B. Thomas Soifer, and Deputy Director, Dr. George Helou, provide daily support and encouragement to the science and development staff at the SSC. Thanks also to David Gallagher, SIRTF Project Manager, for his support of activities at the SSC. The work was performed at the California Institute of Technology under a contract with the National Aeronautics and Space Administration.

Advanced Architecture for the Infrared Science Archive

G. B. Berriman, N. Chiu, J. Good, T. Handley, A. Johnson, M. Kong, S. Monkewitz, S. W. Norton, A. Zhang

Infrared Processing and Analysis Center, California Institute of Technology, Pasadena, CA 91125

Abstract. This paper describes the data mining, catalog-cross comparison, and visualization services available at the Infrared Science Archive (IRSA), NASA's archive node for infrared astronomy data. IRSA is a living archive, which maintains contemporary datasets and continuously develops services to exploit these datasets. Over the past three years, IRSA has devoted most of its resources to support the requirements of the massive 2MASS survey datasets. Given the high volumes of 2MASS data, the services and infrastructure supporting them provide insight in understanding how a future NVO may operate.

1. IRSA's Charter

The Infrared Processing and Analysis Center (IPAC) at Caltech was charged with archiving data sets produced by the Infrared Astronomical Satellite (IRAS). The success of this mission and the demand for its data products made IPAC a leading center for archival research and data distribution, and led directly to the development of the Infrared Science Archive (IRSA) as the archive node for NASA's infrared astronomy missions. IRSA now provides public access to the catalogs and images from the 2MASS and MSX missions, as well as from the IRAS mission. IRSA's requirements are derived on one hand from the specialized needs of projects, and on the other hand from the needs of users analyzing the data. IRSA also holds ancillary catalogs required to allow exploitation of the infrared datasets, including USNOA 2, NRAO VLA Sky Survey (NVSS) and Faint Images of the Radio Sky at Twenty-centimeters (FIRST). A full list of holdings as of August 2000 can be found at the IRSA web site.[1]

2. Access To IRSA Services

While all IRSA services can be invoked via a program interface or via remote HTTP or Java client interfaces, users generally invoke them in server mode through a web client. Processing is performed server-side in the IRSA environment. Results are made available to the user as tables or images that can be

---

[1] http://irsa.ipac.caltech.edu

downloaded. Broadly speaking, the following services can be applied to the data held by IRSA:
1. Completely general catalog queries,
2. Image queries and visualization, with customization of visualization tools applicable to individual missions,
3. Cross-comparison between catalogs, and
4. Statistical representation of large datasets.

IRSA receives on average over 220 requests for data each day, and over 99% of these requests are successfully processed. The architecture of the IRSA services is described by Good (2001).

3. IRSA As A Living Archive

A key feature of IRSA is that it is a *living* archive. That is, by providing robust, contemporary archives and by continuously developing services that have the power to exploit them, IRSA permits the development of new scientific products and opens up new avenues of research. The support provided for IRSA's largest customer, the 2MASS project, has demonstrated the power of such an archive. 2MASS is uniformly surveying the entire sky in three near-infrared bands to detect point sources brighter than about 1 mJy in each band, achieving an 80,000-fold improvement in sensitivity over the first full-sky survey of Neugebauer & Leighton. This deep survey has generated datasets that are by far the largest obtained in any astronomical survey, with roughly 12 TB of images, and an internal catalog of sources now containing over 1 billion entries in an Informix database.

Efficient mining of these huge 2MASS datasets places extraordinarily large loads on IRSA services. Research into special techniques has led to the development of optimized algorithms and software for rapid searches of large databases. As an example of the new science that can be performed with the help of these services, astronomers culled from the 2MASS catalog the very red candidate objects from which the first brown dwarfs were identified (Kirkpatrick et al. 1999, and references therein).

Future surveys will produce data volumes even larger than those generated by 2MASS, and will certainly produce ever more spectacular scientific advances. The services developed by IRSA and the infrastructure to support them therefore provide a window into how a future NVO is likely to function. The remainder of this paper is therefore given over to discussions of special features of IRSA services that permit efficient data mining, and current research at IRSA that will provide the next generation of data mining, visualization and analysis tools.

4. Special Features of IRSA Services For Data Mining Large Volume Datasets

4.1. Catalog Queries and Indexing of Database Tables

Driving the 2MASS data mining requirements is the need to provide efficient and completely general querying methods. Querying has been made efficient in two ways. First, queries are run in parallel fashion across as many processors and

I/O channels as possible. IRSA chose a Sun Microsystems E6500 server and an Informix database because they can be highly optimized for parallel querying. Second, IRSA spatially indexes catalogs using nested hierarchies of increasingly smaller bins. Search mechanisms traverse those branches of the tree to isolate database entries that meet the constraints imposed by the query. IRSA employs three spatial indexes:

1. The Hierarchical Triangular Mesh (HTM), developed at Johns Hopkins to support the Sloan Digital Sky survey, divides the sky into a nested series of equilateral triangles.
2. In magnitude/color space, three dimensional box partitioning with logarithmic steps away from the diagonal.
3. R-tree indexing of image metadata where images cover a large area; R-trees take into account the spatial extent of the elements contained within them.

4.2. Catalog Cross-Comparison and Distributed Queries

Here lies perhaps the greatest challenge to the NVO, and here also lies perhaps the greatest potential for ground-breaking new science. IRSA has developed tools that allow for efficient distributed queries and which handle the complex DBMS functionality involved in cross-comparison of catalogs. The heart of the methodology is the use of three-way joins between tables, and sets of candidate source associations (known as relationship objects). Ma et al. (2000) describe the power of this method in more detail, and future research in this area.

4.3. Image Metadata

IRSA separates images from their associated metadata. The metadata reside in a database catalog, with one record per image. Indexed searches can be made on any parameter, as well as spatially indexed searches based on position. IRSA can therefore efficiently locate images in catalogs it does not hold itself.

4.4. Statistical Representation of Large Datasets

Generally speaking, statistical representations of catalogs or query results are a powerful way of initially studying a large volume of data, allowing a user to quickly refine a search to locate objects of interest, such as those with extreme colors. IRSA provides a service to derive a histogram of a pre-binned representation of the IRSA database.

5. Current Research at IRSA and Development of Next Generation Services

One of IRSA's next generation services is described in detail elsewhere in this volume. Many astronomers have expressed the desire to perform on-the-fly coordinate transformations with the minimum of function calls. Zhang et al. (2001) describe how an Informix datablade solves this problem. A datablade is simply a function or library embedded into the Informix Database Management System. Users simply embed the input and target coordinates into their SQL queries; no further function calls are necessary.

In the next two years, IRSA anticipates that it will ingest the catalogs, images, and spectra from the Infrared Telescope in Space (IRTS) and, in collaboration with the NASA Goddard Space Flight Center, the data sets from the Cosmic Background Explorer (COBE). IRSA will provide integrated access to the Infrared Space Observatory (ISO) data archives, held in Vilspa, Spain. Further in the future, IRSA will archive catalogs from the SIRTF and SOFIA missions. The previous section described research into catalog-cross comparisons. The full value of these data will be realized when they can be combined with distributed multi-wavelength data. Much of IRSA's research and development is therefore aimed at providing a portable, Java-based architecture that will support efficient access to and interaction with multiple, distributed data sets. This architecture, called the On-Line Archive Science Information System (OASIS), is described in detail by Good et al. (2001), and is modeled after the layering methodology employed by Geographical Information Systems (GIS). As part of the OASIS development effort, IRSA is cooperating with STScI, CDS, and ADC to establish an XML output format for catalog search results, and is pursuing similar standards for the transfer of image metadata.

Underpinning the OASIS front-end will be a persistent archive request and management mechanism. IRSA expects to deploy this infrastructure in Spring 2001. It is designed to wrap around existing services, while supporting existing HTML form and CGI technologies. The request manager is an Enterprise Java bean system that uses the WebLogic Applications Server to accept multiple requests from users, control information, maintain state information, and communicate results to users.

Acknowledgments. We thank the NASA Mission Operations and Data Analysis and Science Applications of Information Technology programs and the Digital Sky project for financial support. We thank Drs. G. Helou, W. P. Lee, C. J. Lonsdale, and J. Ma for their technical support.

References

Good, J. C., et al. 2001, this volume, 52

Kirkpatrick, J. D., et al. 1999, ApJ, 522, L65

Ma, J., et al. 2001, in ASP Conf. Ser., Vol. 225 Virtual Observatories of the Future, ed. R. J. Brunner, S. G. Djorgovski, & A. Szalay (San Francisco: ASP), in press

A Parallel, Distributed Archive Template for the VO

Aniruddha R. Thakar, Peter Z. Kunszt, Alexander S. Szalay

The Johns Hopkins University

Abstract. In the proposed Virtual Observatory (VO), there is an urgent need for a prototype distributed archive that a) uses standard interfaces to the outside world, b) contains a parallel and scalable query agent, and c) can serve as a virtual data grid node in the VO. We propose to use the current SDSS Science Archive as a basis for developing such an archive template for the VO. This effort will involve extending the current capabilities of the Science Archive query agent as well as redesigning certain aspects of it. We describe the steps that this effort will entail.

1. Introduction

The multi-Terabyte astronomical archives that are in the process of being built today will give rise to unprecedented sky and wavelength coverage along with an incredibly rich dataset. But the enormous potential for scientific discovery that these archives promise will not be fully realized until they are interconnected in such a way that data from individual archives can be combined, compared, and mined in a seamless fashion. The creation of such a multi-wavelength digital universe-at-your-fingertips is the ambitious and far-sighted goal of the effort to build a Virtual Observatory.

The enormous size of the upcoming archives, along with the multiplicity of standards, software tools, and hardware platforms at the disposal of the scientists that are building these archives, create a daunting challenge with respect to integrating them into a virtual observatory framework. The very task of defining what a virtual observatory is and must provide has consumed several months of discussions and meetings between astronomers. In short, there has been much talk but little action on defining a VO.

What is sorely needed is a prototype for a "VO-ready" archive that can serve as a template for what future archives should (or should not) be, and help to crystallize the concepts and priorities for the VO. We believe that a VO archive must have at least the following features:
- well-defined, standardized interfaces to the outside world;
- a parallel, distributed, and scalable query agent to enable efficient data mining for a large and widespread user community; and
- the proper hardware and software framework to serve as a virtual data grid node.

In order to build such an archive template, we propose to use the Sloan Digital Sky Survey's current Science Archive as a basis and convert it into a VO archive template by making the modifications described below.

2. The SDSS Science Archive

The Sloan Digital Sky Survey (SDSS) is a multi-institution project to build a map of a large part of the northern sky in five wavelength bands (Szalay 1999). The SDSS **Science Archive** (abbreviated SX) is the science database that will result from the survey when it is completed (2005/2006). It is expected to be several Terabytes in size and will contain a catalog of more than 200 million objects and 1 million spectra.

The SX has a client/server architecture that features a lightweight, portable GUI client, a parallel (multi-threaded) distributed server (query agent), and a commercial object-oriented Database Management System (DBMS) - Objectivity (Thakar et al. 2000). It also includes a fast spatial indexing scheme—the Hierarchical Triangular Mesh or HTM (Kunszt et al. 2000), as well as a multidimensional flux index. Although the Science Archive is already considerably optimized for distributed data mining (Szalay et al. 2000), it is still not sufficiently equipped to be a VO data grid node. We aim to take the following specific steps to rectify this and create a prototype VO archive.

3. XML Compliance

A fundamental property of a VO-compatible archive must be the standardization of its interfaces with the outside world. Toward this goal, we recognize the eXtensible Markup Language (XML) as an emerging standard for data interchange on the Internet, and seek to make our archive XML-compliant in a way that will allow data and metadata to be exchanged with any other entity that can decode XML. The attractiveness of XML is that it is flexible and self-contained, so that specific information about reading even the most complex data can be encoded using standard XML's Document Type Definitions (DTDs).

One of the biggest advantages of XML is the wealth of public-domain software and tools that is already available (and rapidly increasing) on the Internet. We list below some of those that are of particular interest to our application:

- **Schema Definition**: These are tools for defining the structure of the data, i.e., the data model. We propose to provide a public XML version of our Abstract, which is essentially a runtime abstraction of our data model, and functions as a type manager and metadata source. Currently available tools include DDML, SOX, and XML Schema.
- **Query Packaging**: These are tools that enable the packaging of queries using XML, such as XML Query, XML-QL, and Quilt.
- **Parsers**: There are also ready-to-use XML parsers available, such as Expat, XML4J, Xerces (Apache).

4. Virtual Data Grid Node

The creation of a data grid will be one of the primary challenges and benefits of a Virtual Observatory. The ability to generate **virtual data**—the complex and often voluminous data that are created on the fly from complex analyses of archival data—efficiently will be crucial for future astronomical research in cosmology and other fields. Virtual Data Grids (VDGs) will be an indispens-

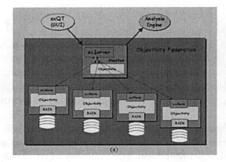 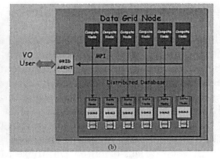

Figure 1. (a) Current and (b) proposed distributed query computation models. The current master/slave model will be replaced by a loosely-coupled, MPI-based massively parallel (MPP) model.

able component of the VO. The GriPhyN project (**Gri**d **Phy**sics **N**etworks, www.griphyn.org) envisages that Petascale Virtual Data Grids (PDVGs) will be necessary in the near future to meet the virtual data needs of the age of multi-Terabyte and Petabyte digital archives. Indeed, these archives **will need to be designed as VDG nodes**. This essentially means that each archive must provide a scalable parallel and distributed framework for executing complex, compute and I/O intensive queries and analysis tasks as close to the archive data as possible so as to minimize network traffic. The GriPhyN proposal contains examples of queries that will become possible (and frequent) with VDGs.

Our current model, although it is parallel, distributed and moderately scalable, is not very well-suited for the large-scale grid computation involving very complex query and analysis tools that is anticipated in a VO context. In order to make it a fully functional grid node, we need to build a massively parallel distributed framework with dynamic load-balancing, resource-scheduling, and message-passing communication between intelligent agents. We propose to reconfigure our current parallel model in order to achieve this objective, as shown in Figure 1. Our master/slave configuration will be replaced by a computational grid of intelligent query agents loosely coupled to a distributed data grid via an MPI (Message Passing Interface)-based communication toolkit like Globus (www.globus.org). It will also be necessary to have a grid agent at the top level that will interface to the outside world and serve as a listener/scheduler for the query agents.

5. Query Language Extensions

Our current query language, SXQL, is a subset of SQL (Standard Query Language) that also includes several object-oriented extensions and astronomical and mathematical macros. We plan to augment the language further so as ultimately to produce a versatile SQL-based scientific query language that includes at least the following features, several of which have already been incorporated into SXQL: the SQL SELECT-FROM-WHERE syntax, aliasing and nesting, the ability to follow links through associations, including language extensions for specifying to-many links, the ability to query on object methods, generic mathematical macro support, specific astronomical macro support, and support for spatial querying using the HTM.

6. Scientific Output

The output from VO queries will often be very complex and voluminous, and will need to be packaged in a self-describing, lightweight format. Again, XML provides the answer. The eXtensible Scientific Interchange Language (XSIL) is an XML DTD for scientific output that is extensible to any discipline (Williams 2000). It contains an extensible object model with a Java API and comes bundled with the Xlook browser. We hope to use XSIL to obtain a flexible, general object transport protocol, a portable ASCII and binary output format, and an ultra-light data format that includes support for binary streams.

7. Concluding Remarks

Although there are significant challenges facing the creation of a Virtual Observatory, the lack of the necessary technology is not one of them. The time is ripe for the implementation of the VO archive template described above. Such a template is sorely needed, and the existing state of software and hardware technology—in terms of storage capacity, network bandwidth, CPU speed, and software standards and technology—makes it achievable.

Within the next few years, the demand for virtual data will see a sharp rise, and the computational power afforded by virtual data grids will be indispensable for scientific research in large-scale structure and other fields within astronomy. Archives like this one will be poised to meet those challenges.

References

Kunszt P. Z., Szalay, A. S., & Thakar, A. R. 2000, in ASP Conf. Ser., Vol. 216, Astronomical Data Analysis Software and Systems IX, ed. N. Manset, C. Veillet, & D. Crabtree (San Francisco: ASP), 40

Szalay, A. S. 1999, Comp. in Sci. & Eng., Mar/Apr 1999, 54

Szalay, A. S., Kunszt, P. Z., Thakar, A., Gray, J., Slutz, D., & Brunner, R. J. 2000, Proc. 2000 ACM SIGMOD on Management of Data, 451

Thakar, A. R., Kunszt, P. Z., & Szalay, A. S. 2000, in ASP Conf. Ser., Vol. 216, Astronomical Data Analysis Software and Systems IX, ed. N. Manset, C. Veillet, & D. Crabtree (San Francisco: ASP), 231

Williams, R. 2000, http://www.cacr.caltech.edu:/XSIL

Tools for Coordinating Planning Between Observatories

Jeremy Jones, Lori Maks

NASA Goddard Space Flight Center, Greenbelt, MD 20771

Mark Fishman, Vince Grella, Uri Kerbel, Dharitri Misra, Vince Pell

CommerceOne Inc., 1100 West Street, Laurel, MD 20707

Abstract. With the realization of NASA's era of great observatories, there are now more than three space-based telescopes operating in different wave bands. This situation provides astronomers with a unique opportunity to simultaneously observe with multiple observatories. Yet scheduling multiple observatories simultaneously is highly inefficient when compared to observations using only a single observatory. Thus, programs using multiple observatories are limited not by scientific restrictions, but by operational inefficiencies.

At present, multi-observatory programs are initiated by submitting observing proposals separately to each concerned observatory. To assure that the proposed observations can be scheduled, each observatory's staff has to check that the observations are valid and meet all constraints for their own observatory; in addition, they have to verify that the observations satisfy the constraints of the other observatories. Thus, coordinated observations require painstaking manual collaboration among staffs at each observatory. Due to the lack of automated tools for coordinated observations, this process is time consuming and error-prone, and the outcome of requests is not certain until the very end. To increase multi-observatory operations efficiency, such resource intensive processes need to be re-engineered.

To overcome this critical deficiency, Goddard Space Flight Center's Advanced Architectures and Automation Branch is developing a prototype called the Visual Observation Layout Tool (VOLT). The main objective of VOLT is to provide visual tools to help automate the planning of coordinated observations by multiple astronomical observatories, as well as to increase the probability of scheduling all observations.

1. Introduction

Planning and executing observations that are coordinated across multiple spacecraft is essential to reaching future space and earth science goals. The current lack of automated tools and interfaces across observatories makes coordinated observing resource intensive for the observatories and consequently limits scientific research. The proposed tools will facilitate the coordinated observing processes necessary to achieve the concept of a "virtual observatory" that will spawn a new era of science data collection.

2. Objectives

A number of advanced visual tools are currently being developed to help the observers and principal investigators of astronomical phenomena in their observation planning. Among early entries in this arena are the Scientist's Expert Assistant[1] and Astronomer's Proposal Tool (APT)[2], which enable the observers to simulate the quality of their observation based upon observing parameters (e.g., target properties, instrument setup, and observatory condition). However, the outcome of the proposed requests is still uncertain as the schedulability of the observation, which is affected by observatory related factors, is still unknown for the requested time period. This problem is magnified many-fold when collaboration among observatories is needed to coordinate the planning of a set of observations through multi-wavelength campaigns that involve variable phenomena, interacting binary systems, and other events of extreme interest to astronomers for the quantitative understanding of galactic and extra-galactic phenomena and for developing realistic physical models.

The main objective of the VOLT project is to provide visual tools to help automate the planning of coordinated observations by multiple astronomical observatories, as well as to increase the scheduling probability of all observations. Thus, these tools will not only provide the users with the required schedulability data, but will also help and guide them in determining the best possible times when the group of observations may be placed in compliance with their coordination goals.

The intended goals of VOLT are:

- The tools will not replicate the software capabilities of the planning and scheduling facility associated with an observatory. Rather, they will interface with these software components, using state-of-the-art communication mechanisms, and retrieve the desired data. Required formatting and normalization of data may be performed by these tools if necessary.
- The coordination of observations will be modeled as a network of temporal constraints that can be solved in a satisfactory manner by use of an appropriate constraint satisfaction engine. Cost-effective solutions will employ free or low-cost search engines.
- If coordinated observations are not schedulable as specified, explanatory help on constraints and constraint relaxation will be provided.
- Emphasis will be on reuse and modularity by making tools easily extensible and configurable to new missions. Following initial application to space-based observatories, later efforts will expand to include queue-based ground observatories, other space sciences missions, and other domains such as earth sciences missions.

---

[1] http://aaaprod.gsfc.nasa.gov/SEA

[2] http://www.stsci.edu/apt/

3. Impact

The primary impact of the tools provided by VOLT will be in the arena of coordinated observations involving two or more observatories, as shown below:

- Coordinated Observations are limited by observatories because of the complexity of the planning process, and the manpower and manual effort required for coordination. VOLT seeks to automate this planning; to enable new types of complex, coordinated observations that are not feasible with current manual methods, and to make coordinated observing a low cost, routine task rather than a special exception.
  Although multi-observatory coordinations involving three to five observatories are envisioned for some studies, currently very few coordinated observations (e.g., 7 - 10% of observations for HST) are requested and take place due to the uncertain, manual and labor-intensive nature of such coordinations. Due to the lack of resources, some observatories (e.g., FUSE) do not support coordinated planning except in extreme cases. Users themselves are also discouraged from requesting three or more coordinated observations due to the complexity involved. These new tools will help both astronomers and observatory scheduling staffs in planning for such complex observations. They will reduce the manual workload, and thus the cost, of coordinated and time-constrained observations.
- Observers with coordinated programs have few tools to will assist them in planning their observations effectively. VOLT will fill this void by applying new technology and innovative solutions for this increasingly important aspect of science planning, not only by automating the procedure, but also by providing visual cues on coordination, and by allowing users to look for alternate solutions. By helping in the planning of observations, these tools can also help in generating more feasible requests even for a single observatory. The pluggable nature of these tools (into other proposal/observation planning tool sets such as APT) will provide an integrated environment enhancing the observation's scheduling probability. The net result will be an increase in science data that would help in the understanding of galactic and extra-galactic phenomena, and benefit a wide variety of astronomical research.

Once VOLT has achieved the above two goals, more astronomers will apply for coordinated observations. Coordinated observations will have become feasible enough that observatories can take a further step in coordination: accepting a single proposal that will apply to all observatories involved in the coordinated program, thereby permitting a given scientific topic to be investigated as a whole, rather than as fragmented proposals that the observer hopes will succeed at each of the separate observatories requested.

4. Acknowledgments

The authors would like to acknowledge the active support of the following persons in developing the VOLT prototype:

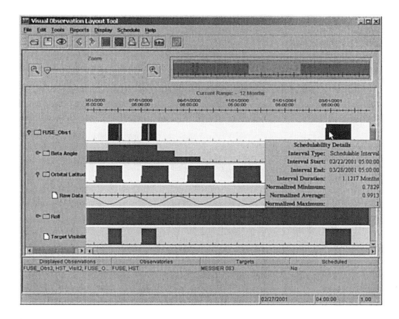

Figure 1. The VOLT prototype schedulability display

Glenn Miller, Beth Perriello, Anuradha Koratkar, Karla Peterson and Peg Stanley: Space Telescope Science Institute - for help in establishing VOLT system requirements.

Bryce Roberts: Johns Hopkins University - for help in interfacing with FUSE observatory data.

References

Peterson, K., Eckert, M., Remage-Evans, N., Hilton, P., Perriello, B., Roberts, B., Smith, E., & Stanley, P. 1999, Proceedings of SPIE, vol. 4010, 279

Misra, D., Bopf, M., Fishman, M., Jones, J., Kerbel, U., & Pell, V. 2000, SpaceOps

The Scientist's Expert Assistant Simulation Facility

Karl R. Wolf, Connie Li

Commerce One, Laurel Maryland

Jeremy Jones, David Matusow

NASA/Goddard Space Flight Center

Sandy Grosvenor

Booz-Allen Hamilton, Seabrook, Maryland

Anuradha Koratkar

Space Telescope Science Institute, Baltimore Maryland

Abstract. In the process of developing an observing program for a given observatory, the planner requires a number of inputs regarding the target and scientific instrument that need to be calculated, found, and/or confirmed. Thus, preparation of a program can be quite a daunting task. The task can be made easier by providing observers with a software tools environment. NGST funded the initial development of the Scientist's Expert Assistant (SEA) to research new visual approaches to proposal preparation.

Building on this experience, work has begun on a new integrated SEA simulation facility. The main objective is to develop the framework for a flexible simulation facility to allow astronomers to explore the target/instrument/observatory parameters and to 'simulate' the quality of data they will attain. The goal is a simulation pipeline that will allow the user to manage the complex process of simulating and analyzing images without heroic programming effort. Tying this into SEA will allow astronomers to effectively come 'full circle' from retrieving archival images, to data analysis, to proposing new observations. The objectives and strategies for the SEA simulation facility are discussed, as well as current status and future enhancements.

1. What is the SEA Simulation Facility?

The Scientist's Expert Assistant (SEA)[1] is a tool designed to investigate automated solutions for reducing the time and effort involved for both scientists and telescope operations staff spent in preparing detailed observatory proposals.

At the past two ADASS conferences (VIII and IX) SEA's Java-based visual target tuning, exposure time calculation, and visit planning capabilities were demonstrated. Since that time, SEA has been embraced by a number of astronomical observatories for inclusion into their observing programs. The Space Telescope Science Institute has already incorporated SEA for production use into its Astronomer's Proposal Tool, while other observatories such as SOFIA are in the initial stages of incorporating it.

A new phase of SEA has now begun: building the new integrated SEA Simulation Facility (SSF). In improving the visualization process to give the user better insight into observation process, the SSF breaks down the observation into elements of the light path. These elements are really software models of various aspects of the light path. The models generally have a variety of parameters that the user can adjust to assist in better understanding the effects of that element in the light path. At any point in the light path, the user can attach one or more visualizations that can be used to observe aspects of the light path at that point.

2. Why a Simulation Facility?

One of the emerging efforts in the astronomical community is the Virtual Observatory (VO). The SSF is working to combine the ability to access existing archives with the ability to model and visualize new observations. Integrating the two will allow astronomers to better use emerging integrated archives of the VO to plan and predict the success of potential new observations.

The SSF provides benefits to a variety of potential users:

- **Observers** can use simulation to:
    1. Effectively determine how various parameters affect their data and scientific objectives
    2. Act as a "Phase 0" tool for the initial "framing" of the observations
    3. Validate proposed observations ahead of time
    4. Support new complex instruments that drive the need for newer visualization tools.
- **Observatory Staff** can use simulation to:
    1. Characterize their telescope, instruments, and detectors
    2. Calibrate instruments with fewer observations.
- **Archive Users** can use simulation to understand the quality and limitations of an archival image.

---

[1] http://aaaprod.gsfc.nasa.gov/SEA/

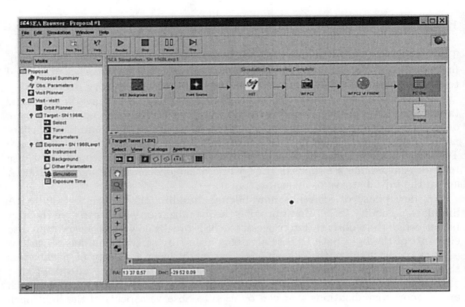

Figure 1. The SEA Simulation Facility Display

3. Highlights of the Design

Project design goals included:
- Testing interactive and innovative ways to look at a proposed observation
- Providing a system that is scalable and supports a variety of models
- Hiding simulation complexity routinely from the user (but providing it when requested)
- Emphasizing scientific fidelity
- Supporting a variety of visualization mechanisms such as imaging and spectroscopy.

The simulation pipeline was envisioned as modeling changes to the photons as they travel from one or more sources though various distortion producing effects before ultimately reaching an instrument's detector. The pipeline consists of two primary element types:

1. **EnergyModels** that modify the photon stream in some way, such as, adding photons, attenuating their rate of production, or distorting them in some other way.
2. **EnergyVisualizations** are attached to an EnergyModel. Their purpose is to visualize the data at that point in the pipeline. Astronomical archive information can also be accessed for comparison purposes.

By default, the pipeline is automatically constructed to model the components of an exposure. Model parameters for the simulation are derived from the exposure specifications. Each component models some part of the exposure's light path. For example, there will be a model representing the background emissions, a model for the observation target (a star, galaxy, etc.), and models for the Observatory, Instrument, Filter and Detector objects defined for this exposure. The user is free to add, replace, modify, or delete EnergyModels and EnergyVisualizations as desired.

The user controls the simulation process by telling the SSF to "render". The rendering process starts at the left most EnergyModel in the pipeline (seen along the top of the display) then progresses to the right as each EnergyModel completes its processing.

What is being processed is an EnergyDataSet. The goal of an EnergyDataSet is to accurately and efficiently contain the simulation data. At the start of the render processing, the EnergyDataSet is initialized then passed from one EnergyModel that can modify it, to the next.

To minimize EnergyDataSet storage requirements and computational load while maintaining quick and easy access to data values, two solutions were implemented:

1. **Smart EnergyDataLayers** that optimize data storage needs by using a sparse matrix and storing double precision values rather than Wavelength arrays unless necessary.
2. **Binning** so that most of the simulation is performed on low resolution data and only a subset of the data is rendered at higher resolution for a more accurate simulation.

4. Current Status and Future Plans

The primary framework for the SSF has been designed and implemented and is now ready for incorporation of existing simulations. Recently released simulation models from the European Southern Observatory are currently being integrated.

Future plans call for:
- Increasing the scientific fidelity of the models
- Prototyping visualization approaches and GUI enhancements.

More information about the SSF is available in a "white paper"[2] on this subject.

References

Koratkar A., Grosvenor S. 1999, in ASP Conf. Ser., Vol. 172, Astronomical Data Analysis Software and Systems VIII, ed. David M. Mehringer, Raymond L. Plante, & Douglas A. Roberts (San Francisco: ASP), 60

Grosvenor S. 2000, in ASP Conf. Ser., Vol. 216, Astronomical Data Analysis Software and Systems IX, ed. N. Manset, C. Veillet, & D. Crabtree (San Francisco: ASP), 695

---

[2]http://aaaprod.gsfc.nasa.gov/SEA/papers/sea_simulation.pdf

Data Information Services at the Infrared Science Archive

John Good, Mihseh Kong, G. Bruce Berriman, Nian-Ming Chiu, Thomas Handley, Albert Johnson, Wen-Piao Lee, Jin Ma

Infrared Processing and Analysis Center, California Institute of Technology, Pasadena CA 91125

Abstract. The Infrared Science Archive (IRSA) at the Infrared Processing and Analysis Center (IPAC) is charged with archiving data from NASA's infrared astronomy projects. Since 1997, IRSA has provided support to the massive datasets generated by the 2MASS project. Services have been developed to perform rapid querying of massive datasets, on-line real time access to a TB-size images database, and cross-correlations between massive datasets. IRSA is building on this work to develop a cross-archive access toolkit that will meet the needs of a National Virtual Observatory. In this paper we describe one aspect of this: OASIS, the On-line Archive Science Information Services astronomical archive integration Java client toolkit, which provides advanced visualization capabilities and distributed data access.

1. Introduction

The National Virtual Observatory (NVO) will encompass data and functionality covering all aspects of astronomical research. No one overarching software "system" will ever adequately provide the range of functionality required. Rather, the NVO should be envisioned as a set of cooperative but reasonably autonomous components, each of which is designed to provide the optimum functionality in a narrow area. The key to NVO success will be to foster an environment where individual development efforts are discrete and intrinsically integrable in the face of a perpetually imperfect understanding of their future use.

As on-line archives become more and more prevalent, one area ripe for development is that of integrated multi-archive data access. Specifically, we are building a client toolkit which supports the ability to view and compose data from various sources and to use this composite knowledge to further the process of data retrieval and analysis.

This paper describes the full range of functionality planned for OASIS. A preliminary version of OASIS was released to the community at ADASS; functionality that existed then will be identified where described here. Parallel work is underway to develop a server-side toolkit with which complex inter-archive composite data access and processing requests can be managed. While these two are envisioned as complementary, with OASIS also acting as a front-end control for the request management environment, they are also independent with OASIS providing access, integration, and visualization for current archive services.

2. OASIS Architecture

The OASIS toolkit marries the kind of functionality found in historical data display programs with the remote-access / dynamic-data-rendering / extensibility model popularized by Web browsers, but in a form customized to the needs of astronomy archival research.

The best way to visualize this synthesis is as follows: A Web browser normally receives and renders HTML with embedded images. When it sees most other datatypes it either fires up a separate plug-in or saves the data to disk. In OASIS, if the data that it sees is an image, source table, sky drawing (like a contour plot), XY graph, etc., it is rendered (or added to an existing display) using astronomy-specific tools. As with browsers, the detailed mechanism by which a user interacts with an archive is defined by the archive (or some other third party) and can be dynamically downloaded to OASIS when needed. OASIS differs from a browser not only in that it intrinsically understands a range of astronomical datatypes but more importantly that it understands their interrelationships and how to render them in composite.

Also similar to browsers, downloaded information (in this case data files or archive access interfaces) can be cached for future reuse or installed formally as packages. There is nothing specific to archive access in this model, so tool builders who wish to add data processing functionality to the mix (e.g., aperture photometry or other image processing) as downloaded or installed components can do so using the same formalism.

The current and future functionality plans for OASIS fall into the following broad categories:

2.1. Visualization

- Sky Map Display. A base image (single image or multi-image composite, either full color or blended images using partial transparency) plus an arbitrary number of overlay layers. Overlay layers consist of drawn symbols, coordinate grids, contour maps, focal plane outlines, etc.
- Table Display. A rendering of tabular data (e.g., catalog subsets) in a spreadsheet/DBMS form. Includes limited client DBMS functionality as well.
- XY Graph Display. For spectra, light curves, histograms, scatter plots, etc.,
- XML/HTML Rendering. Standard URL references are passed to an external Web browser for rendering and any further action. However, some HTML and much of the XML that we expect to handle will contain external references which are meant to be retrieved and rendered using one of the above OASIS tools. For these, the XML/HTML should be handled by rendering components more closely linked to OASIS.

The map display / map layer overlay functionality is already complete, as is basic table display (without any DBMS functionality) and rudimentary XY graphing functionality. A major current effort involves generalizing the map layer interface and creating a "sky drawing" XML dialect to allow easy extension of both OASIS and the data services that feed it by any interested members of the community.

The table display currently allows for "active" cells: URL-like encoding of external references to specific basic data retrieval or Web-based information services. True URL (e.g., NED detailed data for a specific source) are dynamically passed to a "peer" browser. Data references resulting in downloaded data (in contrast to the HTML of the preceding) will be submitted through a "request manager" (see below) and the results handled by one of the built-in renderers (e.g., retrieved images becoming the base for the Sky Map display). A good example of this is one of the modes of the NED (NASA Extragalactic Database) interface, where the currently displayed image is used as the basis for a NED region search (which returns a table of basic source information). Columns in this table are dynamic references to detailed information on the source (detailed source information from NED, journal documentation from ADS, image download references if available, light curve information—an XY graph—from the AAVSO, etc.).

2.2. Archive Request Generation and Request Management

The current OASIS has a small set of custom search interfaces to several image archives (the Infrared Astronomical Satellite, IRAS; the IRAS Galaxy Atlas; the Two-Micron All-Sky Survey, 2MASS; the Hubble Space Telescope Digital Sky Survey, DSS; and the VLA FIRST and NVSS surveys). It also has custom interfaces to the IRSA catalog holdings (and in particular to the terabyte-scale 2MASS source catalog) and to the Vizier holdings at the CDS in France. As was mentioned above, we plan to generalize this interface, allowing open dynamic extension in much the same way that data providers currently build HTML forms to access their own data holdings.

Data retrieved by such services are normally low-level constructs (basic images or tables) and are identified by the return MIME type or by knowledge built into the retrieval component. For images, this is assumed to be FITS format (possibly compressed) and for tabular data will most likely be XML (though the details of this are currently under discussion). We expect that in the future there will be more complex data return structures, containing the above but with an XML "inventory" and possibly packaged in a JAR file. Determining precise details of this will be left to the future NVO project and are easily accommodated in OASIS.

All OASIS data external data requests will be handled through a single "request manager", a multi-threaded object which handles remote submissions and the flow of returning data, either into files or directly to rendering engines (at the user's discretion). A basic version of this is already in place which handles connection to current HTTP/CGI stateless data services. However, in conjunction with the server-side request management infrastructure it would also provide for reconnection to outstanding requests (e.g., when restarting OASIS after exiting and going home for the day) and reacting to asynchronous server-generated signals. This more advanced interaction scenario would be accomplished using RMI and servlet connections to server-side Enterprise Java Beans (EJBs).

2.3. Inter-Component Signaling Events

One of the main reasons for integrating the various visualizations under a single umbrella is to easily support data interrelationships. For instance, a catalog subset can be retrieved based on a displayed image and rendered both as a table and as a scaled (and colored and shaped) symbol layer on top of the image (or any other image of the same part of the sky). Selecting records in the table can also result in the symbols in the overlay being highlighted (and vice versa). The same idea can be extended to highlighting points in a scatter plot based on the table or only making a scatter plot or a histogram using the sources outlined in the image.

The mechanism which makes this possible is an event registration / broadcast mechanism (similar to that used for Java Beans) implemented through a small set of specific event types (e.g., sky location event, sky area definition event, etc.). Individual objects (e.g., map layers) can register to receive such events, though most allow this to be managed through a single "layer control" interface. As with Java Bean events, the event sender does not know (or care) what the receivers do with the event.

Most of this formalism is already in place. The work planned under this proposal in this area is to generalize and externalize the mechanism so that other sets of events can be defined and added by outside groups of users.

2.4. Dynamic Extension

Java class loading functionality makes it simple for outside service / functionality providers to add specialized tools to OASIS (even if only for their own use). However, in order to utilize this we must first provide a set of interfaces that will allow OASIS to initiate the tools and allow the tools access to OASIS internal data (e.g., the current image or table, events, etc.).

For the most part, current information location functionality (search engines, archive home pages, resource lists, etc.) can be used to disseminate information on available extensions and assure quality control. We expect that program-friendly services in this area (along with security and information validation) will be a primary aspect of any primary NVO activity.

2.5. Ancillary Functionality

Much of the underlying functionality needed to build OASIS comes from other sources (FITS image readers, map projection routines, etc.). In addition, we have built and are extending libraries to handle astronomical coordinates (including precession, Besselian-Julian conversion, and proper motion) and XML table parsing.

The CDS Information Hub

Françoise Genova, François Bonnarel, Pascal Dubois, Daniel Egret,
Pierre Fernique, Soizick Lesteven, François Ochsenbein, Marc Wenger,
Patricio F. Ortiz[1]

Centre de Données astronomiques de Strasbourg, Observatoire astronomique, UMR 7550, 11 rue de l'Université, 67000 Strasbourg, France

Abstract. The present status of the CDS information hub, with SIMBAD, VizieR, and ALADIN, is demonstrated, in particular in terms of links with other services such as the ADS, journals, and observatory archives. The XML operational developments, e.g., in ALADIN, are shown, in particular the *Astrores* proposed standard for tabular data. The prototype cross-correlator developed in the framework of the 'ESO–CDS Data Mining project' is also presented.

1. Introduction

The *Centre de Données astronomiques de Strasbourg* (CDS) develops and maintains a set of widely used, interlinked services, in particular SIMBAD, the reference database for astronomical objects, the VizieR catalogue browser, and the ALADIN interactive sky atlas. The CDS services are described in Genova et al. (2000) and the set of companion papers (Wenger et al. 2000; Ochsenbein et al. 2000a; Bonnarel et al. 2000). The services are available from the CDS Web site,[2] together with the Dictionary of Nomenclature, mirror copies of journals and of the ADS, Yellow Page services, documentation, and other services.

2. Interoperability

The CDS services are interconnected, and each service is itself a hub connected to remote reference services.

On one hand, they are part of the astronomy bibliographic network, together with the ADS (Kurtz et al. 2000) and the electronic journals. One can easily navigate from the list of references in a published paper to the ADS, and to the list of objects from this paper in SIMBAD. This link is directly implemented in the journal in some cases. CDS is still more organically linked to electronic publication for tables: it participates to the *Astronomy & Astrophysics* publica-

---

[1] On leave at Osservatorio Astronomico di Capodimonte, Via Moiariello 16, 80131, Napoli, Italy

[2] http://cdsweb.u-strasbg.fr

tion process by preparing and publishing on-line tables in the VizieR catalogue service; it also distributes tables from the AAS and ASP journals.

Links are also being implemented with major data archives. Observation logs can easily be implemented in VizieR, like any tabular data. Active links to the archives, located at the observatory site or in a data center, are installed, and an update mechanism is implemented for evolving logs. ALADIN gives access to these lists of observations through VizieR, and also to archive images with the full ALADIN functionalities when feasible (Bonnarel et al. 2001). Links with archives are implemented in SIMBAD on a case by case basis, with an effort to implement links that actually respond, with HEASARC for objects having a 'high energy' name, with IUE, and soon with ISO, after cross-identification and inclusion of the logs in SIMBAD.

In addition, user data can be used as input in the services: query by lists of objects for SIMBAD and VizieR, integration of user catalogues and images in ALADIN, when using the Standalone version.

3. Standards and Tools

Interoperability between heterogeneous, distributed astronomy information services is based on a few *de facto* standards, such as the 19–digit *bibcode* (e.g., 1999A&A...349..236E; Schmitz et al. 1995) to describe bibliographic references, first defined by NED and SIMBAD, then extended and widely used by ADS, and accepted by the journals. To be able to integrate data from different origins, one must be able to understand their contents. For that purpose, FITS is of course a major standard of astronomy. The completion of World Coordinate System (WCS) coding in FITS (with which ALADIN is fully compatible) is a major step towards standardisation of image astrometry. Another example is the standard description of tables (http://vizier.u-strasbg.fr/doc/catstd.htx), defined by CDS and shared by data centers and journals.

XML is certainly a key tool for metadata management in the future. CDS has lead the *Astrores* consortium, which defined an XML standard for tabular data in 1999 (Ochsenbein et al. 2000b). ALADIN is fully *Astrores*/XML compatible, and VizieR results can be retrieved in XML format (among many others) to be interpreted and used by any other service.

Another problem for interoperability is the maintenance of links. The *Générateur de Liens Uniformes* (GLU; Fernique et al. 1998), first developed by CDS to maintain links between its own services, is widely used for that purpose. A shared dictionary describing links is implemented and updated, and the GLU mechanism allows service providers to use symbolic links in their html pages. These symbolic links are translated to physical links using the GLU dictionary. The GLU dictionary can easily be made compatible with XML syntax, and it can also manage metadata.

4. Towards the Virtual Observatory: Data Mining and Survey Data

The world-wide astronomy web already gives access to many on-line resources, from observational data to published results. One main objective of the Virtual

Observatory is to include survey data, and tools to manage them, in the astronomy web. Survey catalogues can be included in VizieR, where at present time they can very efficiently be searched by position (Derriere et al. 2000). R&D studies are under way to implement the capability to search very large catalogues by other criteria. Remote on–line survey images will soon be accessible through ALADIN, provided that they can be retrieved by an HTTP query, with a FITS WCS description. For instance, ALADIN is already provided by NED as a tool to access their image collection (Bonnarel et al. 2001).

Another aspect of the Virtual Observatory is the need to be able to mine through large amounts of heterogeneous data. This was addressed by the ESO–CDS Data Mining project (Ortiz & Ochsenbein 2001). The project defined a set of *Uniform Content Descriptors* (UCD), to describe hierarchically the contents of tables (i.e., catalogues, tables, surveys, observation logs); and implemented a prototype of fast cross-correlator, able to identify relevant tables in VizieR by their UCDs and to correlate user tables with these tables.

References

Bonnarel, F., et al. 2000, A&AS, 143, 33

Bonnarel, F., et al. 2001, this volume, 74

Derriere, S., et al. 2000, in ASP Conf. Ser., Vol. 216, Astronomical Data Analysis Software and Systems IX, ed. N. Manset, C. Veillet, & D. Crabtree (San Francisco: ASP), 235

Kurtz, M., et al. 2000, A&AS, 143, 41

Fernique, P., Ochsenbein, F., & Wenger, M. 1998, in ASP Conf. Ser., Vol. 145, Astronomical Data Analysis Software and Systems VII, ed. R. Albrecht, R. N. Hook, & H. A. Bushouse (San Francisco: ASP), 461

Genova, F., et al. 2000, A&AS, 143, 1

Ochsenbein, F., Bauer, P., & Marcout, J. 2000a, A&AS, 143, 23

Ochsenbein, F., et al. 2000b, in ASP Conf. Ser., Vol. 216, Astronomical Data Analysis Software and Systems IX, ed. N. Manset, C. Veillet, & D. Crabtree (San Francisco: ASP), 83

Ortiz, P. F. & Ochsenbein, F. 2001, in Proceedings 'Mining the Sky', Conference held in Garching, July 31 -August 4, 2000, in press

Schmitz, M., et al. 1995, Vistas in Astron., 39, 272

Wenger, M., et al. 2000, A&AS, 143, 9

The PyRAF Graphics System

M. D. De La Peña, R. L. White, and P. Greenfield

Space Telescope Science Institute, Baltimore, MD 21218

Abstract. This paper describes the features, plans, and design for PyRAF graphics. PyRAF is an alternative CL for IRAF based on Python. Since IRAF tasks depend on the CL to manage all graphics, any CL replacement must implement a means of handling IRAF graphics.

We have developed graphics kernels for PyRAF written completely in Python that are capable of working with IRAF graphics tasks (including interactive tasks). We have added capabilities such as multiple graphics windows, a scrollable message input/output region, ability to recall previous plots and "undo" features, and automatic focus handling.

The design of the PyRAF graphics system makes use of Python's object-oriented features. We describe the design employed to isolate the details of IRAF's underlying graphics system from PyRAF, making it easier to support multiple kernels, including the ability to use the IRAF Graphics kernel tasks.

1. PyRAF Graphics Features

PyRAF is a new CL for IRAF (Greenfield & White 2000) that has been developed to allow writing scripts in Python that can run IRAF tasks and to allow enhancements to the interactive CL environment. The goal of the interactive environment was to retain the interface and syntax of the original IRAF CL to the maximum extent possible or sensible. Since IRAF expects the CL to handle all graphics, PyRAF must handle IRAF graphics.

The approach taken to the graphics system embodies a number of significant departures from that used by the IRAF CL for graphics. Like IRAF, we wish to retain the ability to use multiple graphics kernels. However, IRAF interactive graphics devices are largely terminal-based, whereas we decided to base our initial interactive kernel on a GUI library (through Python's Tkinter, though other multiplatform GUI libraries are available). Nevertheless, this does not preclude us from emulating the terminal-based graphics devices in the future. Basing graphics on a GUI library allowed us to add a number of features not easily added to the existing IRAF devices, such as dropdown menus, multiple graphics windows, and a scrollable status region.

Integrated with the menus is labeled access to all past plots in the current session, navigation through past plots via short-cuts (next, back, first, and last), the ability to create a new graphics window, plot edit features (undo, redo, refresh, and delete plot), ability to save the metacode to a file or print it (through existing IRAF kernels), and help for PyRAF graphics. A history

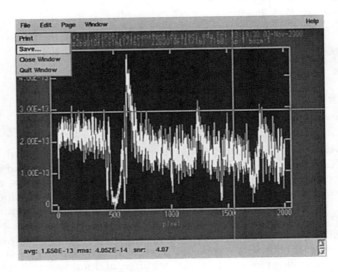

Figure 1. The basic PyRAF graphics window incorporates additional functionality accessible via a menu bar at the top and a buffer at the bottom of the window which contains a history of I/O messages.

buffer, implemented as a "status line," maintains a log of all input/output messages associated with the particular graphics window. The PyRAF system is capable of handling multiple graphics windows for a single PyRAF session. The graphics windows can be resized at any time, and window contents are subsequently resized accordingly. Most importantly for those who are longtime users of IRAF, PyRAF provides access to nearly all of the IRAF interactive features. The basic PyRAF graphics window is shown in Figure 1.

2. Planned Features

The initial release of PyRAF will contain most of the graphics features necessary for a viable system, but we envision further enhancements to the graphical environment. Some of the planned improvements for PyRAF are: alternative kernels to support graphics and image display, an enhanced "status line" buffer for easier history access, improved text rendering, optional balloon help (*aka* tool tips), a "preferences" menu (with options for color, printer choice, font size, etc.), an optional toolbar, additional IRAF interactive features ("capital" letter commands controlling roaming and zooming), use of GUIs for IRAF interactive commands versus special characters, keyboard accelerators, and the ability to plot Python data arrays to graphics windows (akin to IDL).

3. Open Issues

While we are attempting to retain most of the current IRAF CL's graphics features in PyRAF, there are some that have uncertain utility. We may decide not to duplicate such features, particularly if there appears to be little demand for them. These include the ability to run IRAF kernels interactively, support for stdgraph devices such as xterm and xgterm, and some of the IRAF "capital" letter interactive commands.

4. Design Issues

One of the high-level goals of the PyRAF design was to eliminate any direct dependence on IRAF libraries (PyRAF only uses IRAF executables). The PyRAF interactive graphics kernel thus is written entirely in Python with no reuse of IRAF code. Since Python is interpreted, one issue was that of efficiency. This was alleviated in part by basing the graphics on OpenGL, yet most of the action takes place in Python. While graphics rendering is perceptibly slower on slower workstations (e.g., Sparc 4s), the speed is not objectionable. On newer machines, the difference is rarely noticeable.

An important objective of the PyRAF graphics design was to make it easy to support multiple interactive graphics kernels as well as non-interactive ones. While designing a kernel class that has a simple interface is straightforward, there are some tricky issues dealing with the fact that one graphics kernel may be asked to switch to another in midstream while retaining the appropriate metacode information and state for the new kernel to start. Multiple graphics windows introduce further complications (the solution was to instantiate a kernel for each window).

OpenGL provides powerful plotting capabilities (and a great deal that is not needed for simple 2-D plotting). Most of the IRAF metacode plotting primitives are quite easy to render in OpenGL. The only exception is text rendering. For simplicity and portability, the initial OpenGL graphics kernel relies on a simple stroked font implementation. Handling keyboard focus properly when multiple interactive windows are available (including an image display window) also presents special problems and required a few C routines to provide Xlib focus manipulation functionality typically absent from most GUI toolkits. Finally, a full-screen cursor is rendered in software and requires careful handling of when it is and is not enabled. Python exception handling is the key to robust management of the full-screen cursor and other graphics state information.

A Unified Modelling Language (UML) diagram depicting the high-level design of the PyRAF graphics system is shown in Figure 2. GkiKernel is the base class for all graphics kernel implementations and is used to provide a standard interface to the IRAF process which communicates with IRAF tasks. The first level subclasses are responsible for handling the interactive graphics, switching between graphics kernels, and invocation of the IRAF builtin kernels. Specifically, GkiInteractiveBase is the base class for *interactive* kernels and implements the supporting functionality (e.g., menu bar, status line message buffer, page caching, etc.). The specific interactive classes are GkiOpenGlKernel and GkiTkinterKernel. GkiOpenGlKernel is the OpenGL graphics kernel implementation which uses OpenGL (or Mesa) in combination with Tkinter (a Python version of Tk) to render the plots. Alternatively, GkiTkinterKernel is the Tkinter graphics kernel implementation which uses only Tkinter to render the plots. GkiProxy is a proxy base class which implements the GkiKernel interface and

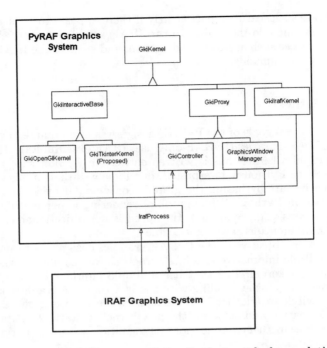

Figure 2. A UML diagram of the fundamental class relationships present in the PyRAF graphics system.

allows switching between different graphics kernels; GkiController is a GkiProxy which selects the active graphics kernel as directed by commands in the IRAF metacode stream. GraphicsWindowManager is a GkiProxy for the active graphics window which also acts as the manager for multiple graphics windows. GkiIrafKernel is a GkiKernel that routes metacode to an IRAF executable. Finally, IrafProcess handles the control and communication between the PyRAF and IRAF tasks and provides a unified interface to the IRAF subsystem.

5. PyRAF System Plans

A public beta version is currently available and the first release should be available by summer 2001. It is worth noting that PyRAF runs on IRAF-supported platforms without any changes to the IRAF system. For further details, users are encouraged to visit the PyRAF web site at http://pyraf.stsci.edu.

References

Greenfield, P. & White, R. L. 2000, in ASP Conf. Ser., Vol. 216, Astronomical Data Analysis Software and Systems IX, ed. N. Manset, C. Veillet, & D. Crabtree (San Francisco: ASP), 59

AstroRoute

P. Fernique

Centre de Données astronomiques de Strasbourg (CDS), Strasbourg, France

D. Durand

Canadian Astronomical Data Centre (CADC), Victoria, Canada

Abstract. The Internet is now the most commonly used medium for the exchange of data between data centres. However the quality of this network of networks is completely outside the astronomical community's control: the routing rules for each country and the technologies deployed depend on various interests generally far from astronomical concerns. In order to get actual measurements of the fluctuating network flow rates between some astronomical Web sites, the CDS has developed a tool called AstroRoute. Its goal is to supply, on an hourly basis, a measurement of the network quality between astronomical Web sites over the world.

1. AstroRoute

AstroRoute is a system to get measurements about the connectivity and the quality of the network between astronomical web sites. It is an initiative and a development of the CDS (Centre de Données astronomiques de Strasbourg) with the help of D. Durand from the CADC (Canadian Astronomical Data Center). Several other institutes collaborate : the Observatoire de Paris/Meudon, ISO, ATNF (Sydney), NOAO, STScI, and CFHT. This project started in June 2000.

2. AstroRoute Working Principle

There exists a lot of network tools to test network quality, but the main problem is that the tests are generally done from one single site—the tester's own host machine. In order to obtain more reliable results, AstroRoute is based on a set of distributed agents which test synchronously a list of targets in order to measure the network connectivity and the bandwidth between any pair of agent/web site in the AstroRoute system. So, it becomes possible to know the authority that should be contacted in case of network problem or bad performances (local, campus, and national network instance) and to build convincing arguments for future upgrades.

To implement this principle, AstroRoute is based on two elements: a main site and some agents. The main site controls all agents: it synchronizes their clock reference, sends the list of astronomical Web sites to be tested and compiles their results. Each agent tests individually these web sites and sends regularly a report to the main site. The resulting data set can be browsed on the main site with a classical web interface.

2.1. A Couple of Technical Points

The agent and main site software are written in Perl in order to keep a good portability. The system is designed to take into account bandwidth less than 200 kilobytes per second. The reports sent by the agents to the main site use a simple SMTP mail. By default, the tests are done hourly, and the reports are sent daily. The result browser is a classical HTML application. It is also possible to use a more powerful Java applet.

3. The Network Measurements

AstroRoute should provide network measurements reproducing a current web usage. This is why the tests are simple HTTP URLs providing a static file or image. With these tests AstroRoute records three measurements for each pair agent/web-site:

- The delay required to establish a network connection between the agent and the site. It means the time required for the DNS address resolution and the time to open an HTTP TCP session.
- The HTTP flow rate. This measurement requires a test URL providing a file larger than 100 KB.
- The failure frequency.

	Agent(s)		
	NOAO us	CADC ca	CDS fr
	123 tests	414 tests	681 tests
ADAC.Tokyo.jp	1.3s/24.0k/4%	0.8s/18.7k	0.7s/15.3k/2%
ADC.Greenbelt.us	0.2s/112.0k/2%	0.3s/59.7k	0.3s/>60.1k/1%
ADS.Cambridge.us	1.2s/1%	0.2s/1%	0.5s/1%
ATNF.Sydney.au	3.8s/15.6k/5%	1.2s/19.1k/1%	1.6s/17.2k/2%
CADC.Victoria.ca	2.2s/64.6k/2%	0.0s/>258.5k/0%	1.4s/35.0k/1%
CDS.Strasbourg.fr	3.0s/36.1k/5%	0.9s/22.0k/0%	0.0s/>225.4k
CFHT.hawaii	0.8s/54.5k	0.2s/53.9k/0%	0.4s/32.4k/0%
ESO.Garching.de	0.6s/30.6k/6%	0.5s/24.7k/0%	0.5s/>76.8k/2%
Gemini.Arizona.us	0.3s/>195.3k/4%	0.5s/52.5k/7%	0.4s/38.4k/14%
IPAC.Pasadena.us	0.2s/>138.0k	0.1s/>109.8k	0.3s/41.0k/0%
ISO.Vilspa.es	0.6s/27.9k/4%	0.7s/26.1k/2%	0.3s/>80.8k/2%
IUCAA.Pune.in	0.7s/8.7k/7%	1.3s/10.2k/6%	1.8s/7.7k/4%
ObsPM.Meudon.fr	0.5s/4.6k/6%	0.6s/20.7k/0%	0.2s/>65.5k/1%
STScI.Baltimore.us	1.3s/98.2k	0.3s/62.4k/0%	0.4s/69.9k/0%
XMM.Leicester.uk	0.5s/10.5k/5%	0.5s/>11.9k	0.5s/>14.2k/1%

Table 1. Excerpt of agents/Web sites pair matrix: time connection, flow rate in kilobytes/sec, and success rate.

4. Global Trends

AstroRoute has been running for six months, and interesting global trends can be seen:

- **Various impact of the peak hours:** The difference between the peak hours and the off-peak hours is very important between European sites and US sites. The bandwidth can be divided by a factor of two or three at 17hr GMT. On the other hand, the access quality to Japan (ADAC) or Australia (ATNF) is not really dependent of time.

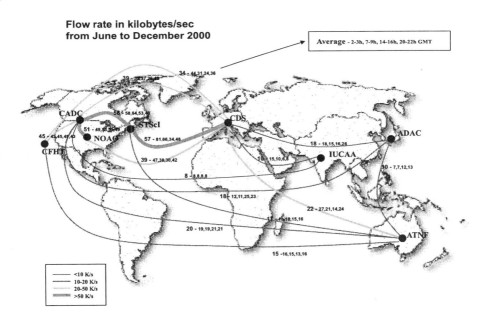

Figure 1. Bandwidth world map.

- **Increasing global quality in "switchback:"** Globally the bandwidth of Internet grows but this longterm tendency results from alternation of regular decrease and sudden improvements (Figure 2).
- **Protectionism of routing policy:** Network agencies seem to keep the bandwidth for local hosts since their network is loaded. For example, it is easy to reach the ISO Center at Vilspa (Spain) from Strasbourg (France) even if European network is loaded, but it is very difficult for an American host to do the same thing at the same time. We can find other examples for the US routing policy specially concerning the Eastern sites.

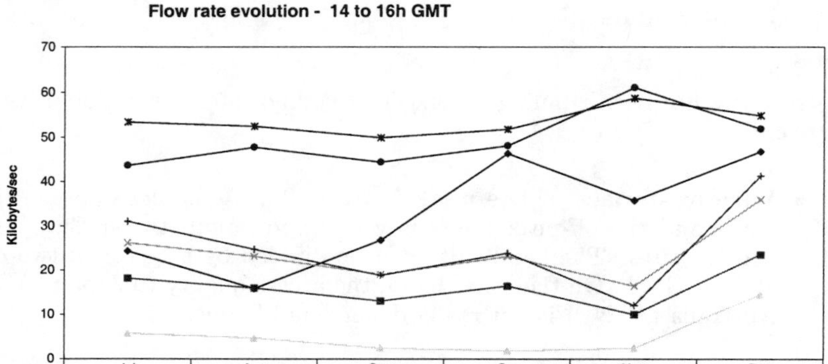

Flow rate evolution - 14 to 16h GMT

	Jul	Aug	Sep	Oct	Nov	Dec
CDS - STScI	24,3	15,8	26,8	46,4	35,7	46,8
CDS - ADAC	18,1	15,9	13	16,4	10	23,5
CDS - IUCAA	5,7	4,7	2,4	1,8	2,4	14,5
CDS - CFHT	26,2	23,1	19	23,1	16,4	35,9
CADC - STScI	53,4	52,5	49,9	51,8	58,7	54,8
CADC - CFHT	43,7	47,8	44,4	48,1	61,1	51,9
CADC - CDS	31	24,7	18,8	23,8	12	41,3

Figure 2. Flow rate evolution.

5. Conclusions

The AstroRoute results can be accessed from CDS via the AstroRoute main page[1]. The objective is now to extend the set of participating sites to obtain meaningful world-wide long term statistics, critical to assess the needs for this key infrastructure in the context of rapid development of interconnection between astronomical on-line resources.

---

[1]http://simbad.u-strasbg.fr/astroroute.pl

Hera: Building a Web-Based Analysis Environment at the HEASARC

Christina W. Heikkila[1], William D. Pence, Thomas A. McGlynn[2], Peter Wilson[3], James Peachey[1], Ning Gan[1], Edward Sabol[1], Keith Arnaud[4]

NASA/Goddard Space Flight Center, Greenbelt, MD 20771

Abstract. We are developing protocols which will allow full scientific data analysis to be performed within suitably configured Web browsers. Currently, although the Web is critical to both the data discovery and publishing aspects of the scientific enterprise, it is used far more sparingly in the gritty details of data analysis. The data analysis environment does not share the advantages of immediate access to new resources, quick and convenient downloads, and the support from commercial vendors that data distribution currently enjoys. We believe we can extend existing protocols to provide a complete analysis environment that users will access through their Web browsers. We will implement these protocols within the HEASARC's FTOOLS and XANADU analysis environments to provide a fully functional system to users. We anticipate these protocols would be extensible to other software environments.

1. Introduction

Hera is designed to bring data analysis to the web. It breaks down artificial barriers between data analysis tools and the data archive, between user data and archive data, between resources at the data center and resources on the user's machine. Currently, users of the HEASARC FTOOLS and data archive must install a copy of the FTOOLS at their site, download any data needed from the HEASARC archive, then begin analysis on their local machine. Using Hera, the user could actually use FTOOLS on archive data all on the HEASARC servers, without installing the entire suite of FTOOLS, and without downloading the data files from the archive.

The HEASARC's Hera program has just begun in the last six months, and has been awarded AISR funding for the next three years. We have been exploring the enabling technology and have developed an initial prototype.

---

[1] Emergent IT

[2] Universities Space Research Association

[3] Apple Computer, Inc.

[4] University of Maryland

2. Goals

Hera should allow seamless use of software and data resources at the user's local site and the archive. Ideally, the user should not care whose machine is actually running a task, and the user's space on the HEASARC servers should seem like just another local disk.

We plan to offer a scripting capability, where users can put together sequences of analysis tools to be repeated. These scripts will allow a mix of local and server tools. Using this scripting capability, Hera will provide simple "cookbook" tools to allow astronomers (or even expert high-school students) to perform routine data analysis quickly.

We will use Hera as a real service at the HEASARC. Thus we have made implementation choices (e.g., Tcl for the user interface) that reflect current HEASARC architecture. However the protocols and concepts used within Hera should be portable to other systems and languages.

3. Examples of Potential Hera Use

The Integral mission anticipates that response matrices required to do spectral analysis of observations from some instruments may be several gigabytes in size. Either users must clog the web downloading these enormous files, or build a copy of the extensive Integral packages needed to generate these files—with no guarantee that they will run on the user's machine. Using Hera, users will be able to analyze archival or even their own proprietary data without placing such a burden on their local machines.

Hera makes it possible to do real data mining in the HEASARC archives. The user has access to both the full archive and the full software system. Tasks can be run over large numbers of retrieved datasets. For example, a user looking at faint variable X-ray sources might first use the Web to query the HEASARC catalogs for long observations by ROSAT or ASCA. He or she could then use Hera to run a script on each of the resulting observations and download the results to a local machine.

An optical astronomer unfamiliar with high-energy analysis can use Hera scripts to guide her analysis. Using Hera, existing HTML cookbooks can not just tell her what tasks to run, but start them as based on simple web form selections.

With a little care, the previous example can be extended to build educational pages which allow college or even high-school students to interact dynamically with real data in the HEASARC archive.

4. Current Status

Using Hera, users can initiate analysis tools on the HEASARC servers or on their local machine (currently limited). The FTOOLS are presented in a set of hierarchical menus. Choosing one brings up a window where the user can set parameters, including the input and output files. A new window is created showing any text output of the FTOOL, and a graphical window shows any images or plots generated.

Every user is given their own personal space on the HEASARC servers, where FTOOLS results files, status files, and preference files are stored. Symbolic links are used as pointers to files in the HEASARC archive which the user has marked for later analysis. Users can also transfer data files to and from their personal space at HEASARC. Hera provides users with a full set of file management tools to manage their user space at HEASARC.

5. Future

Despite the progress we have made so far, there is still much work to be done. In particular, we do not yet have scripting capabilities available—these will allow a user to string together a series of tools to be run on the same data or on multiple data files. We plan to create cookbook recipes as Hera scripts so users can easily analyze large sets of observations. These scripts would also be valuable as teaching aids—they could be integrated in a college or high school exercise involving analyzing data.

We need to create a system of metadata, so that Hera will know which FTOOLS can accept as input which types of FITS files. The FTOOLS use a parameter file which gives some limited metadata for each tool, but describing HEASARC software and data sets is a substantial element of the work needed for Hera. XML looks like a natural candidate for handling this metadata.

The FTOOLS should be able to recognize that the versions of tools installed on the user's machine are out of date, and offer to download newer versions. Some users may choose not to install a local copy of the FTOOLS at all, and simply run all analysis on the HEASARC servers.

We plan to make the first public release of the Hera system by the end of 2001. We welcome suggestions and comments—please contact:
- Bill Pence : pence@milkyway.gsfc.nasa.gov, or
- Tom McGlynn : tam@lheapop.gsfc.nasa.gov.

MAISON: A Web Service of Creating Composite Images On-the-fly for Pointing and Survey Observational Images

M. Watanabe[1] and K. Aoki[2]

Japan Science and Technology Corporation (JST), Tokyo 102-0081, Japan

A. Miura

Institute of Space and Astronautical Science (ISAS), Kanagawa 229-8510, Japan

N. Yasuda

National Astronomical Observatory of Japan (NAOJ), Tokyo 181-8588, Japan

S. Uno

Nihon Fukushi Univ., Aichi 475-0012, Japan

Abstract. MAISON[3] (*Multi-wavelength Astronomical Image Service On-line*) is a Web broker service which allows users to retrieve different images of the same field-of-view (FOV) from separate image servers. Through MAISON, users can readily preview a composite image created on-the-fly from these multiple images. Given a successful development and release of the seminal version, we are currently developing a new version of the MAISON system which will be equipped with several new features.

1. Overview of the MAISON

A flow chart of a single MAISON session is illustrated in Figure 1. The session consists of two systematic queries from a client to remote image servers via the MAISON server.

1.1. The First Query: Getting an Image List

At the first query which is launched at the top page, users are requested to specify the celestial area and remote image servers from which the images are

---

[1] Postal address: PLAIN, Institute of Space and Astronautical Science, Kanagawa 229-8510, Japan

[2] Postal address: ADAC, National Astronomical Observatory of Japan, Tokyo 181-8588, Japan

[3] http://maison.isas.ac.jp

MAISON: Creating Composite Images On-the-fly   71

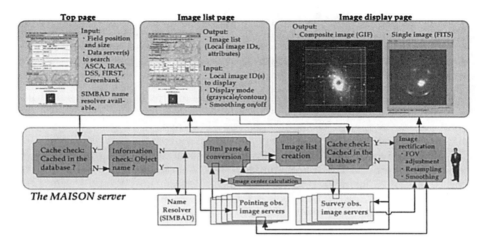

Figure 1. A flow chart of a MAISON session. Three top layer panels represent client browser windows, while a middle layer panel underneath them shows operations of the MAISON server. The MAISON server accesses remote servers which are illustrated underneath the MAISON server layer.

to be retrieved. What is returned in this query is a list of images (not the images themselves) which meet the user's input request and are available at the specified remote servers.

The MAISON system affiliates the remote servers unilaterally, i.e., a remote server is affiliated *as is* with the MAISON system. Because of this unilaterality a condition is imposed on the remote server that it should be a TCP/IP-based server dealing with FITS images that hold popular WCS records. The condition, however, is a quite ordinary one for a standard astronomical image server and therefore it does not actually limit the remote servers available to the MAISON system. Similar to most of other astronomical image servers, the MAISON server accepts either an object name or celestial coordinates in order to specify a celestial area. The object name is resolved into coordinates using a SIMBAD name resolver client routine (Wenger et al. 2000) provided by CDS. The user's initial input parameters are converted into a CGI variable 'QUERY_STRING' that is suitable to the remote servers' specific CGI-form. This conversion is conducted in reference to a conversion table created and maintained off-line manually. The MAISON server then submits the QUERY_STRING to the specified remote servers. Pointing observation images are searched for within a finite area on the sky centered at the users' specified location, thus generally have different FOVs. Because of this we need to know the real center of the individual pointing observation images prior to obtaining the survey images comparable with the pointing images. Accordingly, the submission to the pointing observation image servers always needs to precede that to the survey observation image servers. The HTML information returned is parsed in a manner prescribed specifically in the conversion table. The parsed information is composed into a single HTML list file of the available images and then displayed on the client's browser.

1.2. The Second Query: Getting a Composite Image

At the second query, the users are requested to select from the image list a pair of images to put together into a composite image. These images are then retrieved from their respective hosting remote servers, rectified, composed, and finally displayed on the client browser in GIF format. A single image may be displayed as well using a FITS image viewer provided as a Java applet.

Given the users' input parameters for the second query, the MAISON server converts them into a QUERY_STRING in the same manner as does in the first query. MAISON has a local cache database for recent queries and images. Accordingly, if the QUERY_STRING is found in the database, the locally stored images are used in this session. Otherwise the query is submitted to the specified remote servers. The pair images are then retrieved from their respective hosting remote servers and stored on the MAISON server's working disk space. The images are rectified in reference to their WCS information so as to have exactly the same FOV. The rectification is performed using the WCSTOOLS library (Mink 1999).

The rectified images are then put together into a single GIF image using the PGPLOT[4] library, with one image illustrated in a gray-scale and the other superposed on it as a contour map. A grid and labels for equatorial coordinates are also drawn in the GIF image. The GIF image is finally displayed on the client browser using an 'image' tag of the HTML. In the FITS image viewer invoked for displaying a single image, users may change a color map by mouse-dragging and a magnification of the image by a menu selection in this applet window.

2. Development of the New MAISON

The currently released version of MAISON has several issues of improvement and reinforcement to get a better performance, thus leading us to plan a whole reconstruction of the MAISON. To this end we have already started a development of the new MAISON system. Substantial aspects of the reinforcement has been enlightened by the excellent Aladin system (Bonnarel et al. 2000). Main issues of the improvement and reinforcement include:

- *The image viewer*
  All the images are to be treated in the FITS format and will be displayed in a single image viewer (Figure 2). The image viewer will be coded as a Java applet to allow users to do a prompt interactive operation on the images. More than two images may be selected for the composite image display.
- *Catalog data*
  A catalog data server VizieR (Ochsenbein et al. 2000) has been mirrored at NAOJ/ADAC and will be affiliated with the new MAISON system. The catalogued objects will be superposed on the image viewer.

---

[4]http://astro.caltech.edu/~tjp/pgplot/

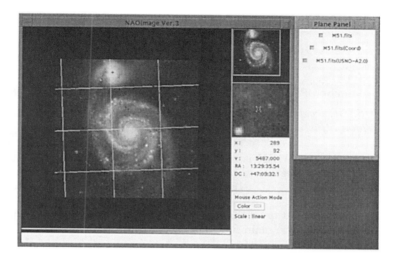

Figure 2. The image viewer for the new MAISON system (under development). The viewer uses FITS files as an input images. The viewer is equipped with subpanels similar to those of SAO image, and any numbers of gray-scale images and contour map images may be selected and overlaid as if they are drawn on 'transparent sheets'. The sheet may be a chart of the catalogued objects in the sky area.

- *DSS wide-field server*
  A DSS image server maintained at NAOJ/ADAC has been featuring a locally modified getimage which is capable of generating an image up to $6°$-square FOV (Taga et al. 2001). The output pixel scale can be set accordingly larger and therefore helps to transfer DSS data significantly faster than the original DSS image. The DSS wide-field server will be affiliated with the new MAISON server.

Acknowledgments. We are grateful to the members of the ISAS/PLAIN and NAOJ/ADAC and all the collaborators who have worked to improve the astronomical data archival environment in Japan. The present work is partly supported by Japan Science and Technology Corporation (JST).

References

Bonnarel, F., et al. 2000, A&A, 143, 33

Mink, D. J. 1999, in ASP Conf. Ser., Vol. 172, Astronomical Data Analysis Software and Systems VIII, ed. David M. Mehringer, Raymond L. Plante, & Douglas A. Roberts (San Francisco: ASP), 498

Ochsenbein, F., et al. 2000, A&A, 143, 23

Taga, M., et al. 2001, to appear in the Report of the National Astronomical Observatory of Japan (in Japanese)

Wenger, M., et al. 2000, A&A, 143, 9

The ALADIN Survey Integrator

F. Bonnarel, P. Fernique, F. Genova, O. Bienaymé, and D. Egret

CDS, Observatoire Astronomique de Strasbourg, 11, rue de l'Université, F67000 Strasbourg

Abstract. ALADIN is a powerful tool for data interpretation, which allows one to integrate survey images and catalogues with reference information from SIMBAD, NED, VizieR, and with images from observatory archives.

Recent evolution of the ALADIN software and of the image database are presented. The main facts are the following:

- The AladinJava standalone version has been distributed, allowing one to use local catalogues and images as input with the full functionalities of ALADIN;

- AladinJava is now able to decompress images in the hcompress format. Decompression for multi-resolution compressed images is also currently included in collaboration with the CEA;

- The image database now includes the whole DSS2 set (Red and Blue) in addition to the previously implemented DSS1 and MAMA images;

- The server is now fully rewritten in order to allow access to a larger heterogeneity of survey images. 2MASS and then DENIS images will soon be implemented in the system.

1. Introduction

The current decade (1995–2005) is marked by several instrumental and observational trends in astronomy. First of all the outset of the 8–10m class telescopes led, beside the scientific results, to the building of huge archives of on-line data. Secondly space missions went on providing data avalanche in more various wavelengths. And last but not least, the sky survey domain is rapidly growing and evolving. Together with the completion of the digitization of all the old-generation sky Schmidt surveys and the appearance of non-optical surveys we are now facing the large optical digital surveys.

This led to the emergence of the new Virtual Observatory concept, widely discussed during two dedicated conferences, held in Pasadena (Virtual Observatories of the Future) and Garching (Mining the Sky) during year 2000. The goal assigned to VO(s) is to make the data easily and rapidly accessible to the whole astronomical community, and to provide dedicated public tools to make usage of these data easier. Presently different partners, such as Observatory archives, survey teams, and data centers have to provide building blocks of a distributed VO.

CDS faced this need by making Aladin (Bonnarel et al. 2000; Fernique & Bonnarel 2000) evolve from a cross-identification facility towards a distributed data integrator, dealing with images, catalogues, survey data, observation logs, and compilation databases.

2. Aladin: From New Cross-identification and Observation Preparation Tools ...

The AladinJava applet is basically a help to test and modify the automatic cross-identification strategies, by allowing overlays of VizieR, SIMBAD and NED sources directly on reference images. From the beginning, the Aladin database gave access to DSS1 images (1.7 arcsec sampling) from the STScI CD-ROM set and to other high sampling (0.67 arcsec/pixel) SERC and ESO Schmidt survey scans, made in Paris at CAI with the MAMA facility (Guibert 1992).

In March 2000 the AladinJava standalone version was released. It allows the same kind of overlays for user catalogues in Tab Separated Value or XML/Astrores format and for user images with WCS FITS headers. Coordinate grids, and Field of Views of various telescopes (currently: CFH12K, XMM, and HST) together with 1.5 deg "low resolution" views have been added in both applet and standalone version in December 2000. The Java applet is now used as reference image facility in NED and at CFHT—discussion are under way with other groups, such as Isaac Newton Group.

In the meantime the spatial resolution and spectral coverage of the whole sky has been improved by the inclusion of the red and blue colors of the DSS2 (Lasker & STSCI Sky-Survey Team 1998) in the Aladin database. The I color of DSS2 will be included in 2001. The total amount of available Schmidt survey data is now larger than a terabyte.

3. ... to Observatory Archive and Survey Data Integration

AladinJava now goes further by providing within the same portable interface beside links to the full content of SIMBAD, VizieR, and NED databases access to ground or space observatory archives—currently IUE spectra or HST and VLA FIRST archive images. The latter are directly usable in the interface, and AladinJava allows to shift from archive image data to reference images and conversely. Access to these archive images is made faster by the integration of a Java HDECOMP module in the interface, which allows to read and decompress images transferred in the much more compact hcompress format. More efficient multi- resolution methods such as MR1/mr-comp are currently included, in collaboration with CEA (Murtagh et al. 1998). The December 2000 release goes further by allowing usage of any on-line image database in FITS WCS, or catalogue server in TSV or XML, through a short ASCII description (in GLU syntax; Fernique et al. 1998). This functionality has been tested with the SUPERCOSMOS Web server.

In addition, inclusion of survey catalogues in VizieR integrates them within Aladin. Currently USNOA, DENIS, and 2MASS catalogues are available. APM and GSCII will be soon. In the meantime a prototype of a new version of the server has been developed which allows management of near-IR and optical CCD

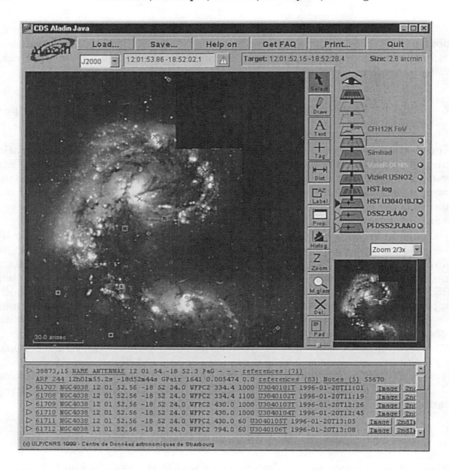

Figure 1. HST image of the Antennae, with overlays from SIMBAD and NED. DSS1 images are on the stack.

survey pixels. With this new server Aladin will be able to deliver DENIS and 2MASS images—this is planned for the first semester of 2001. CFH12K fields reduced at the Paris TERAPIX data center (Radovich et al. 2000) will come later, and may be followed by the MEGACAM survey data.

All these new features make Aladin a real prototype of an Observatory archive and survey data integration tool (Figure 1).

References

Bonnarel, F., Fernique, P., Bienaymé, O., Egret, D., Genova, F., Louys, M., Ochsenbein, F., Wenger, M., & Bartlett, J. G 2000, A&AS, 143, 33

Fernique, P. & Bonnarel, F. 2000, in ASP Conf. Ser., Vol. 216, Astronomical Data Analysis Software and Systems IX, ed. N. Manset, C. Veillet, & D. Crabtree (San Francisco: ASP), 71

Fernique, P., Ochsenbein, F., & Wenger, M. 1998, in ASP Conf. Ser., Vol. 145, Astronomical Data Analysis Software and Systems VII, ed. R. Albrecht, R. N. Hook, & H. A. Bushouse (San Francisco: ASP), 446

Guibert, J. 1992, in Digitised Optical Sky Surveys, ed. H. T. MacGillivray & E. B. Thomson (Dordrecht,Boston MA: Kluwer Academic Publishers), 103

Lasker B. M., & STSCI Sky-Survey Team 1998, in American Astronomical Society Meeting 192 (Washington: AAS), 6403

Radovich, M., Mellier, Y., Bertin, E., Bonnarel, F., Boulade, O., Cuillandre, J. C., Didelon, P., Missonier, G., Morin, B., & McCracken H. J. 2001, in A new era of Wide Field Astronomy, IAU Symposium, ed. R. G. Clowes (San Francisco: ASP), in press

Murtagh, F., Starck, J., & Louys, M. 1998, International Journal of Imaging Systems and Technology, 9, 38

Astronomical Data Analysis Software and Systems X
ASP Conference Series, Vol. 238, 2001
F. R. Harnden Jr., F. A. Primini, and H. E. Payne, eds.

Visual Exploration of Astronomical Documents

Soizick Lesteven, Philippe Poinçot, Fionn Murtagh[1]

CDS, Observatoire astronomique de Strasbourg, 11, rue de l'Université, 67000 Strasbourg, France

Abstract. The CDS bibliographical map is a tool for organizing astronomical text documents into a meaningful map for exploration and search. The system is based on the Self Organizing Map (SOM) algorithm that automatically organizes documents into a two-dimensional grid so that related documents appear close to each other and general topics appear in well defined area.

After the determination of optimal parameters for the SOM's learning process, we have developed a graphical WWW interface which allows the visualization of the document distribution. It shows the localization of documents related to given topics (keyword queries). The map is clickable and provide links to the documents. Recent developments include detailed map of small areas, full text indexing, automatic labeling, ...

Some applications will be presented. One map is available for interactive use on the Web (http://simbad.u-strasbg.fr/A+A/map.pl).

1. Introduction

The continually increasing quantity of textual data requires constant effort in order to update storage and access methods so that the totality of information is easily accessible. Scientific publications are no exception. Astronomy is a good example in view of the enormous mass of data collected by modern satellites and large ground-based facilities, and the numerous scientific articles which result from such data.

The Centre de Données astronomique de Strasbourg (CDS) collects and organises different types of astronomical information (Genova et al. 2000). In particular, the CDS offers on-line access to some bibliographic data.

This article presents ongoing work on the CDS bibliographical maps. These maps are a tool for automatically organizing collections of astronomical text documents and for displaying them in order to facilitate the mining and retrieval of information (Poinçot 1998, 1999). The map is clickable and provides links to the on-line documents. The system is based on the Self-Organizing Map, or Kohonen Map (Kohonen 1995). In the article we present new developments, including detailed map of small areas, full text indexing, and automatic labeling.

---

[1]School of Computer Science, Queen's University Belfast, Belfast BT7 1NN, Northern Ireland

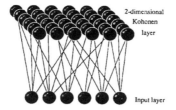

Figure 1. The topology of the Kohonen self-organizing map network.

2. The Self-Organizing Map (SOM)

The SOM is an algorithm used to visualize and interpret large high-dimensional data sets. These maps are one of the artificial intelligence techniques, and more precisely an unsupervised neural network. The topology of the Kohonen SOM network is shown in Figure 1. This network contains two layers of nodes: an input layer, and a mapping (output) layer in the shape of a two-dimensional grid. The input layer acts as a distribution layer. The number of nodes is equal to the number of attributes associated with the input. Each node of the mapping layer also has the same number of attributes. Thus, the input layer and each node of the mapping layer can be represented as a vector which contains the number of features of the input. The mapping nodes are initialized with random numbers. Each actual input is compared with each node on the mapping grid. The "winning" mapping node is defined as that with the smallest Euclidean distance between the mapping node vector and the input vector. The value of the mapping node vector is then adjusted to reduce the Euclidean distance. In addition, all of the neighboring nodes of the winning one are adjusted proportionally. After all of the input is processed (usually after tens of repeated presentations), the result should be a spatial organization of the input data organized into clusters of similar regions.

3. Application to Bibliographic Classifications: Creation of Bibliographical Map

We applied the method to the classification of articles published in *Astronomy and Astrophysics* in the period 1994 up to now (9450 articles). The attributes were based on the bibliographic keywords. We kept only the 269 keywords that appear in at least five different articles. The documents characterized in this way constitute a set of 9450 stimuli to be applied to the network. Learning through 50 iterations (heuristically determined) gives good results for the principal map (15×15 units).

We have adapted the use of the SOM to our own needs. Documents located at a map edge have neighbors at the other side of the map. It is then possible to reconfigure the map without losing the similarity of closely clustered documents. At the end of the training of a map, the number of documents assigned to each node is known. Because it is much easier to visualize the colours of an image than a matrix of numbers we transformed the table of numbers into an image. For this image the colour scale indicates qualitatively the number of documents per

Figure 2. The secondary maps.

node. The map is manually labeled to locate on it the different themes associated with the nodes. The user can select one node of the map, by clicking on the picture, to obtain some information about the articles located in it: the number of documents and the keywords describing them appear on the right side of the interface. The user can also access the article content (title, authors, abstract) and all the facilities provided by the CDS bibliographical service (including a link to ADS and to the on-line full paper when available). The map can also be reached by keyword queries and bibcode queries (standard definition of a document). One map is available for interactive use on the World-Wide Web (http://simbad.u-strasbg.fr/A+A/map.pl).

4. Recent Developments

4.1. Secondary Maps

The size of the SOM map has a strong influence on the quality of the classification. The smaller the number of nodes in the network, the lower the resolution for the classification. The fraction of documents assigned to each node correspondingly increases. It can then become difficult for the user to examine the contents of each node when the node is linked to an overly long list of documents. However, there is a practical limit to the number of nodes: a large number means long training. A compromise has to be found. That is why, for each "over-populated" node of this map, termed the *principal map*, another network, termed *secondary network* or *map*, is created and linked to the principal map. Each secondary network is trained using the documents associated with the corresponding node and its surrounding nodes of the principal map (Figure 2). In this way, a map is created with as many nodes as necessary, while keeping the computational requirement under control.

4.2. Full Text Indexing

To overcome the limitations due to the representation of the documents by keywords, we use all textual information present in the documents (title, abstracts, keywords, ...). We built our own full text indexing tool. All the words are kept and counted, empty words (without meaning) and the less frequent ones are eliminated. A truncature (Porter 1980) is applied to reduce morphologic variants of a word to a single index term. The program allows to keep word associations (2 or 3 words). The first applications show that the two-word associations are meaningful. It is now possible to create bibliographical maps with a full text indexing and to take into account documents for which keywords are not available. It seems that the full text indexing is on one hand noisier due to the automatic indexing (the context is not taken into consideration) but on the other hand more precise because it does not rely on a limited list of keywords.

4.3. Automatic Labeling

For map interpretability, the different themes associated with document/node assignments have to be indicated. Although our maps have a relatively limited number of units, it is important to automate the annotations. Because it is impossible to characterize all nodes without overlapping annotations it is preferable to select a limited number of nodes for characterization. The selection is made according the density peaks and the cross-positions of two peaks. When the position is defined, the process examines the words associated with the documents assigned to the peak and writes the most frequent one beside the peak. When different words have the same occurrence, the automatic choice (the first one) is not automatically the best one. Furthermore, our system makes it possible to classify the words in various categories. It is then possible to annotate the map according to the different classes, corresponding to a precise application.

5. Conclusions

We developed a visual and efficient system to explore astronomical documents based on the SOM. Other data collections, notably catalogue information, have already been processed in the same way, and a cartographic user interface tool has been set up to allow catalog selection (http://vizier.u-strasbg.fr/).

Now that new tools have been developed, a number of new applications are possible. Creation of a bibliographical maps for the SIMBAD objects, for an ADS query, for a colloquium organisation, ...

References

Genova, F., et al. 2000, A&AS, 143, 1
Kohonen, T. 1995, Self Organizing Maps (Berlin: Springer-Verlag)
Poinçot, P., et al. 1998, A&AS, 130, 183
Poinçot, P., 1999, PhD Thesis "Classification et recherche d'information bibliographique par l'utilisation des cartes auto-organisatrices, applications en astronomie"
Porter M. F. 1980, Program, 14(3), 130

The CARMA Data Viewer

Marc W. Pound

University of Maryland

Rick Hobbs, Steve Scott

Owens Valley Radio Observatory

Abstract. The CARMA Data Viewer (CDV) is a CORBA and Java based real-time data viewer for radio interferometric arrays. CDV was designed to solve the problem of a common data viewer for interferometric observatories (the Berkeley-Illinois-Maryland Association array and the Owens Valley Radio Observatory array) with differing visibility data (amplitude and phase) formats and differing hardware and technical specifications. In the coming years, the BIMA and OVRO arrays are to be merged into a common array called CARMA.

We chose CORBA because it is available on many platforms with bindings for many programming languages, enabling remote object support in a heterogeneous computing environment. We chose Java for its cross-platform support and its rich set of lightweight graphical objects, in particular the JTable.

Because of its observatory-independent data model built on top of CORBA, CDV can in principle be used with any interferometer (e.g., SMA, ALMA, VLT).

1. CARMA

CARMA (the Combined Array for Research in Millimeter-wave Astronomy) is a university consortium of Caltech, U. C. Berkeley, the University of Illinois, the University of Maryland, and the University of Chicago. CARMA will combine the existing Owens Valley Millimeter Array, the BIMA Millimeter Array, and new antennas from the University of Chicago at a high altitude site in California. The combined array will consist of six 10.4m, nine 6.1m, and six 3.5m antennas. CARMA's 21 antennas will have unique multi-scale imaging capabilities.

2. Data Format

A visibility is a complex number representing the amplitude and phase correlation between a pair of antennas (a baseline). We defined a single CORBA object, called an ObsRecord, which could encompass simply the different visibility formats, independent of the peculiarities of either telescope array. An ObsRecord contains an array of visibilities records, one for each unique antenna

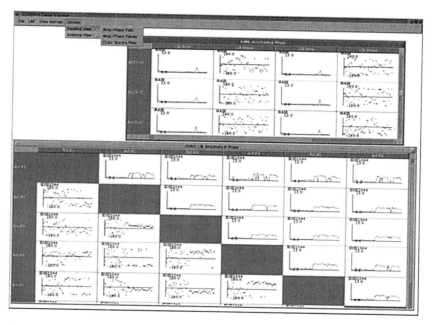

Figure 1. View of the CDV workspace showing different data representations for sideband averages (LSB & USB), which are selected using the Options Menu. Upper window shows baseline-based amplitudes (red trace) and phases (blue dots) vs. time from BIMA. Lower window contains antenna-based plots of amplitude and phase vs. time from OVRO.

pair, measured at a given time from an astronomical source. The visibility data may be multi-channel or single channel and are tagged with frequency, position, velocity, and time information. The channel data are represented in individual spectral windows, which are not necessarily contiguous in frequency. Each observatory produced software to create ObsRecords from its own visibility format and place them on a server using standard CORBA methods. The client-side data viewer can then connect to a server at either observatory (or both at once).

3. The Viewer

CDV is based on the Java 2 JTable class, with a Data Model that matches the ObsRecord specification. Each cell in the JTable contains the data for a particular antenna pair. The phases and amplitudes are displayed in the cells as a function of time (for single channel data; see Figure 1) or frequency (for multi-channel data; not shown). The cell update rate is tied to the observation integration time, typically about 10 seconds.

CDV is a workspace application, with individual data viewers running as windows inside the workspace. Multiple Data Sources (i.e., observatories) may

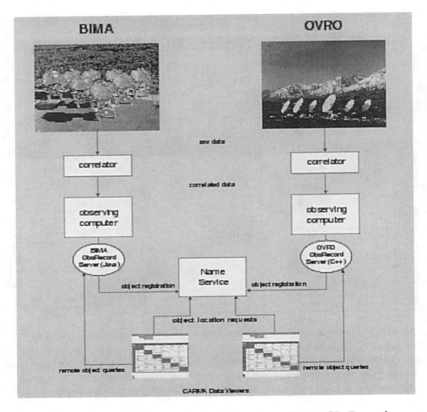

Figure 2. A diagram representing the data flow. ObsRecords are created from the correlated data on each array and registered with a CORBA Name Service. Client viewers can fetch ObsRecords from either observatory by first querying the Name Service for the ObsRecord location, then using the ObsRecord public CORBA interface to obtain its member data.

be monitored simultaneously, and a variety of representations of the data are possible.

Future enhancements will include server and client side caches for low-bandwidth operation and the ability to sort and filter the data (e.g., by time, frequency, baseline separation) using the Java 2 Comparator classes.

4. CORBA Implementation

4.1. Overview

CORBA allows convenient access to remote objects. For the CARMA Data Viewer, the remote objects (ObsRecords) contain a rich hierarchy of header and data that are updated on timescales of a few to tens of seconds. The data alone for a 45 baseline BIMA continuum-only ObsRecord is about 13.5 KB. For 4 spectral windows with 512 channels, the size increases to 200 KB. With 15 baselines, the OVRO data scales to one third the size of the BIMA data. Directly accessing the remote ObsRecord with standard accessor calls to update the Data Viewer would require hundreds of remote method invocations.

4.2. Bottleneck

The latency, overhead, and data transport time can all make significant contributions to performance and must be examined. Latency for calls over the general Internet is typically 100 to 200 msec. There is negligible latency over a 100 Mbps LAN. The overhead in a CORBA call that transfers no information is very small—approximately 1 msec. For transferring data (large CORBA objects) over the Internet, measured transfer speeds were at least 60% of the nominal T1 connecting observatory sites to the Internet. Note that for the largest objects (200 KB), the transfer time is in excess of one second. From measured performance components, the Internet latency (100 to 200 msec) can severely limit CDV if many remote method calls are required per update.

4.3. Solution

The solution adopted was to create a structure that contained all the data items for an ObsRecord. The remote method then returned this structure, effectively moving all of the data for the object in one transaction. The structure can then be used to form a local cloned copy of the remote object. The measured overhead is about 6 msec per transaction for the OVRO continuum data packet (3.5 KB) and should scale with packet size.

4.4. Overall Performance

Using this technique of moving a large structure, it is possible to update a large ObsRecord every few seconds with reasonable latency and CPU loads of a few percent or less on modern CPUS.

4.5. And finally...

Electronic versions of the actual poster in a variety of formats are available at http://www.astro.umd.edu/~mpound/adass/2000/cdv.html

The Design of Solar Web, a Web Tool for Searching in Heterogeneous Web-based Solar Databases

Isabelle Scholl

Institut d'Astrophysique Spatiale, CNRS/Université Paris XI, Bâtiment 121, F-91405 ORSAY, France

Abstract. Today, scientists interested in solar data have access to a large variety of data collections (coming from space or ground-based observatories), but only available with different and heterogeneous software. Moreover, there is no common definition for keywords describing the data. Consequently, this makes the pre-analysis period long and difficult.

This paper presents the design and the architecture of the 'Solar Web Project' which aims to provide scientists interested in solar data with collaborative software for browsing distributed and heterogeneous solar databases using a unique web-based interface.

The project is developed at MEDOC (Multi-Experiment Data and Operations Centre), located at the Institut d'Astrophysique Spatiale (Orsay, France) which is one of the European Archives for SOHO (selected by ESA) and for TRACE data. JAVA and XML are technologies being used for designing and developing this tool.

1. Introduction

Today scientists interested in solar science are facing large amounts of data and many heterogeneous tools to identify and analyze them. Observations often arise from collaborative programs involving several observatories (space and/or ground-based). Cross-data analysis starts then with a heavy task: to identify and to retrieve data around several archive centers, each of them using different user interfaces, vocabularies, and semantics.

The tool developed for the SOHO archive[1] (Scholl 1998) at MEDOC (Scholl 1999) is a first step in the process of simplifying the access to heterogeneous data. SOHO is a spacecraft that has on-board 12 instruments, each of them having their own catalog and data in their own format. The software developed to browse the SOHO catalog and to retrieve files (including a web-based graphical user interface) was designed with open concepts in mind: this tool had to be generic enough to be able to easily manage 1) homogeneously all instrument catalogs and 2) other solar observations than those from SOHO (TRACE data are already available with this tool).

The Solar Web Project (Scholl 2001) ensues from: the acknowlegment that scientists have no unique or generic tool available for browsing several heteroge-

---

[1] http://www.medoc-ias.u-psud.fr/archive/

neous and worldwide distributed archives, the fact that an archive center should not store all mission data (data should be located near the relevant scientific experts), and the need to extend the SOHO archive access system to other solar archives in order to help scientists in their data search phase.

2. System Design

The aim of the 'Solar Web' tool is to provide users with one unique interface to access several heterogeneous and distributed archives. This assumes the development of a software with capabilities of sending multiple queries over the Internet, waiting for remote results before merging them and finally displaying them on one single view. But it also assumes that these archives are coordinated in order to standardize communications and catalogs.

The project is divided in 2 main parts: 1) scientific work: definition of a common model based on a standardized vocabulary for describing a 'Generic Solar Observation', and 2) technical work: software design and development.

The first step is to defined what a 'Generic Solar Observation' means. This phase consists in classifying a few parameters which are sufficient for the purpose of selecting the data. Around twenty keywords are grouped in 4 categories: General Parameters, Observational Technical Parameters, Scientific Parameters, and Data Management Parameters.

The purpose of this model is to help communications between existing archives. By using such a model, existing archives will be able to communicate in a standard way, even if some of them would probably not have all this information or will have to convert their own keywords to this model.

3. System Architecture

The architecture of the 'Solar Web' tool is based on a collaborative system. There are 2 main components:
- the 'Application Server'. Its role consists in accepting connections from clients and managing their queries (local and remote). It is installed under the HTTP server tree.
- the 'Web-User Interface' or client module.

The specificity of the 'Application Server' resides in its capability to communicate with other kinds of servers. It can communicate with databases using four different methods (see Figure 1):

1. the server directly accesses the local database (DB) using the JDBC API,
2. the server directly accesses registered remote databases that offer a direct JDBC access,
3. the server accesses existing cgi scripts that query their local databases (located on remote web-servers) using an HTTP GET method, and
4. the server communicates to any other 'Application Servers' that query their local databases.

Items 2 and 3 are free of developments on remotes sites.

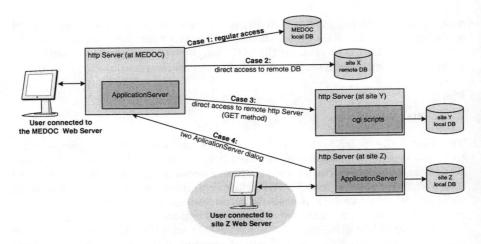

Figure 1. General System Architecture.

4. Software Architecture

This software is based on a 3-tier and distributed architecture. It has been designed to be open to new kinds of data (see Figure 2).

On the server side, the 'Application Server' fulfils the following roles:
- A Session System which manages sessions with clients using a unique ID
- A Query System composed of:
  - a Query Generator (QG) preparing SQL queries for a specific DB. Each DB structure is implemented in dedicated classes and provided by the QG as an API, and
  - a Query Processor which runs queries, gets answers, formats results with a DataBase Manager for local DB access and a Query Router for remote DB access.
- A Cache System which manages buffers associated to clients

On the client side, the applet is dedicated to the powerful query capture and result visualization. The client (excepted for the dialog with the server) and the query generator are the only business specific components. These two modules will be available as an API of the generic software to implement extensions.

Exchanges between the main server and remote applications are made using either a native application protocol to simplify development, or XML with a standardization objective: the 'Generic Solar Observation' is implemented as an XML DTD.

Security is a major concern in this project. Permitting direct remote accesses is always a potential security risk for any organization. A secure protocol for communications between remote servers is currently under study.

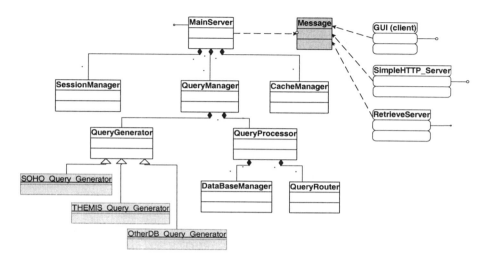

Figure 2. Software Architecture.

5. Conclusion

'Solar Web' is currently a stand alone development effort from MEDOC. Such another project exists, the Virtual Solar Observatory (Hill 2001), but it is still at a very early stage. Discussions are in progress with potential partners. Tests will be made with Bass2000 (Themis data, at Tarbes, France) to experiment with the communication between MEDOC and an existing remote web-server. The status of the project can be found at http://www.medoc-ias.u-psud.fr/archive/solarweb/. We are open to suggestions for new collaborations.

References

Scholl, I., et al. 1998, in ASP Conf. Ser., Vol. 172, Astronomical Data Analysis Software and Systems VIII, ed. David M. Mehringer, Raymond L. Plante, & Douglas A. Roberts (San Francisco: ASP), 253

Scholl, I. 1999, in Proceedings 8th SOHO Workshop, (ESA Publ. SP-446) (Paris 1999), 611

Hill, F. 2001, in ASP Conf. Ser., Vol. 225, Virtual Observatories of the Future, ed. Robert J. Brunner, S. George Djorgovski, & Alexander Szalay (San Francisco: ASP), in press

Scholl, I. 2001, in ASP Conf. Ser., Vol. 225, Virtual Observatories of the Future, ed. Robert J. Brunner, S. George Djorgovski, & Alexander Szalay (San Francisco: ASP), in press

Quick-look Applet Spectrum 2

Markus Dolensky[1], Benoît Pirenne

*Space Telescope - European Coordinating Facility,
Karl-Schwarzschild-Straße 2, D-85748 Garching, Germany*

Scott Binegar, Molly Brandt, Nial Gaffney

*Space Telescope Science Institute, 3700 San Martin Drive, Baltimore,
MD 21218*

Christophe Arviset, Jose Hernandez

*ISO Data Centre, European Space Agency, Space Science Department,
Astrophysics Division, Villafranca del Castillo, PO Box 50727, 28080
Madrid, Spain*

Abstract. *Spectrum 2* is a new quick-look applet for spectra. It serves as part of the ST-ECF HST archive web interface. Prior to display the ESO Fits Translation Utility[2] (FTU) is used to homogenize plots of data from HST instruments STIS, FOS and GHRS. *Spectrum 2* supports manipulations of multiple spectra. It will also be integrated within STScI's new archive browser StarView 6. A stand-alone version of *Spectrum 2* is available for download[3].

This article concentrates on the software architecture and interfaces of this java utility since it is composed of various components mainly developed by ESA-IDC and ST-ECF.

1. S/W Architecture

Figure 1 shows *Spectrum 2* in context of the ST-ECF HST archive web interface (http://archive.eso.org). On user request preview spectra are retrieved from the archive DB. The ESO Fits Translation Utility (FTU) adds special purpose FITS keywords in order to homogenize plots of spectra from HST instruments STIS, FOS, and GHRS. The resulting FITS stream is transmitted to the web client. The web browser hosts the java applet *Spectrum 2*. The applet reads the data stream using a FITS I/O package developed by McGlynn/USRA and then parses certain keywords in order to define a plot (Figure 3). Currently there exists a parser for ISO data and one for HST spectra. The ISO parsing

---

[1] Affiliated to the Astrophysics Division, Space Science Department, European Space Agency

[2] http://archive.eso.org/ftu/

[3] http://archive.eso.org/java/sp2/

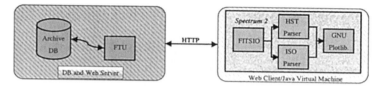

Figure 1. Software components involved in the client/server environment of *Spectrum 2*.

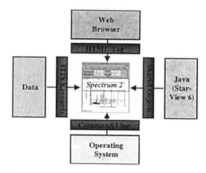

Figure 2. The various machine interfaces of *Spectrum 2*.

capability has been implemented first. This happened in cooperation with the ISO Data Centre.

For the sake of browser support the GUI is based on the Abstract Windowing Toolkit (AWT) rather than on the more powerful Swing libraries. AWT is part of the Java Virtual Machine that executes *Spectrum 2* and therefore does not require any installation on the client side. The actual plot is rendered with Brookshaw's Graph Class Library. A user does not need any knowledge about the internal data structure of HST observations to perform a quick on-line evaluation of individual or associated spectra.

2. Interfaces

Spectrum 2 provides various machine interfaces (Figure 2). It accepts parameters from the command line in application context and from the HTML <APPLET> tag in applet context. It is capable of parsing FITS headers in order to extract plot attributes and finally it can communicate with other Java components via shared objects by which means it can be hooked up to StarView 6 from STScI.

3. Future Work

- The first release will be further tailored to support the visualization and on-line analysis of associations of spectra: Binning and features for the manipulation of individual spectra within an association of spectra are desirable.

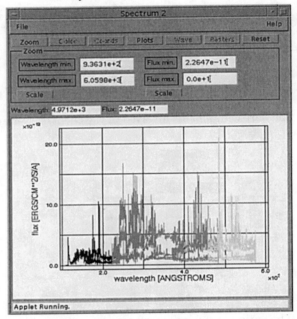

Figure 3.   *Spectrum 2* displaying 16 associated FOS spectra of Eta Carinae (PI Davidson).

- *Spectrum 2* software releases will be made available for download[4].
- Integration within archive browser StarView 6: There will be a static interface comparable to the interface with image viewer JIPA (Dolensky, Mayhew, & Kennedy 1998), but it may be turned into a dynamic Bean component at a later stage.

References

Dolensky, M., Mayhew, B., & Kennedy, B. 1998, in ASP Conf. Ser., Vol. 172, Astronomical Data Analysis Software and Systems VIII, ed. David M. Mehringer, Raymond L. Plante, & Douglas A. Roberts (San Francisco: ASP), 454

---

[4]http://archive.eso.org/java/sp2/

The New JCMT Holographic Surface Mapping System - Implementation

Ian Smith, Fred Baas, Firmin Olivera, Nick Rees

Joint Astronomy Centre Hawai'i HI 96720

Youri Dabrowski, Richard Hills, John Richer, Harry Smith

Mullard Radio Astronomy Observatory, UK

Brian Ellison

Rutherford-Appleton Laboratory, UK

Abstract. The James Clerk Maxwell Telescope (JCMT) on the summit of Mauna Kea in Hawaii is a 15 meter telescope which operates in the sub-millimeter region of the spectrum (in practice from about 2 mm to 350 microns). Operation at such short wavelength requires the dish to have a surface accuracy of 20 microns RMS. The primary reflector surface consists of 276 panels each of which is positioned by 3 stepper motors. In order to ensure the highest possible surface accuracy we are embarking upon a project to improve the mapping of the surface.

The mapping of the primary reflector surface is achieved by "holographic" techniques. Currently a 94 GHz radiation source, located on a nearby Telescope (United Kingdom Infra-Red Telescope) is used to illuminate the JCMT surface. The orientation of the telescope is scanned in two coordinates to measure the beam pattern. From this it is possible to determine the path length of the radiation to, and hence the position of, each panel. A new instrument has been deployed in order to achieve greater accuracy and produce maps at a significantly faster rate. The instrument uses two main frequencies, 80 and 160 GHz, and is frequency agile around these frequencies to allow for the elimination of spurious signal paths.

1. Introduction

The James Clerk Maxwell telescope (JCMT) primary reflector is made up from 276 reflective panels. Each of these panels is attached to three stepper motors that allow the panels to be individually tilted in order to achieve the best shape for the highest antenna gain (Smith 1998).

To maintain and correct the parabolic shape of the primary reflector, a panel position detection system is employed. A 94 GHz radiation source located on the hill above (at the UKIRT) illuminates the JCMT, and its receiver detects this signal. By scanning the JCMT across the beam in a raster pattern whilst moving the secondary mirror up and down, the phase of the signal is recovered. This produces a whole-dish map of the vertical position and tilt of each panel (with 3 micron resolution and 12 micron repeatability).

2. Reasons for a New System

Whilst the current system provides the necessary information to maintain the surface of the JCMT, it suffers from several problems. One of the largest problems is that it takes about two hours to fully map the surface. Apart from the time aspect, this introduces errors given the fact that the temperature of the telescope structure changes over the period of the mapping. It is well known that temperature changes have a significant effect on the position of the panels. The only time when the temperature is stable over this sort of time period is the middle of the night, so to produce the best maps, normal observations have to be halted.

When moving panels, it is necessary to re-map the surface to check that the moves were sufficient and correct, and sometimes the panels need a further adjustment with a third map required. At $\sim$2 hours per map, this can consume an entire observing shift.

Other problems include spurious reflections from the protective Gortex membrane and other structures, as well as a growing maintenance burden from the aging system in place.

3. RxH3 Description

A new system described in the JCMT Holography Receiver RXH3 Preliminary Design Paper[1] has been constructed and was commissioned in December 2000. Still undergoing tests prior to calibration, its main features are: (1) a map of the surface can be produced in about 15 minutes, (2) measurement repeatability of 5 microns RMS, (3) dual frequency (80 & 160 GHz) system (using two frequencies provides a choice of either fast low resolution or slower high resolution mapping, and stepping around these frequencies provides the ability to solve and eliminate signals from spurious paths like reflections from the protective membrane), (4) dual signal system, main and reference beam, which produces both phase and amplitude of the signal, akin to a holographic measurement system and (5) a VME real-time computer data acquisition system for fast mapping and time stamping for accurate registration of the data.

4. RxH3 System Components Description

4.1. Components Located At UKIRT

Real-Time Source Computer, Sx3 This is a PC with a PC-LabCard multi-IO and real-time clock card. This provides control over the PTS frequency synthesiser via a parallel interface. Frequency stepping patterns are received by

---

[1]http://www.jach.hawaii.edu/JACdocs/JCMT/rxh3/d-1-1/d-1-1a.doc

the PC via an RS232 interface, and when commanded, the PC runs the given pattern continuously at a 1 KHz rate. This is achieved by controlling the PTS synthesiser via a parallel interface.

PTS Synthesiser   This digitally controlled frequency synthesiser can produce an output in the range 250 MHz to 350 MHz, and when coupled to the 80 GHz and 160 GHz signals generators produces controllable signals in the range 80.25 - 80.35 GHz and 160.5 - 160.75 GHz. This unit will step in frequency by 5 MHz, at a rate of 1 msec as per the pre-programmed frequency pattern. It will also blank the signal during each hop in order to synchronise the data acquisition system at JCMT, much like the blanking signal in a television signal.

Source   This is a dual frequency system with 80 and 160 GHz channels orthogonally polarised by a wire grid.

4.2. Components Located at JCMT

Optical Relay   This consists of two curved mirrors mounted on the backing structure of the JCMT. Their purpose is to fold the reference beam, which passes through a hole in the JCMT primary reflector, into the reference input of the RxH3 receiver.

Receiver   The receiver consists of two channels, reference and signal, each with two mixers (80 GHz and 160 GHz) to produce a 300 MHz IF. These two IF signals are then passed through a dual channel correlation receiver and each channel produces a sine and cosine output. The signal blanking is also detected and used to generate a TTL sync pulse to drive the DAU (Data Acquisition Unit).

DAU   The DAU is a VME-based 68060 computer with a Wind River's VxWorks operating system and applications coded in the Anglo-Australian Telescope's DRAMA C programming environment[2]). The DAU consists of: (1) a VME XYCOM XY566 16 channel programmable 12 bit A/D converter, (2) a VME XYCOM XY240 TTL I/O card and (3) a VME Bancom BC635VME RIGB card.

4.3. Software Components

The SYSCON (System Controller) DAU and DHS (Data Handling System) components have been written as DRAMA tasks. These tasks and the tcl Control Script communicate with each other using DRAMA messages.

Control Script   The Control Script runs on a Unix workstation, coordinating all activity between systems and issuing STEP and CENTRE commands to the telescope. The concept of a map is understood only at this level, with other systems operating on a row by row basis. The script determines all required parameters for the requested map and generates it by iteratively issuing ROW commands to collect raster data.

---

[2] http://www.aao.gov.au/drama/html/dramaintro.html

SYSCON This DRAMA task runs on a Unix workstation to control the DAU and DHS. In preparation to take data, SYSCON sends the Sx3 control PC a frequency pattern to be executed repeatedly and sends INITIALISE and CONFIGURE commands to the DAU and DHS DRAMA tasks. It instructs the DHS to open a FITS data file. Upon request, SYSCON will send a ROW command to the DAU and instruct the DHS to receive data, monitor the telescope position parameter, TSPOSN (time and position stamps), interpolate the collected TSPOSNs and write data to disk. When all rows are collected, SYSCON instructs DHS to close the data file.

DAU This VME software consists of a VxWorks task and a DRAMA task that enables the A/D card to acquire data from the RxH3 receiver analogue channels. The frequency band is selected, and the Rx state is monitored by the TTL I/O card. Up to 20000 samples can be acquired by the A/D for each row at a rate of 1 K samples per second. Each time the data are sampled, the time card is read, and the time-stamp, TTL status byte and data are stored. Data acquisition is normally synchronised to both the telescope "On Source" (start of row) pulse and the digitisation SYNC pulses generated by the Rx. Once ROW data are collected, the data set is sent to the DHS task as a DRAMA bulk message.

DHS This DRAMA task accepts bulk data messages from the DAU, receives a stream of telescope positions, generates interpolated telescope positions and writes the data to disk.

4.4. Conclusion

RxH3 will quite literally revolutionise the maintenance of the dish as well as achieve significant improvements in surface accuracy.

The ability to map the surface in 15 to 30 minutes will permit the rapid correction of aberrant panels. It will be possible to make many adjustment iterations within just a few hours, and errors introduced by thermal drift in the current system will be minimal.

A surface accuracy of 5 microns RMS will be a significant improvement over that currently achievable and will significantly improve telescope efficiency.

Acknowledgments. RxH3 is a joint development project between MRAO Cambridge UK, RAL UK and the Joint astronomy Centre in Hawaii.

References

Smith, I. 1998, Proc. SPIE, 3351, 190

The 4-m International Liquid Mirror Telescope Project (ILMT)

J. Poels[1], E. Borra[2], J. F. Claeskens[1], C. Jean[1], J. Manfroid[1], F. Montfort[1], O. Moreau[1], Th. Nakos[3], J. Surdej[1], J. P. Swings[1], E. van Dessel[3], B. Vangeyte[1]

Abstract. The working principle of liquid mirror telescopes (LMTs) is first reviewed along with their advantages and disadvantages over classical telescopes. For several reasons (access to regions near the south galactic pole, galactic center, good image quality, etc.), an excellent site for such an LMT is the Atacama desert. A deep (B~24 mag) LMT survey at latitudes near -22° – -29° will cover ~90 square degrees at high galactic latitude and be especially useful for gravitational lensing studies, for the identification of various classes of interesting extragalactic objects (cf. clusters, supernovae, etc. at high redshift) and for subsequent follow-up observations with 8 m-class telescopes. A short description of the handling of data products is also presented.

1. Telescope Technical Description

The surface of a rotating reflecting liquid takes the shape of a paraboloid which is the ideal surface for the primary mirror of an astronomical telescope. The focal length F of the mirror is related to the gravity g and the angular velocity of the turntable ω by means of the relation $F = g/2\omega^2$. The container and the bearing rest on a three-point mount that aligns the axis of rotation parallel to the gravitational field of the Earth (Figure 1). The container must be light and rigid. A thin layer (0.5 mm to 1 mm) of mercury is then spread on the container.

Figure 2 shows the entire telescope system. Comparing the LMT to a conventional telescope, we see that they are similar with the exception of the mount. The top parts, consisting in a focusing system and a detector, are identical, but there is some cost saving in the upper end structure since it does not have to be tilted. The largest savings accrue due to the simple tripod mount.

While an LMT can only observe a zenith strip of constant declination, its observing efficiency compared to that of a classical telescope is very high (no slew, no field acquisition, no lost readout time). The tracking will be done with the Time Delay Integration technique (TDI, also known as the drift scan technique), and low-resolution spectroscopy can be carried out with interference

---

[1] Institute of Astrophysics and Geophysics, Liège University, Belgium

[2] Laval University, Canada

[3] Royal Observatory of Belgium

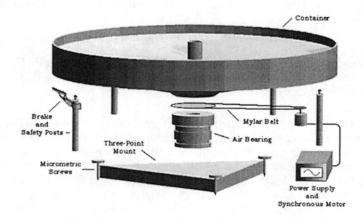

Figure 1. Exploded view of the basic liquid mirror telescope setup.

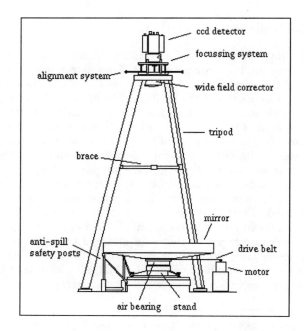

Figure 2. Entire liquid mirror telescope system.

filters. A semi-classical on-axis glass corrector capable of about $30' \times 30'$ degree field will be used to remove the TDI distortion. With a classical corrector, the TDI technique degrades the images since the technique moves CCD pixels at a constant speed in a straight line, while fixed-sky elements appear to move at different speeds along slightly curved trajectories. This latitude-dependent deformation is zero at the equator and increases with increasing latitude.

2. Science with the International Liquid Mirror Telescope

The observational strategy for studies of gravitational lensing effects with a LMT consists in first surveying a sky area as deeply and broadly as possible for interesting targets (e.g., quasar candidates using color and variability criteria) and then selecting gravitational lens candidates among them. The field of view is primarily determined by the number and/or the size of the thin CCDs placed at the LMT prime focus. For the case of multiplely imaged quasars, we find that direct imagery with the 4 m International Liquid Mirror Telescope (ILMT) will lead to the detection of approximately 50 new gravitational lens systems ($\Omega_0 = 1$, $\Lambda_0 = 0$). The natural possibility of photometrically monitoring these at daily intervals with a great accuracy offers a unique opportunity to define a sub-sample of interesting lenses with reliable geometrical parameters, time delay measurements and/or micro-lensing signatures for further astrophysical and cosmological studies. Such a survey will also provide unique data for studies of the galactic structure and stellar populations, including the detection of micro-lensed galactic objects, accurate measurements of stellar proper motions and trigonometric parallaxes useful for the detection of faint red, white and brown dwarfs, halo stars, etc. The ILMT will be located in the Atacama desert and should be operational in 2002. With the ILMT field of view at its geographical latitude, the Earth's rotation will scan the Galaxy from the Southern Pole to the bulge and central regions. Very precise photometric and astrometric data for millions of stars will be obtained in the drift scan mode night after night, permitting the detection of microlensing events toward the galactic bulge.

3. Data Analysis and Computational Aspects

3.1. Data Acquisition

The overall ILMT software architecture can be described briefly as follows: The acquisition of the $4\,\text{K} \times 4\,\text{K}$ CCD data flux relies on C++ software running on a WinNT4 platform housing custom electronic PCI cards with real-time multi-channel acquisition capabilities. This system (a Pentium PC) will be located in a booth that will be part of the telescope building. The collected data stream will then be sent through a microwave data link to a lower altitude remote site where the main processing and analysis will take place. This will allow dedicating a whole system for reliable data acquisition, temporary storage, archiving, observation scheduling and network transactions, thereby enabling data reduction to be performed at the same data rate in a remote data processing location. The integration of CCD captors, TDI mode CCD controller, data acquisition PCI electronic cards, WinN4 PCI software drivers and data acquisition application software is currently undergoing testing. Completion of quality assurance tests related the whole data acquisition chain is expected in early summer 2001.

3.2. Data Processing and Reduction Pipeline

From the data analysis point of view, such a project requires sophisticated algorithms and a massive, reliable computation and storage infrastructure in order to generate an exhaustive catalog of detected sources. Fortunately, modern computer science and artificial intelligence techniques (i.e., pattern recognition using neural networks, fuzzy logic, decision trees, etc.) can permit accurate categorization of the objects and of their photometric and astrometry history (time series). Furthermore, management and analysis of catalog/image databases will be accomplished with powerful data-mining tools. With an appropriate on-line data analysis policy, the survey is likely to yield many short duration events/targets that in turn could be of major interest to large southern hemisphere observatories, such as the ESO/VLT, and to the Virtual Observatory project. Powerful computational resources are needed for the demanding CPU-intensive processes and large number of database transactions. Sun Microsystems has been chosen as hardware manufacturer partner for its UltraSparc III processor technology (running the Solaris 8 Unix-like operating system). Ideally, one TB of Fiber Channel RAID storage should be considered as a minimum after one year of ILMT telescope operation. Storage requirements would increase if an array of such ILMT's should be deployed (with individual telescopes being dedicated to specific spectral bands). On the database side, investigations are still underway to determine which RDMS software vendor best fits the ILMT project's needs. At the time of this writing, Oracle, Informix and Sybase are being considered as potential RDMS's. Given the amount of public domain astronomical image processing software available, it is clear that some de facto standards have already emerged (e.g., SExtractor, IRAF, MIDAS, Drizzle, etc.). Integration of as many existing software solutions as possible seems desirable in building the ILMT data processing pipeline. Another alternative would be to adapt an existing data reduction pipeline to meet the project's requirements: this possibility is still under assessment.

4. The ILMT Project in Detail:

A more detailed description of the ILMT project is available at: http://vela.astro.ulg.ac.be/lmt/.

Part 2. Education & Public Outreach

Hands-On Universe: A Global Program for Education and Public Outreach in Astronomy

Michel Boër, C. Thiébaut

Centre d'Etude Spatiale des Rayonnements (CESR/CNRS), BP 4648, F 31028 Toulouse Cedex 4 France

Hugues Pack

Northfield Mount Hermon School, Northfield, Massachusetts, USA

C. Pennypaker, M. Isaac

University of California at Berkeley, USA

A.-L. Melchior

DEMIRM/CNRS, Observatoire de Paris, France

S. Faye

Lycée Jacques Decour, Paris, France

Toshikazu Ebisuzaki

RIKEN, Tokyo, Japan

Abstract. Hands-On Universe (HOU) is an educational program that enables students to investigate the Universe while applying tools and concepts from science, math, and technology. Using the Internet, HOU participants around the world request observations from an automated telescope, download images from a large image archive, and analyze them with the aid of user-friendly image processing software. This program is now in many countries, including the USA, France, Germany, Sweden, Japan, and Australia. A network of telescopes has been established, many of them remotely operated. Students in the classroom are able to make night observations during the day, using a telescope in another country. An archive of images taken on large telescopes is also accessible, as well as resources for teachers. Students deal with real research projects, e.g., the search for asteroids, which resulted in the discovery of a Kuiper Belt object by high-school students. Not only does Hands-On Universe give the general public access to professional astronomy, it also demonstrates the use of a complex automated system, data processing techniques, and automation. Using telescopes located in many countries over the globe, a powerful and genuine cooperation between teachers and children from various countries is promoted, with a clear educational goal.

1. Introduction

The advantages of using astronomy in the classroom are several. The Universe and its objects have to be learned in most of the curricula, either as part of physics courses, or on its own. Astronomy may provide a good application of many of the concepts developed in physics and mathematics curricula. Astronomy is at the intersection of many areas of the knowledge, either in fundamental and applied sciences. It is a good illustration of the usefulness of an interdisciplinary approach. It is also a good illustration of the emergence of science over the past centuries, and of the idea (still developing) of the place of mankind in the Universe. Many of the astronomical concepts have been developed in various historical areas (Mesopotamia, Egypt, ancient Greece, Arabic countries, Occidental world...).

The goal of the Hands-On Universe (HOU) program is to promote the use of astronomy within the high and middle schools, and to enable students to use data or to request their own observations from professional or dedicated observatories. HOU has historically been developed at the University of California Berkeley, as a curriculum program. In this framework, dedicated software to analyse astronomical images has been written. Now, HOU has been extended to more than nine countries over four continents, and promotes the use of a global network of telescopes.

In this paper we present the main features of the HOU program, and its global approach.

2. Main Teaching Goals of HOU

As explained above, beside astronomy one of the main goals of HOU is to use the concepts and data acquired in Astronomy to introduce the scientific notions in physics and mathematics, for high-school students. As an illustration these concepts may be introduced using astronomical data: In physics, the notion of speed of light may be illustrated by reproducing Romer's experience. In mathematics, notions such coordinates, maps, and transformations may be illustrated using celestial reference systems. Astronomical observation brings these abstract transformations to real life. In image and signal processing, concepts like contrast may be approached when pupils work with actual images. An astronomical observatory has many complex instruments, and if automated, it is a good illustration of automation for a technology course (e.g., housekeeping). Of course, an actual exploration of solar system bodies brings "life" to an introduction to astronomy.

Though HOU has as main goal teaching at the high and middle school levels, many demonstrations have been performed for the general audience, e.g., at the open days of the U. C. Berkeley, at the Science Museum of Tokyo, at the SITEF technological exhibition in Toulouse, and at the Villette Science Museum in Paris.

A specific computer program has been developed to simplify access to most of the functions available in astronomical data processing software. It has been written for PC and Macintosh type computers, and gives a clear interface for pupils. The data processing functions include various color tables, computation

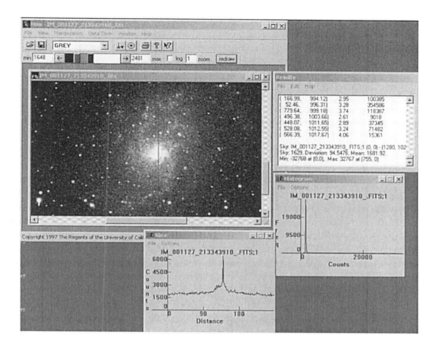

Figure 1. The Messier 33 galaxy displayed using the HOU software, and featuring the "sky" and "slice" tools (image from TAROT, Boër et al. 2001)

of sky level, aperture photometry, slice extraction, histogram, axes plotting, etc. Several of these features are illustrated in Figure 1 on an image of M33 taken by the TAROT telescope (Boër et al. 2000; Boër et al. 2001).

3. Global HOU

Global HOU (G-HOU) is an open network of teachers, high-schools, EPO, and research institutions, whose goal is the use of astronomy in the various curricula. Several resources are shared, like exercises, images, and telescope time.

Several telescopes are giving time to the G-HOU network, e.g., the TAROT instrument in France, the Keio observatory in Japan, the Katzman observatory in the USA, etc. Some of them are professional research instruments, while several others have been developed primarily for educational purposes. Most of these telescopes are able to be remotely operated. This global network is widely spread in longitude (and also in latitude) and has the advantage that pupils may access telescopes during normal classroom hours, while it is the night at the telescope site. Several successful sessions have been made, e.g., during the open days of U. C. Berkeley, the SITEF exhibition in France, or within the framework of the astronomy class of the Tokyo Science Museum. They all bring the powerful (and somewhat magical) possibility of seeing the night sky of other countries. This feature, together with the use of archival, or pre-requested data, open the possibility for students to perform their own scientific program. As an example, for a study of the colors of stars, a sequence of multi-color observations may be requested and the HR diagram built directly from the data by the pupils. Another striking example is the discovery of a Kuiper belt asteroid by the astronomy class of Northfield Mount Herman School.[1] This discovery has been confirmed by another high-school class, showing the level of cooperation reached within the G-HOU program. More generally speaking, G-HOU is a means for children from various countries and languages to exchange and collaborate.

There are, however, some difficulties because educational programs may be quite different in various countries. For instance, there is no dedicated "astronomy" course in France (and in most countries), while there quite often is one in the USA. This means that astronomy has to be used as a tool to introduce other concepts in physical sciences or mathematics, e.g., distance, color, and temperature. The use of astronomy, with "real data" at hand, has proven to be very effective to stimulate pupil interest in science courses, including unpopular areas.

4. Conclusion

We presented the Hands-On Universe program.[2] The most interesting features of this program are the use of actual and archived data within the classroom, and the ability of students from various countries to build their own research program and to work with "their" data.

Several exercises have been designed by the teachers participating in the program, most of them aimed at introducing the notions and concepts used in physical sciences, technology and mathematics courses. Last but not least, through the use of telescopes located in many countries over the globe, a powerful and genuine cooperation between teacher and children from various countries is promoted, with a clear educational goal.

References

Boër, M., et al. 1999, A&AS, 138, 579
Boër, M., et al. 2000, in ASP Conf. Ser., Vol. 216, Astronomical Data Analysis Software and Systems IX, ed. N. Manset, C. Veillet, & D. Crabtree (San Francisco: ASP), 115
Boër, M., et al. 2001, this volume, 111

---

[1]http://www.nmhschool.org/astronomy

[2]http://hou.lbl.gov/global/

Space Science Education Resource Directory

Carol A. Christian, Keith Scollick

Space Telescope Science Institute, Baltimore, MD USA

Abstract. The Office of Space Science (OSS) of NASA supports educational programs as a by-product of the research it funds through missions and investigative programs. A rich suite of resources for public use is available including multimedia materials, online resources, hardcopies and other items. The OSS supported creation of a resource catalog through a group lead by individuals at STScI that ultimately will provide an easy-to-use and user-friendly search capability to access products. This paper describes the underlying architecture of that catalog, including the challenge to develop a system for characterizing education products through appropriate metadata. The system must also be meaningful to a large clientele including educators, scientists, students, and informal science educators. An additional goal was to seamlessly exchange data with existing federally supported educational systems as well as local systems. The goals, requirements, and standards for the catalog will be presented to illuminate the rationale for the implementation ultimately adopted.

1. The Challenge

For the last decade, NASA has invested the creation of a wealth of Space Science educational and informational resources derived from its flight missions and related research. These materials are available from a variety of sources including universities, other educational institutions and NASA Centers. A wide range of users (teachers, students, museum staff, the general public, etc.) are keen to find and access these resources.

Content providers (i.e., the Space Science flight mission personnel, individual researchers and NASA staff), desire to know what kind of resources exist and how materials currently in production complement each other and need a consolidated mechanism for making their own products visible and easily deliverable to the public. NASA as a stake-holder is interested in determining the level of return on its support to education and outreach efforts. These objectives have been addressed through the creation of an education resource directory meeting the varied needs of the different customers of Space Science education.

2. Directory Development Strategy

The directory provides a robust infrastructure spanning all of NASA's Space Science education endeavor. It is to be a single source (i.e., a "one stop shop")

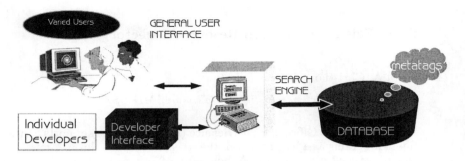

Figure 1. Basic architectural scheme for Space Science directory.

for access to online, hardcopy, and physical media. The model is based upon the successful *amazon.com* interface because we need to address a wide range of user needs in a similar way. More sophisticated attributes of the directory interface will include resource recommendations, user feedback mechanisms, thumbnail sketches of resources and personal preference lists and notifications. Also, providers and stake-holders wish to generate various reports, probe inventory information and track resource usage with other custom interfaces to the directory. There was also a strong motivation not to re-invent needed infrastructure, but to use existing methods where possible.

Additional directory requirements were culled from the end users as well as providers and NASA agency personnel. These requirements include:

- handling multimedia and varied physical media
- insuring quality control and assurance of the content, upholding the level of excellence expected of NASA
- testing materials for utility, functionality, and pedagogy and certifying scientific and technical accuracy
- protecting intellectual property rights
- enabling programmable interfaces for various directory uses, including inventory resolution and various methods of payment

3. Implementation

3.1. Creation of infrastructure and core functionality

The basic architecture on the user side developed in Phase I (Fig. 1) involves the creation of the core database that contains all descriptions of available resources, the overlying search engine, other software applications and an interface layer. The end user interface layer is initially a generic web interface intended to eventually accommodate more individualized features. Developers use customized interfaces to extract more detailed information from the system as well as generate reports.

Registration of materials occurs as content providers enter relevant information through a form based web interface (not shown). Upon entry, information on resources is cached. Subsequently, cursory initial reviews and more substantial evaluation processes serve as a gate to the registry process. Resources needing revision are identified to content developers for rework, and suitable resources are registered in database in order to insure resource integrity. Resources are reviewed for accuracy, relevance, usability, and pedagogical approach and other criteria. Automatic processes generate appropriate information to export information and translate Space Science information for submission to external systems (Eisenhower, Gateway to Education Materials, NASA EDCATS, NASA CORE, SpaceLink, etc.)

In Phase II, tools for user customized interfaces similar to web interfaces such as "My Netscape" and others will be developed. Also the issues of inventory currency and location are being addressed, and methods derived from e-commerce applications must clearly be considered. Phase III involves implementation of user profiles to provide custom recommendations, email notification, cell phone notices and the like.

4. Metatags

The heart of the Directory is a meta-tag scheme used to thoroughly describe the educational resources. After examination of a number of candidate systems, the Dublin Core Standards were adopted. In particular, the national consortium, Gateway to Education Materials (GEM)[1] has defined particular schema relevant to education materials. These established efforts represent years of considerable investment that could be easily adapted to insure that the NASA Space Science directory contains schema (controlled vocabulary and accepted, standard values for meta-tags) relevant to its discipline including astronomy, space physics and solar system physics nomenclature. By adopting these standards, the directory is as widely distributable as possible. The GEM consortium was formed and produced an infrastructure that has an expansive view of its target audience, beyond what NASA requires. In addition, GEM continues to do active research in the area of meta-tags while keeping in lockstep with Dublin Core developments. In this way the NASA directory is continually kept in line with the state-of-the art in information technology education cataloging.

5. Aknowledgements

This work is supported by a contract, NAS5-26555 to the Association of Universities for Research in Astronomy, Inc. for the operation of the Hubble Space Telescope and the Origins Education Forum. We acknowledge the extensive work accomplished by our colleagues at University of California Berkeley, Goddard Space Flight Center, Jet Propulsion Laboratories, and the Lunar and Planetary Laboratories in the establishment of the directory. Valuable discussions with the GEM Consortium were critical to the directory implementation.

---

[1] http://www.thegateway.org/

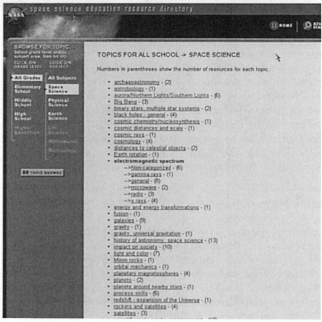

Figure 2. Upper graphic: General User Interface for the Space Science Directory. Lower graphic: User interface demonstrating browsing for specific Space Science resources. See http://teachspacescience.stsci.edu

Hands-On TAROT: Intercontinental Use of the TAROT for Education and Public Outreach

Michel Boër, Carole Thiebaut, Alain Klotz

Centre d'Etude Spatiale des Rayonnements (CESR/CNRS), BP 4346, F 31028 Toulouse Cedex 4 France

G. Buchholtz

Institut National des Sciences de l'Univers, Division Technique (CNRS), Meudon, France

A.-L. Melchior

DEMIRM/CNRS, Observatoire de Paris, France

C. Pennypaker, M. Isaac

University of California at Berkeley, USA

Toshikazu Ebisuzaki

RIKEN, Tokyo, Japan

Abstract. The TAROT telescope has for primary goal the search for the prompt optical counterpart of cosmic Gamma-Ray Bursts. It is a completely autonomous 25cm telescope installed near Nice (France), able to point to any location of the sky within 1–2 seconds. The control, scheduling, and data processing activities are completely automated. In addition to its un-manned modes, we added recently the possibility to control the telescope remotely, as a request of the "Hands-On Universe" (HOU) program of using automatic telescopes for education and public outreach. To this purpose we developed a simple control interface. A webcam was installed to visualize the telescope. Access to the data is possible through a web interface. The images can be processed by the HOU software, a program specially suited for use within the classroom. We used these feature during the open days of the University of California Berkeley and the Astronomy Festival of Fleurance (France). We plan regular use for an astronomy course of the Museum of Tokyo, as well as for French schools. Not only does Hands-On TAROT gives the general public access to professional astronomy, it is also a more general tool to demonstrate the use of a complex automated system, the techniques of data processing and automation. Last but not least, through the use of telescopes located in many countries over the globe, a powerful and genuine cooperation between teachers and children from various countries is promoted, with a clear educational goal.

1. Introduction

There have been several attempts at using astronomical data in the classroom, in general within the framework of physics, mathematics, and/or astronomy courses. Using directly a telescope in the college backyard has many advantages, mainly that children themselves practice astronomy with a telescope. However, several problems may arise:

- Except for the Sun, astronomical observations take place at night, making them somewhat difficult to accommodate on a regular basis, both for pupils and teachers.
- Many schools are located in town, and do not have any dark area where to locate a telescope at night.
- Teachers are typically not experienced astronomers.
- Having a telescope in the school requires some care in handling and maintaining it.
- Not all colleges can afford a telescope with (or even without) a CCD camera.

To that purpose, the Hands-On Universe program (Pennypaker et al. 1998; Boër et al. 2001) has been initiated to use astronomical data within the classroom. Telescope time is exchanged within the HOU network, in order to enable the use various telescopes over the world. Most of them may be remotely controlled, allowing them to be used at night.

2. TAROT, an autonomous observatory

The prime objective of the *Télescope à Action Rapide pour les Objets Transitoires* (TAROT; Boër et al. 1999; Boër et al. 2000; http://tarot.cesr.fr), is the real time observation of cosmic Gamma-Ray Bursts (hereafter GRBs). TAROT is a 25cm telescope, with a full autonomous control system, and able to point to any location over the sky within 1–2 seconds. Figure 1 displays the functional diagram of TAROT. In normal operations, the *MAJORDOME* (Bringer et al. 2000) computes the schedule and sends observation requests to the Telescope Control System, which takes care of the various housekeeping, points the telescope, and activates the CCD Camera. As soon as the data are taken, they are pre-processed, with dark, bias, and flat-field subtraction, cosmic ray removal, astrometric reduction, and a source list is built. The requests for observations are now sent via the web. Should a GRB alert occur (from the HETE-2 satellite), the present observation is interrupted, and the telescope slews immediately to the position of the GRB source.

The interfaces with the users, beside the "alert" connection with the GCN, are as follows:

- The main interface has now been rewritten as a web form. The user is requested to write the coordinate of the source, the number (up to 6), duration, and filter(s) (6 positions) of the frames he/she wants. A unique identifier is attributed to the request as soon as the user validates it. The new request is taken into account by the *MAJORDOME* software at the next start of the scheduling process, at least once a day or whenever

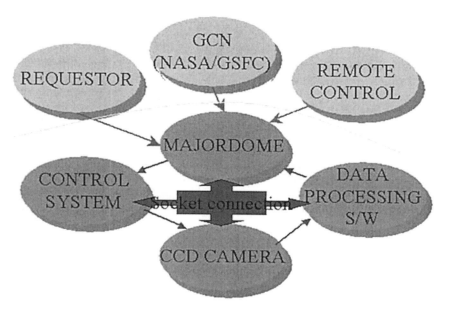

Figure 1.   Functional modules and external interfaces of TAROT.

some event interrupts the operation at night, e.g., rain. Since the form is available through the web, the user can use any computer system.
- As a request of the HOU program, we included a direct remote interface, written in the Java language, again to avoid any preference to a particular operating system. The user can operate directly the telescope, provided the *MAJORDOME* accepts input from this interface. Any kind of image can be acquired from this interface. However, one of the telescope operators has to log in the system to allow operations through this interface.

3.   Discussion

We tested the various TAROT user interfaces at several public demonstrations. They proved to be very reliable. During the day, the presence of a webcam enables the user to see the immediate reaction of an instrument located at several hundreds or thousands kilometers from him. At night, images are available through the web within one or two minutes, on a page which includes the image in JPEG format, and the FITS header. Optionally, the sources from the USNO A2.0 catalog can be superimposed on the image (Thiébaut & Boër 2001), an asteroid chart can be requested, and the DSS can be extracted using a preformatted SKYVIEW query.

Since the prime goal of TAROT is doing science, we still prefer that users from schools either use frames from the scientific program (including frames acquired during the last night), or send requests to the batch interface, reserving the direct remote interface for demonstration purposes during the day or at night. We plan also to enhance this interface par allowing the *MAJORDOME* to schedule in advance the blocks of nights allowed for a use in direct access mode.

We found also that what seems evident to the astronomer has to be explained to general audiences, e.g., phenomena like saturation of frames, angles expressed in hours, and that a telescope located in the northern hemisphere has some difficulties to look at e.g., the Magellanic Clouds (this has also to be explained to several astronomers), or that the accessible sources in the sky vary from winter to summer. To cope with these last points, we plan to have a more interactive and pedagogical interface. In any case, this exercise of porting a system devoted to a somewhat specialized audience to the general public proved to be a very interesting and rewarding adventure for the TAROT team.

Acknowledgments. The TAROT program is funded by the Centre National de la Recherche Scientifique, Institut National des Sciences de l'Univers (CNRS/INSU).

References

Boër, M., et al. 1999, A&AS, 138, 579
Boër, M., et al. 2000, in ASP Conf. Ser., Vol. 216, Astronomical Data Analysis Software and Systems IX, ed. N. Manset, C. Veillet, & D. Crabtree (San Francisco: ASP), 115
Boër, M., et al. 2001, this volume, 103
Bringer, M., et al. 2000, Experimental Astrophysics, submitted
Pennypaker, C., et al. 1998, in Proceedings of the Misato International Symposium on Astronomical Education with the Internet, ed. M. Okyudo, T. Ebisuzaki, & M. Nakayama (Tokyo: Universal Academic Press, Inc), 45
Thiebaut, C. & Boër, M. 2001, this volume, 388

A Japan-U.S. Educational Collaboration: Using the Telescopes in Education (TIE) Program via Intelsat

P. L. Shopbell, E. Hsu

Caltech/JPL

N. Kadowaki

Communications Research Laboratory, Japan

G. Clark

Mt. Wilson Institute

Abstract. In 1993, a proposal at the Japan-U.S. Science, Technology, and Space Applications Program (JUSTSAP) workshop lead to a series of satellite communications experiments and demonstrations, under the title of the Trans-Pacific High Data Rate Satellite Communications Experiments. These experiments were designed to explore and develop satellite communications techniques, standards, and protocols in order to determine how best to incorporate satellite links with fiber optic cables to form high performance global telecommunications networks. This paper describes a remote astronomy experiment executed during the second phase of this program, which established a ground- and space-based network between Japan and several sites in the U.S. This network was used by students, scientists, and educators to provide real-time interaction and collaboration with an automated 14-inch telescope and associated data archive. We find that such networking can provide educators and students with a unique collaborative environment which greatly enhances the educational experience.

1. The Network Layer

The Remote Astronomy experiment made use of a network encompassing geostationary satellites and an extensive ground-based fiber optic network across three countries, connecting seven endpoints. This connectivity was achieved using the NASA Research and Education Network (NREN) in the U.S., the CA∗net3 network in Canada, Intelsat #702 between Canada and Japan, and bits of optical fiber connections for the "last mile" at several endpoints. A range of bandwidths were used, from standard T-1 (1.44 Mb/s) to ATM-based OC-3 (155 Mb/s). The Intelsat bandwidth was kindly donated and, as such, set our overall network performance limit at DS-3 speeds (45 Mb/s). The network infrastructure is illustrated in more detail in Figure 1. (See also Kadowaki et al. 2001.)

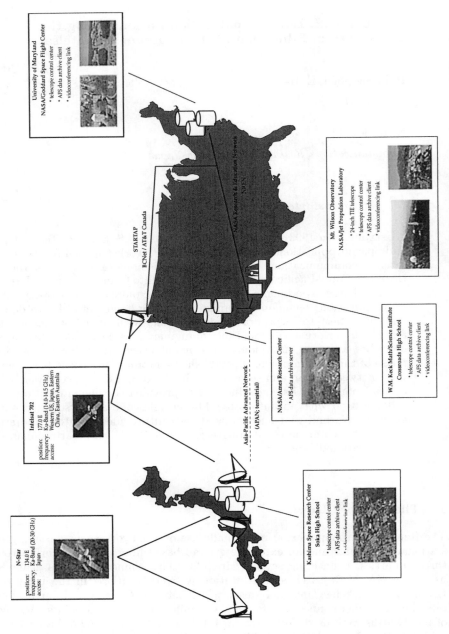

Figure 1. An overview of the U.S.-Japan network used for the Remote Astronomy experiment.

2. The Application Layer

Telescope Control – The experiment employed a fully-automated 14-inch telescope and CCD camera at Mt. Wilson Observatory. This telescope is part of the Telescopes in Education (TIE) project, an award-winning program initiated and maintained by one of us (GC), which allows school children around the world to remotely operate the telescope and obtain their own pictures of astronomical objects. The telescope is controlled via a commercial digital planetarium software product, *TheSky*. This software includes tools for controlling local and remote telescopes and cameras. For the Remote Astronomy experiment, the *TheSky* software package was installed at four endpoints of the network: Kashima Space Research Center (KSRC) in Japan, JPL, Crossroads High School in California, and the University of Maryland (UMD).

Data Archiving – In order for users at each site to obtain the acquired data simultaneously, we used a shared disk protocol known as the Andrew File System (AFS). AFS provides a robust network file service, with server and client software freely available for numerous operating systems. AFS servers were installed on systems running Windows NT at both the Ames Research Center (ARC) and JPL. AFS clients were installed on each of the endpoint observing systems at KSRC, JPL, Crossroads High School, and UMD.

Videoconferencing – Numerous remote observing projects, including one executed by JPL and Caltech in the first phase of the JUSTSAP experiments (Cohen et al. 1998; Shopbell et al. 1998), have pointed to the communication requirements of the partners in shared observing sessions. We therefore installed a standard H.323 videoconferencing package, *CuSeeMe*, on each of the endpoint observing systems. Associated video cameras and microphones were installed where needed. The *CuSeeMe* package includes video and command-line interfaces; it does not include a shared whiteboard or more complex interactive tools.

3. Experiment Operation

The experiment was comprised of three two-hour observing sessions, on June 29, June 30, and July 7, 2000. At the start of each session, participants at all four observing stations contacted each other through a public *CuSeeMe* reflector. The satellite delay affected comprehension of the audio stream to a far greater degree than the video stream; a standard teleconference phone line was therefore used for the audio portion of the final session. The chat mode of *CuSeeMe* was found to be very useful for transmitting detailed information where errors could not be tolerated, such as target names and coordinates. When word was received from Mt. Wilson that the 14-inch telescope was ready for use, the four participant sites connected to the telescope using the *TheSky* software.

As the lead astronomer, one of us (PLS) developed a detailed observing plan in advance, including an overall science goal for the session, choices of target objects, and comparison images from the Internet. The scientific themes of our sessions were "The Types of Galaxies" and "The Lives of the Stars." The former session concentrated on imaging a number of galaxies of differing types, which we then compared to HST images of nearby and very distant galaxies, in an attempt to explore the common classifications of galaxies. The latter session concentrated on imaging a number of gaseous nebulae of different types, representing both the birth and death sites of stars.

Each session began with a brief introduction to the subject, presented by an astronomer. The sessions then became heavily observational in nature, with most discussion centering around the data being obtained by the telescope. Periodically as needed, control of the telescope camera was taken by one of the participant sites, who then moved the telescope to a desired object and obtained an image of it. The acquired image was automatically placed in the shared AFS directory, giving all participants simultaneous access to the data. The process of taking an image was relatively rapid, required approximately 10 minutes per exposure. During that time period, the participants not controlling the telescope would examine the previous image, discuss its relevance to the scientific goals, and monitor the progress of the current exposure. Comparisons were made with published imagery from other telescopes and the Internet. Each session ended with a brief summary of the observations and their relevance to the scientific goals for the session, again presented by the guiding astronomer. A final question-and-answer period provided seeds for discussion and thought after the end of the collaborative session.

4. Conclusions

The widespread adoption of the Internet by research and educational institutions has created vast opportunities for collaboration and interaction on a global scale. Satellite technology can provide a crucial piece of this new model by enabling high-bandwidth networking to virtually any location in the world. As collaborative applications become more complex, and as more simultaneous streams are involved (e.g., video, audio, whiteboard, AFS data, telescope commands), the required network capacity also increases rapidly. The current bottleneck against more widespread use of satellite technology in collaborative research and education is the relative difficulty in building the network. Finally, we note that collaborative software is advancing rapidly; researchers and educators should familiarize themselves with the powerful capabilities available now and in the near future. Such tools provide a means for technical and social interaction at levels previously impossible. Modern collaborative software, together with global satellite networking, can bring together remote resources—both hardware and human—for impressive results in research and education.

References

Cohen, J. G., Shopbell, P. L., & Bergman, L. 1998, Remote Observing with the Keck Telescope Using the ACTS Satellite, final report for NASA grant #BK-509-20-42-00-00

Kadowaki, N., Gilstrap, R., & Foster, M. 2001, Space Comm., in press

Shopbell, P. L., Cohen, J. G., & Bergman, L. 1998, in ASP Conf. Ser., Vol. 145, Astronomical Data Analysis Software and Systems VII, ed. R. Albrecht, R. N. Hook, & H. A. Bushouse (San Francisco: ASP), 348

Using the Net for Education and Outreach in Astronomy

Leopoldo Benacchio

Osservatorio Astronomico di Padova, Vicolo dell'Osservatorio, 5-35122 Padova, Italy

Abstract. This paper consists of two parts. In the first is briefly described the "Catch the Stars in the Net!" education and outreach via the Internet project, giving also an account of some important lessons learned about the way the Network changes the rules of the game. In the second part the special Web module "The Universe at Your fingertips," developed for visually impaired and even completely blind Web users, is described.

1. The Catch the Stars in the Net! Project

Catch the Stars in the Net! is an education and public outreach program carried out by the Padua Astronomical Observatory, with the support of the main Italian telephone and Internet Company: Telecom Italia. The Project is in fact based on the Web and the so-called "new technologies" of communication (international edition[1] or the Italian one.[2]

A new working method, more complex and time consuming than the traditional one, but also more interesting and rich in results was adopted. The project produces the Web site, but every initiative, studied and realised by the astronomers, is first of all proposed to a focus group. Typically, we use schools if educatin is concerned, and a group of "typical" users for public outreach. The initial version is given to these users, from which suggestions, criticism, and suggestions are collected and used for the production of the final release. Only after this procedure, a module or an initiative becomes public in the Web site.

In other words, the Web site only represents the "visible" part of a more complex project of education and popularisation of our Science. With the site, we want to satisfy the interests of the Internet surfer who wants to learn some astronomy concepts, or who is simply curious about these themes. At the present, several levels are satisfied: from Primary School children to Secondary School students, to adults who want to "know more" about the sky.

The Project received awards from the public and the critics (e.g., 1998 New Media Prize; 1999 Kidscreen Digital Kids).

---

[1] http://www.astro2000.net

[2] http://www.lestelle.net

2. The Website

The site is composed of educational or popular "modules," organised in different levels of complexity and depth, in order to meet the specific interests of the user. The site is accessible to everyone, but each module is designed with a particular user in mind (i.e., teachers, students, less than 10 years old children, teens, curious and simply interested). A careful modality of exposition of contents and themes (i.e., education, outreach, history, curiosity/news, and communication) is used in the design of the site.

The interaction and collaboration with the public and, above all, with schools (students and teachers) is very important during all the working phases. Thanks to this collaborative work we are also able to understand future directions and extensions of the Project. Catch the Stars in the Net! takes advantage of the potentialities offered by the so-called new technologies applied to education and popularisation. The topics covered and the methodology of exposition have been carefully studied and realised, according to the possibilities of the Net, with the aim of making the site educationally valid, but at the same time interesting and fascinating for the user. Multimedia, interactivity, and hypertext have in fact given their contribution to the creation of the Web Site. In this way, even if rigorous in content and innovative in the teaching method, the site is very attractive and stimulating for anyone who wants to know astronomy or simply has some curiosity about it.

For what concerns teaching, we have to say a word about the Virtual Planetarium, a module of the Project that is nowadays an example and reference in the teaching of sciences via the Internet in our country. The Virtual Planetarium is in fact used by a lot of schools in their scientific curricula and by individuals. Using this site as a real interactive course of astronomy, students can experience a new way to learn. It is a lively way that surpasses the traditional limitations of texts. This gives the opportunity to learn the arguments following a favourite way and with an adequate rhythm to individual capabilities. You learn by yourself—that is a very good way to learn.

The site isn't static, but follows the continuous growth and evolution of both astronomy and new computer technologies, making necessary frequent updates of the contents. At present, between various initiatives that are accessible to everyone, we have on-line:

a) The Virtual Planetarium is an interactive astronomy course via Internet for 10–18 years old students. It contains 27 educational chapters, in which there are the main themes of astronomy, from the Earth and the theory of gravity to stellar evolution and cosmology. The course contains more then 800 text pages, hundreds of pictures, movies, and animations, and also virtual experiments, all specifically build for the project. The Virtual Planetarium can be used as a self-learning tool and as an aid to teachers who can adapt it to the scholastic curriculum and to the peculiarity of each class of students. The chapters are independent, so that they can be explored one by one and then appropriately inserted in the annual curriculum. A peculiar teaching method is used in the teaching of Science (Karplus Cycle), by cutting a priori misconceptions. Wide space is given to the "teacher page," containing the experiences of teachers that have adopted the Virtual Planetarium in the scholastic curriculum.

b) Voyage into the Cosmos: from Galileo Galilei to Interplanetary Probes is the Web server of the astronomical exposition, which took place in Padua and Rome in 1997, and which had more then 70.000 visitors in six months. The site contains a great exposition of many items of the modern astrophysics and a view of current astronomical instruments. The history of the most important astronomical discoveries, from Galileo to the present, as well as the main missions of cosmic exploration is covered. A particular part is given to the so-called "new astronomy," that is the observations in non-optical spectral bands, the research about neutrinos, gravitational waves, and cosmic rays.

c) Ask the astronomer, where the user can ask an astronomer every kind of question, using a simple form. The answers are collected, argument by argument, in a proper page. The user is tempted in this way both to visit the site often and to collaborate actively in its development.

d) Request a videoconference (initiative for Italy only). This initiative, completely free to the user, offers the possibility to schedule a videoconference about a fixed theme or by "question-answer" according to his requirements.

e) Astronomy for all (initiative for Italy, some parts also in English), initially born as a deepening of Voyage into the Cosmos, consists of two sessions dedicated to different users. The first is Starchild, adapted from a NASA initiative and dedicated to 6–14 years old children. There are few basic concepts (gravitation, appearance and morphology of planets, etc.) with suitable and selected pictures. The second is instead dedicated to people who want to "know more." This site offers a rigorous and deep view about all the themes of modern astronomy. There are a lot of pictures and movies coming from NASA, Hubble Space Telescope Institute, and other famous scientific institutes.

f) The Specola Museum: a tour to the Observatory (also in English). The building in which our Observatory is hosted is an ancient tower raised in 1117. At first it was a castle, and then a fortress for hundreds of years. In 1767 the big tower became an Astronomical Observatory. In this site you can read the history and make a virtual tour of what is still considered one of the finest Observatories of the 18th century.

g) Under the Moon shadow (soon translated into English), completely dedicated to eclipses, and particularly to the one of August 1999, this site was created to satisfy interest about this astronomical event. The contents included 1) information about "when and where" the eclipse will happen, 2) "what happens and what to see" during the phenomenon, and 3) "how to observe and to take pictures" in safety. There are also many pages about the origin of eclipses, as well as interesting historical information, ancient myths and legends about them.

h) Constellations on the Net (soon translated into English), is dedicated to the 88 official constellations and shows their characteristics and other information about them. Maps, visibility time from Italy, position, mythological history, and information are given for every constellation. There are also pictures by which the users can learn to distinguish them on the sky.

i) Serravalle's clock describes the surprising discovery of a tower clock dial of the 14th century, in 1993 during the repair of the "Torre Campanaria" (Main bell tower) of the Municipal Palace in Vittorio Veneto (Treviso-Italy). The goal is to spread this discovery and simultaneously to stimulate the user's interest toward other themes, objects of future developments such as the "time" and the periodic astronomical phenomena used in its measure.

3. The Sky at Your Fingertips!

In the project website, the first scientific outreach Web site completely aimed at visually impaired users of the Net is now available. The only one at this moment in the Net, the site is a classical journey around planets, comets, and galaxies to discover the wonders of the Universe. A main characteristic of this part must be stressed: this site is not "adapted for" or "accessible to" visually impaired or blind people, but it is especially developed for these users. Each page contains an astronomical image, a professional one, converted to tactile form, which may be printed through a Braille printer, and a detailed description of the image itself. The conversion is made by means of a piece of software developed at the Department of Electrical and Computer Engineering at the University of Delaware (see http://www.ece.udel.edu for software and literature). Images can be produced in a tactile format on paper via an embosser, a typical hardware for visually impaired people that allows the user to print in Braille format. While touching the printed image the user can listen to the textual description via a vocal synthesiser, a hardware or software device generally present in the PC's used by blind Internet users.

The site contains an overview of sky objects: the Sun and Solar System planets, stars, and galaxies. The physical phenomena associated with the chosen astronomical image are also included. The functional texts, that are linked to the images, are easy and clear, and they are the description of the tactile images rather than a standard explanation.

The co-operation of a blind person has been essential to the implementation of this site, in order to overcome, rather than technical problems, many specific ones arising from the different way blind people perceive distance, depth, size, and shape of objects.

"The Sky at your Fingertips" is a small contribution towards the philosophy of Web design for all. The site, also available in Italian, can be considered as the first experiment of its kind, and we are going to improve it according to the users' suggestions.

Part 3. Sky Surveys

The Uccle Direct Astronomical Plate Archive Centre UDAPAC—A New International Facility for Inherited Observations

J.-P. De Cuyper[1], E. Elst, H. Hensberge, P. Lampens, T. Pauwels, E. van Dessel

Royal Observatory of Belgium, Ringlaan 3, B-1180 Uccle, Belgium

N. Brosch[1]

Wise Observatory, Tel Aviv 69978, Israel

R. Hudec[1]

Astronomical Institute, Cz-251 65 Ondrejov, Czech Republic

P. Kroll[1]

Sonneberg Observatory, Sternwartestr. 32, D-96515 Sonneberg, Germany

M. Tsvetkov[1]

Sky Archive Data Center, Institute of Astronomy, 72 Tsarigradsko Shosse Blvd., BG-1784 Sofia, Bulgaria

Abstract. An international facility to store, catalogue and digitise photographic plates is under development at the Royal Observatory of Belgium in Uccle-Brussels. The creation of such a facility requires a well-organized effort and substantial resources. However, the cost is not exorbitantly high if one takes advantage of the new generation of commercial, photogrammetric, flatbed scanners, which use the latest CCD technology. These scanners are relatively inexpensive, very fast, and comparable to dedicated scanners like PDS instruments in geometric and radiometric precision.

1. Goals and Objectives

The international workshop "A European Plate Centre" took place in March 2000 at the Royal Observatory of Belgium in Uccle (ROB), one year after the international workshop "Treasure Hunting in Astronomical Plate Archives", held at Sonneberg Observatory, Germany (Kroll, la Dous, & Bräuer 1999; la Dous 1999). At the ROB workshop, it was decided to attempt to create, within 5 years,

---

[1] Member of core team

a scientifically useful digital archive, based on observations selected from half a million photographic plates to be stored at the ROB in Uccle. More information on the project is available at the UDAPAC web site http://udapac.oma.be.

2. Scientific Rationale

Many astronomical objects are exhibit either secular changes (e.g., proper motions, stellar evolution), periodic variability (e.g., binaries), or repeated outbursts at irregular intervals (e.g., novae). Many changes take place on timescales that are too long to be covered at all adequately by modern observing, however sophisticated the equipment. Innumerable questions—concerning, for instance, long-term trends in dynamical and astrophysical evolution, pre-outburst conditions of cataclysmic events, apsidal motion, or the causes of newly-discovered stellar variability of a longer period and larger amplitude than previously suspected—can only be answered by examining archival material that spans many decades. The preservation of past records is therefore vital to this science. User-friendly digital archives are an essential research tool to complement modern observations.

The world's collection of photographic images represents the costly output from more than a century of devotion and skill. The collection is already nominally in the public domain, but as a universal resource it is seriously under-exploited. The main reasons are : (a) lack of information in digital form about the plates, and (b) lack of digital versions of the observations.

The astronomical community itself must be responsible for creating a user-friendly on-line database of calibrated digital images from this resource. The creation of such a database requires a well-organized effort and substantial, but not exorbitantly high, resources.

3. Existing Situation

The locations and contents of most of the relevant photographic archives are already documented in the *Catalogue of Wide-Field Plate Archives* (Tsvetkov 2000)[2]. That facility is an essential first step in locating the plates needed for a given task, such as multi-scale analyses of specific targets.

Although most plate archives are now 'closed', i.e., no-one is actively in charge and there are no regular loan arrangements, all have log-books and card-catalogues which contain the relevant information about the exposures. A few observing logs are already on-line.

Some observatories (particularly ESO and to a lesser extent Kitt Peak) did not routinely request observers to return their plates. Consequently much of the plate collection remains widely dispersed even though it is no longer needed by the original observers.

Commercial photogrammetric scanners are now relatively low-cost, and are capable of precise geometric and radiometric performances that make them com-

---

[2]http://www.skyarchive.org/catalogue.html

petitive with PDS instruments, especially since the archive will include plates of a wide range of size, quality and accuracy.

4. Strategies to Achieve the Goals and Objectives

The main goal is to create a unique resource for scientific research that exploits the long time-base of these archived astronomical observations. The proposed astronomical data centre at the ROB will safeguard a substantial fraction (about 25%) of astronomy's total heritage of direct photographic observations in the world, and will undertake the digitization and processing of a wide selection of particularly time-sensitive observations. The group has identified a number of key project tasks whose completion constitute stepping-stones towards the main goal. These are outlined below.

4.1. Storage Conditions

The appropriate control of humidity and temperature is to be planned and costed with expert advice. New plate envelopes might be needed for some collections.

4.2. Rapid Key Projects

A small sub-set of research projects, that can be executed relatively quickly from new scans of old plates, is needed in order to demonstrate convincingly the soundness of the project and thence to attract funding to establish it firmly. Each scientific project must be able to compete on merit in the post-photographic age, with time variability as one of the fundamental features. Particular emphasis is to be placed on the complementarity of the archival material, and links with modern high-cost projects are to be explored. As far as possible, scientific proposals should link to research programmes at the host observatory (ROB). They must also promote innovative ideas for using an archive as a central research tool. Plate archives with contents suitable for rapid key projects are to be identified as ready for immediate transfer.

4.3. Comprehensiveness of Plate Survey

The WFPA list is to be completed, by determining which plate stores can be sent and which are still in active use, together with a list of declarations of intent.

Observatory directors or custodians of plate archives will be invited to consider donating redundant equipment, and to participate in the transport responsibilities. A worldwide request is to be issued for plates in individual collections to be returned for optimal conservation and/or digitization.

4.4. Scanner Selection

Test plates are to be circulated to commercial scanners to evaluate their adequacy for this project, with parallel tests run on special-purpose scanners (e.g., at Sonneberg, Cambridge, Edinburgh). Commercial photogrammetric scanners prove to be very fast (8–20 min for a $10'' \times 10''$ plate at $15\,\mu$m pixel size) and of enough geometric and radiometric precision to make them cheaper alternatives to the special-purpose machines. A full evaluation of their accuracy, repeatability and stability is taking place.

4.5. Archive Software

Adequate indexing of content, quality and other details is essential for the development of an automatic request scheme for scans of located plates. Existing query software is to be adapted as far as possible. The database must provide the data in a standard format such as FITS, and must employ efficient and suitable means of data compression.

Existing scanner related software from other domains like areal photography and electron microscopy can be used to build a user-friendly, fully documented database.

4.6. Funding

Support is to be sought for specific facets of the project, such as :
1. travel for scientific visits and meetings;
2. outside consultants;
3. hardware: refurbishment, shipping, instruments, office equipment;
4. manual tasks suitable for casual employment;
5. preservation of historic scientific material;
6. a "virtual observatory" and public archive resource for research;
7. a set of scientific projects at the cutting edge of modern science.

Acknowledgments. J.-P. De Cuyper carried out this research in the framework of the project "IUAP P4/05" financed by the Belgian Federal Scientific Services (DWTC/SSTC).

References

la Dous, C. 1999, wgss_newsletter 11, 26,
    http://www.skyarchive.org/wgss_newsletter
Kroll, P., la Dous, C., & Bräuer, H.-J. (eds.) 1999, 'Treasure-Hunting in Astronomical Plate Archives', Proceedings of the international workshop held at Sonneberg Observatory March 4-6, 1999, Historica Astronomiae, 6, (Verlag Harri Deutsch, Thun und Frankfurt am Main)
Tsvetkov, M. 2000, Catalogue of Wide-Field Plate Archives, Sky Archive Data Center, Institute of Astronomy, Sofia, Bulgaria,
    http://www.skyarchive.org

Providing Improved WCS Facilities Through the Starlink AST and NDF Libraries

D. S. Berry

Centre for Astrophysics, University of Central Lancashire, Preston, Lancs, UK, PR1 2HE, Email: DSBerry@uclan.ac.uk

Abstract. The AST library provides a comprehensive range of facilities for creating and using world coordinate systems that describe astronomical data. This paper gives a brief over-view of these facilities and a description of the integration of these facilities with the Starlink "Extensible N-Dimensional Data Format" (NDF), which stores bulk data in the form of N-dimensional arrays of numbers. Most of the major Starlink applications have been updated to take advantage of the facilities provided by this integration of AST and NDF.

1. NDF—The Data Format

NDF is a data format for storing arrays of numbers with associated ancillary information. It is the native data format used by Starlink packages, and is based on the Hierarchical Data System (HDS), which allows data to be stored in the form of arbitrary structures of arrays and scalars, using character, integer, logical or floating point primitives.

The following components are recognized within an NDF structure:

DATA – An N-dimensional array of pixel values representing the spectrum, image, etc. stored in the NDF. This is the only mandatory component; all the others are optional.

TITLE – A character string for general use as a heading for such things as graphical output; e.g., "M51 in good seeing".

LABEL – A character string to be used on the axis of graphs to describe the quantity in the NDF's DATA component; e.g., "Surface brightness".

UNITS – A character string describing the physical units of the quantity in the NDF's DATA component.

QUALITY – An array of the same size and shape as the DATA array holding a set of unsigned bytes. These are used to assign additional "quality" attributes to each pixel (for instance, whether it is saturated, or part of an active area of a detector).

VARIANCE – An array of the same size and shape as the DATA array representing the measurement errors or uncertainties associated with each individual DATA value. If present, these are always interpreted as variance estimates.

AXIS – An array of 1-dimensional look-up tables, one for each pixel axis, used to transform pixel coordinates into an arbitrary user-defined rectangular coordinate system.

HISTORY – A structure used to keep a record of the processing history which the NDF undergoes.

MORE – A structure in which applications may store a collection of arbitrary HDS objects (which may include other NDFs). The interpretation of these objects is left up to the software which creates them.

WCS – A new component which provides facilities for storing an AST *FrameSet*. This holds a description of an arbitrary collection of coordinates frames, together with (potentially non-linear) mappings for converting between each of the frames, and the pixel coordinate frame.

2. NDF—The Data Access Library

Starlink distributes a library of routines which provide access to all the components of an NDF structure. The library is documented fully in *Starlink User Note 33*, available from the Starlink librarian, ussc@star.rl.ac.uk, or from the Starlink WWW site http://www.starlink.rl.ac.uk/. It includes facilities for doing *on-the-fly* data conversion so that a wide range of non-NDF data formats can be read and written transparently by NDF-based software. The library is written mainly in FORTRAN, but additional C and Perl interfaces are available. Since NDF is based on HDS, it transparently converts NDFs written on machines which have a different byte-ordering from that on which the data are read.

3. AST—A Library for Handling World Coordinate Systems in Astronomy

The AST library provides a comprehensive range of facilities for attaching world coordinate systems to astronomical data, for retrieving and interpreting that information and for generating graphical output such as coordinate grids based upon it. Documentation is available in *Starlink User Note* 210 (FORTRAN) and 211 (C), available from the Starlink librarian or from the Starlink WWW site. One of the design goals of the AST library was to make it "stand-alone"—it does not depend on any other Starlink libraries (with the exception of the SLALIB positional astronomy library). Although it is object-based, it has procedural interfaces for both C and FORTRAN.

One of the most important AST objects is the *FrameSet*, which represents an arbitrary collection of coordinate *Frames*, with associated *Mappings* describing how to convert positions from one Frame to another (see Figure 1). An AST Frame represents a coordinate system and can be:

- A simple Cartesian coordinate system (e.g., pixel coordinates).

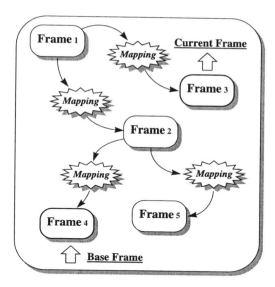

Figure 1. An AST FrameSet is a network of coordinate Frames with interconnecting Mappings. One Frame is nominated as the "Base" Frame (typically the pixel coordinate Frame), and another as the "Current Frame" (an R. A./Dec. Frame for instance).

- A spherical coordinate system (e.g., celestial coordinates). "SkyFrames" have built-in knowledge of how to convert between the main celestial coordinate systems.
- A combination of the above (e.g., an R. A./Dec./Wavelength system).

An AST *Mapping* is a recipe for transforming position between Frames. Sub-classes of Mapping are available which provide basic linear transformations, a wide range of spherical projections, generalised algebraic expressions, and user-defined transformations. Mappings may be combined together to produce compound Mappings (see Figure 2).

Graphical facilities within AST include production of entire annotated coordinate grids from an arbitrary Mapping (any discontinuities and singularities in the Mapping are handled transparently). Any graphics package capable of drawing straight lines and text can be used with AST. A simple interface module needs to be written for each new graphics package. AST comes with a pre-written interface for PGPLOT.

Objects can be saved and retrieved as a set of "*keyword=value*" strings from any external data system capable of storing lines of text, but special support is provided for storing AST objects in the form of FITS header cards. FrameSets can be stored and retrieved using several different popular encodings of WCS information within FITS headers, including the standard proposed by Calabretta & Greisen[1].

---

[1] http://www.aoc.nrao.edu/~egreisen/

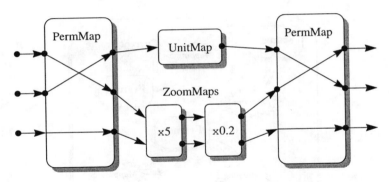

Figure 2. AST Mappings can be combined together to form compound Mappings. These compound Mappings can be simplified to remove unnecessary processing. The above Mapping for instance would be simplified into a 3 axis UnitMap.

4. Putting it all Together: NDF + AST = WCS

The current version of the NDF access library has been modified to include a new standard item, the "WCS" component, which attaches an arbitrary collection of world coordinate systems to the data.

Each NDF now has a "current coordinate Frame" which can be Pixel coordinates, "*Axis*" coordinates (defined by the AXIS component of the NDF), or any other coordinate Frame which has been imported into the WCS component (e.g., R. A./Dec., focal plane position, offset from the field centre, etc.). Data reduction commands report positions in the current Frame of the NDF being processed, and expect the user to supply positions in the same Frame. User commands exist to change the current Frame in any NDF.

Most of the main Starlink packages (GAIA, KAPPA, CCDPACK, POLPACK, ESP, CONVERT, etc.) have been modified to take advantage of the facilities provided by AST and the WCS component. These include:

- Propagation of WCS through the processing chain
- Display of WCS grids
- Automatic data alignment based on WCS
- User-interaction in any preferred coordinate Frame

Through these packages, the Starlink Software Collection now provides a powerful, uniform interface for importing, modifying, exporting and displaying world coordinate information.

Acknowledgments. This project was funded by the Starlink Project,[2,3] Rutherford Appleton Laboratory, Didcot, Oxfordshire, UK, OX11 0QX.

---

[2]http://www.starlink.rl.ac.uk/

[3]ussc@star.rl.ac.uk

Astronomical Data Analysis Software and Systems X
ASP Conference Series, Vol. 238, 2001
F. R. Harnden Jr., F. A. Primini, and H. E. Payne, eds.

A System for Low-Cost Access to Very Large Catalogs

K. Kalpakis[1], M. Riggs[1], M. Pasad[1], V. Puttagunta[1]

Computer Science & Electrical Engineering Department, University of Maryland Baltimore County, Baltimore, MD 21250

J. Behnke

NASA Goddard Space Flight Center, Greenbelt, MD 20771

Abstract. Many new and some old astronomical catalogs contain data for very large numbers of objects. To conduct their studies, researchers must have rapid access to those catalogs. At the same time, the monetary cost of achieving fast access should not come at the expense of resources that would be better used to support the actual scientific studies. We demonstrate how to achieve fast access to data on a low cost desktop for a very large catalog using the Informix Object-Relational database system. We report on experimental results from the development of a solution for efficiently indexing the USNO-A2.0 catalog, which has approximately 500 million objects. The solution offers significant performance improvements over some existing methods. We also describe an extension of Informix that enables users to apply their IDL scripts to data stored in Informix using SQL. This extension brings the powerful data-analysis and visualization capabilities of IDL within Informix.

1. Introduction

Several astronomical catalogs, such as Tycho-2 and GSC-I, contain data for very large numbers of (mostly static) stellar objects. In this study, we focus on the USNO-A2.0 catalog. USNO-A2.0 contains 526,280,881 stars, listing Right Ascension, Declination (J2000, epoch of the mean of the blue and red plate), and blue and red magnitude for each star.

While those data can be made available to public and astronomical communities all over the world via the WWW, there are several problems to be dealt with because of their sheer bulk. To conduct any kind of research on such datasets, researchers must be provided with rapid access mechanism(s). At the same time, costs of achieving fast access should not come at the expense of resources that could be better used to support the actual scientific studies. Low cost solutions that achieve efficient and effective storage management and rapid access to (query processing of) such large catalogs is of vital importance to the research community.

---

[1]Supported in part by NASA under contract number NAS5-32337 and cooperative agreement NCC5-315.

We demonstrate how to achieve fast access to data on a low cost desktop for a very large catalog using the Informix Dynamic Server, an extensible Object-Relational database system (ORDBMS). Further, we illustrate how to enhance the functionality of the system by extending the ORDBMS with a statistical/scientific-computing package, IDL.

2. System Summary Description

In designing our system, we wanted to ensure that the following types of spatial queries could be processed efficiently without a significant hardware cost:

Spatial Window Find all stars within a user-defined bounding rectangle,

Spatial OR-Window Find all stars within at least one of two user-defined bounding rectangles,

Spatial Multi-Join (catalog correlations)

> **Spatial Chain-Join** Find all stars in a "chain-joined" sequence of catalogs (join catalog A with B, and B with C)
>
> **Spatial Star-Join** Find all stars in a "star-joined" sequence of catalogs (join catalog A with B, and A with C)

Spatial Self-Join (catalog mining) Find all stars that are within a specified box centered on each star.

The spatial window and spatial multi-join queries are very critical and users expect them to be processed online (e.g., response times in the order of couple of seconds). Spatial self-joins over large spatial datasets usually require a very large computation time and we do not consider them for online processing at this time.

Hardware and System Software We used is a PC with an AMD Duron 650Mhz processor on a FIC model AZ11 motherboard with an IDE ATA-66 and an Ultra/ATA-100 controller, 256MB of PC100 RAM, and two IBM Ultra/100 7200rpm IDE disks (model DTLA307045). The cost of our hardware platform is well below $2,000. The machine is running the Linux Mandrake version 7.0 Operating system, with the updated Linux Kernel version 2.4.0-test4, and is also running the Informix Dynamic Server 2000 version 9.20 UC-1, configured with our custom datablade (which we call *SimpleShape*, and custom data loading software.

Custom Datablade SimpleShape provides a 16-byte opaque User Defined Type (UDT) storing 4 small floats, full R-tree index support including bottom-up index building, and full B-tree index support based on 2nd order Hilbert curves. It defines two other UDTs with appropriate I/O functions and constructors: the *SimplePoint* to store the coordinates of 2-D points, and the *SimpleBox* to store the lower left and upper right corner points of a rectangle. Each star in the USNO-A2.0 catalog comes packaged as three integers: RA (right ascension), DEC (declination), and MAG (a coding for the field, and the read and blue magnitudes). We create a table with schema (**coordinates SimplePoint, magnitude Integer**) to store the catalog. The data are stored in the database in their original (compressed) format, while user-defined functions convert them into more readily usable values. The RA and DEC fields are stored in a SimplePoint value. To further simplify access to the data, we define additional utility User Defined Functions (UDFs). An example SQL spatial window query is:

```
SELECT Ra(coords), Dec(coords), Red(mag), Blue(mag), Field(mag)
FROM catalog
WHERE Within(coords, monetbox(-3.22, 22.45, 1.23, 22.78) );
```

Custom Loader Due to a bug in the regular table load command for the PC platform, and the lack of an Infomix high performance loader for Linux, and in order to avoid time-consuming string conversions during loading, we developed a custom high-performance loader. Our loader uese the Virtual Table Interface (VTI) of Informix which maps a binary file into a database table. An example of its use is:

```
CREATE TABLE monet(ra INT, dec INT, mag INT)
USING vti_load(file='/tmp/zone0000.cat');
```

IDL Extension for the Informix Server To enable quick prototyping of powerful data-mining applications we also developed an extension to Informix that enables users to apply their IDL scripts to data stored in the database. The IDL extension allows users to tap the data analysis and visualization capabilities of IDL through Informix and SQL. Our implementation is based on UDFs and the RPC mechanism. We define two UDFs, the **idl_exec** that evaluates a user provided IDL script for each qualifying tuple, and the **idl_agg** which evaluates an aggregation defined via three IDL scripts over groups of tuples specified by standard SQL expressions. We also define utility and casting UDFs to access data returned by IDL and to pass data to IDL from the database server. For example, the following SQL computes three cluster centers of an $m \times n$ array of data stored in a database table using the IDL CLUST_WTS function, and then returns them as a virtual table of cluster centers:

```
SELECT toList( idl_agg( ROW(ra,dec),
ROW( ''count=-1'', ''count=count+1 & if count eq 0 then
y=[$ 1, $ 2] else y = [[y]], [$ 1, $ 2]]'',
''w=CLUST_WTS(y, N_CLUSTERS=3); $ RETURN w'' ) ) ) ::
list( ROW(x float, b float) NOT NULL) FROM monet;
```

3. Experiments

We compared the performance of our SimpleShape datablade and customized loader with that for two other datablades for Informix, the Geodetic and the Shapes2 datablades, and also with the performance of a traditional relational database (e.g., no user-defined indexes, no first-class citizen UDTs and UDFs, and no user-defined loaders). We created four database schema, Relational, Geodetic, Shapes2, and SimpleShape, to store USNO-A2.0 data. The experiments for the SimpleShape schema were conducted on the PC platform described earlier, while the others were conducted on a Sun UltraSparc 60 with 512MB of RAM and two 8GB and two 4GB SCSI disks. In all the experiments, we used random query windows of size 0.0666 × 0.0666 decimal degrees. We can see from Table 1 that the SimpleShape datablade and custom loader reduced the loading and indexing times with respect to the other two datablades by more than a factor of five, while reducing the amount of disk space used by more than a factor of three. Further, they were both substantially better than the traditional relational approaches in terms of loading and indexing spatial data. Moreover, as we can see from Table 2, our custom datablade reduced the number of page reads for spatial window and multi-join queries substantially over the traditional relational approach, and by more than a factor of two with respect to the Geodetic and Shapes2 datablades. For Spatial self-joins, the SimpleShape is dramatically better than the Geodetic and Shapes2 datablades, while is competitive with relational approaches (with respect to page reads; the SimpleShape gives substantially better performance on response time and page reads over relational methods as the number of stars increases). For more detailed experimental results and analysis please refer to (Kalpakis et al. 2000) or see http://www.csee.umbc.edu/~kalpakis/monet for updated information.

Table 1. Loading and Indexing All 526M USNO-A2.0 Stars.

Schema	Loading Time[a]	Indexing Time[a]		Table Size	Index Size	
		B-tree	R-tree		B-tree	R-tree
Relational[b]	3 days (46 secs)	6 days		13GB	56GB	
Geodetic[b]	14 days (234 secs)	90 days	25 days (448 secs)	190GB	66GB	104GB
Shapes[b]	10 days (162 secs)	57 days	15 days (256 secs)	140GB	86GB	77GB
SimpleShape	1 day (27 secs)		1 day (8 secs)	15GB	14GB	20GB

[a]The times in parentheses are measurements for 100K stars.

[b]Estimate for 526M stars based on regression from measurements for up to 140K stars.

Table 2. Performance for Various Queries and Schemas for 60K Stars.

Query	Number of Page Reads				Elapsed Time
	Relational	Geodetic	Shapes2	SimpleShape	SimpleShape
Spatial OR-window	22	35	44	14	0.163 secs
Spatial Self-Join	4440	1005023	60443	6775	3072.000 secs
2-Chain Spatial Join	23	49	87	19	2.377 secs
3-Star Spatial Join	24	50	43	12	9.532 secs

References

Kalpakis, K., Behnke, J., Pasad, M., & Riggs, M. 2000, Performance of Spatial Queries in Object-Relational Database Systems, NASA/CESDIS Technical Report TR-00-226.

An Observer's View of the ORAC System at UKIRT

Gillian Wright, Alan Bridger, Alan Pickup, Min Tan, Martin Folger

UK Astronomy Technology Centre, Royal Observatory, Blackford Hill, Edinburgh, EH9 3HJ, UK

Frossie Economou, Andy Adamson, Malcolm Currie, Nick Rees, Maren Purves, Russell Kackley

Joint Astronomy Centre, 660 N. A'Ohoku Place, Hilo, Hawaii 96720, USA

Abstract. The Observatory Reduction and Acquisition Control system (ORAC) was commissioned with its first instrument at the UK Infrared Telescope (UKIRT) in October 1999, and with all of the other UKIRT instrumentation this year. ORAC's advance preparation Observing Tool makes it simpler to prepare and carry out observations. Its Observing Manager gives observers excellent feedback on their observing as it goes along, reducing wasted time. The ORAC pipelined Data Reduction system produces near-publication quality reduced data at the telescope. ORAC is now in use for all observing at UKIRT, including flexibly scheduled nights and service observing. This paper provides an observer's perspective of the system and its performance.

1. Introduction

The UKIRT ORAC project aimed to improve the observing efficiency and publication rate at UKIRT. It did this by making it simpler to prepare observations in advance, by giving observers feedback on their observing as it goes along, and by producing near publication quality reduced data at the telescope. In May–Aug 2000, integration and commissioning with the entire UKIRT instrument suite was completed, following the successful commissioning run with the near-IR camera UFTI in October 1999. The ORAC software has now replaced all of the software at UKIRT that interacts with observers.

ORAC is an integrated suite of software which provides a modern observing interface covering three key areas: advance preparation and specification of observing parameters (ORAC-OT); sophisticated scheduling and sequencing of the instrument and telescope (ORAC-OM); and automatic data reduction producing close to publication-quality images and spectra in real time (ORAC-DR). More detailed descriptions of the ORAC software have been presented in Bridger et al. (2000a, 2000b), and Economou et. al (1999) and references therein. This paper provides an overview of the system from a users point of view.

2. Preparing to Observe

Software which enables the astronomer to specify prior to observing precisely how their observations should be obtained reduces instrument and telescope set-up time during observing. A good remote preparation tool also allows observations to be prepared in the UK, or elsewhere, to a level such that other people know exactly what to do to obtain the data efficiently during "service" and "queue-scheduled" nights. The ORAC Observing Tool (ORAC-OT) was developed from the Gemini Observing Tool (Wampler et. al 1997), and thus inherited much of its functionality. An important feature for the observer is a visual Position Editor which is used to specify guide stars, and check positions and offsets. The Position Editor shows the area around a target which can be reached by the UKIRT guider and areas that are vignetted, as well as the science field of view.

The UKIRT use of the ORAC-OT is heavily based on a number of Library Programs which contain templates for all the common observing techniques at UKIRT, as well as fully defined standard star observations. For observers using the tool without the assistance of staff astronomers, we added intelligence to the OT components that specify instrument configurations, and have also found that a check or validation facility for observations is invaluable. For example, when specifying the configuration of the grating spectrometer CGS4, observers enter the desired grating, central wavelength for the spectrum, and an estimated source brightness, and then may use default buttons to automatically select the recommended order, on-chip exposure time, and other parameters, based on simple table-driven rules. All of these can be easily over-written if the observers choose to use their own settings instead. This level of help has made preparing to observe much easier for most people, but still allows expert users freedom to use non-standard setups. Although preparation now consists of completing a template for each source, errors are still possible. The validation facility both confirms that an observation or science program is executable (e.g., coordinates can be reached by UKIRT) and warns if it may not be scientifically complete (e.g., absence of normal calibrations). The ORAC-OT will be released for remote use at users home institutes in the near future. In the meantime it is used at the JAC to prepare before going to the summit to observe, so that UKIRT staff can gauge user reaction, and fine tune the automation of value specification. Users have found the combination of template libraries and default keys easy and efficient to use.

3. Observing with ORAC

The ORAC-OM is run at the telescope on a dedicated Solaris workstation and enables the observer to select which observation in their program is to be carried out and then to control the progress of that observation's execution by the system. The ORAC-OT is also run at the telescope so that any last minute adjustments to parameters can be made e.g., if the observing plan changes in response to newly acquired data, or the seeing changes sufficiently to require changes to the on-chip exposures. ORAC-DR is started up and it then reduces the data automatically as they appear. When the observer sends an observation

An Observer's View of the ORAC System at UKIRT

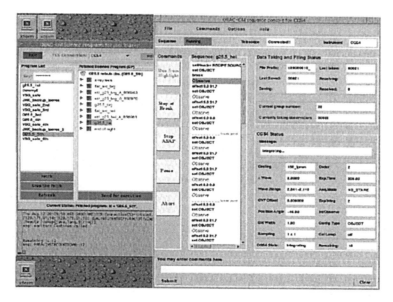

Figure 1. The ORAC-OM in use at UKIRT

for execution its "astronomer-friendly" form becomes a real executable sequence of commands. These include the instrument aperture, used to automatically adjust the telescope pointing for changes in the instrument or instrument read area, as well as the telescope slewing commands. The automated inclusion of the instrument aperture has helped to make changes between instrument configurations, and target acquisition in general, much faster.

An important part of the ORAC-OM is the user interface to the sequencer software which controls and sequences the instrument set-up and telescope positioning, and instructs the instrument when to take data. A real time display of the executable sequence of telescope and instrument actions is provided and observers can start/stop/abort observing at any point. The display also provides feedback as to sensible stopping places which ensure completed reducible data sets - in the example in Figure 1 these are after every set of four offsets, marked by a dotted line. The observer can request a stop at this "break point" rather than keeping track of data frames manually. This OM sequence console is also responsible for showing you the status of the instrument and the data acquisition.

4. On-line Data Reduction

The purpose of the ORAC Data Reduction system is to offer observers near real-time data reduction, and hence to let them easily assess the quality of their data while observing. This results in more efficient use of the telescope (stop when you have sufficient signal to noise) and better quality data (take another standard /longer exposures if need be). Rather than consisting of large packages dedicated to each instrument, ORAC-DR uses Starlink's general-purpose data-reduction software, and drives it via a modular pipeline. The pipeline manager detects when a raw data frame becomes available, and reduces it using the recipe specified in the data header. ORAC-DR is described in more detail in Economou et al. (1999, 2001). Each observation defined with the OT contains a recipe name through which the user selects how the data should be reduced. A set of jittered frames taken with one of the UKIRT cameras could be reduced using a recipe called JITTER_SELF_FLAT, or perhaps JITTER_SELF_FLAT_APHOT if photometry of the point sources is desired. Although the reduction steps are automated, the results, including intermediate steps, are available for interactive analysis and display (slices, spectral extraction, seeing measurements, etc.).

5. Conclusions

The ORAC software suite has achieved its goal of making observing at UKIRT easier, and has been well received by observers. Efficiency gains are of course difficult to quantify, although we are in the process of acquiring relevant statistics, and impact on the publication rate could take a long time to assess. Almost all users have found ORAC to be more efficient and easier, and a variety of aspects of the system have contributed to this. For example users no longer spend time looking for suitable guide stars, switching instruments is much faster and fewer manual actions are needed overall. Initial assessments indicate that on average about a 20% gain in efficiency has been achieved.

Acknowledgments. We would like to thank the UKIRT staff scientists and visiting observers for their help as ORAC was commissioned and brought into full time use. Scientists and software engineers at the UK-ATC and UKIRT made many valuable comments during the development phases. Finally, but by no means least, we thank all the groups who provided software that has been incorporated into our system.

References

Bridger, A. Wright, G. Economou, F. Tan, M. Currie, M. Pickup, A. Adamson, A. Rees, N. Purves, M. & Kackley, R. 2000a, Advanced Telescope and Instrumentation Control Software, ed. Lewis, H., Proc. SPIE 4009, 227

Bridger, A. Wright, G. Tan, M. Pickup, A. Economou, F. Currie, M. Adamson, A. Rees, N. & Purves, M. 2000b, in ASP Conf. Ser., Vol. 216, Astronomical Data Analysis Software and Systems IX, ed. N. Manset, C. Veillet, & D. Crabtree (San Francisco: ASP), 467

Economou, F. Bridger, A. Wright, G. Jenness, T. Currie, M. & Adamson, A. 1999, in ASP Conf. Ser., Vol. 172, Astronomical Data Analysis Software and Systems VIII, ed. David M. Mehringer, Raymond L. Plante, & Douglas A. Roberts (San Francisco: ASP), 11

Economou, F. Jenness, T. Cavanagh, B. Wright, G. Kerr, T. Adamson, A. & Bridger, A. 2001, this volume, 11, 314

Wampler, S. Gillies, K. Puxley, P. & Walker, S. 1997, Telescope Control Systems II, ed. Lewis, H., Proc SPIE 3112, 246

ASTROVIRTEL: Accessing Astronomical Archives as Virtual Telescopes

F. Pierfederici, P. Benvenuti, A. Micol, B. Pirenne, A. Wicenec

ST-ECF, Garching bei Munchen, Germany

Abstract. We present here ASTROVIRTEL[1]: a project supported by the European Commission within the "Enhanced Access to Research Infrastructures" action. ASTROVIRTEL is already being used by European astronomers as a Virtual Telescope, enabling them to access a huge amount of astronomical data with the support of the ASTROVIRTEL personnel. At the same time, operating ASTROVIRTEL—that is, being involved in the definition of the user requirements, and in the implementation of the necessary tools— is an ideal way to get acquainted with the scientific drivers and the technology required to build a Virtual Observatory.

1. Introduction

The ASTROVIRTEL Project is supported for a three year period by the European Commission (EC), within the "Enhanced Access to Research Infrastructures" action of the "Improving Human Potential & the Socio-economic Knowledge Base" section of the EU Fifth Framework Programme. It is managed by the Space Telescope European Coordinating Facility (ST-ECF) on behalf of ESA and ESO and is aimed at improving the scientific return of the ESO/ST-ECF Archive.

European users can exploit ASTROVIRTEL as a Virtual Telescope, retrieving and analyzing large quantities of data with the assistance of the Archive operators and personnel. In addition to serving the scientific community directly, the Project will be used to define the scientific requirements for a more comprehensive and sophisticated multi-wavelength Virtual Observatory.

2. ASTROVIRTEL

Although individual grants cannot be funded under the current scheme ("Enhanced Access to Research Infrastructures"), we believe that the strength of ASTROVIRTEL is to be found in the support, both in terms of manpower and of computing power, that it is able to offer to its users. Once a submitted proposal has been accepted by the scientific review panel and has been judged

---

[1] http://www.stecf.org/astrovirtel/

technically feasible, its investigators are invited to ESO for a period of one to two days.

The aim of this preliminary visit is to understand the requirements of the scientific program. Once that has been achieved, work on the program starts. The role of the ASTROVIRTEL team at this stage is similar to that of a team of assistant astronomers at an observatory on a mountain, i.e., to offer help and consulting in data mining and retrieval, data reduction, and software development. Moreover, investigators can store and process their data using the computing facilities available at ESO, if the tasks are too demanding for the resources of their home institutions.

Another strength of the ASTROVIRTEL project is its ability to help investigators access, retrieve and process data residing in archives and data bases not housed at ESO. These include ISO, CFHT, ING, and MAST.

3. Accepted Proposals

The ASTROVIRTEL Cycle I Call-for-Proposal deadline was June 15, 2000. By that date a number of proposals were submitted, out of which five were accepted by the science review panel and by the technical feasibility team. The accepted proposals, listed below, are publicly available on the web from the ASTROVIRTEL home page:
- D. Burgarella, "A quantitative Approach of Rest-Frame Ultraviolet Morphology",
- D. Egret, "Multi-wavelength Cross-identification of Star Catalogues towards the Magellanic Clouds",
- G. Hahn, "Search for Precovery Images of Near-Earth Asteroids in the ESO Schmidt Plate Archive",
- S. Smartt, "From galaxy formation to supernovae - stellar populations in resolved galaxies",
- C. Zwintz, "Asteroseismology with the HST Fine Guidance Sensors".

The present cycle will be followed by two other calls for proposals; one every year, with deadlines, probably, at the beginning of the summer. It is intended that work on accepted proposals will be concluded in one year's time, in order to avoid work-load pile-ups.

4. Resources Already Available in House

The following resources are already available to ASTROVIRTEL PIs:
- ESO/ST-ECF archive[2]: 1 million observations (HST and ESO data), 8.0 TB of scientific data, growth rate: 5 TB/year, one big robotized jukebox with 1100 DVD slots and 6 DVD drives, two jukeboxes with 670 DVD disks each, on the fly calibration pipelines for HST data, and on the fly calibration of VLT data (soon).

---

[2]http://archive.eso.org

- Catalogues and survey data access: GSC1 and GSC2,[3] Tycho-2,[4] Hipparcos,[5] USNO A2.0,[6] IUE final archive (INES),[7] DSS1 and DSS2 red, blue and IR,[8] the HST Hubble Deep Fields, ESO Imaging Survey (EIS), NTT SUSI Deep Field (NDF) and SOFI Infrared Images of the NTT Deep Field, Science Verification and Commissioning Data from the VLT and WFI, HST NICMOS and STIS parallel observations, ESO Schmidt Plates Collection, and ESO Lauberts and Valentijn Images.
- Computing power: two Beowulf systems (1 master + 9 nodes each, 10 times faster than state of the art quad-processors servers on WFI (wide field imager) reduction pipelines, 8 x 36 GB fiber channel RAID each, Gigabit network switches), state of the art SUN workstations, and Alpha VMS systems.
- Manpower: dedicated personnel to assist ASTROVIRTEL PIs, in-house development of data mining tools, search and retrieval of datasets not available in-house (ISO, CFHT, MAST, etc.), support for reduction and analysis of the data, and ASTROVIRTEL can take advantage of the extra expertise available within ESO.

5. Goals & Future

One important goal of the ASTROVIRTEL project is to enhance the ESO/ST-ECF archives by making tools developed for any of its approved proposals publicly available. This will empower the whole astronomical community with new and more powerful browsing/data mining tools, dedicated reduction/analysis pipelines, and real cross-archive interoperability.

Another important goal of ASTROVIRTEL is to understand the requirements of tomorrow's Virtual Observatories. This implies an understanding of the scientific drivers of a Virtual Observatory as well as of the technology it needs. Handling in a technically efficient and scientifically meaningful way several interconnected multi-wavelength (and multi-instrument) archives is an ambitious task that constitutes the core of the VOs. ASTROVIRTEL aims at tackling this very problem by making use of the experience gained during its years of operation.

Acknowledgments. We are grateful to Martin Kornmesser for the nice pictures he contributed to our poster.

---

[3] ©Space Telescope Science Institute

[4] ©Copenhagen University Observatory and ESA, 2000

[5] ©ESA, 1997

[6] ©US Naval Observatory, 1998

[7] ©ESA, 1999

[8] ©1993, 1994, AURA, Inc. all rights reserved

Astronomical Data Analysis Software and Systems X
ASP Conference Series, Vol. 238, 2001
F. R. Harnden Jr., F. A. Primini, and H. E. Payne, eds.

PICsIM – the INTEGRAL/IBIS/PICsIT Observation Simulation Tool for Prototype Software Evaluation

J. B. Stephen and L. Foschini

Istituto TeSRE/CNR, Via P. Gobetti 101, 40129 Bologna, Italy

Abstract. The INTEGRAL satellite is an observatory-class gamma-ray telescope due for launch in early 2002. It comprises two main instruments, one optimised for imaging (IBIS) and the other for spectroscopy (SPI). The PICsIT telescope is the high energy (150 keV - 10 MeV) plane of the IBIS imager and consists of 8 individual modules of 512 detection elements. The modules are arranged in a 4 x 2 pattern, while the pixels are in a 16 x 32 array. This layout, which includes a dead area equivalent to one pixel width between each module, together with the event selection procedure, which (in standard mode) does not allow the identification of coincidences between separate modules, leads to a non-uniformity of the background which is significantly different for single-site events and for multiple energy deposits. Other sources of background variations range from the separate low energy detector, situated immediately above the PICsIT plane, to the large mass of the SPI telescope at a short distance to one side. The algorithms for performing all the imaging and spectral deconvolution for PICsIT are currently being produced for delivery to the INTEGRAL Science Data Centre. In order to maximise the information which may be extracted from the PICsIT data, we have designed a prototyping environment which consists of a GUI to a highly structured and modular set of procedures which allows the easy simulation of observations from the data collection phase through to the final image production and analysis.

1. Introduction

The INTEGRAL high energy gamma-ray satellite consists of two primary instruments – a high resolution spectrometer (SPI) and a wide spectral and angular coverage imager (IBIS). IBIS achieves the goal of being able to create images of the gamma-ray sky over the energy range of $\approx 15\,\mathrm{keV}$ – $10\,\mathrm{MeV}$ with good angular resolution and large field of view by utilising two position sensitive detection planes, one solid state (ISGRI) optimised at lower energies and the other using scintillator technology (PICsIT) for high energy detection, in conjunction with a coded aperture mask. There are many possible sources of disturbance in the imaging process for the PICsIT detector, ranging from the presence of ISGRI between PICsIT and the coded aperture, through the modular form of the detection plane construction to the masses of the other instrumentation nearby.

2. The PICsIM Environment

The PICsIM simulation tool is written entirely using the IDL[1] language and runs under both Windows NT and Linux operating systems. The tool itself provides only minimum functionality, mainly in the displaying of the images produced, however it is the interface which allows the developer straightforward and consistent control of the algorithms he is creating. This is performed by using a rigid system of data and algorithm storage which allows the operator to use the tool to identify the set of procedures available for use at any one time (e.g., for deconvolution of the image or definition of the accumulation parameters), and to select the required action. In this way it is also possible to bypass the data simulation step (which is itself a limited Monte-Carlo process which does not take into account all physical processes, and to use the results of a detailed Monte-Carlo simulation which is also being developed at the TeSRE institute (Malaguti et al. 2000) merely by placing the event files so produced into the event file directory and ensuring that a run-ID and associated statistics file is associated with them. Furthermore, although to date the various procedures used to define the imaging process are actually IDL source code, in principle they could also be pre-compiled C or FORTRAN programs called by the main IDL procedure.

The entire PICsIM simulation, display and evaluation package is controlled through use of IDL widgets. The top level is the main control GUI which then activates one of the other second-level GUIs on user command. The various commands currently available are detailed below:

- SIMULATE This command activates the GUI which allows the simulation of an observation. The simulation widget interface is shown in Figure 1. The source definition procedures (user-defined) are located in the relevant directory and displayed, as are the instrument characteristic procedures, which describe how the hardware is functioning, and the observation strategy (e.g., hexagonal pointing, staring etc.) Once a valid combination is selected a script file is created and the simulation automatically launched, creating a series of event files, one for each pointing, with a run-id assigned which is unique for every observation.
- ACCUMULATE The event files form the input for this GUI which is shown in Figure 2. The user can select the files which he wants to accumulate by run-id (i.e., observation), and further detail may be imposed by selecting also by sub-pointing. The effective time may also be changed and the energy channels of the accumulation may be defined by means of selection of a user written procedure. Details of a particular observation may be obtained (these are held in the associated statistics file)
- DECONVOLVE The deconvolution GUI allows the user to select between various methods of deconvolution (provided as IDL procedures in the appropriate directory) for the accumulated data files.
- CLEAN This GUI screen shows all the options open (again in terms of IDL procedures) to the user for cleaning the resulting image, or filtering the accumulated data files before deconvolution.

---

[1]IDL 5.3, Research Systems Inc., Boulder, Colorado USA

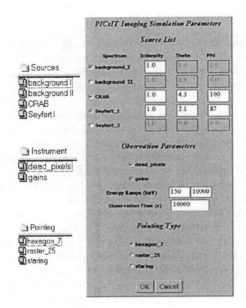

Figure 1. The GUI to the 'SIMULATE' sub-menu

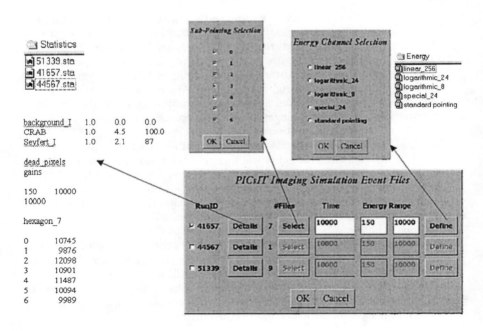

Figure 2. The GUI to the 'ACCUMULATE' sub-menu

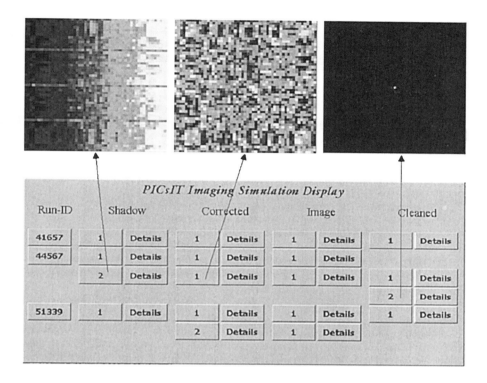

Figure 3. The GUI to the 'DISPLAY' sub-menu

- DISPLAY This GUI provides a workspace within which the operator can display and perform limited analysis on any accumulated shadowgram, deconvolved and/or cleaned image or sum of images. Figure 3 shows how the operator can display the original detector image, the corrected shadowgram and the cleaned sky image of a point source.

3. Conclusions

The PICsIM simulation environment is a useful tool within which the algorithms are being developed for use in the final data analysis data package at the ISDC. Its highly modular form and easy widget-driven interface allows a rapid evaluation and comparison of new procedures to be made.

References

Malaguti, G., Ciocca, C. and Di Cocco, G. To be published in Proceedings of the 4th INTEGRAL workshop, Alicante, Spain, September 2000

Astronomical Data Analysis Software and Systems X
ASP Conference Series, Vol. 238, 2001
F. R. Harnden Jr., F. A. Primini, and H. E. Payne, eds.

An Enhanced Data Flow Scheme to Boost Observatory Mine-ability and Archive Interoperability

Alberto Micol

Space Telescope European Coordinating Facility, European Space Agency

Paola Amico

European Southern Observatory

Abstract. The major astronomical observatories in the world, Gemini, HST, VLT, NGST, etc, invested or plan to invest large amount of human resources and money to build archive facilities that support their data flow. The classical Data Flow Scheme allows its users (calibration/quality control scientists, principal investigators, archive scientists) simple files retrieval and access to basic ambient and calibration data, leaving other valuable information totally unexplored.

Is there more to exploit in a large data archive/data flow? Is it possible to improve the Data Flow Scheme in order to foster the mine-ability of an archive, making, at the same time, the every day life of the quality control scientist easier? What are the common missing steps for an archive/observatory to be miner-ready? We will answer all these questions and suggest a newer approach for a data flow scheme, where the Data Quality and the Archive can be seen as two different clients of the same sub-system: the Observatory Data Warehouse.

1. Introduction

HST and VLT insiders have great familiarity with the concept of Data Flow System (DFS): it is a closed-loop software system, which incorporates various subsystems that track the flow of data all the way from the submission of proposals to storage of the acquired data in the Science Archive Facilities. Typical DFS components are: Program Handling, Observation Handling, Telescope Control System, Science Archive, Pipeline and Quality Control. All these components produce various sorts of data with different formats and "handling" rules. Therefore, the information flow among the subsystems suffers from "hiccups". Ultimately, some data may be lost in the process and the referential integrity within the DFS may be compromised.

ST-ECF and CADC have always been busy in trying to patch the current HST Data Flow but unfortunately, as we will see later, only a posteriori. The nitty-gritty details of on-the-fly calibration of science HST data, the jitter extraction pipeline, the WFPC2 associations, and the FOS associations are examples of patching the engineer-oriented HST Data Flow. Those systems have introduced a previously missing, basic, scientifically-oriented description of the HST datasets.

It is thanks to this a posteriori effort of reconstructing what actually happened during the observations that higher-level, ready-for-science data products are now immediately available to scientists. Indeed, cosmic ray free co-added images, mosaics of dithered WFPC2 observations, and combinations (at least for a first visual inspection) of FOS spectra are generated on-the-fly upon demand.

In the case of the VLT, soon after the completion of the observing phase of a service mode programme, the PI receives a complete data package, which includes, among other, zeropoints and science frames free of instrument signatures. The archive, and therefore the future user of the science data, does not receive the same information and therefore the work of the quality control scientist, at the back end of the data flow, is partially lost.

2. The Classical Data Flow

After Phase 1 and 2 of the proposal preparation, the observations are scheduled and then executed. Both telemetry and science data are acquired and stored, while some reduction pipeline produces quick look products and a Quality Control team inspects the data, but usually only a few measurements are taken and passed to the PIs. For example, the PIs of VLT programmes receive a Quality Control report, which compares the user requirements in terms of seeing, fraction of lunar illumination, moon distance and airmass with the true values measured on site and stored in the ambient database.

These are certainly necessary steps. But are they sufficient? What is described by an investigator in his/her proposal (pointing information, S/N ratios, image quality etc) doesn't necessarily agree with what the observatory has been able to achieve at run time[1][2]. Discovering what actually happened during the observations is usually left to the PIs. Furthermore, such effort is then lost since no feedback is given to the Observatory. Moreover, an archive scientist will later have to go through the same reduction for his/her own study. Again no feedback will be provided to the archive facility.

Any data-miner will have to go through those steps again and again. This observatory does not qualify as miner-ready.

---

[1] An HST example: the observed dither pattern can differ from what was actually requested, due to errors (usually at the sub-pixel level) in the positioning of some optical element (e.g., the lack of repeatability of the STIS Mode Selection Mechanism), or because of jitter, or some other effects (aberrations, deformations). Hence the offsets as from the proposal DB (and/or the WCS in the science data header) are not necessarily the correct ones to be used to combine the images. Some other more reliable source of pointing information (e.g., Observatory Monitoring System jitter files in the case of HST) must be used, or otherwise, direct measurements on the images (via cross-correlation techniques) are required.

[2] A VLT example: users do not receive any information on the wind speed and direction with respect to where the telescope points, an important piece of information when excellent image quality is required. Another example comes from the requirement, typical of infrared observations, to monitor the level and the variations of the sky background in bands over 2 microns and, eventually, to measure if it correlates with the mirror temperature and/or the external temperature.

3. A Better Data Flow Scheme

As highlighted in the previous paragraph, there are two main problems in todays' DFS implementations: (1) lack of interoperability within the various DF subsystems (2) insufficiently detailed description of the observations.

Two steps are necessary to overcome these limitations:

(a) The adoption of a Data Warehouse[3] to control the various DFS activities. The information flows among DF subsystems via the data warehouse. The advantages of having it at the centre of the DFS are multiple. Among them, it guarantees a homogeneous access to the information created by any DF subsystem; it may also be the place used to develop and integrate tools to check for referential integrity. (b) The introduction of a new DFS component, let's call it "Characterisation" step, responsible for any data manipulation/reduction to extract all the parameters, which are useful for a thorough description of what actually happened during the observations. It should consist of a set of reduction pipelines to measure (and compare) those parameters requested in phase 2 (e.g., offsets of dithered frames, S/N of spectra, image quality, etc).

These Characterisation tasks should be executed at a later time than the Data Quality ones since they require a better calibration, using improved reference files and software typically unavailable at the time of the observations. This special activity should be carried out some time (1 year ?) later[4].

In the end the data warehouse should not only contain Phase 2 and scheduling information, but also:

1. Parameters to be used for instrument trend analysis. (e.g., PSF, noise properties, bias level trends, etc.)
2. Parameters to be used to calibrate the data (e.g., zeropoints)
3. Parameters requested by the PI during phase 2 (e.g seeing, fraction of lunar illumination, etc). Comparison of requested and obtained values will be used for Data Quality assessment.
4. Parameters used to scientifically characterise the observations (Limiting magnitudes, background properties, object detections along with rough preliminary object properties measurements, density of point/extended sources for images, etc.)

---

[3]The term *data warehousing* generally refers to combine many different databases across an entire enterprise and its application is rather general and not at all confined to scientific databases. Data warehousing emphasizes the capture of data from diverse sources - the DF subsystems - for useful analysis and access, but does not generally start from the point-of-view of the end user or knowledge worker who may need access to specialized, sometimes local databases. The latter idea is known as the data mart. These data-marts (files, spread-sheet, DBMS, etc) are used in the day to day operations and allow insertions, updates, deletions or, in other words, all those operations that are not possible in the data warehouse itself. Once consolidated, the information is extracted and translated from the data mart and sent to the data warehouse for permanent storage.

[4]It could be combined with the production of Preview data (1992), which pipeline is run just after the data exits the proprietary period, to ensure the best quality (better calibration s/w, better calibration files) of the products, and to respect the data reservedness. The fact that the same data is processed twice -the first time to assess the data quality the second one to extract parameters useful for a scientific charactersation of the archive contents - helps in understanding better the data, their problems, in discovering and resolving possible inconsistencies.

5. Some preview products (imagettes, binned spectra, histograms, etc) could be generated.
6. A layer with pre-compiled and up-to-date statistics of some of the parameters listed above.

While certain parameters are already measured (mainly 1 and 2 above), others (some of 3, and mainly 4, 5 and 6) are not part of the current Data Flow Systems. Though 1 and 2 above are stored in the so-called calibration database, and though 3, 4, 5 and 6 above could end up into a "characterisation database", more is to be gained by integrating those two aspects within the observatory data warehouse. Having all this information on-line will greatly improve the way an instrument scientist or an archive scientist works.

The mine-ability of the system is greatly enhanced since engineers and scientists, both inside and outside the observatory, will have homogeneous access to information like: **(a)** ready-to-use measurements, **(b)** ready-to-view preview products, **(c)** a scientific view on the archive as opposed to the standard, sterile catalogue (observation log), **(d)** a quality control view on the archive, (trend analysis techniques and instruments/telescope health checks could benefit from monitoring parameters such as the noise levels of detectors, the measured resolution versus time and slit width, the image quality, etc.), **(e)** a superior level of abstraction, since at this level the underlying complexity of the various subsystems that collected the necessary information must have been removed.

Without this level of abstraction, it will be difficult to achieve effective interoperability among archives.

4. Conclusions

We highlighted the typical problems which HST and VLT Data Flow Systems are facing today. Dispersing the information into several subsystems that are not interoperating is the immediate cause of glitches and inconsistencies, which, for the intrinsic heterogeneous nature of the DFS, are then difficult to identify and repair. We claim that a central repository of the information produced by all the various DF subsystems will greatly help to reach smoother operations.

Industry is facing the same kind of problems; indeed data warehousing is one of the hottest industry trends. The astronomical community should try to benefit from that effort.

Up to now an archive user, being an external user or an instrument scientist, has been able to browse through an observation log representing basically Phase2 information. The aim of introducing a characterisation step is to provide not only better information on what actually happened during the observation, but also to provide a higher level interface to the archive: a miner-ready interface which doesn't need nor want to know the details of the particular DFS, but which can help the scientist in identifying the data s/he needs.

A good DFS must be able to remove its own signature. A good Observatory (not only the archive) must be miner-ready.

INES Version 3.0: Functionalities and Contents

E. Solano, R. González-Riestra, A. Talavera[1], F. Rodríguez

Laboratorio de Astrofísica Espacial y Física Fundamental (LAEFF), P.O. Box 50727, 28080 Madrid, Spain

A. de la Fuente, I. Skillen[2], J. D. Ponz, W. Wamsteker

Villafranca Satellite Tracking Station (VILSPA), P.O. Box 50727, 28080 Madrid, Spain

Abstract. We describe the functionalities and contents of Version 3.0 of the IUE Newly Extracted Spectra (INES) System, which has been developed jointly by ESA and LAEFF, and which has been operational at the INES Principal Centre (LAEFF) since August 2000. At the time of writing, it is being distributed to the National Hosts.

1. Introduction

The IUE Newly Extracted Spectra (INES) System was developed by the ESA-IUE observatory at VILSPA in order to make IUE data available in a simple and efficient way. The system design was driven by the concept of delivering fully calibrated data, ready for analysis, with minimum development and maintenance costs (González-Riestra et al. 2001). Special attention was given to making the graphical user interface, tuned for the occasional users of the archive, display useful scientific information in a simple form.

INES has been operational since November 1997. In this paper we will summarize the main functionalities of the INES Version 3.0. This release includes new features such as a built-in name resolver, homogenization of object names and coordinates, query by list of objects, and an improved data viewer facility, including errors, quality flags and updated bibliographic references. An overall description of the system capabilities is given in the next section. A detailed user guide to the system can be found in the INES Newsletter (2000, http://iuearc.vilspa.esa.es/Ines_PC/Newsletter.pdf).

2. Functionalities

The functionalities of Version 3.0 of the INES data distribution system are outlined below.

---

[1] Presently at the XMM Science Operations Centre, VILSPA

[2] Presently at the Isaac Newton Group, La Palma

Archive Search: The INES archive contains 110033 entries. The query to the access catalogue is made by means of an HTML fill-in form with permits the Archive to be queried by object name, coordinates, object type, observing date, instrumental parameters and object or image list (see Figure 1). Four predefined output fields are available, emphasising General, Observation, Variability and Pointing information, and each may be output in HTML, ASCII or as tab- or comma-separated values. The output fields may be ordered by date and time of the observation, coordinates, camera and image number, object type and object name. The system allows one to select either the Principal Centre at LAEFF, Madrid, Spain (http://ines.vilspa.esa.es) or the Mirror Site at CADC, Victoria, Canada (http://ines.hia.nrc.ca).

INES version 3.0 has a built-in name resolver utility which permits one to query the Archive using any of the object names provided by SIMBAD. The Name Resolver gives 168571 identifications for the 9494 astronomical objects contained in the INES access catalogue.

Results from Search: The following utilities are provided in HTML output format (Figure 2):

Links to Publications: Associated with each spectrum is the number of publications which have made use of it. By clicking the link, the reference of the publication and links to the abstract and/or the full paper are obtained through the ADS facility. In addition to this, ADS also includes direct links to the INES archive. The IUE publication catalogue includes 38812 images referenced in 2103 scientific articles published before January, 2000.

Data Previews: A browse plot of a spectrum including bad pixels and flux errors can be generated by clicking on the corresponding link. A panel summarising the observation is displayed next to the plot, and the full FITS header can be listed from there. For a high resolution spectrum, zoom plots of 30 Å of selected regions may be generated transparently on the Principal Centre/Mirror server by entering the desired central wavelength. A copy of a browse or zoom plot can be saved as a GIF file. (Figure 3).

FITS Header Display: Links are provided to display the FITS primary and binary table headers of each requested low resolution or re-binned spectrum.

Data Retrieval: Spectra may be retrieved individually or in groups. Multiple retrieval of concatenated, high resolution spectra can be restricted to a specific wavelength interval. Spectra are delivered as FITS files. Single spectra are retrieved uncompressed from the appropriate repository: Principal Centre/Mirror Site or National Hosts.

For multiple retrieval, it is possible to include/exclude individual spectra. Inclusion/exclusion of files by type is also possible. Multiple spectrum retrieval generates a packed file in either tar or ZIP format. Compression of packed files is also possible and recommended for network efficiency, in particular when downloading large data sets.

On-line Help: Help on a specific keyword can be obtained by simply clicking on it.

On-line Access to Project Documentation: A detailed description of the spacecraft, the IUE Final Archive and the INES System is given in the on-line project documentation. The information is stored in PDF format files to be easily browsed.

Access Statistics: The distribution package also includes some tools to help in the administration of the National Hosts. These tools, implemented in a set of Perl scripts, allow one to generate access statistics for the National Host with different views and selection of the time period, and to monitor the status of the network by using "ping" and "traceroute" in an interactive form.

HelpDesk: The Principal Centre includes a Help Desk facility, based on "JitterBug", to channel questions and to provide continuous support to users of the Archive.

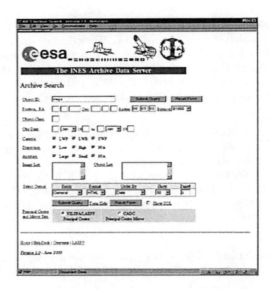

Figure 1. Search capabilities of the INES System (see Section 2 for details).

References

González-Riestra, R., Cassatella, A., Solano, E., Altamore, A., & Wamsteker, W. 2000, A&AS, 141, 343

González-Riestra, R., et al. 2001, this volume, 156

INES Version 3.0: Functionalities and Contents 155

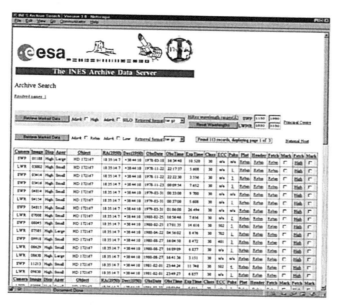

Figure 2. Result of the search displayed in Figure 1. The different sources of data are indicated by the grey areas (Principal Centre/Mirror Site) and the white fields (National Host). Data are retrieved in a transparent way for the end user regardless of the repository.

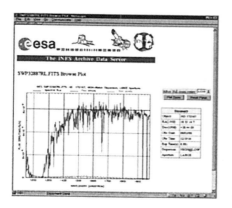 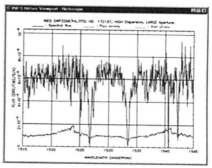

Figure 3. Data previewing capabilities supplied within the INES system. **Left panel:** high resolution concatenated spectrum re-binned to the low dispersion wavelength scale (González-Riestra et al. 2000). **Right panel:** The active box (HiRes 30 Å zoom centre) allows the choice of part of the high resolution data.

Astronomical Data Analysis Software and Systems X
ASP Conference Series, Vol. 238, 2001
F. R. Harnden Jr., F. A. Primini, and H. E. Payne, eds.

INES: The Next Generation Astronomical Data Distribution System

R. González-Riestra, E. Solano, A. Talavera[1], F. Rodríguez, J. García, J. Martínez, B. Montesinos, L. Sanz

Laboratorio de Astrofísica Espacial y Física Fundamental (LAEFF), P.O. Box 50727, 28080 Madrid, Spain

A. de la Fuente, I. Skillen[2], J. D. Ponz, W. Wamsteker

Villafranca Satellite Tracking Station (VILSPA), P.O. Box 50727, 28080 Madrid, Spain

Abstract. The IUE Archive was the first astronomical archive to be made accessible on-line, back in 1985, when the World Wide Web didn't even exist. The archive stores more than 110000 spectra which span nearly two decades of Ultraviolet Astronomy. The IUE Newly Extracted Spectra System (INES), a complete astronomical archive and its associated data distribution system, was developed with the goal of delivering IUE data to the scientific community in a simple and efficient form. Data distribution is structured into three levels: a Principal Centre at LAEFF (Laboratory for Space Astronomy and Theoretical Physics, owned by the Spanish National Institute for Aerospace Technology) and its Mirror at CADC, a number of National Hosts (currently 22), and an unlimited number of end users. The INES Principal Centre can be reached at http://ines.vilspa.esa.es.

1. Introduction

Archives are essential components of all astronomical space observatories. Their holdings are in most cases unique, and they allow the re-use of the data for many different scientific purposes. Astronomical space archives are excellent tools for the study of variable phenomena, and provide reference information for the planning and calibration of new missions. Moreover, archives are a precious source of data for many didactic purposes.

From its beginning, the ESA-IUE project at VILSPA has made special efforts to define the mission archive and to distribute data to the scientific community worldwide. In 1985 the IUE ULDA (Uniform Low Dispersion Archive) became the first on-line astronomical archive.

---

[1]Presently at the XMM Science Operations Centre, VILSPA

[2]Presently at the Isaac Newton Group, La Palma

INES: Next Generation Astronomical Data Distribution

Taking advantage of the expertise gained with ULDA and the distribution system for IUE Final Archive data, the ESA-IUE Observatory began to develop INES in early 1997. INES is a complete astronomical archive and the associated mechanism for data distribution. Its goal is to make accessible the complete dataset obtained by the International Ultraviolet Explorer during nearly twenty years of operations to the scientific community in a simple, user-friendly and efficient form (Wamsteker et al. 2000).

In early 1998, ESA decided to put an end to its involvement in the IUE project, and to deliver the IUE archive to the scientific community. In May 1998 the Science Programme Committee of ESA selected LAEFF (Laboratory for Space Astrophysics and Theoretical Physics, owned by the Spanish Institute for Aerospace Technology INTA) to become the INES Principal Centre. During the period January 1999–June 2000, ESA and LAEFF developed jointly the "INES Transfer Programme", after whose completion LAEFF assumed full responsibility for the maintenance and development of the INES system. Thus, in the coming years, LAEFF will be the repository of the largest and most complete set of data in the ultraviolet domain.

2. The INES System

The INES system has the following design goals:
- to deliver **fully calibrated data** in a form that does not require a detailed knowledge of the instrumental characteristics,
- to decrease the volume of information in the archive, excluding intermediate products and technical information without direct relevance to the user,
- to apply state-of-the-art technology in terms of data distribution techniques, but minimizing the development effort by using tools already available,
- to reduce the costs associated with the archive support.

2.1. The Distribution System

The distribution system has a three-level structure to avoid single points of failure in the system and to reduce the chances of "network bottlenecks" impacting performance:
- A **Principal Centre** at LAEFF, with a **Mirror Site** at the Canadian Astronomical Data Centre (CADC). Both sites contain the master archive and provide the data not available at the National Hosts (high resolution and spatially resolved low resolution data, see below). The Principal Centre is responsible for new software releases and the update of the catalogues.
- A number of **National Hosts**, containing the access and publications catalogues and the full set of low resolution and re-binned spectra. At the time of writing, INES is installed at the following National Hosts: Argentina, Austria, Belgium, Brazil, Canada, China (P.R.)[3], ESO, France,

---
[3] Local access only

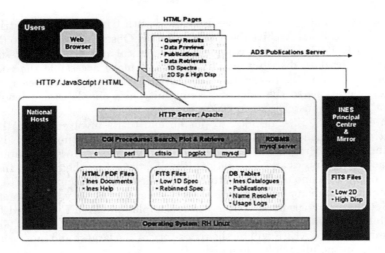

Figure 1. Implementation Scheme of the INES System.

Israel, Italy, Japan, South Korea, Mexico, Morocco, The Netherlands, Poland, Russia, Spain, Sweden, Taiwan, the United Kingdom and the United States. [4]
- an unlimited number of **end users**, who access the archive via the WWW using any standard browser.

2.2. The Data

The INES archive contains the following set of IUE data:
- Low dispersion spectra, re-extracted from the IUE Final Archive files with improved algorithms (Rodríguez-Pascual et al. 1999) (1.2 GB),
- High dispersion spectra, re-binned to the low dispersion wavelength scale (González-Riestra et al. 2000) (0.6 GB),
- High dispersion spectra, with orders concatenated and correct wavelength scale (Cassatella et al. 2000, González-Riestra et al. 2000) (12.5 GB).
- Low dispersion spatially-resolved spectra, as available in the IUE Final Archive (16.8 GB).

High dispersion and bi-dimensional spectra reside both at the Principal Centre and its Mirror. Low dispersion and re-binned data are distributed to all National Hosts. Requests for retrieval of the latter types of data are resolved locally at the National Host, and requests for high dispersion or bi-dimensional spectra are forwarded to the Principal Centre or its Mirror, from where the data are delivered to the user in a transparent way. All the data are delivered in FITS

---

[4]The minimum hardware required for a National Host is: a server with an Intel Pentium or similar CPU, 32 MB RAM, 5–6 GB hard disk, a permanent Internet connection, and a CD-ROM reader. The major software components used in the National Host installation are (see Figure 1): the Linux Operating System (the current installation has been tested in RedHat Linux 6.2. and 7.0), PERL (version 5.005-03) together with the Data Base Interface and Data Base Driver modules, MySQL (version 3.22-32-1), CFITSIO, and PGPLOT (version 5.2.0).

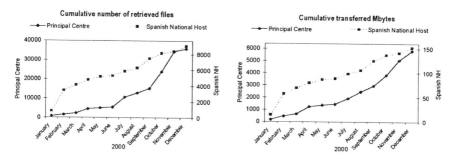

Figure 2. Usage statistics of the INES Principal Centre and the Spanish National Host in 2000.

format and can be read into standard image processing packages (e.g., MIDAS, IDL, IRAF) without special-purpose software.

INES has been operational for three years now. The first release of the system (INES 1.0) was installed in VILSPA in November 1997. INES 2.0, which represented a major upgrade of the system, was installed in August 1999, and distributed to the National Hosts in November of that year. INES 3.0 was installed for testing at the Spanish National Host in August 2000, and is being distributed at the time of writing. The main functionalities of this last release of the system are described by Solano et al. (2001). The usage of the archive has increased substantially along the last year (see Figure 2). The number of files retrieved from the Spanish National Host and the Principal Centre was 26000 in 1999, growing up to 45000 in 2000. The total volume of data transferred increased from 2.3 to 6.0 GB in the same period.

References

Cassatella, A., Altamore, A., González-Riestra, R., Ponz, J. D., Barbero, J., Talavera, A. & Wamsteker, W. 2000, A&AS, 141, 331

González-Riestra, R., Cassatella, A., Solano, E., Altamore, A. & Wamsteker, W. 2000, A&AS, 141, 343

Rodríguez-Pascual, P. M., González-Riestra, R., Schartel, N. & Wamsteker, W. 1999, A&AS, 139, 183

Solano, E., González-Riestra, R., Rodríguez, F., Talavera, A., de la Fuente, A., Skillen, I., Ponz, J. D. & Wamsteker, W. 2001, this volume, 152

Wamsteker, W., Skillen, I., Ponz, J. D., de la Fuente, A., Barylak, M. & Yurrita, I. 2000, Ap&SS, 273, 155

Astronomical Data Analysis Software and Systems X
ASP Conference Series, Vol. 238, 2001
F. R. Harnden Jr., F. A. Primini, and H. E. Payne, eds.

The OaPd System for Web Access to Large Astronomical Catalogues

Lucio Benfante, Alessandra Volpato, Andrea Baruffolo, Leopoldo Benacchio

Osservatorio Astronomico di Padova, Vicolo dell'Osservatorio, 5 - 35122 Padova, Italy

Abstract. At the Padova Astronomical Observatory (OaPd), we are developing a system for managing large sets of astronomical data, such as the GSCII catalog that will be released in the near future. In this paper we describe the different parts of the retrieval system: the Database Management System (DBMS) containing the data and the meta-data of the catalogs, the CORBA services, the Java Servlets and finally the Java Applet Client. A brief description of the hardware and of the support software is also presented. The result is a system that allows one to access a set of astronomical catalogs using heterogeneous clients (i.e., Astrobrowse, StarCat). The first public release is expected in January 2001.

1. Introduction

At the Padova Astronomical Observatory (OaPd), we are developing a system for managing large sets of astronomical data, such as the soon-to-be-released GSCII catalog. The main purpose of the system is to allow a powerful and efficient querying activity on large astronomical catalogues through the Internet by using a simple Web interface. The only requirement for the user is to have a java-enabled Web browser. Figure 1 shows the overall architecture of the system as described in Baruffolo, Benacchio & Benfante (1999). In the following sections we briefly describe both the hardware and software components of the system.

2. Database Server

The catalogue data and meta-data are managed by the Informix Dynamic Server (IDS), an object-relational Database Management System. The IDS is enhanced by the PosAstro DataBlade. This OaPd-developed module provides functionality that is required to handle spatial information for astronomical objects. Functionalities and features available in the current version of the PosAstro DataBlade allow one to create tables containing astronomical coordinates objects that can be queried by content. The DataBlade supports the creation of R-Tree based indices on coordinates objects (Guttman 1984, Baruffolo 1999) that are used by the database query engine to optimize execution of queries.

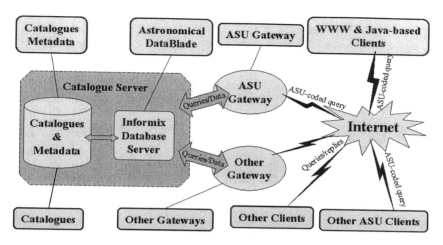

Figure 1. Overall architecture of the system

3. CORBA Services

The system is mainly based on Java code and platform-independent protocols. Nevertheless the use of some legacy C code was necessary. To allow the access to that code in a language-neutral, platform-independent and distributed manner, we embedded it into CORBA objects[1]. The C code that is now available as CORBA services is:

- the C version of the SLALIB for string decoding, sexagesimal conversion, coordinates and astrometry functions (Wallace 1999),
- the library for querying the NED[2] name resolver,
- the library for querying the Simbad[3] name resolver.

4. Servlets

The server services are provided to the clients by a set of HTTP Java Servlets (Davidson & Coward 1999). The main servlets are the ASU Servlet, the Catalogue Metadata Servlet and the Name Resolver Servlet.

The ASU Servlet provides the main query service for the system. It receives a standard ASU query (Albrecht et al. 1996) and, after retrieving data from the Database Server, it returns the requested catalogues data to the client in the selected format. Any client following the ASU specification (e.g., ESO's Skycat[4]) will be able to submit queries and retrieve results. The current allowed output

---

[1] http://www.omg.org/technology/documents/formal/corbaiiop.htm

[2] http://nedwww.ipac.caltech.edu

[3] http://simbad.u-strasbg.fr

[4] http://archive.eso.org/skycat

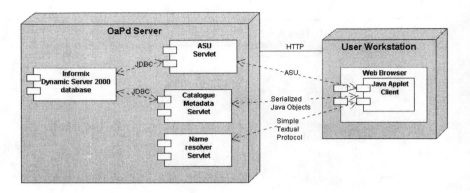

Figure 2. Java Applet Client communications

formats are HTML with tables, Tabbed Separated Values (TSV) and Starbase format (Roll 1996). Support for FITS and XML output will be provided at a later date, if necessary. The servlet uses the CORBA services for name resolving and for astronomical coordinates manipulation (via Slalib).

The Catalogue Metadata Servlet provides the client with the meta-data of the catalogues available in the database. Examples of these meta-data are the list of catalogues, the list of fields of each catalogue, the units of a field, etc. The servlet communicates with its clients by exchanging serialized Java objects through an HTTP connection. To make this communication easier, the applet utilises a servlet-specific Java library providing a simple interface for the Metadata Servlet services.

The Name Resolver Servlet is an HTTP wrapper to the CORBA name resolving services. Its textual communication protocol is very simple. It receives as input the name of the name resolver to query (NED or Simbad) and a name of an astronomical object to resolve. The output will be the resolved coordinates of the object or an error code.

5. Java Applet Client

The user interface is implemented with a Java applet to allow portability and service via the Web. The applet is a thin client for the OaPd catalogue server. It is structured as a "wizard" that guides the user in forming the queries and checks its input. The query execution is done on the server side. For requesting server-side services, the applet contacts the servlets through the HTTP protocol. Figure 2 shows the interaction among the applet and the servlets.

6. Hardware & Software

The server-side part of system runs on a COMPAQ AlphaServer DS10 466 MHz, equipped with 640 MB of RAM. The operating system is Compaq Tru64 UNIX (ex Digital UNIX). The data, meta-data and indexes of the catalogs are managed by the Informix Dynamix Server 2000. The final projected disk space is 250 GB.

The server-side Java objects are executed by a Compaq Java 2 Virtual Machine (VM)[5]. For the client-side Java Applet a Java 1.1 VM has been ordered. The servlet engine is Tomcat[6], mounted on an Apache Web Server.

The CORBA objects are handled by the ORBacus[7] Object Request Broker for C++ and for Java.

References

Guttman, A. 1984, ACM SIGMOD Conference, 47

Albrecht, M, et al. 1996, "Astronomical Server URL", http://vizier.u-strasbg.fr/doc/asu.html

Roll, J. 1996, in ASP Conf. Ser., Vol. 101, Astronomical Data Analysis Software and Systems V, ed. G. H. Jacoby & J. Barnes (San Francisco: ASP), 536

Baruffolo, A. & Benacchio, L. 1998a, in ASP Conf. Ser., Vol. 145, Astronomical Data Analysis Software and Systems VII, ed. R. Albrecht, R. N. Hook, & H. A. Bushouse (San Francisco: ASP), 382

Baruffolo, A. & Benacchio, L. 1998b, SPIE Proc., 3349, 274

Baruffolo, A., Benacchio, L. & Benfante, L. 1999, in ASP Conf. Ser., Vol. 172, Astronomical Data Analysis Software and Systems VIII, ed. David M. Mehringer, Raymond L. Plante, & Douglas A. Roberts (San Francisco: ASP), 237

Baruffolo, A. 1999, in ASP Conf. Ser., Vol. 172, Astronomical Data Analysis Software and Systems VIII, ed. David M. Mehringer, Raymond L. Plante, & Douglas A. Roberts (San Francisco: ASP), 375

Davidson, J. D. & Coward, D. 1999, "Java Servlet Specification, v2.2", Sun Microsystems, http://java.sun.com/products/servlet/

Wallace, P. T. 1999, Starlink Project, Starlink User Note 67.45

---

[5] http://www.compaq.com/java/alpha/

[6] http://jakarta.apache.org/tomcat/

[7] http://www.ooc.com

Astronomical Data Analysis Software and Systems X
ASP Conference Series, Vol. 238, 2001
F. R. Harnden Jr., F. A. Primini, and H. E. Payne, eds.

The Submillimeter Array Data-Handling System

Jun-Hui Zhao, Takahiro Tsutsumi

Harvard-Smithsonian Center for Astrophysics, 60 Garden St., Cambridge MA 02138

Abstract. We report on the basic design and current status of the data-handling system for the Submillimeter Array (SMA). Components of this system currently under development include the data storage format, archive, and off-line data reduction software.

1. Introduction

The Submillimeter Array (SMA) is under construction at Mauna Kea (Moran 1998). The SMA's first fringes from observations of celestial sources were obtained with two antennae on September 29, 1999. A year later, the first phase closures were successfully achieved on Uranus. A synthesis image of this planet at 230 GHz was made from the observations using the SMA's first three elements. As the SMA correlator comes on-line, the maximum data production rate will approach 2.75 MB per sampling. For a typical integration time of 10 seconds, the daily data production rate of the SMA would be 20 GB/day. In this paper, we present the design of the data-handling system and report the status of the software development in support of data reduction and analysis for SMA users.

2. Software Design and Development

Figure 1 shows an overview of the architecture of the SMA on-line data-handling software. Communication between the data-handling computer *Smadata* (a Sun Ultra 60 running Solaris) and the real-time system (the SMA correlator *Crates* and a control computer *Hal9000*) is accomplished with remote procedure calls (RPC) via a local network (100 Meg/sec Ethernet). *Smadata* is a central host of the data-handling server (smadata-svc), performing the post-correlator data processes such as data formatting, on-line correction, and flagging. In addition, this data computer also hosts the servers for data archiving, database management, data replication and HTTP.

RPC server and Data Format The RPC server *smadata_svc*, developed in C, provides several data services to process the data received from real-time computers *Crates* and *Hal9000*. The cross-correlation data from the SMA correlator and ancillary data are organized and stored in a number of FITS tables following the FITS-IDI standard (Diamond et al. 1997;Flatters 1998). During

The Submillimeter Array Data-Handling System

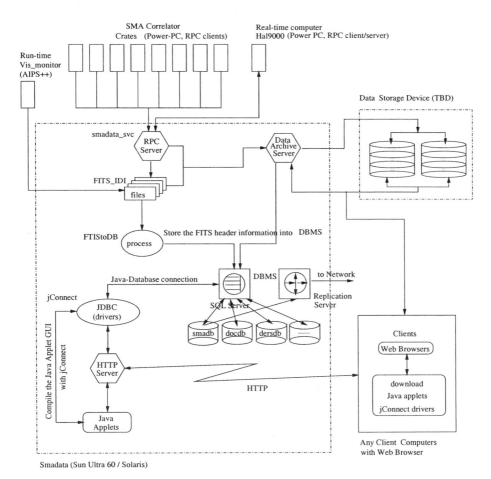

Figure 1. The SMA Data-Handling Software Architecture. The host computer is Smadata, a Sun Ultra 60 running Solaris. The RDBMS is Sybase. The JDBC utilizes jConnect from Sybase. This configuration is for the primary site currently located on Mauna Kea. Eventually, this system will be moved to the SMA headquarters in Hilo. There already exists a dedicated network link (45 MB/s) between Mauna Kea and Hilo.

an observing run, a visibility data monitor (Vis_monitor, under development in AIPS++) will provide a handy, run-time imaging facility for data quality control. At the end of each observing run, a single portable FITS-IDI is produced. The SMA FITS-IDI can be directly read into the AIPS environment and is ready for off-line data analysis.

On-line Archive and Storage A Sybase SQL Server relational database management (RDBM) system is being used at the SMA for various types of data management. Its relatively low cost (compared with other commercial packages such as Oracle) is suitable to the size of a project like the SMA. With standard ANSI SQL (Structure Query Language), the software functions supported by Sybase also meet our requirements for archiving documentation and data management.

With this commercial software, we are also developing an on-line archive system to handle SMA interferometer data. The FITS-IDI files will be stored in mass storage. The header information of the FITS tables in each FITS-IDI file along with the file location is archived in the SMA astronomical database (SMADB), which is managed by the Sybase server. The preliminary design of this system is illustrated in Figure 1. At the termination of each observing run, the RPC server *smadata_svc* triggers a process, FITStoDB, which extracts all the header data from the FITS-IDI files and converts them to the database in Sybase.

The SMA Astronomical Database (SMADB) The database model is based on the data structure of the FITS-IDI file. Ten relational Sybase tables are needed to model the SMADB. Table 1 (RUN_LOG) contains the general information for each observing run. The information about the correlator that generates the visibility data is included in Table 2 (CORR). The mandatory keywords for each FITS-IDI file are stored in Table 3 (FITS_KEY). The general information on FITS tables in each FITS-IDI file is stored in Table 4 (TAB_NM). The parameters for frequency setup, source coordinates and velocities are stored in Tables 5 (FREQ), Table 6 (SOUR), and Table 7 (VELO). The information regarding the array geometry is saved in Table 8 (ARR_GEO). The information on the visibility data can be found in Table 9 (VIS), and byte-size and location of each FITS-IDI file are stored in Table 10 (DFILE).

Data Replication A primary data archive system is located at the Mauna Kea site Eventually, it will be shifted to the SMA Hilo base facility. Most SMA users are located at two remote institutes, CfA in Cambridge (Massachusetts) and ASIAA at Nankang (Taipei). Due to the large volume of SMA visibility data, users and applications at these sites would suffer unacceptable delays in receiving complete data sets and would also generate a large amount of network traffic if they could only access data from the primary site. To avoid this problem, the current design includes replication of the data on the local systems.

User Interface A JDBC driver, Sybase's jConnect, has been installed in the Server host computer Smadata. The basic configuration for the SMA On-Line Archive System is illustrated in Figure 1. JDBC provides standard Java API codes that allow us to develop a specific Java Applet GUI (Graphical User Interface) to communicate with SMADB via the SQL server. The data computer also hosts an HTTP server. This Server provides a port for outside clients to download Java Applets and therefore to establish a connection with the database server. As soon as the client/server connection is established, the data transaction can proceed via the network.

3. Hardware for Data Storage

As the SMA becomes fully operational (with all 8 antennae and a full set of MIT/SAO correlators), data storage will become a major issue for the on-line data archive system described in the previous sections. We will inevitably need a high capacity mass storage system.

We continue to investigate hardware devices for data storage, including a DLT library or DVD-R juke-box. However, we have a temporary solution for keeping the visibility data on-line during the construction and testing phase. The current storage hardware system is implemented with several multipack disks attached to the data server *Smadata* (Ultra 60) while either the DLT library or DVD-R juke box is being considered.

4. Off-line Data Reduction Software

Three primary interferometric data reduction environments (AIPS++, AIPS, and Miriad) are chosen by the SMA staffs for off-line data reduction. Utility codes for calibrations are under development in support of the SMA specifications.

Acknowledgments. This paper is based on SMA Technical Memo 138. We thank the SMA staff for their many helpful comments and discussions in the course of the software development.

References

Diamond, P. J., Benson, J., Cotton, W. D., Wells, D. C., Romney, J. D. and Hunt, G. 1997, VLBA correlator Memo No. 108 (NRAO)

Flatters, Chris, 1998, AIPS Memo No. 102 (NRAO)

Moran, J. M. 1998, in Advanced Technology MMW, Radio, and Terahertz Telescopes, Ed. Thomas G. Phillips, Proc. SPIE Vol. 3357, 208

Zhao, J.-H., Mailhot, P., and Tsutsumi, T., 2000, SMA Technical Memo No. 138 (SAO)

See Spike Run... Run, Spike, Run

Leslie Zimmerman

Space Telescope Science Institute, 3700 San Martin Drive, Baltimore, MD 21218

Abstract. No, Spike is not man's best friend, but it is one the best friends that our planning and scheduling teams have. Spike (Science Planning Integrated Knowledge Environment), is an integral part of the HST planning and scheduling system. It provides long range planning information to the scheduling team; it provides bright object, timing, and orientation information to the observation planning team; and it provides graphical description of the observations and their constraints.

In testing Spike, we verify that Spike retains the knowledge of all of its old tricks, as well as the new ones. In other words, with each release of HST Spike we have to verify all or some combination of the functions Spike performs. Here, we present our strategies for regression testing the following functions of Spike: generating graphical observation descriptions, generating alerts for bright objects, relating timing and orientation link information, and long-range planning.

Graphical User Interface (GUI)

The GUI graphically displays the constraints that are computed by Spike and stored in the description files and in the database. To test this, we exercise a subset of the menu items in the GUI and compare the display output to the corresponding values in the database or description files.

The windows that describe where the constraint suitabilities intersect ("constraint windows"), as well as the windows that show where the observation may be scheduled on the long-range plan ("plan windows"), are graphically displayed in the GUI with color-coding. The individual constraints on the observation may be viewed separately to allow one to investigate cases where the constraint suitabilities do not intersect.

In addition to the graphical display of the constraint windows and plan windows, the GUI also allows easy access to most of the Spike functions, such as loading a proposal (to calculate its observations' constraints) and running the scheduler to plan where the observations may be scheduled (which generates "plan windows").

Bright Object Alert System (BOAS)

BOAS searches the database for bright objects associated with observations that have just been processed. BOAS then generates alerts for those bright objects.

The alert files are sent to the Contact Scientist for review. Only when the Contact Scientist has determined that the bright objects pose no health and safety threat to the instruments can the observation complete processing and be set ready for flight scheduling on the telescope.

BOAS processes all bright objects that have entered the database since the last time BOAS was run. We test BOAS by selecting several observations with bright objects in the database and setting the BOAS time_stamp to be before the date that the objects were entered into the database. We then compare the alert files and alert database entries to those that were generated in the previous version of BOAS. Unless there were changes made to the functionality of the bright object alert system, the files and database records should be identical.

Link Set Generation

Spike translates observer specified special requirements that link observations together by timing or orient constraints into database entry commands. These commands are written to "assignment files", which are used to populate the database with the description of the linked sets of observations.

To test this, we use a set of proposals that use a variety of the different types of timing and orient links that can be requested. We then use Spike to generate assignment files for those linked observations. If there are no changes to the functionality, then these assignment files will match those generated with the previous version of Spike.

Long Range Planning

We use a specific set of proposals with a variety of the different special requirements that can be requested. These proposals are then loaded into Spike. When Spike loads the proposals, it generates its own descriptions of the observations contained within those proposals and writes those descriptions to "description files". It also generates windows, which are the times during the plan that those observations may be performed. These "constraint windows" are an intersection of all of the constraints for an observation within the plan's start and end times. The constraints include sun and moon avoidance, target visibility, guide star availability, and timing/orient links.

After the proposals have been successfully loaded, the Spike scheduling command is run to create a long-range plan. This scheduler uses a particular set of criteria (which are given weights by the user) to find the best places for the loaded observations to schedule with respect to one another in order to make the most efficient long range plan. Once the scheduler finishes, the "plan windows", places where the observations may be scheduled, are written to the database.

To verify that nothing has changed in the functionality of Spike with respect to generating a long range plan, we need to satisfy ourselves that Spike has described and scheduled the observations in the same way. We do this by comparing the plan windows that we have just generated to those generated by the previous version of Spike. If the plan windows are identical, we can assume that the descriptions of the observations are also identical. Otherwise the different description would change where Spike would try to schedule the observation.

CIA V5.0—the Legacy Package for ISOCAM Interactive Analysis

S. Ott

ISO Data Centre, Astrophysics Division, Space Science Dept. of ESA, Villafranca, P.O. Box 50727, 28080 Madrid, Spain

R. Gastaud

DAPNIA/SEI-SAP, CEA/Saclay, F-91291 Gif sur Yvette Cedex, France

B. Ali[3], M. Delaney[1,4], M-A. Miville-Deschênes[5], K. Okumura[1,5], M. Sauvage[2] & S. Guest[1,6]

Abstract. The ISOCAM Interactive Analysis System (CIA) (Ott et al. 1997; Delaney 2000) was developed to support the calibration of ISO-CAM, (Cesarsky et al. 1996) the infrared camera on board ESA's Infrared Space Observatory (ISO) (Kessler et al. 1996) and to perform its astronomical data processing. The development, a collaborative effort involving several institutes, was led by ESA and began in mid-1994. Currently the system is used by 70 institutes, including the ISO Data Centre at VilSpa and the CAM consortium. CIA is generally available to the astronomical community and runs, using IDL 5,under DEC VMS Alpha, Solaris, DEC Unix, Debian (PC) Linux and HP/UX. CIA can be obtained electronically[7]. With the end of ISO's post-operational phase in December 2001, the legacy version, CIA V5.0, will be released. We present the latest calibration results and algorithmic improvements, give examples of CIA's current data processing capabilities and outline the foreseen scope of the legacy version.

---

[1]ISO Data Centre, Astrophysics Division of ESA, Villafranca del Castillo, Spain

[2]CEA, Saclay, Gif-sur-Yvette, France

[3]Infrared Processing and Analysis Center, JPL and Caltech, Pasadena, USA

[4]Stockholm Observatory, Salsjöbaden, Sweden

[5]Institut d' Astrophysique Spatiale, Orsay, France

[6]Rutherford Appleton Laboratory, Chilton, Didcot, Oxon, England

[7]http://www.iso.vilspa.esa.es/archive/software/CIA/CIA_form.html

CIA V5.0 Legacy Package for ISOCAM Interactive Analysis

1. Introduction

The ISOCAM Interactive Analysis System (CIA)[8] was developed to support the calibration and operation of ISOCAM, the infrared camera on board of ESA's Infrared Space Observatory (ISO)[9].

The system is mainly IDL-based. CPU intensive tasks are coded in C++ and FORTRAN. It consists of 250,000 lines of code, 150 MB of documentation and 80 MB of calibration products.

Currently the main effort is spent on improving the system's functionalities in processing ISOCAM astronomical data without compromising its ability to continue dedicated calibration analysis. As the next version 5.0 will be the legacy version, purging of obsolete routines and upgrading the documentation is also a priority.

2. Algorithmic Improvements for CIA V5.0

The latest algorithmic improvements for ISOCAM include: sky-cube deglitching, a very powerful technique to reject the effects of cosmic ray impacts for datasets with highly redundant pointings (a sky pixel has been observed several times) (Ott et al. 2000); improved distortion correction, including circular variable filters, and an improved projection algorithm, resulting in higher quality mosaics (see Figures 1 & 2); dedicated treatment of solar system objects; improved error propagation; improved deglitching for strong point sources; use of distorted PSFs for simulations; revision and upgrade of photometric tools; and use of improved pointing information.

3. System Improvements for CIA V5.0

Major foreseen system improvements (apart from bug-fixes and the purging of obsolete routines) are: improved on-line help: new HTML based help and search for multiple topics; multiple copies for widgets; tool to recalibrate ISOCAM astrometry for wheel jitter; and upgrade to IDL V5.4.

---

[8] CIA is a joint development by the ESA Astrophysics Division and the ISOCAM Consortium. The ISOCAM Consortium is led by the ISOCAM PI, C. Cesarsky.

[9] ISO is an ESA project with instruments funded by ESA member states (especially the PI countries: France, Germany, the Netherlands and the United Kingdom) and with the participation of ISAS and NASA.

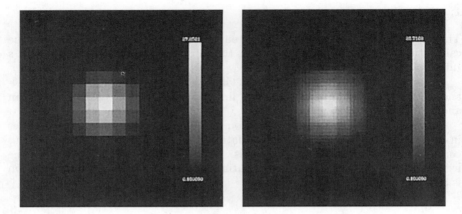

Figure 1. ISOCAM raster observation of HIC96441 at $4.5\,\mu$. The star has been observed 81 times crossing the whole detector. Left image: Resulting mosaic without distortion correction. Right image: Previous distortion correction.

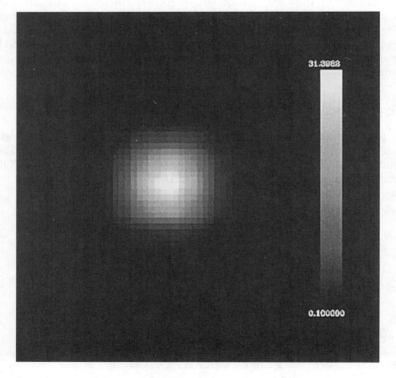

Figure 2. Mosaic of HIC96441 using the new distortion correction. Note the better reconstruction of the PSF.

4. New and Updated Contributed Packages

The following contributed packages are expected to be delivered together with CIA V5.0:

- BESD. *B*aseline *E*xtraction and *S*ky *D*eglitching is a tool to extract very faint sources for observations with many redundant pointings (Metcalfe et al. in preparation);
- PRETI. *P*attern *RE*cognition *T*echnique for *I*SOCAM is a tool using multiresolution analysis for faint source processing (Starck et al. 1999);
- SLICE V1.0. *S*imple and *L*ight *ISOC*AM *E*nvironment is a powerful tool to correct ISOCAM's long-term transient and variable flat-field (Miville-Deschênes et al. 2000);
- ISOCAM Parallel processing package. This package is tailored to analyse CAM parallel data (Ott et al. in preparation).

Acknowledgments. We would like to thank our many contributors:
A. Abergel, B. Altieri, J.-L. Auguéres, H. Aussel, J-P. Bernard, A. Biviano, J. Blommaert, O. Boulade, F. Boulanger, C. Cesarsky, D. Cesarsky, P. Chanial, V. Charmandaris, R-R. Chary, A. Claret, A. Coulais, C. Delattre, F-X. Désert, T. Deschamps, P. Didelon, D. Elbaz, Y. Fuchs, P. Gallais, K. Ganga, G. Helou, M. Heemskerk, M. Kong, F. Lacombe, D. Landriu, O. Laurent, P. Le Coupanec, J. Li, L. Metcalfe, M. Perault, A. Pollock, P. Roman, D. Rouan, M. Rupen, J. Sam Lone, R. Siebenmorgen, J-L. Starck, D. Tran, R. Tuffs, D. Van Buren, L. Vigroux, F. Vivares, T. Võ and H. Wozniak.

References

Cesarsky, C., et al. 1996, A&A, 315, L32

Delaney, M. ed. 2000, ISOCAM Interactive Analysis User's Manual, Version 4.0[10], ESA Document Reference Number SAI/96-5226/Dc

Kessler, M., Steinz, J. A. et al. 1996, A&A, 315, L27

Miville-Deschênes, M-A., et. al. 2000, A&AS, 146, 519

Ott, S., et al. 1997, in ASP Conf. Ser., Vol. 125, Astronomical Data Analysis Software and Systems VI, ed. G. Hunt & H. E. Payne (San Francisco: ASP), 275[11]

Ott, S., Metcalfe L., & Pollock, A. 2000, in ASP Conf. Ser., Vol. 216, Astronomical Data Analysis Software and Systems IX, ed. N. Manset, C. Veillet, & D. Crabtree (San Francisco: ASP), 543[12]

Starck, J-L., et al. 1999, A&AS, 138, 365

---

[10]http://www.iso.vilspa.esa.es/manuals/CAM/users_manual_v4_0.ps.gz

[11]http://www.iso.vilspa.esa.es/science/pub/adass.ps

[12]http://www.iso.vilspa.esa.es/science/pub/2000/cosmic_ray.ps.gz

The Multimission Archive at the Space Telescope Science Institute

Paolo Padovani[1], Damian Christian, Megan Donahue, Catherine Imhoff, Timothy Kimball, Karen Levay, Marc Postman, Myron Smith, Randy Thompson

Space Telescope Science Institute, Baltimore, MD 21218, USA

Abstract. We present an overview of the Multimission Archive at the Space Telescope Science Institute (MAST). The Hubble Data Archive has expanded to provide easy on-line access to non-HST data. MAST includes the following: IUE, EUVE, Copernicus, ORFEUS, ASTRO HUT, WUPPE, and UIT data, and VLA FIRST data. MAST is also the active archive site for the Far Ultraviolet Spectroscopic Explorer (FUSE), launched in June 1999, and provides access to the Digitized Sky Survey. We discuss the relevance of MAST for "data mining" studies, its literature links, and the features of the World Wide Web interface. We finally present our plans for expansion, which include the science public archive of the Sloan Digital Sky Survey, the new version of the Guide Star Catalog, and "on-the fly" creation of advanced data products.

1. Introduction to MAST

The Multimission Archive at the Space Telescope Science Institute (MAST) is NASA's optical/UV science archive center. MAST, established in October 1997, builds upon the infrastructure developed for the Hubble Space Telescope archive but expands this service to support nine additional missions (see below).

Our data holdings include eight space-based missions, three of which (HST, EUVE, and FUSE) are currently active as of January 2001. We also provide archival services for two ground-based sky surveys: the Digitized Sky Survey and the VLA 20 cm radio survey known as FIRST. The combined MAST data volume exceeds 12 TB, making it one of the most significant astronomical collections available on-line today. All MAST data can be accessed at the MAST home page[2].

The scientific value of MAST comes, in part, from the rich and varied astrophysical phenomena that dominate the optical/UV regions of the electromagnetic spectrum and from the convenient and user-friendly implementation of our World Wide Web (WWW) archive interface. The utility of the MAST

---

[1] Affiliated to the Astrophysics Division, Space Science Department, European Space Agency

[2] http://archive.stsci.edu/mast.html

supported missions is demonstrated by the scientific literature publication rate of well over 1000 papers per year which make substantial use of these data.

2. The MAST Holdings

MAST includes the following missions:

HST Hubble Space Telescope (1100–25,000 Å), which contains over 250,000 exposures of $\sim 30,000$ individual astronomical sources. Spectroscopic and imaging data are available from 7 widely-used instruments;

IUE International Ultraviolet Explorer (1200–3350 Å), which contains more than 104,000 spectral images of $\sim 10,000$ individual astronomical sources;

Copernicus (OAO-3) far- (900–1560 Å) and near- (1650–3150 Å) ultraviolet spectra of 551 objects;

EUVE Extreme Ultraviolet Explorer (70–760 Å) spectroscopic observations of ~ 400 sources, mostly galactic. EUVE was active through January 2001;

FUSE MAST is the active archive site for the Far Ultraviolet Spectroscopic Explorer, a NASA-supported mission successfully launched on June 24 1999, which is exploring the Universe with high resolution spectroscopy in the far UV (905–1190 Å) spectral region. FUSE is obtaining high resolution spectroscopy in the far-ultraviolet spectral region reaching 10,000 times fainter than Copernicus and superior resolution than HUT. FUSE is part of NASA's Origins Program under the auspices of NASA's Office of Space Science. MAST provides access to both proprietary and public FUSE data.

ASTRO includes three UV missions from the ASTRO 1 and 2 Space Shuttle missions:

- Hopkins Ultraviolet Telescope (HUT) (825–1850 Å), which includes about 500 ultraviolet spectra of more than 300 targets;
- Wisconsin Ultraviolet Photo-polarimeter Experiment (WUPPE) (1400–3200 Å), obtained simultaneous ultraviolet spectra and polarization measurements. It includes 400 observations of roughly 200 targets;
- Ultraviolet Imaging Telescope (UIT) (1200–3300 Å), which contains about 1,600 images of more than 200 targets;

ORFEUS Orbiting Retrievable Far and Extreme Ultraviolet Spectrometers, two UV missions from ORFEUS 1 and 2 Space Shuttle missions:

- Interstellar Medium Absorption Profile Spectrograph (IMAPS) (950–1150 Å), obtained high resolution (R=75,000 for IMAPS-1) objective-grating echelle spectra. The IMAPS archive contains roughly 600 spectra of ten hot stars from the first Shuttle flight;
- Berkeley Extreme and Far-UV Spectrometer (BEFS) (400–1200 Å), which returned high-resolution ($R = 5,000$) far UV spectra of 175 astrophysical objects from the two Shuttle flights. Extreme UV spectra (400–900 Å) were obtained for a subset of the targets.

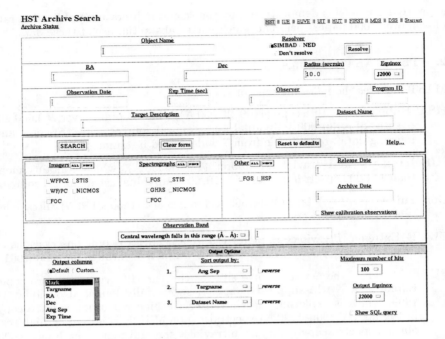

Figure 1. The HST World Wide Web Interface.

FIRST Faint Images of the Radio Sky at Twenty-centimeters, a radio survey at 20 cm (1.4 GHz) of over 10,000 deg^2 down to a flux of 1 mJy. The radio images and the source catalog, currently $\sim 720,000$ entries, are available;

DSS Digitized Sky Survey, digitized photographic plates from the Palomar and UK Schmidt telescopes;

ROSAT As a service to the optical/UV community, MAST provides also access to ROSAT (ROentgen SATellite) X-ray data. The ROSAT Master observations log (ROSMASTER) at the High Energy Astrophysics Science Archive Research Center (HEASARC) is in fact accessible via an interface which is very similar to the other MAST interfaces.

3. The MAST Interface

The MAST listings are available via a simple WWW interface. A sample search page for HST data is shown in Figure 1. Similar interfaces are available for all MAST data. Archival data may be searched by name (resolved by SIMBAD or NED), position, object category, and observation specifics (date, instrument, filters, exposure time, etc.). Previews are available for most MAST missions, allowing users to have a "quick look" at the data before retrieving them.

3.1. MAST Cross-Correlations with Astronomical Catalogs

The potential use of the MAST archive is greatly increased by allowing users to search more than one mission/instrument at a time and cross-correlate the archive holdings with astronomical catalogs. Cross-correlations can be performed using the Hipparcos and Sky2000 stellar catalogs, an active galactic nuclei catalog, the Abell Galaxy Cluster catalog, and any user-supplied list of positions. MAST users can select a sample of astronomical sources based on a range of properties (e.g., redshift, magnitude, radio flux for active nuclei) and then look for the relevant entries in MAST. Work is in progress to expand this facility by using NASA's Astronomical Data Center (ADC) interface. This will allow cross-correlations to be made between MAST and any of the ADC catalogs and tables, opening up new possibilities for the exploitation of MAST data.

4. Literature Links

MAST data are linked to the scientific literature to allow users easy access to MAST-based papers directly from our interface. For most missions the WWW interface returns the papers based on a given dataset/proposal. Active links are provided to the Astrophysics Data System (ADS). We have worked with ADS to provide the complementary service, that is links between astronomical abstracts and MAST data previews, retrieval pages, and observation logs. Literature links to IUE, ASTRO, EUVE, BEFS, and Copernicus papers are already in place, while more than $\sim$ 50% of HST-based papers have been linked with their proposal ID. Work is on-going to complete the HST literature links and to provide the same service for the remaining MAST missions.

5. The Future of MAST

MAST will incorporate additional ultraviolet and optical data in the future, including those from the GALEX and CHIPS missions scheduled for launch in Fall 2001 and Spring 2002, respectively. Furthermore, MAST will host the science public archive of the Sloan Digital Sky Survey (SDSS) and the new version of the Guide Star Catalog (GSC-II). The SDSS will map in detail one-quarter of the entire sky, providing images for more than 100 million sources and redshifts for more than a million galaxies and quasars. GSC-II will include proper motion and color information, in addition to accurate coordinates, magnitudes and classification, for all objects in the sky down to at least 18th magnitude, an estimated 2 billion sources. MAST will further enhance the scientific value of its data holdings by archiving Mosaic Imager data from the National Optical Astronomy Observatories. To fully exploit the multiwavelength parameter space, which is being made available also by the many large surveys completed and under way, MAST will establish closer ties and coordination with other archive centers. MAST will work towards providing the community with science-ready products. This will include data characterization and catalogs of selected HST imaging data, thus enabling the identification of faint optical counterparts at various wavelengths, "on-the-fly" co-added spectra/images for objects with multiple exposures, and combined spectra spanning detector and mission boundaries.

STPOA—The New Pipeline Package for the HST Post-Operational Archive

A. Alexov, P. Bristow, F. Kerber, M. Rosa

Space Telescope - European Coordinating Facility, European Southern Observatory, Karl-Schwarzschild-Str. 2, D-85748 Garching, Germany

Abstract. The "Post-Operational Archive" (POA) project at the Space Telescope European Coordinating Facility (ST-ECF) has recently released an upgraded version of the HST Faint Object Spectrograph (FOS) calibration pipeline. We present an overview of the capabilities of this STPOA release, showing examples of the resulting scientific gain. We discuss the lessons learned from the in-depth study of an HST Legacy instrument, the FOS, and the problems encountered therein, including erroneous header contents, difficulties in attaining and processing telemetry uplink/downlink information, and software environment restrictions.

1. Introduction

Archive research and data mining are now considered vital parts of scientific research, but scientific data integrity has to be checked, improved and validated before massive research projects can be started.

Typically, fewer resources are available for work on legacy instruments than are allocated for work on operational instruments. The HST Post Operational Archive (POA) project, based on an agreement between ESA and NASA, aims at improving the scientific value of HST legacy instrument data. We investigated the feasibility of improving the FOS data archive by resolving important calibration issues with the STSDAS off-line FOS calibration pipeline.

In July 2000, the POA project team released STPOA, a new IRAF external package. It contains 'poa_calfos', a replacement for the standard FOS calibration pipeline, along with other tools for FOS re-calibration.

2. STPOA—The New Pipeline Package

The STPOA external IRAF package is available on-line[1] and can be installed at any IRAF user location. The July 2000 (v1.0) release is already in place and being used at ST-ECF and STScI.

The POA project has officially taken over the user support of the FOS. Our website contains the most up-to-date documentation, technical information and

---

[1]http://www.stecf.org/poa or ftp://ftp.eso.org/pub/stpoa

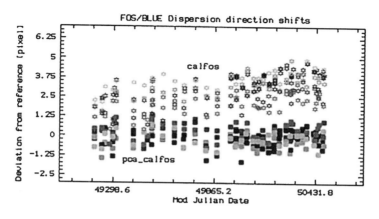

Figure 1. Comparison of wavelength zero-point offsets using 'calfos' (open symbols) and 'poa_calfos' (closed symbols).

software for the FOS. Also, our team operates a new helpdesk; the email address for questions and issues is ecf-poa@eso.org.

In its current version (v1.0), the new 'poa_calfos' calibration pipeline software fixes the zero-point problems for most FOS/BLUE channel data. It takes into account the residual effects from the Geomagnetic Image Motion Problem (GIMP), the offsets introduced by the electronic setup of the detector (YBASE values), and the variations of the ambient temperature in the FOS optical bench.

The ST-ECF HST archive[2] now supports FOS data retrieval with On-The-Fly (OTF) calibration using either the new pipeline 'poa_calfos' or the old pipeline 'calfos'. One can either retrieve re-calibrated FOS data or process the raw data using the new software.

3. Software and Science Improvements

Improvements reduce the uncertainty in the wavelength zero-point from as much as six pixels to less than one pixel. Figure 1 compares the results obtained using the old pipeline 'calfos' (open symbols) and the new pipeline 'poa_calfos' (closed symbols): various gratings are indicated by different grey scales; the time range is April 1993 to January 1997 (the end of the FOS lifetime); 'poa_calfos' corrects the drift in the zero-point track to ±1.25 pixels (or ±0.75 pixels if the data are oversampled, which is not shown here).

An example of the scientific improvement of data, using the new pipeline, is shown in Figure 2. The raw data were taken in Rapid-Readout mode, spanning approximately 40 minutes, observing the same object. A well known spectral line was used to compare the wavelength position to the expected value from literature:

---

[2]http://www.archive.eso.org

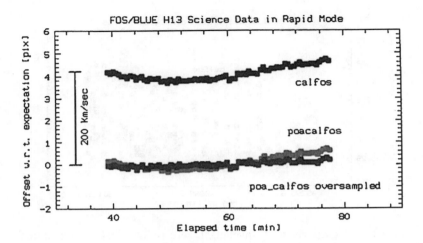

Figure 2. Science Case: wavelength zero-point offsets using 'calfos' (top) and 'poa_calfos' (bottom curved) and 'poa_calfos' oversampled (bottom flat).

- 'calfos' results are offset by ≈ 4 pixels or 200 km/sec;
- 'poa_calfos' decreases the offset to ≈ 0.5 pixels or 50 km/sec; the curvature of the results shows the residual sub-pixel GIMP;
- 'poa_calfos' with oversampling (not available in the current release) can remove the GIMP effects completely .

4. Verification of the On-board GIMP Procedure

In order to perform a verification of the GIMP using on-board information, the uplink HST telemetry data were checked to make sure HST did what it was commanded to do. The downlink engineering telemetry data were checked to match the uplink, as well as help solve the GIMP issue using on-board instrument readouts, info which is not in the science data or FOS data headers.

4.1. HST Telemetry Uplink Check

We needed the HST uplink telemetry data for our POA GIMP investigation. These data were never required to be archived in a database; they were written to 36,000 tapes and stored in a federal vault. The difficulties in attaining even a small fraction of the uplink HST telemetry information and being able to process it, made us realize that if we had to acquire the entire set and re-extract all the necessary information, our re-calibration task would have been close to impossible. A warning for the future! Fortunately we found some uplink telemetry (May '96 to Jan '97) from a colleague. We used the PASS software from STScI to extract the on-board GIMP related information. We thereby confirmed that the uplink GIMP coefficients were properly used by the on-board data processor.

4.2. HST Telemetry Downlink Check

Engineering telemetry downlink data are only routinely archived at STScI, so the ST-ECF HST archive had to be updated for the purpose of doing the on-board comparison. The existing downlink telemetry extraction software was never meant to be used outside STScI or distributed off-site. There is no general purpose engineering telemetry downlink reader; therefore these data are almost inaccessible. The ST-ECF archive team ported this software from VAX/VMS to DEC/Alpha; they modified it to access the parameters we needed. We used the on-board voltage, temperature and magnetometer readings to verify the geomagnetic field models used in 'poa_calfos'. We confirmed that the instrument and on-board GIMP calculations were synchronized.

An important lesson learned from this experience is that one should view the Science and Engineering downlink telemetry as one unit! Uniting them and having the ability to easily access the information would greatly benefit *all* calibration work in the future.

5. Final POA/FOS Release—August 2001

The final version of the Post Operation Archive FOS pipeline is planned for release in August 2001. All 24,000 FOS datasets will be reprocessed with 'poa_calfos' and re-ingested into the STScI archive. OTF re-calibration is (and will be) available at the ST-ECF HST archive for all FOS data. Substantial "cleanup" of the FOS calibration reference file data will be performed. The FOS section of the HST Data Handbook will be updated accordingly. Technical reports on re-calibration issues will be available on our website.

6. Conclusions and a Look Ahead

When the POA work on the FOS is finished, we will have a homogeneous archive for this HST instrument. A researcher can view the FOS data from 1990 to 1997 together, in a coherent manner and confidently make comparisons with data from other sources.

POA work has the advantage of looking at the entire dataset range all at once; this is impossible to do while the instrument is in operation. Calibration issues evolve over time, the "global view" is not available over a short time scale during the "live operational mode" of an instrument. We will apply the expertise gained from our FOS work to the next HST legacy instrument project.

Long running satellite missions need to project ahead in terms of computer hardware/software and portability issues. Post Operation work will be done for most missions; such work will be made easier if all telemetry and pipeline processing software is portable to current technologies and is distributable.

Lessons learned from POA work should also influence the design of calibration procedures and archive architecture for future instruments.

Acknowledgments. We would like to thank the staffs of STScI and Goddard Space Flight Center for their help in unearthing documents and other "dusty" information on the FOS for our analysis.

The SuperCOSMOS Sky Surveys

Mike Read and Nigel Hambly

WFAU, Institute for Astronomy, University of Edinburgh, Blackford Hill, Edinburgh, EH9 3HJ, UK, E-mail: mar@roe.ac.uk

Abstract. The SuperCOSMOS Sky Surveys (SSS) programme is digitising multi–colour (BRI), multi–epoch Schmidt survey plates with the ultimate aim of covering the entire sky. The R-band is covered at two epochs. The object catalogue and compressed pixel data are stored online and are accessible via the WWW. Coordinates, magnitudes, morphological data and proper motions are available for all objects down to B $\sim$ 22, R $\sim$ 20 and I $\sim$ 19. The survey is available on-line at: http://www-wfau.roe.ac.uk/sss.

1. SuperCOSMOS

SuperCOSMOS is an ideal machine for large scale survey work. Important characteristics of the machine include:
- fast scanning time: SuperCOSMOS scans a 320 × 320 mm area of the Schmidt plate in $\sim$ 2.5 hours.
- positional accuracy: the design of the machine and environmental control enable relative positional measurements at a precision of $\sim 0.5\,\mu$m
- good dynamic range and 16–bit digitisation in 10 μm pixels: these are well matched to the image quality of the photographic atlases.

2. Available Services

The SSS data consist of object catalogues to the plate limits from BRI plates including astrometric and photometric calibration. Colours, proper motions, image classification and morphological parameters are available. H-compressed pixel data are available in small FITS file subsets with included object catalogues. In relation to the SSS data, the Wide-Field Astronomy Unit (WFAU) currently offers online queries for:

- small area (up to 15 arcmin) pixel images with attached object catalogue,
- medium–scale (up to 10°) object catalogues with user–specified magnitude range and output format (FITS, ASCII or tab–separated),
- multiple object small pixel images in batch mode.

Whole–sky catalogue access and analysis online is currently being planned.

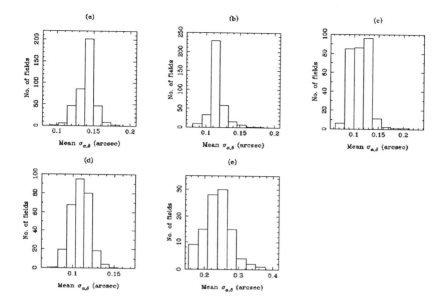

Figure 1. Astrometric residuals from Tycho-2 standards. Five panels show: (a) SERC J/EJ, (b) SERC ER/AAO–R, (c) SERC–I, (d) ESO–R and (e) POSS–I E. Modal values are in the range 0.1 to 0.2 arcsec.

3. Astrometric and Photometric Calibration

Absolute positions are calibrated with respect to the Tycho–2 catalogue while proper motions are zero-pointed on the extragalactic frame using the images of galaxies. The initial photometric calibration is based on the Guide Star Photometric Catalogue I; in due course we plan to recalibrate using GSPC–II.

Figure 1 shows histograms of the mean RMS residual per star per plate in either coordinate for Tycho–2 reference catalogue stars.

For quasar images from Veron–Cetty & Veron (1998), Figure 2 shows distributions of residual X- and Y-displacement (between first and second epoch plates) and indicates that the zero point of proper motions is truly extragalactic.

4. WWW Interface

Most people will interrogate the South Galactic Cap (SGC) database via the World Wide Web. An interface has been set up to enable subsets of the large database to be extracted. The primary data format is FITS for both images and binary tables of object catalogues, which enables porting of the data into established software packages. Object catalogues are additionally available in plain text format for readability. Basically, the user specifies the region of interest (up to 15 arcmin for images and up to several 100 square degrees for object

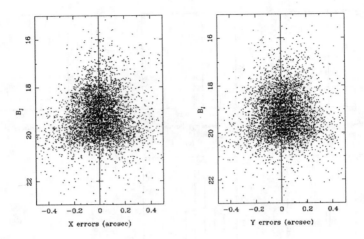

Figure 2. Distributions of residual X- and Y-displacement as a function of B_J magnitude in QSO images for the entire South Galactic Cap survey.

catalogues) and a primary colour; FITS files are then efficiently extracted and sent back to the user's home institute for browsing and manipulation.

5. Accuracy of Parameters

As is usual with measurements from wide field Schmidt plates, the external astrometric and photometric accuracies of the data are limited by position- and magnitude-dependent systematic errors. However, the internal accuracy (i.e., when comparing image data in restricted position and magnitude ranges) is unaffected by this, and for some astronomical applications it is the internal accuracy that is important. Typical numbers for the astrometric and photometric precision for well–exposed stellar images are as follows:
- $\sigma_{\alpha,\delta} \sim 0.05$ arcsec; $\sigma_{B,R,I} \sim 0.05$ mag; $\sigma_\mu \sim 5$ milliarcsec/yr (internal)
- $\sigma_{\alpha,\delta} \sim 0.3$ arcsec; $\sigma_{B,R,I} \sim 0.3$ mag; $\sigma_\mu \sim 10$ milliarcsec/yr (external)

6. Completeness and Reliability

Completeness and reliability of image classification of the SGC data have been assessed by comparison against external datasets. We find that the completeness is essentially 100% to within 1.5 mag of the nominal plate limits (B_J=21.5, R=20.5, I=17.5). For the B_J data we find that the image classification is > 90% reliable to $B_J \sim 20.5$. The limiting factor for reliable image classification is the resolution of the photographs coupled with the typical angular size of galaxies as a function of magnitude: at B_J=20, the typical scale size of a galaxy is comparable to the resolution.

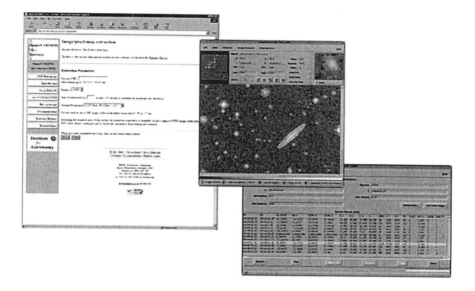

Figure 3. Extraction example using the GAIA/SkyCAT software.

7. Example

In the example shown in Figure 3, a 10 arcmin region has been extracted in the B band from a south galactic pole field containing a rich cluster of galaxies.

The data have been displayed using the GAIA/SkyCAT software. In this way, it is possible to browse the data by clicking on objects in the image display or the catalogue list to highlight them in both. All the catalogue parameters for a given object can then be examined (e.g., position, shape, brightness and classification).

8. Timescales

A 5000-square degree area centred on the SGC is available online now. By early 2001 we will have the entire southern sky available online in both J and second epoch R. The four–colour southern sky survey will be complete by mid–2002. It is hoped that we will then expand the survey into the northern hemisphere to cover the entire sky using the Palomar first and second epoch surveys.

References

Veron-Cetty, M.-P. & Veron, P. 1998, "A Catalogue of Quasars and Active Nuclei", Garching, ESO Scientific Report Series Vol. 18

Part 4. Software Development Methodologies & Technologies

SOFIA's CORBA Experiences: Instances of Software Development

John Graybeal[1], Robert Krzaczek[2], John Milburn[3]

Stratospheric Observatory for Infrared Astronomy, NASA Ames Research Center, MS 207-1, Moffett Field, CA 94035

Abstract. Developing data systems for special purpose applications—like one-of-a-kind telescopes—is a singular, if not idiosyncratic, process. Developers must master and wisely use rapidly changing software technologies to produce systems faster, better, and cheaper, meanwhile keeping up with iterative requirements and schedules. Architectural standards such as CORBA may help—or may lead to slow, hard to change, and expensive data systems.

The Stratospheric Observatory for Infrared Astronomy (SOFIA) will use CORBA in several different environments—the airborne mission systems (MCS), the ground support system (DCS), and a Facility Science Instrument (FLITECAM). A review of CORBA development experiences on the MCS reflects the challenges and choices made, while comparison with other SOFIA implementations shows the variety of CORBA applications and benefits.

1. Introduction

1.1. The Stratospheric Observatory for Infrared Astronomy

The Stratospheric Observatory for Infrared Astronomy (SOFIA)[4] is a major infrared and submillimeter observatory scheduled to begin operations within two years. A joint collaboration of NASA and DLR (the German Air and Space Agency), the Boeing 747-SP aircraft will carry the 2.5-meter telescope at or above 12.5 km, where the telescope will collect radiation primarily in the wavelength range from 0.3 micrometers to 1.6 millimeters. With a 20-year operational lifetime, SOFIA is designed to maximize astronomical value per unit cost, as compared to the other observing platforms such as balloons or satellites. To meet this goal, it must be a particularly flexible, long lasting and economic platform for conducting research. The SOFIA data systems will be key to achieving these objectives.

---

[1] Logicon Sterling Federal

[2] Rochester Institute of Technology, Center for Imaging Studies

[3] University of California, Los Angeles, Astronomy Department

[4] http://sofia.arc.nasa.gov/

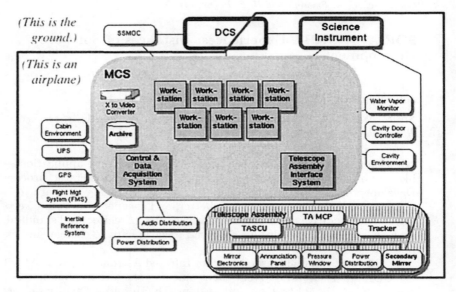

Figure 1. SOFIA's Software Components

1.2. SOFIA Software Overview

SOFIA has five principal software development environments:
- Science Instruments (SIs),
- the Mission Control Subsystem (MCS),
- the Telescope Assembly (TA),
- the Data Cycle System (DCS) and
- custom dedicated subsystems (for example, the water vapor monitor).

Different organizations develop software components in each environment, and the subsystem interfaces must be carefully negotiated and defined. As might be expected, functional requirements differ for each of the components, and development standards differ for the various environments. These variations all affect whether, and how, CORBA is used in SOFIA software.

In the following subsections we describe each of SOFIA's software components, so that their use in SOFIA—and CORBA's use in each component—will be clearer. Figure 1 provides graphical context for these descriptions.

1.3. Science Instruments

SOFIA's numerous Science Instruments, or SIs, are developed by science teams according to specifications provided by Universities Space Research Association (USRA), the prime U.S. contractor for SOFIA. The observatory will change SIs as often as every week, thereby supporting a wide variety of scientific investigations over its 20-year life cycle, while using the latest science instrumentation.

There are two classes of science instruments, Facility Science Instruments (FSIs) and Principal Investigator Science Instruments. FSIs will be operated by the observatory, and as such must undergo more thorough review and meet more challenging development criteria. The First Light Infrared Test Experiment Camera (FLITECAM) is an FSI designed to perform critical checkout and integration phase activities for SOFIA.

On SOFIA, science instruments control the observatory, but only partially. Scientists using SOFIA will command science operations through an SI's user interface, and the SI in turn commands some aspects of the observatory through the MCS. But other observatory controls must be performed by observatory staff using their own interfaces. (The Data Cycle System, described below, will eventually provide a consistent interface for many of the science instruments, and perhaps even for the observatory itself.)

1.4. The Mission Control Subsystem

The Mission Control Subsystem (MCS) is the command and control heart of SOFIA. It communicates with the observatory subsystems (including the TA), coordinates their actions, and provides an interface for science instruments, scientists, observatory staff, and other users to interact with the observatory. To make this interface as accessible as possible, the MCS implements an ASCII-based command language, available over a standard TCP/IP socket interface.

The MCS is a highly distributed system with a high-speed network of nine workstations, most running the Solaris operating system. The MCS must achieve reliable, high throughput for both SOFIA mission housekeeping data and science commands. (Science data does not go through the MCS but is maintained internally by all SIs and also by the DCS for the FSIs.) The MCS must provide enough configurability and flexibility to serve as the SOFIA baseline data system for 20 years of science operations (Papke et al. 2000).

1.5. The Telescope Assembly

The Telescope Assembly (TA) performs control and pointing for SOFIA's telescope and related components. It is being developed by a consortium of companies under the direction of the Deutschen Zentrum für Luft- und Raumfahrt e.V. (DLR, the German space agency). Although the TA's subsystems communicate with each other to some degree, the MCS is responsible for coordinating their work to perform science effectively.

1.6. The Data Cycle System

The Data Cycle System (DCS) provides an observatory-level, science-oriented interface to SOFIA. On the ground it facilitates science interactions for all parts of the SOFIA data life cycle. During flights the DCS standardizes science functions available through the SOFIA science instruments. For both of these functions, the DCS must be capable of rapid reconfiguration to address a wide variety of science functions and interfaces.

The first DCS implementation will provide basic functions for the SOFIA Facility Science Instruments. The DCS will eventually support all science instruments that take advantage of it.

1.7. Custom Dedicated Subsystems

SOFIA's custom dedicated subsystems include the water vapor monitor, the Cavity Door Control Subsystem, the Environment Control Subsystem, and the Mission Audio Distribution System. These are dedicated systems performing specialized functions in support of the SOFIA mission. Most software is implemented within dedicated embedded systems, and if a subsystem communicates with the MCS, it usually uses a subset of the SOFIA Command Language.

2. Key CORBA Attributes

Briefly, for those unfamiliar with CORBA (Common Object Request Broker Architecture), it is a standard which describes certain "middleware" products. These products provide an infrastructure on top of which application features may be developed. CORBA specifies a certain set of features useful in an object-oriented, distributed, multi-processor environment, including:
- object registration, location, and activation,
- naming and trading services (find objects based on their names and properties),
- parameter marshaling and De Marshalling (send data between systems),
- event notification and
- object life cycle management, security, and transactions.

In summary, CORBA's features generalize typical object-oriented behaviors (access, method calls, etc.) from a single-processor environment to a distributed environment.

One feature not built into CORBA is low-latency and real-time distribution of data. Systems with real-time constraints would typically be designed to avoid the need for data marshaling services that help convert data objects to different language and operating system formats in a distributed system. Such systems typically use consistent languages and operating systems.

3. SOFIA Environments Not Using CORBA

Some of the SOFIA software architectures have compelling rationales for not using CORBA. The Telescope Assembly and custom dedicated subsystems do not incorporate CORBA at all. Those subsystems' software is analogous to firmware—largely stable code with specific functionality, typically implemented on single embedded CPUs with limited interface complexity. Given stable system requirements, pre-established stand-alone development environments and no need for on-the-fly component associations, CORBA would add little value.

Many of the science instruments have environments similar to the "embedded subsystems" just described. They are also fairly stable systems, with unchanging connectivity needs, and often use only one or few processors and a single programming language and operating system. While the functional complexity of many of these science instruments (and other non-CORBA systems) is quite high, functional complexity alone does not imply the need for CORBA. For most science instruments, it is more appropriate to build the software for the expected needs, designing only the specific flexibility that has been defined, and then make changes in the developed software on an as-needed basis.

4. SOFIA Products Using CORBA

By reviewing the three SOFIA software products that do use CORBA—DCS, MCS, and the FLITECAM FSI—we identified a set of attributes that contributed to the decision to use CORBA. Not surprisingly, these attributes (see Table 1) are well matched to CORBA's intended environment.

Table 1. CORBA Motivators

Factor	DCS[a]	MCS[a]	FLITECAM[a]
Number of Operating System Types	•••	••	••
Number of Processors (i.e., CPUs)	•••	•••	••
Number of Supported Languages	•••	•	••
Object Orientation of the System	••	••	•••
Amount of Data Distribution	•••	••	•••
Tolerance for Communication Latency	•••	••	•
Need for On-the-fly Component Connections	•••	••	••
CORBA Choice	**ILU**	**TAO**	**Visibroker**

[a]Minor, moderate, and strong factors in CORBA selection are represented by one, two, and three dots, respectively.

Another attribute common to the three SOFIA software products using CORBA is that at least one member of each development team had experience with CORBA or a similar standard. In some cases the experience was limited, and most of the subsequent training was on-the-job, but having some sense of the technology was a consistent starting point.

As seen from Table 1, the DCS, MCS, and FLITECAM each chose a different CORBA product (ILU, TAO, and VisiBroker), with characteristic strengths and weaknesses discussed below. CORBA fit the three systems to different degrees and for different reasons, but each chosen CORBA implementation appeared well suited to its intended use.

CORBA Features Used The principal CORBA features used by these three systems are shown in Table 2. For all three systems the list is quite short, due in part to the systems' early development phase and their developers' CORBA experience. However, this also reflects the fact that while CORBA is very powerful, for any given project only a subset of its features may be useful. In the end, a given data system may not leverage as much of the entire CORBA package as might be expected.

4.1. Data Cycle System (DCS)

The Data Cycle System will support science operations over the entire 20-year operational lifetime of SOFIA, operating more or less continuously throughout that time. (Although it has ground-based and airborne components, we focus

Table 2.   CORBA Features used by SOFIA Systems

Function	DCS	MCS	FLITECAM
Name Service/Registration	–	YES	YES
Data (De-)Marshalling	YES	–	YES
Language Portability	YES	–	–
Remote Object Services	YES	–	YES
Connections to Remote Objects	YES	YES	YES

here on the ground-based components.) It will provide an interface for scientists to gain access to SOFIA (e.g., via proposal preparation) and its archived data.

With these requirements in mind, the DCS was designed to collect software on the fly, at user request. For a given data pipeline, algorithms from sites such as Ames, Goddard, and Australia might be combined to reduce archived raw data; while for a proposal, submission components that submit data, automatically review it and provide scheduling metrics could be combined. Even more impressive, new software components can be developed, tested and integrated while existing versions of those components remain fully available to users.

DCS CORBA Motivators   Since DCS designers were already familiar with other inter-process and inter-machine communication systems such as Remote Procedure Calls (RPC), moving communications from a procedural abstraction to CORBA's object abstraction was straightforward. Additionally, they anticipated a requirement to support many implementation languages: whereas one instrument team's chosen language might be Research System's IDL, another instrument builder might favor Java or C++.

The DCS will be a widely distributed system, not only spanning local high-speed networks but also taking advantage of possibly distant computer resources. A flexible and robust communications layer was therefore a crucial element of the DCS, and this is a CORBA strength.

DCS CORBA Implementation Process   The DCS group performed a trade study with RPC, DCE, CORBA and DCOM, choosing CORBA as the best fit for their requirements. Following a small proof of concept using the Perl scripting language, they began building key products and designing their interfaces with the CORBA product ILU (Inter-Language Unification). (Much of the underlying DCS software to date has been developed in C, an interface for which ILU had implemented early on.)

Benefits of ILU for DCS   Promising features of ILU included its free source code (valuable when troubleshooting), multiple underlying protocols, and its multiple language support (a strong plus for the DCS). The flexibility of the product was also impressive, as ILU has easily supported a wide range of functions.

DCS CORBA Current Issues/Regrets   The fact that ILU does not currently support the CORBA 2 standard may eventually become a problem for the DCS. Due to its "Erector set" approach (picking tools and features to use as needed), there is more internal machinery to consider. Finally, it seems likely that using a single CORBA implementation between the MCS and DCS will considerably simplify SOFIA's software maintenance effort, and this will be more carefully evaluated in coming months.

DCS CORBA Status and Future The DCS team has already implemented a sophisticated prototype that addresses certain DCS requirements using multiple computers. Many capabilities must be added to produce the real DCS system, and other CORBA features are likely to be needed for that product. Investigations will be performed to identify and evaluate potentially useful features and to consider the tradeoffs going from ILU to the MCS CORBA product.

4.2. Mission Control Subsystem (MCS)

The job of the MCS can be summarized as routing and displaying data, accepting, routing and responding to commands, performing calculations on data (for example on coordinate systems) and coordinating the observatory and its subsystems. This involves a lot of communication across multiple machines, the state of which may change at any time for reasons both planned and unforeseen (e.g., at 41,000 feet cosmic rays noticeably affect standard workstation memory, with unfortunate consequences for operating system reliability). The MCS is therefore built on the Jini model, with self-registering and self-discovering software components that make connections to each other as needed to perform their allocated functions.

The MCS hardware architecture has evolved over its first two years of development. Whereas multiple processor types (Sun SPARC, Motorola PowerPC) and operating systems (Unix, VxWorks) were originally expected to co-exist, it appears Sun processors with Unix (and its real-time features) may be sufficient to meet SOFIA's needs.

MCS CORBA Motivators Initially, the MCS team was not motivated to use CORBA at all. However, CORBA seemed to address several concerns of the MCS, especially given the likely need to support multiple platforms and operating systems. The ability to connect different processes at run time (and in any order, if the custom-developed code supports it) was particularly appealing. One staff member had overview knowledge of CORBA, while another was very interested in using it, and the team hoped to get significant code reuse by adopting it.

MCS CORBA Implementation Process A "hero programmer" on the MCS team developed a demonstration application using Visibroker in a little over three weeks. The programmer had no previous experience with CORBA, but was very interested in using it. Based on the evident functionality, the project leaders decided to proceed using CORBA tools.

The team purchased the Visigenix VisiBroker product in order to make use of its excellent documentation (the planned CORBA tool had no documentation available during its beta phase), and the first three months of MCS implementation used VisiBroker. Once The ACE ORB (TAO) was formally released, the team replaced the Visibroker software with TAO, encountering only minor issues. Partly reflecting this experience, the team encapsulated MCS CORBA use as much as possible, so that another product, or custom code, can replace it if necessary. As a result of this encapsulation and the limited CORBA use, the MCS team needs only one or two experienced CORBA developers, whereas three or four would be more typical for a twelve person team.

Benefits of TAO for MCS   Like ILU, TAO is freely distributable and its source code is available (Graybeal et al. 2000). It has been extremely stable, with no bugs observed since its first release. It complies with many of the features in CORBA 2.4, the most recent standard. An especially valuable feature is TAO's low communication the—overhead TAO developers took into account real-time data distribution issues during TAO's development.

MCS CORBA Current Issues/Regrets   The MCS does not take advantage of CORBA's many features, notably event channels, data management, and other cross-platform features. (The lack of an access-by-value mechanism caused significant latency problems for the data-driven MCS.) Despite the team's focus on rapid data transmission, it is still possible that CORBA and/or TAO introduces too much latency to these communications.

Another potential issue is the criticality of the CORBA Name Server in the MCS system. A failure of this component would force the entire MCS to be restarted, and a suitable backup mechanism has not yet been designed.

MCS CORBA Status and Future   The MCS currently uses only a few specific CORBA functions but thoroughly depends on those functions. Future work includes addressing the issues identified above and might either increase CORBA use or eliminate it entirely.

As a specific example of the alternatives, consider the MCS data handling mechanism. Initially, the MCS design rejected CORBA's data marshaling/demarshaling capabilities: the team didn't want to access data across machines. The lack of a pass-by-value feature meant up to six handshakes were required for each data item, and the packing/unpacking process seemed likely to be too slow for the system's requirements. Since the MCS needs to know and manage data structures anyway in order to input and output data, CORBA wasn't expected to add much value.

However, in the current MCS implementation the data structure is unpacked and repacked on every machine in the communication path, and the original format of the data is ignored. On top of that, pass-by-value features are becoming available in CORBA. As a result the current design assumes the overhead burden of CORBA but gains little from its data distribution and management capabilities. This design will be revisited as MCS development goes forward, to take into account lessons learned and improvement in the tools.

4.3. First Light Infrared Test Camera (FLITECAM)

The First Light Instrument Test Experiment Camera (FLITECAM) has a relatively complicated architecture with many components that are selectable during FLITECAM operation. This architecture is designed to maximize flexibility, component reusability and distributability to an unusual degree, making it more like the highly distributed MCS or DCS than a classic monolithic system for a science instrument.

FLITECAM CORBA Motivators   FLITECAM has a multi-tier distributed architecture with many small servers. The goals of this design were to support remote instrument control, and to allow redistributing processes to other CPUs as desired. It is an intensely object-oriented architecture, with a heavy Java emphasis, but support for C++ is also necessary. Finally, the developer wanted flexibility in his selection of Java development environments, which meant development and testing might take place on any platform.

FLITECAM CORBA Implementation Process   At the outset, the FLITECAM software developer had no CORBA experience. He initially compared CORBA to a solution using RMI/JNI (Remote Method Invocation/Java Native Interface) by developing prototypes using each protocol. He chose CORBA as being the easier to use, especially for certain programming tasks such as exception handling. Immediately thereafter he started building his final product, using a Windows-based development environment.

Benefits of VisiBroker for FLITECAM   Visibroker's integration with the chosen development environment, Java IDE (JBuilder 3.5) running on Windows, is extremely good. The developer found that development on Windows and operation on Solaris is feasible using this tool. The Visibroker CORBA solution has proven more powerful and simpler to use than RMI/JNI. Earlier platform independence issues appear to be fully resolved.

FLITECAM CORBA Current Issues/Regrets   An early problem with Visibroker's version compatibility (Solaris vs. Windows platforms) caused some worry about future inconsistencies, but those have declined. The developer also feels some need for more in-depth understanding of CORBA's capabilities.

FLITECAM CORBA Status and Future   Product development for FLITECAM is well under way and will soon begin core software integration with an external deliverable (containing many C++ modules). In addition, CORBA-related development will focus on pursuing a new event channel model.

5. Conclusions and Recommendations

5.1. Whether to Use CORBA

Does it make sense to use a distributed object services product such as CORBA? Most projects will fall clearly on one side or the other of the CORBA benefit-vs-risk equation. As this paper describes, run-time application coordination in an object-oriented, distributed environment will benefit from CORBA middleware, a conclusion tempered to some degree by whether or not the project has hard real-time requirements. If a data system runs strictly on a single computer, is likely to remain fundamentally stable for several years or already must support well defined interfaces between all of its subsystems, CORBA usefulness will be limited at best.

If the decision is not so clear, the authors believe that a bias in favor of trying CORBA is appropriate. This conclusion is based on four points:

- a reasonably competent software developer can learn and apply CORBA basics in a short period of time (less than a month);
- only essential or useful CORBA elements need be adopted;
- developers can replace CORBA features that prove unsatisfactory;
- it is difficult to evaluate CORBA's value without using it.

Of course, developers should expect to re-evaluate their initial decision once they have some CORBA experience, but changes of heart will not be catastrophic. With the standard's maturity and the release of public domain CORBA products, deciding about CORBA is only slightly more consequential than any other development tool decision.

5.2. Tips When Introducing CORBA

These lessons may be helpful when beginning a CORBA installation.

Seek Out Good Documentation As an example, the TAO CORBA documentation is more readable than the OMG documentation, and the documentation that comes with the VisiBroker product is also good. By now there are also many books on the various CORBA implementations.

Obtain Training As Needed Developers unfamiliar with CORBA may benefit from a short training, say three to five days. Once using the CORBA product, they will quickly go beyond most training classes.

Have Expertise at Hand When something doesn't work, it will be important to have someone to ask. While products such as TAO have responsive mailing lists, the MCS team also benefited from occasional access to CORBA experts.

Encapsulate CORBA Services In a good object oriented design, not every object will need CORBA's services; and those that do should inherit those services where possible. As a result, the developers of most of the data system's objects should not be writing CORBA code directly, and the system should minimize the number of objects using CORBA services. This approach is especially useful if one is unsure whether or not CORBA is the right tool for a data system's development.

References

Graybeal, J., Brock, D., & Papke, B. 2000, Proc. SPIE Vol. 4009-17, "The Use of Open Source Software for SOFIA's Airborne Data System"

Papke, B., Graybeal, J., & Brock, D., 2000, Proc. SPIE Vol. 4014-35, "An extensible and flexible architecture for the SOFIA Mission Controls and Communications System"

Software Engineering Practices for the ESO VLT Programme

Giorgio Filippi, Paola Sivera, Franco Carbognani

European Southern Observatory, Karl-Schwarzschild Str.2, D-85748 Garching bei München, Germany

Abstract. With the first light of its fourth telescope unit, the European Southern Observatory is successfully completing the commissioning of the main telescopes of the Very Large Telescope program, featuring four 8-meter telescopes. Two of these are already available to the scientific community.

The VLT Software Control group is proud of the fact that VLT commissioning has never suffered delays due to software. A decisive factor in achieving this was the use of software engineering practices, and this paper reports on the experiences of applying these practices to the VLT Control Software and building the group culture as a simple but effective set of standards and tools.

Main areas of interest are analysis and design methodology and tools, software configuration control, software testing, problem reporting and tracking, and documentation. The paper will also discuss what didn't work and changes that are needed and planned.

1. The VLT Project

The European Southern Observatory (ESO) Very Large Telescope (VLT) project consist of four 8-meter telescopes, fifteen instruments, and the VLT Interferometry (VLTI). The VLT Control Software project provides installation of all control and monitor functions can be characterized as follows:

- 3-4 millions lines of code
- 80 workstations for operations, ∼40 for development
- 100-120 software people involved
- 25-30 sites involved in total (12-15 at any given time)
- coding begun in 1991; planned project completion: 2003
- incremental installation and operation.

The VLT Software has been and will be reused on other projects (e.g., NTT, 3.6, 2.2, VST, .etc.).

2. Main Ideas

In 1991 when the VLT development began, the question was whether a traditional, basically no-rules approach would suffice, or whether a more structured software engineering approach was needed for a project of its size and characteristics. In retrospect, it is easy to identify the decisions responsible for the successful approach:

- The software engineering (SE)/quality assurance (QA) approach was endorsed at the outset of the project and dedicated resources (one full-time person) were allocated.
- In order to allow many people to work at many different sites in a productive way, standards were defined and put in place. Whenever possible, standards were enforced by providing tools, not simply paper definitions.
- VLT Common Software was defined to collect software solutions that could be used across the project. To be effective, the Common Software has been handled like a commercial software product: non-regression testing, backward compatibility, documentation, installation of the operating system (OS) and of third party tools (GNU, etc.) included as part of the VLT Common Software, and distribution to internal and external developers as releases once or twice per year.
- To support diverse needs, SE tools have been selected to allow development loops at different frequencies. The same tools have been used to support both twice-per-year Common Software releases and daily releases of new versions for the integration and commissioning of telescope and instruments.
- The Code Management tool, more than just supporting software development, became a major integration tool.
- Automatic testing was used extensively to validate new releases.

At the beginning very little was available; hence SE had to be put rapidly in place while the project was developing (see Table 1) in order to provide timely standards and tools for early project needs. It was found that by paying attention to the proper synchronization between the two processes, both SE and project needs could be met. This may be of interest for organizations or groups that intend to start but are afraid that SE practices have to be fully in place before starting. Of course this is preferable, but the other approach can also work.

The SE approach applied to the VLT project can be grouped in the following areas that are discussed below: Software Life Cycle, Documentation, Development Environment, (Automatic) Testing, Configuration Management, Releases, and Problem Reporting.

3. Software Life Cycle

The Software Life Cycle provides a reference model for project activities. It defines the phases of the project and the deliverables that should emerge from each of them. In view of the many interesting discussions of the pros and cons of various models described in the literature, it is important to remember that the S/W life cycle is a tool, not an end in itself! The SW Life Cycle has been described in the VLT SW Management Plan (ESO document VLT-PLA-ESO-00000-0006).

For the VLT we used the traditional waterfall model (sequence of phases: requirement, analysis, design, implementation, test, etc.) in an "incremental"

Table 1. SE and VLT Milestones.

Year	Development	Milestone
'91	Start of a SE/QA role in the software (S/W) team	VLT SW Requirements
'92	Basic development environment Documentation standards	VLT SW Specifications
'93	OS and development tools	Start Coding
'94	First VLT Common S/W release S/W Problem Reports	First external distribution
'95	Code management (cmm) Automatic test support (tat)	First Field Test (NTT)
'96	S/W process audited	Wide external distribution
'97	Use & tune	Installation at Paranal
'98	Use & tune	UT1 First Light, First external Instrument (FORS)
'99	Use & tune	UT1 inauguration, UT2 first light
'00	Use & tune	UT2 inauguration, UT3 and UT4 first light

way, i.e., identical sequences were started at different times for the different parts of the project. Every release of each major package repeated a similar sequence of phases.

4. Documentation

The S/W life cycle defines the types of documents to be produced at each stage. The main types we used were: requirement specification, functional specification (high level design), design specification (detailed design), user manual, and test procedure.

The documents followed the numbering system defined at the VLT level and all had the same layout. A document template was provided for each of the major document types.

Whenever possible, documentation was extracted from the code. This was applied mainly to interface specifications (.h files, command tables, etc.) and for "man pages" used both in the design phase as detailed description of each function and later as user documentation. The "man page" is maintained in the file that contains the corresponding code.

5. Development Environment

A standardized development environment is an essential requirement for managing different people, in different places, who are all developing software for subsequent integration. The cornerstones of the development environment have been:

- A standard variable setup obtained using normal Unix mechanisms (.cshrc, .login) and a combination of project-wide, machine-specific, and user-

specific files. Variable definitions are consistent with other standards (makefile, tools, etc.) and help to reinforce their usage.
- The definition of "Software Module" to group sets of files that provide well-defined functionality. Each software module has a unique name that becomes the stem for all named items that are part of the module (files, variables, etc.). The files are organized in a standard directory structure. The software module is the smallest Configuration Item (CI) used to build the software and therefore the basic concept for integration and configuration management.
- A directory structure as a flexible, controlled mechanism for integrating code and distributing it. It defines different areas (development, integration, release) with similar structure but different responsibility. By selecting the area via environment variables (e.g., INTROOT, VLTROOT) and by consistently applying the precedence rules in all other tools (makefile, PATH setting, etc.), the directory structure provides a flexible tool to support parallel work on different stages of the project.
- Naming conventions and coding rules that speed up communication during integration and reduce maintenance cost later on. This also saves time by providing default answers to many trivial questions.
- A controlled set of development tools (OS, GNU tools, etc.) is essential to document and standardize the installation and configuration of the OS(s) used. The same applies to all tools used to develop the software (VLT uses the GNU family, TCL/TK and some commercial products). The goal is to be capable at any moment of building a known configuration.
- Standard makefiles made up by a project-wide set of rules, centrally stored and managed, to which every developer has only to add the module-specific part: in the simplest case, the names of the files to be treated, or a complete makefile if need be. Implementation based on features of GNUmake is strictly correlated with the environment variable set-up and directory and software module conventions.

6. Testing

The most important aspects of testing are that it be automatic (to the extent feasible) and that it be conducted independently of the person whose code is under test. Test software is made of two parts: (1) a standard framework to provide the general features of creating the needed test environment (processes, communication, etc.), sequentially running a set of test programs, scripts, etc., and gathering the output, and (2) after having accounted for what can change from one execution to the other (e.g., time stamps, machine names, etc.), a means of comparing results with a reference execution.

The developer has to provide at least one test executable (from C/C++ programs and shell, tcl, etc. scripts), ordered in a so-called TestList. Because the test tool compares the current output with a previously generated one, test programs can be quite simple: execute an action and print out the result. No interpretation with error-prone "if-then-else" provisions are needed in the test program itself. Minimising test program complexity is important both to limit the cost and improve reliability. Make the automatic test affordable, and not a nightmare.

Execution of the test is then accomplished via a standard command: "cd <moduleName>/test; tat". The needed environment is created (every time!) to allow the test to be executed with the same initial conditions. All tests are executed and the result is compared with a reference file. PASSED or FAILED is the only test output. PASSED means the result of the current run is identical to the one defined as reference. FAILED means that something was different and requires that an investigation begin. A very compact output is necessary when tests are used for non-regression during release integration.

Despite the "magic" word "automatic", testing is not without its costs. Dedicated resources, both people and computers, are needed for this task in addition to the work of each developer in preparing and using the test software. The size of test software has been of the same order as delivered software.

In addition to the internally developed tool, commercial software products, like Purify, have been used.

7. VLT Common Software Releases

The VLT Common Software provided the building blocks for all applications and has been used by all internal and external development teams. To be effective it has to be readily available and reliable. This has been achieved by treating it like a commercial product rather than as a scientific prototype.

Key aspects have been: identify all components (OS, tools, ESO-developed S/W), maintain a reasonably paced upgrade cycle (6-12 months), document both in man pages and on paper, enforce backward compatibility, and impose systematic non-regression testing before release.

8. Configuration Management

The Configuration Item is the software module, and a system is made up of a list of pairs "moduleName"-"version". Normally one page is enough to specify a build for a system made up of several thousands of files.

One central archive, accessible to all sites and all people, was essential. The archive is the coordination point: if a software unit is in the archive it means that it is *ready* to be used, because it has been *tested* and all files are *consistent*.

The supporting tool (cmm) is a thin layer of procedures on top of an RCS archive. Its basic rules are:
- All files belonging to a module are handled at the same time;
- Only one person at time can modify a module;
- Branches are supported;
- There is only one archive;
- A client-server protocol implements all operations;
- Commands consist of a very simple set.

Figure 1 gives the total number of transactions executed per month over the last three years. "Archiving" accesses are when a new version of a module is created, "copying," when a read-only copy is retrieved. Note the high number

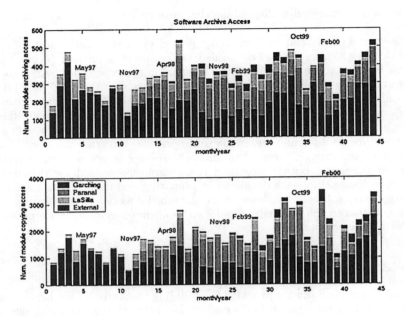

Figure 1. Total CMM Archive accesses by all sites.

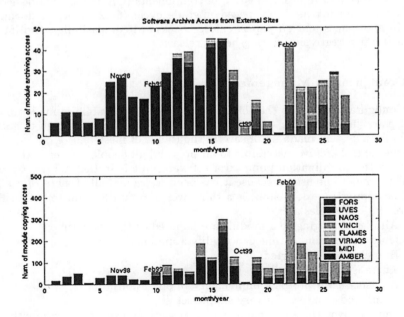

Figure 2. CMM Archive accesses by European Consortia sites.

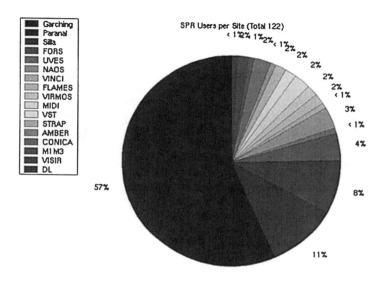

Figure 3. Software Problem Report Users per site.

of read-only copies, especially at the end of the integration process of a release or during first telescope integration and commissioning. This is a sign that regeneration from scratch was the rule. Figure 2 gives the same data but limited to transactions serving non-ESO sites, mainly European institutes that are part of one of the several Instrument Consortia providing the VLT instruments.

9. Software Problem Reports (SPRs)

Tracing bugs and modifications is a necessary complement to Configuration Management. For VLT software, a Web-interfaced central database to submit, query, and modify SPRs has been implemented using the commercial tool Action Remedy. This application supports the following work flow:

- Any user can submit an SPR. Depending on the area, some people are automatically informed.
- Every week or two, the S/W Configuration Control Board discusses all new SPRs and either assigns a responsible person and deadline or rejects it.
- The responsible person produces the fix and after having archived the modified module(s), closes the SPR.
- All status changes are broadcast to all interested parties (a Cc field allows people to follow up).
- Comments can be added by anyone at any time.

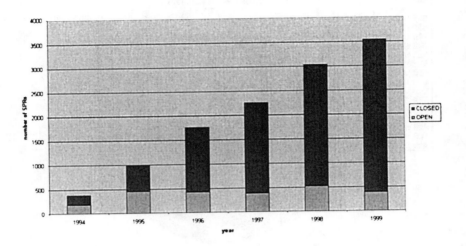

Figure 4.   Number of Software Problem Reports per year.

Figure 3 gives the total number of SPR users, i.e., people that submitted a problem or proposed a change. In principle this could be the number of software people that participate in the development and integration of the project. A bit less than 60% were generated by people from the development team (either at the Garching headquarters or on assignment at the Observatory), about 20% by technical support staff based in the Paranal Observatory, and the last 20% by people from Consortia.

Figure 4 give the cumulative number of SPRs each year and shows a continuous increase of the total number due to new software made available each year. Except for the early days, the number of open SPRs at any time was always below a critical level (about 500 SPRs) that was more or less what the team could deal with between releases.

Figure 5 gives the answer to the most important question: Is the system becoming stable, or will a maintenance team bigger than the development team be needed to maintain it? A healthy system should show a peak corresponding to the first integration, followed by a decline. In Figure 5, three different areas are compared over a three-year period: the Common Software had its "glory days" earlier and is declining as a sign of stability, while the Telescope Control Software shows its peak at the integration and commissioning of the first Unit Telescope (1998, UT1 first light). The third group includes other software packages that were at the initial stages of their life cycles in the years shown.

10.   User Reaction

In addition to crude numbers, it can also be instructive to report people's reactions. Initial feelings that ranged from indifference to hostility and rejection

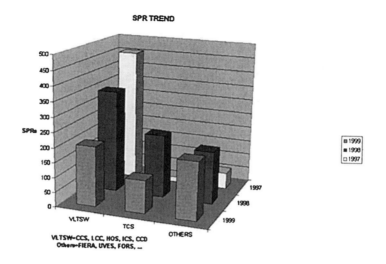

Figure 5.   Software Problem Report trends.

have completely faded away a few years later, and the SE has become a strength (and the pride) of the software team. In addition to supporting the work, SE also provided quantitative results, often needed as a way to present work done to upper management.

The established SE system was audited by an external team (in 1996) that found no deficiencies in the approach and in the implementation. Beside this, it has been positively recognized by the integration and commissioning teams of both Paranal and La Silla

A July 2000 survey asked those who had used the VLT Common Software to identify the three best and three worst aspects of the systems. Of a total of 22 responses collected, 15 people mentioned SE practices as one of the best items, while none mentioned SE among the worst aspects. Table 2 contains these reported opinions.

11. The Future

All SE practices described are currently in place and part of the daily routine. Beside keeping the SE tools up to date with the new versions of the underlying free software, there are a few major areas of development. First, the Object Oriented approach is now the rule and analysis and design have to be tuned for that. Unified Modeling Language (UML) and Use Cases are replacing the current practices for the early phases. A second area is use of the Web as the repository for documentation. This tool became widely available in the middle of the VLT S/W project and was not employed as vigorously as it should have been. Third, a stronger "testing" culture will be promoted: more support will be given to developers in terms of both tools and training in writing good test

Table 2. User opinions of VLT Software Engineering.

"Best" Rank	Characteristic
1	Programming standards
1	S/W Dev. Environment + Standards (Makefile,
1	S/W engineering (configuration, SPR, programming
1	S/W engineering, standards
1	The development environment (makefile, file
1	standard for programming, directory structure,
1	std module structure - INTROOT/VLTROOT concept
2	Configuration Management
2	Standard module structures with makefile and
2	Lots of documentation
3	Documentation
3	Test support
3	modular concept
3	cmm, vltMakefile
3	software configuration control and vltMakefile

software. Last but not least, it is time to close the loop with metrics: We have to measure in a quantitative way what we are doing and define numeric goals.

12. Conclusion

Overall, the VLT-SE approach has been a blend of ideas inspired by existing standards and some pragmatic down-to-earth choices. From the implementation point of view, standards have been using commonly available free software (RCS, GNUmake) with minimal "home made" wrappers, allowing an easy and cheap way to serve the quite widespread community that worked on the VLTI.

Standards and practices have been enforced more by means of such tools, than by "police inspections." With the exception of a few key areas in which deviations were simply not permitted, it has been the voluntary behavior of developers that has determined the results.

In summary:

- SE is necessary for a big project and should be used project-wide.
- It comes at a reasonable cost (but not for free).
- It can be accepted (and liked) by developers.
- It can be implemented in gradual steps: it is not necessary to "have it all", but using all that one has, consistently and continuously is essential!

Exploiting VSIPL and OpenMP for Parallel Image Processing

Jeremy Kepner[1]

Massachusetts Institute of Technology Lincoln Laboratory Lexington, MA 02420

Abstract. VSIPL and OpenMP are two open standards for portable high performance computing. VSIPL delivers optimized single processor performance while OpenMP provides a low overhead mechanism for executing thread based parallelism on shared memory systems. Image processing is one of the main areas where VSIPL and OpenMP can have a large impact. Currently, a large fraction of image processing applications are written in the Interpreted Data Language (IDL) environment. The aim of this work is to demonstrate that the performance benefits of these new standards can be brought to image processing community in a high level manner that is transparent to users. To this end, this talk presents a fast, FFT based algorithm for performing image convolutions. This algorithm has been implemented within the IDL environment using VSIPL (for optimized single processor performance) with added OpenMP directives (for parallelism). This work demonstrates that good parallel speedups are attainable using standards and can be integrated seamlessly into existing user environments.

1. Introduction

The Vector, Signal and Image Processing Library (VSIPL[2]) is an open standard C language Application Programmer Interface (API) that allows portable and optimized single processor programs. OpenMP[3] is an open standard C/Fortran API that allows portable thread based parallelism on shared memory computers. Both of these standards have enormous potential to allow users to realize the goal of portable applications that are both parallel and optimized.

Exploiting these new open standards requires integrating them into existing applications as well as using them in new efforts. Image processing is one of the key areas where VSIPL and OpenMP can have a large impact. Currently, a large fraction of image processing applications is written in the Interpreted

---

[1]This work is sponsored by DARPA, under Air Force Contract F19628-00-C-0002. Opinions, interpretations, conclusions and recommendations are those of the author and are not necessarily endorsed by the United States Air Force.

[2]http://www.vsipl.org/

[3]http://www.openmp.org/

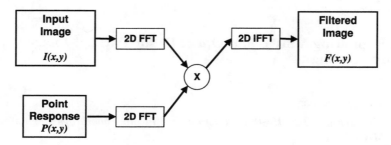

Figure 1. **Basic 2-D Filtering.** FFT implementation of 2-D filtering which performs the mathematical operation: $F(x,y) = \int \int P(x',y') I(x-x', y-y') dx' dy'$

Data Language (IDL[4]) environment. The goal of this work is to show that it is possible to bring the performance benefits of these new standards to the image processing community in a high level manner that is transparent to users.

2. Approach

Wide area 2-D convolution is a staple of digital image processing (see Figure 1). The advent of large format CCDs makes it possible to literally "pave" with silicon the focal plane of an optical sensor. Processing of the large images obtained from these systems is complicated by the non-uniform Point Response Function (PRF) that is common in wide field of view instruments. This paper presents a fast, FFT based algorithm for convolving such images. This algorithm has been transparently implemented within IDL environment using VSIPL (for optimized single processor performance) with added OpenMP directives (for parallelism).

The inputs of image convolution with variable PRFs consist of a source image, a set of PRF images and a grid which locates the center of each PRF on the source image. The output image is the convolution of the input image with each PRF linearly weighted by its distance from its grid center. The computational basis of this convolution is addition with interpolation of 2-D overlapping FFTs (see Figure 2). Today, typical images sizes are in the millions (2 K x 2 K) to billions (40 K x 40 K) of pixels. A single PRF is typically thousands of pixels (100 x 100) pixels, but can be as small 10 x 10 or as large as the entire image. Over a single image a PRF will be sampled as few as once but as many as hundreds of times depending on the optical system.

There are many opportunities for parallelism in this algorithm. The simplest is to convolve each PRF separately on a different processor and then combine all the results on a single processor. This approach works well with VSIPL, OpenMP and IDL (see Figure 3). At the top level a user passes the inputs into an IDL routine that passes pointers to an external C function. Within the C function OpenMP forks off multiple threads. Each thread executes its convolution using VSIPL functions. The OpenMP threads are then rejoined and the

---

[4]http://www.rsi.com/

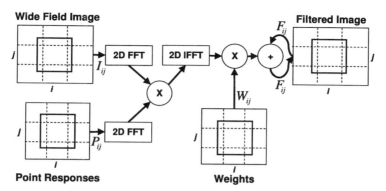

Figure 2. **Wide Field Filtering.** FFT implementation of 2-D filtering for wide field imaging with multiple point response functions. Each portion of image is filtered separately and then recombined using the appropriate weights. The equivalent mathematical operation is:
$F_{ij}(x,y) = W_{ij}(x,y) \int \int P_{ij}(x',y') I_{ij}(x-x', y-y') dx' dy'$

results are added. Finally a pointer to the output image is returned to the IDL environment in the same manner as done by any other IDL routine.

3. Results

This algorithm was implemented at Boston University[5], on an SGI Origin 2000 (64 300 MHz MIPS 10000 processors with an aggregate memory of 16 GB). IDL version 5.3 from Research Systems, Inc. was used along with SGI's native OpenMP compiler (version 7.3.1) and the TASP VSIPL implementation. Implementing the components of the system was the same as if each were done separately. Integrating the pieces (IDL/OpenMP/VSIPL) was done quickly, although care had to be taken to use the latest versions of the compilers and libraries. Once implemented the software can be quickly ported via Makefile modifications to any system that has IDL, OpenMP, and VSIPL (currently these are SGI, HP, Sun, IBM, and Red Hat Linux). We have conducted a variety of experiments that show linear speedups using different numbers of processors and different image sizes (see Figure 4). Thus, it possible to achieve good performance using open standards underneath existing high level languages.

---

[5] http://scv.bu.edu/

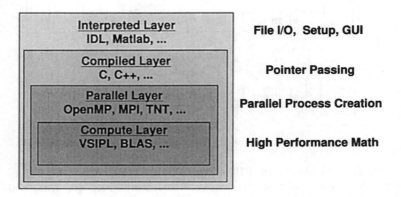

Figure 3. **Layered Software Architecture.** The user interacts with the top layer, providing a high level of abstraction for high productivity. Lower layers provide performance via parallel processing and high performance kernels.

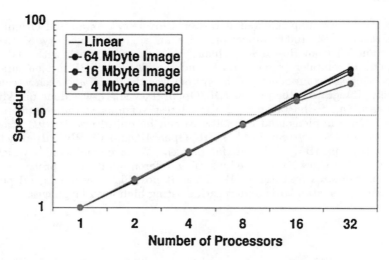

Figure 4. **Parallel Performance.** Measured speedups of wide field 2-D filtering application on a shared memory parallel system (SGI Origin2000). Results indicate linear speedups are achievable using open software standards underneath high level programming languages.

CORBA as an Interoperability Tool for Astronomy

Marc Wenger

Centre de Données astronomiques de Strasbourg, France

Laurent Frisée

Haute Ecole Renequin Sualem, Liège, Belgium

Abstract. Object oriented programming is becoming increasingly important. At the same time, protocols for managing network classes are developing. Independent from any vendor, the CORBA protocol is a major player in this domain, making it a good candidate for managing interoperability between astronomical data providers and other applications. The protocol allows class providers and users to operate in different and independent environments with CORBA managing all communications in a transparent way.

This paper presents the prototype developed at CDS using CORBA for interconnecting different services in a heterogeneous context of several operating systems and different languages. Ease of development and problems encountered are assessed.

1. CORBA

Client/server architecture needs tools to manage communication between the different components. In traditional programming, sockets and Remote Procedure Calls are employed. In object oriented programming, CORBA has become a standard tool that offers remote execution of object methods for specifically designed CORBA classes. It is an interface specification designed by the Object Management Group (OMG[1]), a consortium linking both commercial companies and academic institutions from the object world. It ensures a standard independent from any vendor. Many CORBA environments are available, some commercial like Orbix or Visibroker, others free like omniORB, ORBACUS, and many others.

A CORBA class is defined by its interface, specifying in the Interface Data Language (IDL), the methods, their parameters and return values to be used on the client side and to be implemented on the server side. An IDL compiler produces the *stub* code needed by the client programme to call the object method implemented on its side, and the *skeleton* code interfacing the implemented methods on the server side.

---

[1] http://www.omg.org

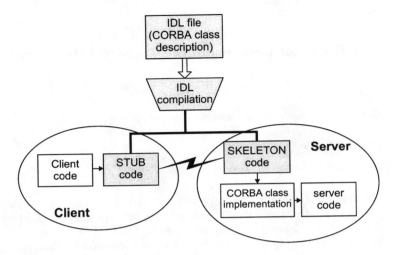

Figure 1.   CORBA architecture

Once the stub is available on the client side and the CORBA object is created, the methods are called in the same way as if they were *normal* objects defined and used in an application. The server developer implements the methods, i.e., writes their code as for any class. Figure 1 shows this architecture.

One of the main characteristic of CORBA is its ability to work well in an heterogeneous world: client and server may run on different hardware under different operating systems, have their applications written in different languages and be implemented in different CORBA environments from different providers.

This makes CORBA particularly well suited for astronomy where every data provider and user has their own computing environment. CORBA allows communication in every situation and facilitates the development of interfaces for interoperability between data providers.

2.   Interoperability in Astronomy

There are already many interactions between applications in astronomy: services like NED, ADS or CDS/SIMBAD have developed packages allowing client/server queries. Such packages generally require use of a specific language and have a proprietary protocol.

It is obvious that the need for interoperability will increase in the future: Virtual Observatories, Data Mining tools, Remote Observing and Data integration systems are developments that will require interaction between heterogeneous services to improve their functionality and efficiency (Graybeal 2001).

Before using CORBA for interoperability on a large scale, the CDS began by experimenting with it for internal services that have long been interoperating through classical client/server architecture.

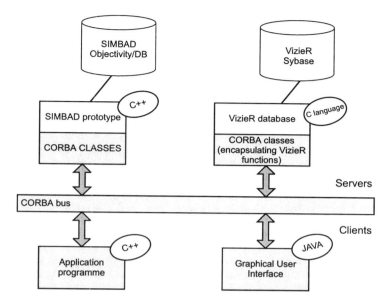

Figure 2. CORBA prototype at CDS

3. The CORBA Experiment at CDS

The experiment consisted of two server extensions (see Figure 2): (1) addition of CORBA classes to the SIMBAD prototype using Objectivity/DB (written in C++ so that the required CORBA classes were easy to develop; Wenger et al. 2000) and (2) encapsulation of the VizieR interface (written in C and using the Sybase relational DBMS; (Ochsenbein et al. 2000)) for querying catalogues by identifier.

Two clients were developed in this experiment: a simple server package written in C++ for querying SIMBAD through the CORBA class, and a graphical user interface written in JAVA to query both SIMBAD and VizieR through the CORBA classes.

This prototype incorporated several key features:

- Use of legacy code in different languages, demonstrating the ease of encapsulation via CORBA classes.
- Connection with database management systems of different DBMS technologies – Sybase as relational and Objectivity/DB as object oriented. (This revealed interaction problems between a DBMS and CORBA.)
- Development of different clients using different languages, namely C, C++ and JAVA, to assess the heterogeneous implementation of CORBA.

4. Lessons Learned

The entire system was developed and implemented within a few months. Use of different subsystems in the prototype uncovered three interesting technical prob-

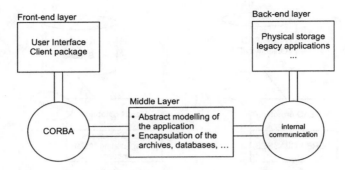

Figure 3.   Three-tier architecture

lems: (1) an incompatibility between the compilers required by a CORBA tool and Objectivity/DB, (2) an interference between the Standard Template Library (STL: template classes provided only as C++ header files) and a mini-STL used by one CORBA environment (MICO) and (3) a "buggy" interaction between the threading mechanisms used by the CORBA tools and by Objectivity/DB.

Bypassing these problems required moving from one CORBA tool to another and also implementing a three-tier architecture – this also allowed better independence between the server components (see Figure 3), leaving CORBA classes between fully mastered pieces of code and helping to solve some incompatibilities with pre-existing packages.

5. Next Steps

A follow-up to this experiment could consist of defining some standard classes for querying each concerned service. Query language, parameter syntax and return values build the standard. The XML language could be an important element of this standard for exchanging data through CORBA classes. Distributing an IDL file is enough to allow users to access the CORBA services.

Generalization to more applications would be a first step towards the definition of business classes in Astronomy like those that already exist in other disciplines such as Medicine (CORBAmed) and Biology (Life Science Research).

References

Ochsenbein, F. et al. 2000, A&AS, 143, 23
Wenger, M. et al. 2000, in ASP Conf. Ser., Vol. 216, Astronomical Data Analysis Software and Systems IX, ed. N. Manset, C. Veillet, & D. Crabtree (San Francisco: ASP), 247
Graybeal, J. 2001, this volume, 189

Specifics on a XML Data Format for Scientific Data

Ed Shaya,[1] Brian Thomas

Raytheon ITSS/NASA/ADC

Cynthia Cheung

NASA/ADC

Abstract. An XML-based data format for interchange and archiving of scientific data would benefit in many ways from the features standardized in XML. Foremost of these features is the world-wide acceptance and adoption of XML. Applications, such as browsers, XQL and XSQL advanced query, XML editing, or CSS or XSLT transformation, that are coming out of industry and academia can be easily adopted and provide startling new benefits and features. We have designed a prototype of a core format for holding, in a very general way, parameters, tables, scalar and vector fields, atlases, animations and complex combinations of these. This eXtensible Data Format (XDF) makes use of XML functionalities such as: self-validation of document structure, default values for attributes, XLink hyperlinks, entity replacements, internal referencing, inheritance, and XSLT transformation. An API is available to aid in detailed assembly, extraction, and manipulation. Conversion tools to and from FITS and other existing data formats are under development. In the future, we hope to provide object oriented interfaces to C++, Java, Python, IDL, Mathematica, Maple, and various databases. http://xml.gsfc.nasa.gov/XDF

1. Introduction

The eXtensible Data format is an XML-based language for describing and encapsulating scientific data. Its primary purpose is to be the mathematical and computer science kernel for other, more fully-featured, discipline-oriented formats. With the advent of global acceptance of XML as the mark up language for data, now is the time to rethink our data formats. It is a time when cross disciplinary interchange of data is greatly needed. At NASA there is a pressing need for earth scientists, solar physicists, astronomers, and biologists to communicate and to share data and processes. XML provides a common set of rules for documents that are a great boon to interoperability. A data format grounded

---

[1]Physics Department, U. of MD

in XML will not only be understandable to other scientists but could also be crafted so it can be viewed by the public with common browsers.

XML is a very important breakthrough in documentation and metadata expression. It brings in or formalizes the concepts of self-validation, self-description, recursive variable substitution, separation of presentation from information, hierarchical structures, the multi-file document, and standard multilangual character sets and tagging. It continues to grow in popularity. It is not likely that any of these important concepts in documentation will ever disapear.

The development of XML techniques now will, in the end, result in great cost benefits. Tools designed to use XML will likely be useable by other groups with minor adjustments. This is because all XML data are parsed with standard parsers and converted to a standard Document Object Model (DOM), which has standard application programming interfaces.

But we can go one step further in interoperability by accepting a common XML language specifically for data description. The present lack of a common object model of data and data description has made it very difficult to develop tools useful to a wide audience. Cross-disciplinary search is nearly impossible with the current technology.

2. The eXtensible Data Format (XDF)

An XDF document describes a structure of related scientific data. Our goal is to develop an XML language that reasonably covers the needed mathematical and computer science constructs in a generic data model. Discipline specific descriptors are left to specific languages that can inherit XDF capabilities. This allows the essential and common data structures to be represented in a consistent manner independent of the scientific specialty involved.

In XDF, data are stored in a manner that accurately reflects scientists' views of data. Most scientific numeric data can be seen as an object assembled from objects of two catagories: (1) a simple parameter set to a single value, possibly plus or minus infinity, or a range of values, possibly of infinite extent (e.g., $x = 3.1$ or $0 < y < 180$), or (2) gridded samples of scalar or vector fields embedded in an N-dimensional space (field arrays). These include tightly sampled grids such as spectra, images, animations, or time-series measurements. And, they include sparsely sampled fields such as interferometric u-v maps, event detection, or sets of pointed telescope observations.

Examples of N-dimensional spaces include physical space, projected space, time, wavelength, frequency, energy scales, or some other parameter space. Complex numbers can be considered as vectors in the complex plane.

These two basis objects can be assembled in several ways, into lists of lists or field arrays. A record is a list of values for a selection of parameters for a particular item or target. A list of these records is a table.

Because observational data always has some finite resolution, it is always gridded (sometimes variably). Therefore both field arrays and tables can be represented similarly as ordered N-dimensional data cubes. Software can take advantage of this similarity by sharing input/output methods between the two. In fact much of the data handling can be similar: subsetting, taking cross-sectioning, etc. However, the fundamental difference between the two categories is the fact that tabled properties rarely form a continuous space and therefore interpolation and analyses depending on interpolation are not sensible.

One of the key concepts in object oriented methodology is that data should be wrapped with the information necessary to read it and to make it useful. The XML language allows for this by including in data documents either references to applications or code in the form of ECMAscript or Java. An XML document can have references to files containing data and different types of data files can be handled by different applications. This not only allows input and preliminary processing to be self directed, but it also allows some of the data to be generated on the end users' machines. Eventually XDF will include functions for calculating values along axes, and if necessary, calculating positional information for every grid point. When applications are referenced or embedded within the data documents, it greatly reduces the learning curve needed to begin working with scientific data.

3. Examples

Here is a simple example:

```
<XDF>
   <structure name="structure A">
      <array name="array I" description="reduced data">
         <units>&Jansky;</units>
         <dataFormat>
            <float width="10" precision="5"/>
         </dataFormat>
         <axis name="x" axisId="x">
            <valueList count="512" start="23.32" step="2.22"/>
         </axis>
         <axis name="y" axisId="y">
            <valueList count="1024" start="12.11" step="2.26"/>
         </axis>
         <read readId="format">
            <for axisId="x">
               <for axisId="y">
                  <readCell/>
                  <skipChar/>
               </for>
            </for>
         </read>
         <data href="imagefile1"/>
      </array>
      <array name="array b" description="flat field">
         <unitless/>
         <dataFormat>
            <integer width="5"/>
         </dataFormat>
         <axis axisIdRef="x"/>
         <axis axisIdRef="y"/>
```

```
            <read readIdRef="format"/>
            <data href="imagefile2"/>
        </array>
    </structure>
</XDF>
```

One sees that the data description takes only a few minutes to understand. This is a structure with two arrays of data. Each is a 2-D image with axes "x" and "y" and the axis values are specified with start and step attributes. The second array reuses the information from the first one for axis and read instructions. The read section tells one to increment the x value after each row of y is read. The dataformat is F10.5 for the first array and I5 for the second with a single ignorable character between numbers. The units are Janskys. An entity &Jansky; is automatically replaced from an entity list with its SI equivalent. All XDF data end up in SI units in this way.

In our data model the axes are critically important. High resolution images can be subsections of the low resolution ones. Understanding how to do that comes from the fact that the axes data are placed in a well established manner. Axes from one array can point (using axisIdRef mechanism) to ones in other arrays to indicate that they are parallel or aligned. Arrays can be merged or appended to other arrays. Each component of vectors indicate which axis they line up with.

High dimensional tables are possible. A group of tables with the same layout that differ essentially by the value of some parameter would be represented by adding another dimension specifying the varying parameter. Data that need to be split at variable sizes can be segmented with each segment having a name. Field headings can also be grouped by higher level field names.

Other features include parameters with scoping, and inheritability. Another paper (Thomas, Shaya, & Cheung 2001) describes the FITSML language which has astronomical keywords and inherits XDF properties.

4. Future Versions

The XDF is being used at the Astronomical Data Center where it is getting a good solid testing of a large and disparate set of data. It is working out well but we welcome any suggestions from the scientific community for improvement or new ideas to be included in future versions. We expect new versions to come out every so often. There will be no problems with earlier version documents because applications can easily include XSLT scripts that transform earlier versions to newer ones immediately before processing. We feel it is quite possible that in a short time one data format will take you from observation, through XML manuscript creation http://xml.gsfc.nasa.gov/article, to XML query of archives.

References

Thomas, B., Shaya, E., & Cheung, C. 2001, this volume, 487

Declarative Metadata Processing with XML and Java

Damien Guillaume, Raymond Plante

University of Illinois Urbana-Champaign, Urbana, IL 61801

Abstract. Metadata processing is an essential part of the Astronomy Digital Image Library (ADIL) and the BIMA Image Pipeline because of their complex and evolving data models. In order to choose the best technical solutions for these systems, a number of techniques have been evaluated; XML, Java, XSLT, DOM and Quick are described here.

An increased use of declarative (as distinguished from procedural) knowledge is found to improve the stability and flexibility of such systems.

1. Introduction

The Astronomy Digital Image Library[1] (ADIL) and the Berkeley Illinois Maryland Association (BIMA) Image Pipeline[2] are two projects carried out at the National Center for Supercomputing Applications (NCSA). Both projects make a heavy use of metadata, i.e., information about the processed data. The metadata is small in size compared to the processed data, but can be fairly complex, and the models can evolve with time as new features become available. The eXtensible Markup Language (XML) is well suited for encoding metadata because it supports flexible complex structuring, structure validation, and a format that can be read directly by both people and machines. Java is adapted to XML data because it is a solid object-oriented language, and many software packages for XML are written in Java.

This article explains a number of ways to use XML and Java to process metadata that have been tried for the ADIL or the BIMA Image Pipeline. It also highlights the value of using declarative knowledge, as opposed to procedural knowledge, to process the metadata.

2. Declarative vs. Procedural Knowledge Representation

The difference between declarative and procedural knowledge is somewhat subjective because of the possibility of converting one into the other. However, some relative comparisons of a number of characteristics show a clear difference between the two. In this article "declarative" and "procedural" will be used in the sense defined by the comparisons of Table 1.

---

[1] http://adil.ncsa.uiuc.edu/

[2] http://monet.astro.uiuc.edu/BIP/index.html

Table 1. Declarative vs Procedural Knowledge

DECLARATIVE KNOWLEDGE	PROCEDURAL KNOWLEDGE
Easy to validate	Hard to debug
Glass-box	Black-box
Explicit	Obscure
Data-oriented	Process-oriented
Ability to use knowledge in ways that the system designer did not foresee	Extensions are dangerous for stability
Slow (requires interpretation)	Fast (direct execution)
May require high-level data types	simple Data types can be used

3. Using XSLT

Extensible Stylesheet Language Transformations (XSLT), a part of the Extensible Stylesheet Language (XSL)[3], is a language to define XML tree transformations. XSLT 1.0 stylesheets are purely declarative, in the sense described in Table 1. Since they describe the transformations using only constants (without variables), a loop variable is not possible with XSLT. An XSLT processor is used to read the XSLT stylesheet and transform XML files into other XML files or HTML[4] files.

XSLT is very powerful for making tree transformations such as moving a subtree, changing the structure, adding subtrees, or formating the result. Some operations, however, are difficult or even impossible. Impossible operations often occur when a variable is needed, but some simple data processing operations can make the stylesheet very hard to read (even worse than a piece of code). A good way to solve these problems is to define some new high-level operations in a programming language like Java and use them in the stylesheet (though this might not be possible with any XSLT processor).

At some point, it becomes necessary to use some procedural knowledge. This limit can be pushed by using the right high-level data types. With XML already reasonably well advanced, XSLT provides a way to define a number of operations in a declarative way. One hopes that progress in the definition of XML (e.g., schemas, better data types) will improve XSLT rapidly.

---

[3] http://www.w3.org/Style/XSL/

[4] http://www.w3.org/MarkUp/

4. XSL Transformations in the ADIL

The ADIL is using the Astronomical Markup Language[5] (AML), an XML language, to store the metadata. The main operation of the ADIL server is to display some metadata in HTML, and XSLT is very well suited for this. Since not all the metadata is displayed and some metadata require complex transformations prior to display, it made sense to separate the transformation into two steps: first, the creation of an intermediary XML file containing only the information to display, and then the transformation of this XML file into HTML, with the addition of the user interface.

Since the complex transformation needed to realise these two steps proved impossible with XSLT, a Java program had to be used to transform the project metadata. Alternatively, the complex transformations could have been handled through special purpose Java methods callable from an XSLT stylesheet.

5. APIs for XML

There are two standard Application Programming Interfaces (APIs) for XML: Simple API for XML (SAX)[6] is a de-facto standard for generating parsing events, and Document Object Model (DOM)[7] is W3C[8] recommended for handling XML trees in memory. DOM parsers are often based on SAX parsers, but other possibilities include using a custom interface or creating a new interface based on SAX or DOM. The choice depends on the required performance: applications based only on SAX are faster than the those based on DOM parsers, but application-specific code has to be added to handle XML trees. Custom interfaces added on top of DOM can be very stable and powerful, but the result is very inefficient because of all the intermediate layers.

6. Using a DOM Parser

The declarative information used by a validating DOM parser (apart from the XML files) is mainly the Document Type Definition (DTD), describing the XML language of the documents to process. Using a DOM parser results in procedural knowledge (like a Java program) based on DOM objects. If no custom API is used on top of DOM, the types of the objects usually do not reflect their meaning in the application and cannot contain associated methods (e.g., all data defined as strings, even when numbers are used). Another problem with the DOM API is its complexity: this has led some developers to create alternative interfaces, like JDOM[9], a DOM-like interface simplified and optimized for Java.

---

[5] http://monet.astro.uiuc.edu/~dguillau/aml/

[6] http://www.megginson.com/SAX/sax.html

[7] http://www.w3.org/DOM/

[8] http://www.w3.org/

[9] http://jdom.org/

7. Using Quick

A good alternative to DOM parsers is Quick[10], open-source software initiated by JXML, Inc[11]. Quick transforms XML documents into Java objects automatically with a configuration file that is written in the QJML markup language and contains both the XML schema for the XML documents (XML schemas are next-generation DTDs containing type information) and the mappings between XML elements and Java classes. The mappings describe objects called "Coins" because they have a Java side and an XML side.

Applications using Quick make use of procedural knowledge, as they would with DOM parsers, except that they use Java classes adapted to the application, and the objects can be used in an explicit way. More declarative knowledge is used with Quick because the QJML file contains XML-Java mappings, while DOM parsers only use a DTD. Because this declarative knowledge can be used to replace some procedural knowledge, a utility for Quick (QJML2Java) was written to create Java classes based on the QJML file. These generated classes usually contain fields for the attributes or sub-elements, accessor methods, and some other useful XML-related methods. They can be extended with custom classes so that custom fields and methods are added to the object interfaces, while the link between the generated classes and the QJML file is preserved. If the schema must be updated (e.g., to add a new attribute to an element), the classes can be regenerated.

8. Conclusions

Before switching to Java, the BIMA archive was using Perl scripts to handle metadata, and much of the system information was hard-coded. The introduction of more declarative knowledge, by explicitly defining the objects to use, what information they contain, and how to access it, resulted in more stable, more flexible, and (unexpectedly) faster software.

Generating Java classes with an XML configuration file effectively replaces procedural knowledge (hand-written Java classes) with declarative knowledge (a configuration file that can be used to generate the same classes automatically).

With the help of higher-level data types, further improvements in this direction can be expected. Already fairly evolved, XML has greatly assisted the creation of declarative knowledge, as demonstrated by the XSLT language.

---

[10] http://jxquick.sourceforge.net/

[11] http://www.jxml.com/

Astronomical Data Analysis Software and Systems X
ASP Conference Series, Vol. 238, 2001
F. R. Harnden Jr., F. A. Primini, and H. E. Payne, eds.

Funtools: An Experiment with Minimal Buy-in Software

Eric Mandel, Stephen S. Murray, John Roll

Smithsonian Astrophysical Observatory, Cambridge, MA 02138

Abstract. Minimal buy-in software seeks to hide from its users the complexity of complex code. This means striking a balance between the extremes of full functionality (in which one can do everything, but it is hard to do anything in particular) and naive simplicity (in which it is easy to do the obvious things, but one can't do anything interesting). Minimal buy-in acknowledges that design decisions must be made up-front in order to achieve this balance.

Reported here are recent efforts to explore minimal buy-in software through the implementation of Funtools, a small suite of FITS library routines and analysis programs that attempts to hide the complexity of FITS coding. In particular, the use of "natural order" in the Funtools design enables it to "do the right thing" automatically.

1. Introduction

The Funtools[1] project arose out of conversations with astronomers about the decline in their software development efforts over the past decade. A stated reason for this decline is that it takes too much effort to master one of the existing FITS libraries simply in order to write a few analysis programs. This problem is exacerbated by the fact that astronomers typically develop new programs only occasionally, and the long interval between coding efforts often necessitates re-learning the FITS interfaces.

The goal was to develop a minimal buy-in FITS library for researchers who are occasional (but serious) coders. In this case, "minimal buy-in" meant "easy to learn, easy to use, and easy to re-learn next month". Conversations with astronomers interested in writing code indicated that this goal could be achieved by emphasizing two essential capabilities. The first was the ability to write FITS programs without knowing much about FITS, i.e., without having to deal with the arcane rules for generating a properly formatted FITS file. The second was to support the use of already-familiar C/Unix facilities, especially C structs and Unix stdio. Taken together, these two capabilities would allow researchers to leverage their existing programming expertise while minimizing the need to learn new and complex coding rules.

For example, the authors' group at SAO pursues research in X-ray astronomy, where the primary data are stored as rows of photon events in FITS binary

---

[1]An acronym for Fits Users Need TOOLS

tables. Each row consists of information about a single photon event, e.g., position, energy, and arrival time. In order to process photons in an X-ray table, one typically reads the events from stdin into an array of C records, manipulates these event records and then writes the results to stdout:

```
while( (nev=fread(ebuf, sizeof(EvRec), MAXEV, stdin)) ){
  for(i=0; i<nev; i++){
    ev = ebuf + i;
    if( ev->pha == XXX ) ev->energy = NewPHA(ev->pha);
  }
  fwrite(ebuf, sizeof(EvRec), nev, stdout);
}
```

In essence, the goal of the Funtools effort was to implement this code fragment as simply as possible for FITS binary tables.

2. Design Approach

The Funtools design approach was to *minimize the number of public subroutines* in its library. This approach was based on the hypothesis that a few carefully crafted routines would be easy to learn, easy to remember, and easy to use by occasional coders. The aim was to make it easier to "see the forest for the trees" by minimizing the number of trees.

The implications of this approach were two-fold. Firstly, the routines would have to be useful in their basic operation, while containing appropriate hooks to activate more sophisticated functionality. Routines would not be automatically added to the library in order to extend functionality. Instead, optional keyword specifiers were added to the calling sequence of basic routines, taking care to avoid making those routines overly complex along the way.

Secondly, it was decided that routines should act on behalf of the coder, especially with regard to FITS formatting issues (e.g., header generation and extension padding). Since proper formatting of FITS files is one of the most difficult and tedious parts of FITS programming, it is very desirable to have Funtools provide this service automatically. For example, when image or binary table data is written, the associated headers should be generated automatically as needed. In doing so, great care must be taken to ensure that all necessary parameters (including user-specified parameters) are correctly incorporated into the header. Designing such automatic services required maintaining the "state" of processing for a given FITS file, so that the library could "do the right thing" on behalf of the user.

Adhering to these design principles required the imposition on coders of a few rules of "natural order". Of course, order is important for any I/O library: one cannot read a file before it is opened or after it is closed. But in the current case, we decided to make this concept explicit. For example, in order to ensure that headers are automatically and properly generated when an image is written to a FITS extension using stdio, all parameters are required to be written before outputting the image data itself. It was hoped that users would accept such rules if they resulted in services being performed automatically and correctly. Examples of "natural order" are discussed below.

3. Implementation

The implementation strategy was centered on perfecting the code needed by a select few image and table processing algorithms, including the canonical X-ray analysis task: for each X-ray event (row in a binary table), read selected input columns into user space, modify the value of one or more of these columns, and output the results by merging new value(s) with the original input columns. The Funtools library implements this standard X-ray event-processing algorithm in a simple and straightforward manner:

```
typedef struct eventrec{ int pha, energy; } *Ev, EvRec;

ifun = FunOpen("in.fit[EV+,pha=5:9||pha==pi]", "rc", NULL);
ofun = FunOpen("out.fits", "w", ifun);

FunColumnSelect(ifun, sizeof(EvRec), "merge=update",
   "pha",    "J", "r",  FUN_OFFSET(Ev, pha),
   "energy", "J", "rw", FUN_OFFSET(Ev, energy), NULL);

while( (ebuf=(Ev)FunTableRowGet(ifun, NULL, MAXEV, NULL, &nev)) ){
  for(i=0; i<nev; i++){
    ev = ebuf + i;
    if( ev->pha == XXX ) ev->energy = NewPHA(ev->pha);
  }
  FunTableRowPut(ofun, ebuf, nev, 0, NULL); free(ebuf);
}
FunClose(ofun); FunClose(ifun);
```

The Funopen() routine opens the specified FITS image or binary table extension and sets up the required column and spatial region filters. As shown above, the "output" call can be passed an input handle (argument 3) to inherit parameters and other information from the input file. In addition, if the input file is opened in "copy" mode ("c" in argument 2), the user can automatically copy the input extensions to the output file by appending "+" to the extension name ("EV+" in the example). The "natural order" rule here is that the input file must be opened before the output file.

The heart of Funtools processing for binary tables (i.e., X-ray events) is the FunColumnSelect() routine, which specifies how to read columns into a user-defined C struct and/or how to write columns to an output file. Named columns (arguments 4, 8, etc.) are read into the record structure offsets (arguments 7, 11, etc.) They are automatically converted from the FITS data type to the user-specified data type (arguments 5, 9, etc.) In addition, if the output file has inherited the input handle, then the output file also will know about columns having "w" mode. Finally, the optional "merge=[update/replace/append]" string (argument 3) specifies that processed columns will be merged with the original input columns on output. Thus, the same subroutine can be used to set up reading, writing, and/or merging of named columns.

Once columns from the binary table have been selected for processing, the routines FunTableRowGet() and FunTableRowPut() are used to get and put rows. (Similar routines are available for image processing.) FunTableRowGet() reads event rows and stores the selected columns into an array of structs. Buffer space is allocated on the fly if argument 2 is NULL. FunTableRowPut() writes the event rows to the output file. Note that FITS headers are generated and written automatically as needed. To support these automatic services, the natural order rule is that get and put calls must be alternated.

Finally, the FunClose() routine is called to flush and close Funtools files. Note that FITS extension padding will be added automatically as needed (although one can call FunFlush() explicitly). Also note that the remaining input extensions are copied automatically to the output if in "copy" mode. To support these services, the natural order rule is to close the output file before the input file(s) to copy remaining extensions.

In Funtools, much of the complexity of dealing with FITS is hidden. FITS headers and extension padding are written automatically as needed. Output files easily inherit input parameters and other extension information. Copy of input extensions is specified easily on the command line. The price paid for these automatic services is the imposition of some rules of "natural order," although options are available for cases where coders are forced to violate these rules. For example, if input files must be closed before output files, coders can call FunFlush() beforehand to copy the remaining input files explicitly.

4. Discussion

Initial response to the Funtools library has been very positive, with researchers reporting that they often can write programs of considerable complexity in less than a day. Indeed, the early success of Funtools naturally leads to the question of how far a library of this sort can go. Are there limits to its functionality, beyond which a more traditional (i.e., complex) library is needed? Do the rules of "natural order" eventually impose unacceptable restrictions on the library or are they no more than common sense restrictions that everyone can follow? These and other issues can be explored during the further development of Funtools.

Funtools is available at SAO/HEAD R&D Group page.[2] It has been ported to Sun/Solaris, Linux, Dec Alpha, SGI, and Windows (using Cygwin). The Funtools library supports uniform access to FITS tables and images, and to non-FITS arrays and raw event lists. A number of sample programs are also offered; for example, funcnts calculates the background-subtracted image counts in user-specified regions. As always, suggestions and comments are welcome.

Acknowledgments. This work was performed in large part under a grant from NASA's Applied Information System Research Program (NAG5-9484), with support from the (Chandra) High Resolution Camera (NAS8-38248) and the Chandra X-ray Science Center (NAS8-39073). We wish to thank Leon VanSpeybroeck and Francis A. Primini for many helpful discussions.

---

[2]http://hea-www.harvard.edu/RD

Embedded Astrophysics Query Support Using Informix Datablades

A. Zhang, T. Handley

Infrared Processing and Analysis Center, California Institute of Technology, Pasadena, CA 91125

Abstract. The 1.2 billion stars in the Two Micron All Sky Survey (2MASS) working dataset provide significant science opportunities and accompanying database challenges. Effective and efficient access to large datasets such as these is an important service of the Infrared Science Archive (IRSA). By embedding domain-specific query support into a query engine, IRSA provides a significant step toward more efficient queries. This paper describes IRSA's new generation query support in which the Informix server has domain-specific embedded query support, e.g., datablades modules with astronomical functionality. The first IRSA Astronomical datablade developed supports coordinate conversions.

This Astronomical datablade provides scientists, projects and the public with embedded coordinate conversions among common astronomical coordinate systems. Supported conversions include Equitorial, Ecliptic, Galactic, and Super Galatic, including the conversion between Julian and Besselian. This enables data retrieval with no intermediate or client-side processing steps – the user retrieves data from the database as usual. This capability is being deployed to enhance the current IRSA general query support services.

1. Introduction

With dramatic data volume growth in astronomical observations, a new millennium of information retrieval is fast approaching. Such a new era challenges the astronomical community to move to more advanced computer technologies in data mining, data management, data archiving and data analysis. A more creative and efficient way of data retrieval combined with required data analysis is needed. IRSA's astronomical datablade module is one of the solutions for addressing such technical challenges.

2. Datablade Technology

The fundamental datablade module is a software package. It can define any functionality required. Essentially an embedded module is used to extend the intrinsic functionality of Informix Server by implementing user-defined data types and their supporting routines. The science community or new missions are able to define their own data objects and manipulate those database objects using their own analytic methods in a natural, flexible way.

The basic datablade includes a set of Structural Query Language (SQL) statements and a set of supporting code written in an external language such as C. The datablade accepts user-defined database objects that extend the SQL syntax and its commands.

In addition to the above, generally speaking, datablades provide better performance and simpler client-side applications. Datablade modules handle code for manipulation and storing data, so the application does not have to include low-level resources. Furthermore, datablade module routines and data types can be accessed using SQL as other intrinsic functions and data type. Finally, datablade modules are easy to upgrade.

IRSA has extended the Informix database server with an astronomical coordinate conversion capability. Astronomical coordinates are converted and processed within the database server instead of within a client-side application.

3. Coordinate Conversion Datablade

Scientists, researchers and engineers who work in astronomical projects deal with coordinate conversions on a daily basis. In the conventional way, the coordinate related data are just loaded into the database. In most cases, in order to consider the efficiency of data storage, archives ingest coordinates in a common coordinate system, such equatorial J2000. In order to support such processing in a pipeline or data analysis, coordinate conversions are required.

If the coordinates that users need are not stored in the database table, a client-side program is required to accomplish the coordinate conversion. This step cannot be eliminated, since the database serves only as a storage and search machine.

In some cases, database tables are designed to give a certain degree of flexibility by containing additional coordinates. This approach only partially solves the problem and may actually raise other archive issues.

Coordinate conversion is a complicated process. The resulting coordinate pair (RA, Declination) depends on observation time, epochs of the FROM and TO coordinate systems, and various correction conditions. Ingesting more than one coordinate system may meet some users' immediate needs. However, this does not satisfy the general users' demands, since there is no way to ingest all of the coordinate values required. One extra coordinate in a database table, like the 2MASS working database table, would require an additional 20 GB of disk space. Just storing extra columns to meet scientific requirement is not the solution for the problem by its nature. And, overgrown tables will result in serious efficiency and storage problems. However, a database table that does not store extra columns can result in more steps to accomplish a job.

4. Datablade Design and Functionality

The coordinate conversion datablade (CNV) transforms coordinates among common astronomical coordinate systems including the conversions between Julian and Besselian. It also supports the conversion with specified proper motions, unknown proper motion or radio source, or without proper motion. For position angle calculation, the datablade handles both epochs of position angle, which of course may differ. In addition, CNV provides a set of conversion corrections, such as FK4-FK5 systematic correction, elliptic aberration E-term correction, photometric magnitude correction, or any combination.

Whenever alternative coordinates are required, only an SQL statement is needed. Calling a set of SQL functions in either dbaccess, an Informix provided query tool, or in application software does the conversion. Since these functions are part of the server, they are transparent to users. Users only need to know the SQL.

5. SQL Function Calls

Based on the nature of the conversion, the functions in IRSA's astronomical datablade are divided into three subsets with convenient default values: (1) general conversion functions; (2) conversion functions with at least one galactic or super galactic coordinate system; and (3) conversion between galactic and super galactic coordinate systems.

Functions accept input values of RA and declination in decimal degrees or sexagesimal degrees (depending on the type of transformation desired), and output values in either decimal or sexagesimal degrees, according to the functions invoked.

(1) General Conversion Functions – These functions transform any astronomical coordinate into another coordinate with a defined transformation.

(2) Conversion Functions with at Least One Galactic or Super Galactic Coordinate System – This set of functions is used to convert a non-Galactic or a non-Super Galactic coordinate system to a Galactic or Super Galactic coordinate system, or vise versa (in this category, one less argument is required).

(3) Conversion Functions between Galactic and Super Galactic Coordinate Systems – Any function in this set can be performed by the above two sets of conversion functions.

6. Enhanced Archive Architecture Using the CNV Datablade

Currently IRSA provides a rich service for astronomical catalog query and image archive retrieval. When a user wants to perform a positional query or cone search, the speed of searching is dependent on whether the Informix optimizer chooses indexing path. To enhance regional searching, spatial index columns are added to the tables. The table index is built on those tables where the input columns contain ra and dec on coordinate Equitorial Julian 2000. This design adds a constraint, i.e., positional values must be in J2000 Equitorial coordinates.

The architecture of the Data Ingest and Upload service within IRSA for cross-comparison is enhanced by deploying the CNV datablade. The new data ingestion system can convert any astronomical coordinate, whether it is in decimal degree or sexagesimal degree, to Equatorial J2000 system if spatial indexing is required.

7. Astrophysics Query Using Embedded CNV

The Informix server with CNV datablade promotes flexible data processing at IRSA. By launching CNV datablade, users are able to retrieve data in any coordinate system regardless of what is stored in the database. Users are allowed to select proper observation time, input epoch and output epoch. Users are even given the flexibility of selecting different correction terms or conditions, e.g., with or without proper motion.

Uploading tables for cross-idenitification comparisons is one of the important features that IRSA provides. J2000 Equatorial coordinates in decimal degree was previously the only coordinate supported by this application, but with the advent of the CNV datablade, the upload utility is more flexible. Users can now load any coordinates in their list, and IRSA will convert them internally and return the objects of interest in the requested coordinate system.

8. Conclusion and Future Work

The embedded astronomical query which combines a data searching engine with coordinate conversion capability is an efficient tool for the astronomical community. It reduces intermediate steps and makes scientists and engineers' work simpler. Future work includes further optimization of the datablade based on usage experience and full deployment within IRSA and the Space Infrared Telescope Facility.

msg: A Message-Passing Library

John Roll, Jacob Mandel

Smithsonian Astrophysical Observatory, 60 Garden Street, Cambridge, MA 02138

Abstract. As the foundation for the SAO MMT Instrument control software, we have developed an ASCII protocol for passing messages between client and server applications. Server interfaces are described as registered commands and published values. Clients may execute these commands and subscribe to published values. The mechanics of exchanging data between the client and server over a TCP/IP socket are handled by the library. The protocol and its implementing libraries have been designed with simplicity in mind. The simplest server can be written in three lines of code. However, we also have used the msg library to build complex systems with dozens of registered commands and hundreds of published values.

Introduction

Our message-passing library provides an elegant and practical solution to the problem of connecting together the functions of a data system. The API is built on a publish/subscribe metaphor that is easy to use. A server makes public ("publishes") data and commands. Clients request notification ("subscribe") about changes to those data. Clients may also execute the server's commands and actively "set" or "get" the data a server makes public.

The table below represents the code schematic of a service interacting with a client. These are all of the routines that are needed.

Server	Client	
msg_server()	msg_client()	Server and client initialization.
msg_allow()		Server host-based authorization.
msg_register()	msg_command()	Server registers available commands. Clients may call commands.
msg_publish()	msg_subscribe()	Server publishes values, clients may subscribe to value(s) to allow passive notification of changes.
	msg_set()	Clients may actively set or get values that
	msg_get()	have been published by the server.
msg_loop()	msg_loop()	Servers and clients enter the event loop to allow messages to be processed.

With the exception of the routines for specifying the data types on "set/get" calls and some convenience routines for synchronous and asynchronous commands, these routines cover 99% of the msg API. Simple server and client programs are thus very short and easy to code.

Protocol

A message in our protocol is a new-line terminated, variable length ASCII string of the form:

[#] command args\n

where:
- [#] The packet number signals the need for acknowledgment. If omitted or 0, no acknowledgement is expected. Server packets always have even numbers, while client packets have odd numbers.
- command args The message itself consists of all characters between the message number and the terminator. A command may not begin with a numeric character, to avoid conflict with packet numbers.
- \n – A new-line character terminates the message.

Our protocol implementation defines a basic command set to handle the passing of scalar values between clients and servers. Servers can easily add other application specific commands via the msg_register() call. In addition to passing data as command arguments and returned values, application specific commands may also define inter-packet data. Because such data are not necessarily terminated by a new-line, the sender and receiver of this data must each agree to write/read exactly the same number of bytes or they must define their own logical EOF mark.

Telnet Test

The protocol stream is made up of ASCII text. This simplification led directly to our "telnet test" requirement: any changes to the protocol had to preserve support for manual control in a telnet session. Meeting this test soon became an important design goal of our implementation, since it helped us focus on simplicity as a major aim. From a practical standpoint, telnet is an invaluable debugging tool.

```
john@snappy.harvard.edu : !tel
telnet snappy 2001
Trying 192.168.1.100...
Connected to snappy.
Escape character is '^]'.
1 set mode flat
1 ack
2 get camera
2 ack on
3 getcamerascale 1
3 ack 14 14
```

Implementation

We have implemented our message passing protocol in both C and Tcl. The C and Tcl libraries have slightly different API's in order to take advantage of the strengths of each language, but are very similar in overall design. The C version is written to be efficient in large multi-threaded instrument control services. The Tcl version is written to enable quick and simple GUI clients using Tcl/Tk widgets. The latter has also proven useful in creating simple servers, simulators, and protocol translators, where the speed of C is unnecessary and where Tcl's string capabilities are needed. Also, the Tcl implementation allows other Tcl/Tk programs to be used in close cooperation with our control systems. This includes the ds9 image display and control systems at the MMT Observatory.

Another important aspect of the msg implementation is that the library allows systems of clients and servers to recover automatically when an individual component is killed and restarted. This is a very valuable feature when working with complex hardware that is not yet robust itself.

Lab Experience

The msg library has been used in the Hectospec positioner lab over the last four years to develop and calibrate a 12-axis robotic positioner.

Its server systems include:

hecto The robotic positioner control server with several dozen message commands and several hundred status values.

snappy A multi-threaded frame grabber/controller connecting three intensified CCD cameras, which check positional accuracy for the robot in real-time.

stepper A server to control VME-based stepper motor controllers.

epbox A very simple server allowing access to the electronics peripheral box via an RS-232 serial line.

Several GUI interfaces are used to monitor the state of these complex systems by subscribing to the servers' published state values. Many of these values are displayed directly on a GUI as numeric values, while others are translated by the GUI to color and position indicators. The entire system can be run from command scripts or button presses on the GUI. The message protocol updates critical values 20 times a second. Other values are updated only once a second, or less if they are unchanged.

Each server has a companion simulator, written in Tcl. These simulators present the same message interface as the actual hardware control servers. High-level GUI and scripting software can be tested against the simulators without requiring the hardware to be available.

Telescope Experience

The msg library has been used to create an instrument control system for the MMT mini-cam instrument, which is the first of SAO's facility instruments for the converted MMT Telescope. The mini-cam hardware was recently delivered to the MMT and will soon be available for use by astronomers.

Its server systems include:

topper The camera top box controller, including two shutter blades and two large filter wheels.

scanserv A simple barcode scanner, which determines the filters currently loaded into the topbox.

detector The science CCD array controller. The detector controller is multi-threaded, and handles the message passing protocol and the ICE protocol used by the IRAF ccdacq package.

keith The CCD dewar temperature monitor server.

guidecam The guide camera CCD array controller

guide The guide error computation server.

focus The MMT secondary control server.

telescope A protocol translator which talks to the MMT telescope control server.

As in the hectospec lab, the mini-cam system has GUIs to monitor the state of its servers and command scripts to control its operation. Some of the most important benefits of using the message based APIs for our systems came during recent systems integration at the MMT. The telescope protocol translator used to connect the several internal clients and servers to information and control of the telescope was written in a few hours after we arrived at the MMT. The focus server, used to control the telescope secondary with errors generated by the guide error computation server, was written in a few minutes. Here is the source that was added to the Tcl language secondary control program:

```
source msg.tcl

proc FOCUS.focerror { err args } {
    global focus

    set focus [expr $focus + $err]
}

msg_server     FOCUS
msg_allow      FOCUS cfaguide
msg_publish    FOCUS focus focus
msg_register   FOCUS focerror
msg_up         FOCUS
```

DASH—Distributed Analysis System Hierarchy

Masafumi Yagi, Yoshihiko Mizumoto, Michitoshi Yoshida, George Kosugi[1], Tadafumi Takata[1], Ryusuke Ogasawara[1]

National Astronomical Observatory, Osawa, Mitaka, Tokyo 181-8588, Japan

Yasuhide Ishihara

Fujitsu Limited, Nakase, Mihama, Chiba 261-8588, Japan

Yasuhiro Morita, Hiroyuki Nakamoto, Noboru Watanabe

SEC Co. Ltd., 22-14 Sakuragaoka, Shibuya, Tokyo 150-0031, Japan

Abstract. We developed the Distributed Analysis Software Hierarchy (DASH), an object-oriented data reduction and data analysis system for efficient processing of data from the SUBARU telescope. DASH consists of many objects (data management objects, reduction engines, GUIs, etc.) distributed on CORBA. We have also developed SASH, a standalone system which has the same interface as DASH, but which does not use some of the distributed services such as DA/DB; visiting astronomers can detach PROCube out of DASH and continue the analysis with SASH at their home institute. SASH will be used as a quick reduction tool at the summit.

1. Introduction

DASH is an object-oriented data reduction and analysis system for the 8.2 m SUBARU telescope in Hawaii. It is the Observatory's system for managing, reducing, and analyzing the huge amount of astronomical data produced by various Subaru instruments. Design goals include reproducibility and portability of the reduction, seamless operability on heterogeneous computer systems, open architecture, trial-and-error processing, pipeline processing, and research-target-oriented analysis. For these purposes, CORBA is adopted as a distributed object environment (Takata et al. 1998). With CORBA, we can easily connect the system backbone coded in C++, and the GUIs coded in JAVA2 (Figure 1). The details of the DASH system were reported in Yagi et al. (2000) and Mizumoto et al. (2000).

---

[1]Subaru Telescope, National Astronomical Observatory of Japan, Hilo, HI 96720, USA

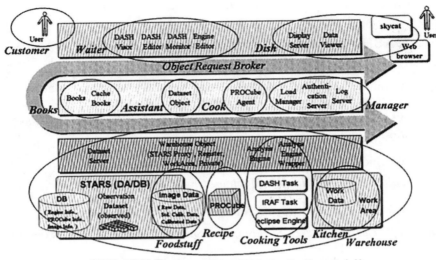

Figure 1. DASH application model (SUBARU restaurant)

2. Observation Dataset and Quality Control

There are three subsystems of the Subaru computer system: the Subaru Observation Software System (SOSS), the Subaru Telescope data ARchive System (STARS), and DASH. The key idea connecting them is the "Observation Dataset" (Kosugi et al. 1998).

The Observation Dataset consists of Dataset rules and an Observation log. Dataset rules define how the observation should be done. For example, it defines how often bias frames should be taken and which commands should be used. Every observer makes his/her observation plan in SOSS following Dataset rules, so that all calibration data to reduce a target should be obtained during the observation. The Observation log and the Dataset rules are sent to STARS with the data. After the observation, DASH reads the Observation Dataset from STARS. In the dataset, a template reduction procedure (skeleton PROCube) is also specified. DASH fills the template with the data, and the target is thus automatically reduced/analyzed to output some data quality parameter, such as seeing size. We can improve the Dataset rules by examining the data quality and feed it back to SOSS.

3. SASH

Just prior to this conference, we released SASH[2], a stand-alone system extracted from a core part of DASH. Because DASH is strongly connected with STARS

---

[2]http://optik2.mtk.nao.ac.jp/SASH

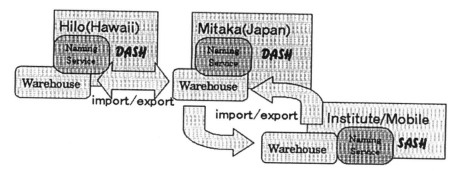

Figure 2. Phase 1: Current System

(Subaru data archiving system), and the SUBARU user/group authentication system, DASH is available only at SUBARU Hilo (Hawaii) and Mitaka (Japan). SASH is designed so that visiting astronomers can continue the data analysis at home institutions and feed the results back to the DASH systems.

For portability, the SASH platform is coded in JAVA2, with some C++ backbones of DASH rewritten in JAVA2. Users may add to the SASH platform using reduction engines such as eclipse or IRAF. There are few UNIX OSs on which JAVA2 1.2.2 is available. We therefore selected Solaris Sparc and Linux for the SASH environment at present. Some engine wrapper scripts written in Bourne shells must be rewritten because of the difference of behavior of some binaries (eg. awk).

4. Connection between DASH and SASH

There are two DASH systems now, in Hawaii and Japan. Each is connected to a local database and archive system, and the DB/DA systems are not synchronized. On the other hand, SASH has its 'database' as files. So, DASHs and SASHs share 'reduction' by exporting/importing warehouses, which contains engines, PROCubes, and data (Figure 2). Users should explicitly detach and attach warehouses. DASH and SASH use their own naming service for local objects.

In the next development phase, we plan is to connect the two DASHs at Hilo and Mitaka via IIOP (Figure 3). The naming service of each system will collaborate with the other to resolve objects so that users can continue the reduction at Hilo on the Mitaka system, and vice versa. If a requested object is not in its local system, the object is transferred from another system through ORB. This function depends on faster Internet speeds between Hilo and Mikata than are currently available.

Our ultimate plan is to connect DASH and SASH via IIOP and to use an interoperable naming service on all systems (Figure 4). Importing/exporting would then no longer be needed. If some data were needed, SASH would automatically connect to the nearest DASH and to retrieve them.

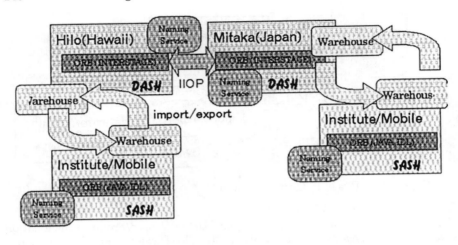

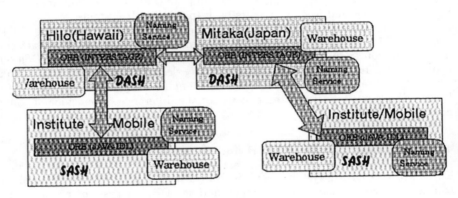

Figure 4. Phase 3: Final System

References

Kosugi, G., et al. 1998, Proc. SPIE, vol. 3349, 421

Mizumoto, Y., et al. 2000, Proc. SPIE, vol. 4009, 429

Takata, T., et al. 1998, Proc. SPIE, vol. 3349, 247

Yagi, M., et al. 2000, in ASP Conf. Ser., Vol. 216, Astronomical Data Analysis Software and Systems IX, ed. N. Manset, C. Veillet, & D. Crabtree (San Francisco: ASP), 510

ADASS Web Database XML Project

M. Irene Barg, Elizabeth B. Stobie, Anthony J. Ferro, Earl J. O'Neil

University of Arizona, Steward Observatory, Tucson, AZ 85721

Abstract. In the spring of 2000, at the request of the ADASS Program Organizing Committee (POC), we began organizing information from previous ADASS conferences in an effort to create a centralized database. The beginnings of this database originated from data (invited speakers, participants, papers, etc.) extracted from HyperText Markup Language (HTML) documents from past ADASS host sites. Unfortunately, not all HTML documents are well formed and parsing them proved to be an iterative process. It was evident at the beginning that if these Web documents were organized in a standardized way, such as XML (Extensible Markup Language), the processing of this information across the Web could be automated, more efficient, and less error prone. This paper will briefly review the many programming tools available for processing XML, including Java, Perl and Python, and will explore the mapping of relational data from our MySQL database to XML.

1. Introduction

The ADASS POC formed a Web Site Working Group (WSWG), chaired by Richard A. Shaw (STScI), whose charter is to define the scope of the existing www.adass.org Web presence. Two of the many issues the WSWG will be addressing are the focus of this paper. These are: 1) the development of an ADASS Conference database, and 2) www.adass.org site management. POC member Betty Stobie, of the University of Arizona (UofA), volunteered to take on the task of building a database of participants, invited speakers, and papers presented at past ADASS conferences. To accomplish this, she enlisted help from the other members of the NICMOS Software Team at the UofA, and this paper chronicles their efforts.

2. Building the ADASS Web Database

MySQL was chosen because it is freely available and is a small efficient database server. Three tables were created within this database:

1. Invited speakers: speaker's name, affiliation, title and special topic.
2. Participants: name and affiliation.
3. Programs: presentations from oral, poster, demonstration, and BOF.

2.1. Data Ingest Tools Used, Perl to the Rescue

Most of the source for the data came from online HTML documents found at former ADASS Conference host sites. HTML documents can be created in many ways, sometimes resulting in poorly formed HTML (i.e., missing end tags). HTML is a structural markup language based on the Standard Generalized Markup Language (SGML) metalanguage. While HTML documents do have hierarchical structure, processing HTML documents using an XML parser would require the following steps:

1. preprocess all HTML documents using something like HTML TIDY[1],
2. convert the HTML documents to XML,
3. write a program to parse the XML document to obtain the required data.

Perl, on the other hand, can reduce the process to a single step. Using the Perl modules HTML::TokeParser[2] and HTML::TableExtract[3], an HTML document can be parsed as arbitrary chunks of text, which is a much simpler technique. A Perl program was written to process online ADASS Conference Programs to extract column data such as author, title, and special-topics. The extracted data was used to populate the MySQL *program* table. Starting with ADASS IX and working backward, we found that our original Perl program had to be modified for each conference year. This was expected, and it was easy to tweak the Perl code to accommodate documentation differences.

3. Web Site Management

Now that the ADASS Conference database has been built, how should it be managed? ADASS is a collaboration between several hosting institutions from North America and Europe. The easiest way to provide access to the www.adass.org Web Database from all of these institutions, is through a Web development platform. When looking for a Web application development platform, the key criteria are:

- ease of configuration,
- management through the Web,
- support for XML.

The object-oriented nature of the XML API's SAX (Simple API for XML) and DOM (Document Object Model) makes processing XML more suited to object-oriented programming languages. Three such development environments were considered:

1. Webware for Python (http://webware.sourceforge.net).
2. Java tools by the Apache XML Project (http://xml.apache.org).
3. Zope (http://www.zope.org).

Of the three development environments reviewed, Zope was the only environment that met all three criteria outlined above.

---

[1]HTML TIDY, written by Dave Raggett.

[2]HTML::TokeParser, written by Gisle Aas

[3]HTML::TableExtract, written by Matthew P. Sisk

3.1. Zope

Zope was the last development environment to be evaluated and the easiest to configure. Zope is an Open Source application server developed by Digital Creations. Because it is written in Python, it is object-oriented and extensible; moreover, Zope has a large, active community of users. Most importantly, Zope provides through-the-Web management.

3.2. Content Management, Data Access and Data Sharing

Central to Zope is the Document Template Markup Language (DTML). It is a variable insertion and expression language that provides "safe scripting" of Zope objects and dynamic HTML content generation. DTML uses a server-side-include syntax (analogous to PHP:Hypertex Preprocessor). It was built from the ground up to use and extend the Web. Zope encourages building a site whose structure maps to the structure of the content. This is best illustrated by the following piece of DTML code.

```
<dtml-var standard_html_header>
<table>
<dtml-in get_program>
<tr><td><dtml-var adassnum></td></tr>
</dtml-in>
</table>
<dtml-var standard_adass_footer>
```

This example calls an SQL query method *get_program* and builds a table by iterating over the resulting record objects using the "dtml-in" tag and inserting the variable values with the "dtml-var" tag. The two variables *standard_html_header* and *standard_adass_footer* are DTML documents located in the Zope hierarchy, and made available through Zope's concept known as "acquisition". This is very useful for centralizing resources to implement a common design across a Web site. Because objects in Zope are hierarchical, URL's map naturally to objects in the hierarchy. For example, the URL:

```
http://localhost:8080/ADASS/Database/program/
```

will access the *index_html* document object located in the *program* folder.

3.3. Web Management and the Zope Security Model

Zope's security model makes it easy to provide developers access to a Web site without compromising security. Zope manages users with "User Folders", which are special folders that contain user information. An example would be to create a separate folder for each ADASS conference year, managed by the conference host. The host site would link its online registration and abstract submission pages to the www.adass.org Zope folder. The host site maintains complete control of that folder, while having access to the common objects in the Zope hierarchy. Other conference specific information would continue to be located and maintained at the host's Web site.

4. XML Support

4.1. ADASS Web Database and XML

Currently the ADASS Web Database provides for XML content through the optional "Download Data" button on any of the database query forms. When querying the database, pressing the "Search" button will render the results into HTML, but by pressing the "Download Data" button, the results will be mapped into XML. The mapping is simple, a table maps to a table element that can contain zero or more row elements. Each row element will contain zero or more field elements.

4.2. Zope and XML

Since Zope is written in Python, program developers have access to the Python XML library through external methods in Zope. Internal to Zope, XML support consists of XML-based protocols such as WebDAV and XML-RPC. There is an "XML Document" add-on and additional XML methods are being developed by Digital Creations.

5. Summary and Conclusions

So, what is the *ADASS Web Database XML Project*? More specifically, what role will XML play in the future of the www.adass.org site? The problems we experienced in building the ADASS Conference database showed us that the old way of delivering Web content must change. The www.adass.org Web site should make all interfaces capable of delivering XML content. When the client is a browser, the XML can be rendered into HTML. When the client is a program, the XML can be delivered directly to that program via XML-RPC or other mechanisms that support XML-over-HTTP protocol. The main role Zope can play is to become the central site for ADASS Web Database management. Requiring future ADASS Host Sites to link their online registration and abstract submission pages to the Zope site would:

- insure consistency of the data structure from year to year;
- insure consistency in the way participant names and affiliations are entered into the database;
- provide up-to-date information on any upcoming conference (like statistics on categories of papers submitted to date, participants, affiliations, etc.).

The WSWG will be looking at other development platforms. We would like to conclude with one final observation. When choosing a development platform for any Web site, the "keep it simple" scenario keeps returning. It is essential to choose a platform that promises to be the easiest for configuration, maintenance, and code development and still manages to be robust. If it's easy to use, it will be accepted by all, but if it's difficult to use, then individual host sites will continue to re-invent the wheel each year.

Acknowledgments. We are grateful to Dr. Rodger I. Thompson, University of Arizona, NICMOS Project, for providing the resources that allowed us to do this project.

Astronomical Data Analysis Software and Systems X
ASP Conference Series, Vol. 238, 2001
F. R. Harnden Jr., F. A. Primini, and H. E. Payne, eds.

DISCoS—Detector-Independent Software for On-Ground Testing and Calibration of Scientific Payloads Using the ESA Packet Telemetry and Telecommand Standards

F. Gianotti and M. Trifoglio

Istituto TeSRE/CNR -Via P. Gobetti 101, 40129 Bologna, Italy

Abstract. The ESA Packet Telemetry and Telecommand Standards satisfy the requirements of a great variety of scientific space missions, thus providing a standard basis for cost-effective and technically compatible developments of on–board and on-ground data handling systems for a wide range of projects. This paper describes the design and the implementation of detector-independent software, running on Unix platforms, for near real-time acquisition, archiving, and basic processing of ESA-based telemetry and telecommand data produced during on-ground testing and calibration of various space-borne X– and γ–ray detectors.

1. Introduction

Every scientific payload that adopts the ESA Packet Telecommand (TC) and Telemetry (TM) Standards uses the TC and TM source packet structures to receive the commands which determine the payload set up and operations, and to send the data generated by its subsystems. These standards dictate the guidelines for the packet structure, formed by a header, a data field and a trailer. They also specify the information to be included in the header and in the trailer in order to allow the verification and decoding of the packet content. Among them, the Application Identifier (APID) is required to identify the subsystem which is the destination or the source of the TC or the TM packet, respectively. In addition, for a given APID, the Type/Subtype keywords identify the specific operating mode under which the data have been produced, and, then, the structure of the data contained in the data field.

Suitable Electrical Ground Support Equipment (EGSE) hardware and software are required in order to test and verify, before launch, all the payload functionality and performances. To this purpose, the EGSE provides the simulation of the relevant missing on–board subsystems, generates and sends to the payload the TC packets, receives and archives all the TM packets. By inspecting in near real time the TM reports and the TM housekeeping packets, the EGSE verifies the correct execution of the TC, and is able to monitor the payload health, as required to support the basic engineering tests. In addition, it is important that EGSE software allow the EGSE operator to easily verify all the different scientific operating modes, and, in particular, to verify and calibrate the scientific performance of the detectors illuminated with X– and γ–ray sources or beams.

245

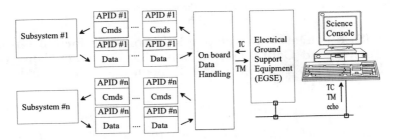

Figure 1. TM and TC packets data flow during ground tests.

2. DESIGN CONCEPT

The use of the ESA TC/TM Standards, allowed us to adopt the common design concept sketched in Figure 1, where the EGSE is limited to the engineering functionality, while the scientific functionality is provided by the Science Console, which receives in near real time, from the EGSE, a copy of all the TC and TM packets. The EGSE was procured by the industry, and the Science Console software was in charge of the scientific team responsible for the instrument acceptance test and calibration.

The DISCoS (Detector Independent Science Console Subsystem) software we present in the following sections, is the part of the Science Console software which was integrated with various detector specific software written in order to unpack the information contained in the TC/TM packets and to perform the quick look on the scientific data. The current version, which is being developed for the AGILE mission (Trifoglio et al. 2000), profits from the experience gained with the previous versions which have been exploited for the XMM–EPIC mission (Trifoglio et al. 1997; Trifoglio et al. 1998), and the INTEGRAL–IBIS mission (La Rosa et al. 1999; Trifoglio et al. 1999).

3. SOFTWARE ARCHITECTURE AND IMPLEMENTATION

The DISCoS software consists of the C programs Monitor, Receiver, Archiver and Provider, which allow the Science Console to acquire, verify, and archive in near real-time the TC/TM packets in one set of files for each measurement setup, and to reconstruct, either in live or in playback mode, the various streams of TM scientific packets pertaining to particular operating modes, i.e those having the same APID/Type/Subtype. Figure 2 sketches how these programs interact together and with the unpacking programs (Processors) and the quick look programs (Quick Look Analysis and Graphical Display) to be written by the DISCoS users.

Once started from a shell script, every second the Monitor updates a screen window once per second with status information and relevant parameters that the concerned programs write on the Shmmon shared memory by using specific routines. In addition, the Monitor program allows the operator to generates fake start/stop TC, to be used by the Science Console in order to close and open the measurements files independently from the TC packets generated by the EGSE.

Detector-Independent Software for Testing ESA Payloads 247

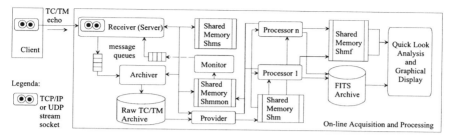

Figure 2. The Science Console software architecture.

The Receiver is the program which interfaces the EGSE through a TCP/IP or UDP socket on an Ethernet LAN 10/100 BaseT. Once started from a shell script, it waits as a Server, and, in the TCP/IP case, establishes with the EGSE a stream socket connection. Hence, using a fork the Receiver generates the Archiver program. As long as the connection is active, the Receiver performs a reading loop. On each iteration, by using a non blocking call it reads from the Monitor task message queue the next fake TC packet, if any; and by using a blocking call, it reads from the LAN in order to acquire the next TC/TM packet. For each packet, the Receiver inspects the header in order to verify the Application Identifier (APID), the Source Sequence Counter (SSC) and the packet length. Hence, by using a blocking call, the Receiver sends through a message queue the packet to the Archiver program, and restarts reading. In case a Start/Stop measurement command is detected, the Receiver waits until the Archiver message queue is empty, then sends a SIGINT signal to terminate the running Archiver, forks a new Archiver, and eventually restarts reading.

Every time it is started, the Archiver opens a new set of raw files: one to contain all TC/TM packets, and one for the TC and the TM Housekeeping only. Their locations and names are automatically derived from the progressive number (Run ID) which identifies the measurement. An additional suffix is used to identify the files containing the data generated during the instrument configuration. By using blocking calls, the Archiver reads each TC/TM packet from the message queue and writes each packet to the local disk using low level C–Unix Unformatted I/O routines. Upon receiving a SIGINT signal from the Receiver, the Archiver completes the reading and archiving of any TC/TM packet from the message queue, closes the raw files, and then terminates itself.

The Provider reads and sorts the TC/TM packets from the raw file. This program is run by a shell script either in live or in playback mode. In the former, the raw file name is derived from the current Run ID and the forthcoming TC/TM packets are read upon receiving the SIGUSR1 signal from the Receiver. In the playback mode, the program starts reading from the raw file, whose name has been provided by the script. For each packet, the Provider verifies the correctness of the packet header, sorts the packet by APID, and writes the packet in the column of the shared memory Shm assigned to the APID. Indeed, this shared memory is managed as a two–dimensional circular buffer capable of containing some hundreds of packets for each APID. Information exchanged with the Provider, through the shared memory Shms, allows each Processor to read new TC/TM packets having the required APID as soon as they are available in the shared memory Shm. A synchronization mechanism guarantees that no TC/TM packet is overwritten until it has been read by the related Processor. Unless a time-out occurs, before overwriting the Provider waits for a SIGUSR2 signal generated by the concerned Processor indicating that the new data have been already read.

In order to allow interfacing with detector-specific programs, the DISCoS software includes a set of C, Fortran and IDL callable routines. They allow the user to write the Processors as separated programs which have access to the monitor window and to the various source packet streams sorted by APID, without having to deal with either the EGSE interfacing or the TC/TM packet acquisition, verification and archiving. Usually, one Processor program is run for each APID. For each sorted stream, the Processor has access to 100% of the acquired packets, regardless of input rate and processing to be performed on the packet. A typical Processor reads the TC/TM packets from the shared memory Shm, verifies and archives the instrument data extracted from the packet data field in a format suitable for further off–line analysis (e.g., FITS format). Other DISCoS routines allows the Processor to communicate these data, through the shared memory Shmf, to another user's program, written in IDL, which is devoted to quick look purposes. Depending on the selected mechanism, on each call the quick look program receives the unpacked data related to either the next or the last packet processed by the Processor. A typical quick look program produces and displays in near real time further data products (e.g., time profile, spectra, and images) allowing the user interaction.

4. CONCLUSIONS

The DISCoS software presented herein has demonstrated its effectiveness and re-usability in the Science Consoles developed for several EGSE and Test platforms which have been designed adopting the ESA Packet Telemetry and Telecommand Standards for instrument data and commands transport structure. The software architecture and implementation have allowed a simple porting from the original HP–UX Unix workstation platform to Linux PC platforms with the GNU Compiler Collection (GCC).

References

La Rosa, G. et al. 1999, Proc. AIP Conference, 510, 693
Trifoglio, M. et al. 1997, Proc. of the Fifth Workshop Data Analysis in Astronomy, ed. V. Di Gesu' et al., World Scientific, 233
Trifoglio, M. et al. 1998, Proc. SPIE, 3445, 558
Trifoglio, M. et al. 1999, Proc. SPIE, 3765, 572
Trifoglio, M. et al. 2000, Proc. SPIE, 4140, 478

CAOS Simulation Package 3.0: an IDL-based Tool for Adaptive Optics Systems Design and Simulations

Marcel Carbillet, Luca Fini, Bruno Femenía, Armando Riccardi, Simone Esposito

Osservatorio Astrofisico di Arcetri, Largo Enrico Fermi 5, 50125 Firenze, Italy

Élise Viard, Françoise Delplancke, Norbert Hubin

European Southern Observatory, Karl-Schwarzschild Strasse 2, 85748 Garching-bei-München, Germany

Abstract. The IDL-based simulation software Code for Adaptive Optics Systems (CAOS) was originally developed to simulate the behavior of generic adaptive optics systems. The modular structure of the software allows the simulation of a great variety of different systems and is particularly suited for the adoption of graphical techniques for the programming of applications. It is actually composed of a global user interface (the CAOS Application Builder – presented in Fini et al. 2001), and a set of specific modules: the CAOS Simulation Package. We present in this paper the last version (3.0) of the CAOS Simulation Package, together with an example of an application to the Large Binocular Telescope interferometer adaptive optics system.

1. Introduction

In the framework of the "Laser Guide Star for 8-m Class Telescopes" *Training and Mobility of Researchers* network funded by the European Union, an IDL-based software package has been developed to simulate generic adaptive optics (AO) systems. The structure of the software is modular. Each elementary physical process such as turbulence in atmospheric layers, propagation of light from source to observing telescope and through the turbulent layers, the wavefront sensor, is modeled in a specific module. The resulting software, called Code for Adaptive Optics Systems (CAOS), is composed of a global graphical user interface (GUI), the CAOS Application Builder (Fini et al. 2001), and a set of specific modules—the CAOS Simulation Package. A list of modules and brief descriptions are presented in Section 2. An example of an application to the Large Binocular Telescope (LBT) interferometer AO system is presented in Section 3. References to more information about CAOS are given in Section 4.

2. The Modules of CAOS Simulation Package 3.0

Table 1 shows a complete list, together with a very brief description, of the modules of CAOS Simulation Package 3.0.

Table 1. Descriptive list of the modules.

Module	Purpose
wavefront generation modules	
ATM - ATMosphere building	to simulate the turbulent atmosphere
SRC - SouRCe definition	to define the observed source
GPR - Geometrical PRopagation	to propagate the light
LGS-specific modules	
LAS - LASer generation	to define the projected laser beacon
NLS - Na-Layer Spot building	to simulate the 3D sodium LGS
wavefront correction modules	
TTM - Tip-Tilt Mirror	the tip-tilt correcting mirror
DMI - Deformable MIrror	the deformable correcting mirror
wf sensing and reconstruction	
TCE - Tip-tilt CEntroiding	to reconstruct the tip-tilt
SHS - Shack-Hartmann Sensor	to simulate the Shack-Hartmann sensor spots formation
CEN - CENtroiding calculus	to compute the SH centroids
TFL - Time FiLtering	to emulate commands time-filtering
REC - REConstruction module	to reconstruct the wavefront
calibration-oriented modules	
CFB - Calibration FiBer	to define a calibration fiber
CSQ - Command SeQuencer	to generate calibration commands
MCA - Make CAlibration data	to elaborate calibration data
other scientific modules	
IBC - Interf. Beam Combiner	to simulate co-phasing of two beams
IMG - IMager module	to simulate image formation
STF - STructure Function	to compute the structure function
WFA - WaveFront Adding	to linearly combine two wavefronts
BSP - Beam SPlitter	to emulate a beam-splitter device
utility modules	
PSG - Phase Screen Generation	to generate turbulent phase screens
DIS - data DISplay utility	to display any kind of input data
SAV - data SAVing utility	to save cubes of data (XDR format)
RST - data ReSTore utility	to restore XDR cubes of data

Using the CAOS Application Builder, a simulation can be built by connecting together the required occurrences of the desired modules, represented by the boxes of Figure 1. The only constraints are those imposed by input/output types. Each module comes with an individual GUI in order to set its own physical and numerical parameters, during the design step or independently at a

CAOS: Tool for Adaptive Optics Design and Simulation

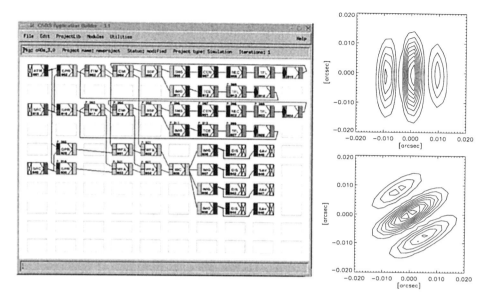

Figure 1. The CAOS worksheet corresponding to the LBT interferometer (left), together with the contour plots of two of the obtained PSFs: R-band, 1″ off-axis, parallactic angle of 0° (up), and 120° (down).

later time. The whole structure of a simulation can be saved as a "project" that can be restored for later modifications and/or parameters upgrading. The IDL code, corresponding to the designed simulation, is written down during the saving of a project, and it can be modified "by hand" in order to be completed with additional tasks not supported by the CAOS package.

3. An Example of an Application to the LBT Interferometer

Let's assume that we would like to simulate high-angular-resolution observations in the red and near-infrared wavelength bands[1], with the LBT interferometer[2] and with AO correction. Figure 1 (left) shows the whole project designed for such a purpose, with a natural guide star (NGS) of 10th magnitude for the AO sensing, either on-axis or 1″ off-axis with respect to the astronomical object. The turbulent atmosphere is modeled with two layers: a ground layer evolving along the baseline and weighted with 30% of the total turbulent energy, and an upper layer (at 10 km altitude) evolving orthogonally to the baseline and weighted with 70% of the total turbulent energy. For both layers the wind speed is 5 m/s. The total Fried parameter r_0 is 20 cm (at 500 nm), and the wavefront outer-scale L_0 is 40 m. Each pupil of the LBT interferometer has its

---

[1] namely R (700±110 nm), J (1250±150 nm), H (1650±=175 nm), and K (2200±200 nm)

[2] 2×8.25 m with a 14.4 m baseline

own AO system made of a tip-tilt (TT) correction loop, and a high-orders (HO) correction loop. The HO loop is made of: a 34×34 Shack-Hartmann lenslet array with 8×8 pixels/sub-aperture and 0.15″/pixel for the sensing, a modal rejection of the Zernike modes over number 231 (20th radial order) during the wavefront reconstruction, and a 35×35 actuators deformable mirror (that corresponds to a projected inter-actuator distance of ∼23.5 cm on the primary mirror) for the correction. The TT loop contains a quad-cell detector (with 0.25″/cell). Both sensings are performed in R, the light from the NGS being split 95% for the HO loop and 5% for the TT loop, assuming an overall efficiency of 60% and a read-out noise of 3 e^-rms. The time-filtering acts in each loop as a pure integrator, and the differential piston is supposed to be perfectly corrected. The scientific CCD – on which the point-spread function (PSF) corresponding to the astronomical object is formed – make 128×128 pixels images[3]. The total temporal history of each simulation run – one per value (0°, 60°, and 120°) of the parallactic angle – is 2.075 s (corresponding to 415 iterations of 5 ms each), but the resulting interferometric PSFs (one per band (four) and per off-axis (two) considered) are integrated over the last 2 s for sake of AO stability. For each of the simulation run, a different realization of the turbulent atmosphere was considered (since each parallactic angle corresponds to a different period of the observing run). Figure 1 (right) shows two of the 3×4×2 obtained PSFs, while Table 2 synthesizes the quality of all the 24 AO-corrected interferometric PSFs obtained in terms of Strehl ratio.

Table 2. Strehl ratios obtained for each of the interferometric PSF.

0″	R	J	H	K	1″	R	J	H	K
0°	.606	.851	.907	.946	0°	.554	.827	.893	.938
60°	.581	.836	.897	.939	60°	.522	.809	.881	.929
120°	.557	.816	.878	.921	120°	.504	.791	.864	.912

4. More Information...

For more information and references on the CAOS Simulation Package and the CAOS Application Builder, see http://www.arcetri.astro.it/caos. See also Correia et al. (2001) for a description of the CAOS-compatible simulation package AIRY (Astronomical Image Restoration in interferometrY).

References

Fini, L., Carbillet, M., Riccardi, A. 2001, this volume, 253
Correia, S., Carbillet, M., Fini, L., et al. 2001, this volume, 404

---

[3] with ∼2.1 mas/pixel in R, ∼3.8 mas/pixel in J, ∼5.0 mas/pixel in H, and ∼6.7 mas/pixel in K

The CAOS Application Builder

Luca Fini, Marcel Carbillet, Armando Riccardi

Osservatorio Astrofisico di Arcetri, Firenze, Italy

Abstract. In order to ease the design and implementation of simulation projects a graphical interface has been built on top of the CAOS Simulation Package. The CAOS Application Builder allows a user to build a simulation program (a *project*) by putting together elementary building blocks and specifying the data flow between blocks. When the project has been defined to the user's satisfaction the IDL code which implements the program is automatically generated.

1. Introduction

The CAOS software system is a set of software tools specifically designed to allow the modeling of any kind of Adaptive Optics system. It was originally developed in the framework of the "TMR Network on Laser guide star for 8 meter class telescopes" funded by the European Community. In this same conference an overall description of the original package (see: 249) and a specialized version dedicated to the deconvolution of multiple interferometric images for the Large Binocular Telescope (see: 404) are also presented.

The Application Builder has been designed in order to provide the scientists with a Graphical Programming Environment in which elementary building blocks could be assembled together to create complex simulation applications in a straightforward manner, so that the user could concentrate on the scientific aspects of his/her problem, while mundane coding problems were managed by some automatic tool.

The functionalities and the overall architecture of the AB are the result of trade-offs among various requirements, and mainly of three principal goals:

- the programming effort for the development of the AB had to be *small*, i.e., much less than the programming effort devoted to the development of the full package,
- the AB must have marginal impact on the structure of simulation programs, i.e., the coding of blocks must not be significantly affected, and the time efficiency at run-time of the code produced must not be degraded significantly,
- the requirements on the coding of single blocks imposed by the use of the AB must not prevent the use of blocks in the traditional way, i.e., as ordinary routines to be called by a user written program.

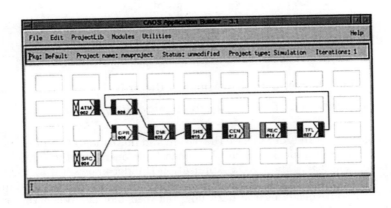

Figure 1. A Project is being built in the Application Builder.

2. How the Application Builder Works

When the AB is launched it appears to the user, as shown in figure 1, as a graphical window (the **worksheet**) provided with a number of rectangular slots and with a number of pull-down menus.

The building blocks, called **modules** (see figure 2) can be selected from a list and placed on the worksheet to build up the simulation program (the **project**); modules can be put into any free slot on the worksheet and then inputs and outputs can be joined by means of **links** which represents the data flow in the program.

Modules are represented as "computational blocks" provided with up to two inputs and up to two outputs; in order to convey the concept of input and output data types, the input and output types are encoded by different colors, so that it is clear that only equally colored inputs and outputs can be joined together.

A few modules have been designed to be "generic type", i.e., they accept input (output) of any type, and the actual type is assigned when they are linked to some typed output (input). This mechanism is useful for general purpose modules (e.g., the data display module) as an alternative to providing a particular typed module to display data values of each particular data type. The generic data display module has been designed so that it can check the data type of its input and use the appropriate section of code to display the value of a particular item.

For any module that requires the specification of run-time parameters, the user can invoke a Graphical User Interface which helps in the definition of values to be used in the execution of the program. At exit the parameter definition GUI creates the required data structures and saves them in the working directory by means of an IDL standard **save** command.

It often happens in the design of a simulation program that the same module must be used more than once, in different parts of the program, e.g., an optical element which is used twice in the optical path.

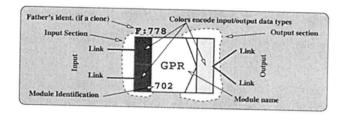

Figure 2. Anatomy of a Module.

This case is handled by the AB by means of "module clones": all modules of the same type share the implementation code, clones also share the set of runtime parameters. When the parameter definition GUI is started for a clone, the parameters which are actually modified are those of the "father" module. This ensures that the two modules correctly represent the same physical element.

A special "feedback" module must be used to "close loops", i.e., to assemble simulation programs which include feedback loops.

3. A Few More Goodies

In order to ease the project building, the AB includes a few more tools.

The **project library**, is a repository of projects which are distributed together with the CAOS software system which can be merged into a project being developed, or can be used as examples for similar ones.

Moreover the CAOS software distribution can be customized by defining **packages**. A package is a subset of the available modules, possibly specialized for a particular area of applications. By selecting a package the user is set into an environment which is exactly suited to the application to be developed.

4. Code Generation

When the project is finished it can be saved on disk. The save operation generates the source code which implements the simulation program together with a textual description of the graphic layout of the project by which the project may be later restored in the AB for subsequent use.

The code is subdivided in two IDL procedures: project.pro, which contains initialization and looping instructions, and mod_calls.pro, containing the sequence of procedure calls corresponding to the project.

As an example here follows a simplified version of the code generated by the project shown in figure 1:

```
COMMON caos_block, tot_iter, this_iter, calibration, signature
ret=src(O_004_00,src_00004_p,INIT=src_00004_c)
ret=atm(O_002_00,atm_00002_p,INIT=atm_00002_c)

;------------------------------------------------ Loop is closed Here
IF N_ELEMENTS(O_027_00) GT 0 THEN O_029_00 = O_027_00
```

```
ret=gpr(O_004_00,O_002_00,O_006_00,gpr_00006_p,INIT=gpr_00006_c)
ret=dmi(O_006_00,O_029_00,O_025_00,O_025_01,dmi_00025_p,INIT=dmi_00025_c,TIME=dmi_00025_t)
ret=shs(O_025_00,O_010_00,shs_00010_p,INIT=shs_00010_c,TIME=shs_00010_t)
ret=cen(O_010_00,O_012_00,cen_00012_p,INIT=cen_00012_c)
ret=rec(O_012_00,O_014_00,rec_00014_p,INIT=rec_00014_c)
ret=tfl(O_014_00,O_027_00,tfl_00027_p,INIT=tfl_00027_c)
```

The project.pro procedure skeleton which is "wrapped" around the above code is shown below:

```
RESTORE, 'Projects/luca1/src_00004.sav'    ; Restore parameters
RESTORE, 'Projects/luca1/atm_00002.sav'
RESTORE, 'Projects/luca1/gpr_00006.sav'
RESTORE, 'Projects/luca1/dmi_00025.sav'
RESTORE, 'Projects/luca1/shs_00010.sav'
RESTORE, 'Projects/luca1/cen_00012.sav'
RESTORE, 'Projects/luca1/rec_00014.sav'
RESTORE, 'Projects/luca1/tfl_00027.sav'

@Projects/luca1/mod_calls.pro              ; Initialization

FOR this_iter=1, tot_iter DO BEGIN         ; Main loop
        @Projects/luca1/mod_calls.pro
ENDFOR
```

5. Implementation

Because the IDL language was selected as the best choice for the implementation of modules, it was also decided to implement the AB in the same language, although not fully suited to this particular task. It is started as an usual script from the IDL prompt, and it is completely independent from the simulation program it has created: the simulation program can and will run independently on the AB itself.

The final version of the AB is made up of some 39 IDL source files for a total of 6700 lines of code (including full documentation of the source code).

References

Fini, L. 1999, Arcetri Tech. Rep. No 5/99

Carbillet, M., Femenia, B., Delplancke, F., Esposito, S., Fini, L., Riccardi, A., Viard, E., Hubin, N., Rigaut, F. 1999, in *Adaptive Optics Systems and Technology*, R. K. Tyson and R. Q. Fugate eds. SPIE Proc. 3762, 378

Carbillet, M., Riccardi, A., Fini, L., Viard, E., Delplancke, F., Femenia, B., Esposito, S., Hubin, N. 2001, this volume, 249

Correia, S., Carbillet, M., Barbati, M., Boccacci, P., Bertero, M., Fini, L., Richichi, A., Vallenari, A. 2001, this volume, 404

Astronomical Data Analysis Software and Systems X
ASP Conference Series, Vol. 238, 2001
F. R. Harnden Jr., F. A. Primini, and H. E. Payne, eds.

Software Fault Tolerance for Low-to-Moderate Radiation Environments

R. Sengupta[1], J. D. Offenberg[1], D. J. Fixsen[1], D. S. Katz[2], P. L. Springer[2], H. S. Stockman[3], M. A. Nieto-Santisteban[3], R. J. Hanisch[3], J. C. Mather[4]

Abstract. The primary intention of NASA's Remote Exploration and Exploration (REE) project is to use commercial off-the-shelf, scalable, low-power, fault-tolerant, high-performance computation in space. Most of the faults caused by the radiation environments in regions of space of interest to REE (Deep Space, Low Earth Orbit) are transient, single event effects. Some of these faults can cause errors at different application levels. System and applications software can potentially detect and correct some or many of these errors. We discuss different software fault tolerance approaches such as replication, voting, and masking with a focus on algorithm-based fault-tolerance. Combined software and hardware approaches such as fault avoidance, redundancy, masking, and reconfiguration are discussed. These approaches allow trade-offs between reliability, power, cost, and computation power for spacecraft in a low-to-moderate radiation environment.

1. Introduction

The Remote Exploration and Exploration (REE) project's global objective is to move ground-based supercomputers to space. This means building a high-performance, reliable, low-power, parallel computer for space applications, using mostly commercial off-the-shelf (COTS) components. REE intends to achieve this goal by using software-implemented fault tolerance (SIFT) techniques that will perform fault detection, isolation, and recovery without compromising reliability or performance. In order to gain experience with SIFT techniques, several generations of test beds are being built to perform fault tolerance experiments. These test beds will be provided with a SIFT middle-ware layer residing between the operating system and the applications, and between the operating system and the hardware. A software fault injector for simulation of radiation faults

---

[1] Raytheon ITSS, 4500 Forbes Blvd, Lanham MD 20706

[2] Jet Propulsion Laboratory, 4800 Oak Grove Dr., Pasadena CA 91109, http://www-ree.jpl.nasa.gov

[3] Space Telescope Science Institute, 3700 San Martin Dr., Baltimore MD 21818

[4] Code 685, NASA's Goddard Space Flight Center, Greenbelt MD 20771

has been generated to test the system and study its performance. REE has the goal of a flight test in 2005.

2. Hardware Fault Tolerance

Hardware fault tolerance can be attempted at different levels and in different ways. *Fault avoidance* can be defined as using better components or radiation hardened components to avoid single event upsets caused by cosmic ray hits. Although this is the only approach chosen by current projects and ranks high in reliability, the cost of radiation hardened components is the main tradeoff. Power consumption is higher and computational power is lower than for systems based on radiation hardened components versus parallel COTS-based systems. *Fault detection* would stop the system when a fault occurs. This has the advantage of identifying fatal errors but at a loss of throughput. *Masking redundancy* means running in the presence of faults. A few processors can run the same program and vote to identify errors in any single processor. Errors can be masked from application software. No software rollbacks will be required to fix errors. *Reconfiguration* means removing failed components from the system. When failure occurs in a component, its effects on the remaining portion of the system can be isolated. We are not concerned about how it failed. We expect to have enough other similar components to take its place. These are typically off-the-shelf processors, disk (for ground-based test beds only), and memory units but the desired reliability units e.g., EDAC, CRC checks, may not be built into these components. A large number of functional units with spares can be used, which will be switched automatically to replace a failing component. One can build desired the level of redundancy through system-level algorithms to coordinate these components.

3. Software Implemented Fault Tolerance (SIFT)

SIFT is a contrasting approach to a fault tolerant multi-processor. *Fault isolation* can be obtained by physical isolation of failed hardware components. *Fault masking* is enhanced through multiple buses or a redundantly linked network over which multiple copies of data are transmitted. Reliable fault-masking uses non-faulty elements to compensate for the effects of faulty elements. *Voting* takes place as a function of the software. Critical tasks in SIFT are required to be iterative tasks such that voting is done on the state of data before each iteration. This guarantees that each processor is working on the same loop. If an error is discovered in the vote on this state of the system, the software attempts to locate and remove the faulty components. This can be done by changing the bus of the presumed faulty processor. If the fault persists after the move, the processor is retired. If the fault disappears, the bus is returned. Thus, the SIFT can configure around the fault. Each processor can make a report of a failed component to the run-time supervisor. Depending on the type of fault, it may not be decidable which processor, or bus, is actively failing. However, the system can continue to run in a fault-masking mode (Banâtre & Lee, 1994).

The clocks in SIFT are not necessarily synchronous. This is a major advantage, as clock synchronization can be a significant added expense.

Fault Tolerance for Low–Moderate Radiation Environments

Figure 1. Sample 10000s NGST image, before and after deglitching (Fixsen, 2000).

Replication ensures reliability but is expensive in terms of hardware or runtime cost. The idea is to take a majority vote on a calculation replicated N times. A hardware solution requires N times the number of processors, as well as a reliable voter. Software solutions require each processor to run N copies of surrounding computations and then vote on the result. This slows down the computation by at least a factor of N. However, the savings from using COTS components could offset the added costs. A candidate application for this approach appears in Figure 1.

3.1. Algorithm-Based Fault Tolerance (ABFT)

In the broadest sense, algorithm-based fault tolerance (ABFT) refers to a self-contained method for detecting, locating, and correcting faults with a software procedure. REE is looking at techniques for numerical linear matrix multiplication, LU decomposition, QR decomposition, single value decomposition (SVD), and Fast Fourier Transforms (FFT). These techniques typically involve augmenting data with a (linear) checksum, whose value is predictably transformed by the linear algorithm. Checksum encoding is embedded in the data.

ABFT can detect, locate, and correct faults by exploiting the structure of numerical operations. REE is focusing on error detection, using result checking. This implies that a simple post-condition exists which can be checked at a cost that is low compared with the cost of the base algorithm. If (transient) errors are detected, the routine can be re-run. Many REEs science applications spend much of their time in core numerical operations; for example, NGST's phase retrieval code[5] (Figure 2) spends ~70% of its time on FFTs.

4. Faults and Errors

The radiation environment causes faults at the hardware level; many of which become errors at the application or system level. About 99.9% of faults are transient single event effects; faults can cause errors that cause the node or application to crash or hang. System software can detect such errors. Restarting the application, rollback, or rebooting will be acceptable solutions in those cases.

Hard to detect errors would change the application data, allowing the application to complete but with incorrect output. Applications can detect most

---

[5]http://www.ngst.nasa.gov/cgi-bin/doc?Id=486

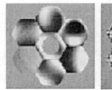

Figure 2. Sample wavefront corrected by 349 actuator deformable mirror and primary mirror segments, using NGST's phase retrieval code.

errors using ABFT, assertion checking, and other application-level techniques. The ability to correct errors with low overhead is a major advantage.

Huang and Abraham (1984) proposed the idea of using properties of the solution as acceptable tests: *"Since we test the intermediate results for correctness with respect to the algorithm, the end solution is correct if the intermediate results are correct. If processor errors occur that do not affect the solution, then they are not errors."* One application of this principle would be to only detect and correct processor errors in excess of either the inherent noise of the data set or the desired accuracy in the end result.

5. Conclusion and Future Work

Combined hardware and software fault tolerance approaches would be most effective for space-borne components in a harsh environment. The combination of COTS and SIFT can dramatically lower the cost and power usage while increasing computation power. *"A system should be as simple as possible in order to achieve its required function—and no simpler"* (Banâtre & Lee).

Future work planned for this study includes making provision for corruption of a counter or corruption of input data during the computation. It is also necessary to develop a better practical understanding of failure rates to have confidence in ultimate reliability; methods for detecting faults during the correction procedures themselves are also needed.

Acknowledgments. These studies are supported by NASA's Remote Exploration and Experimentation Project, which is administered at the Jet Propulsion Laboratory.

References

Huang & Abraham 1984, IEEE TC, 33(6), 518
Wang & Jha 1994, IEEE TC, 43(7), 849
Banâtre, M. & Lee, P. 1994, "Hardware and Software Architectures for Fault Tolerance", Springer Verlag
Fixsen, D. J., et al. 2000, PASP, 112, 1350

Extending the XImtool Display Server Using Components

Michael Fitzpatrick and Doug Tody

National Optical Astronomy Observatories[1], *IRAF*[2] *Group*

Abstract. The IRAF[2] display server, XImtool, has been enhanced to allow arbitrary custom components to be dynamically integrated, permitting the functionality of the main display server program to be extended in arbitrary ways with no change to the core program. Components may include both computational and GUI elements; new components are interfaced as named instances of the Widget Server *Client* class and are fully integrated into the base server using messaging. The compiled portion of a component executes as an external process, allowing components to be written in any language or to use any existing resources. To illustrate the use of this facility, an example is shown that provides real-time pixel value and WCS display using direct access to the actual disk image used to load the display.

1. Introduction

XImtool was originally designed using the IRAF Widget Server architecture (Tody, 1995). This provided a separation of the client application code and the user interface, either of which could be replaced or augmented with little or no change to the other. Using traditional methods, however, adding any major new functionality to the program would still likely involve substantial changes to both the client and GUI code. To solve this, XImtool has been enhanced to allow arbitrary custom components, executing as external processes, to be dynamically loaded at runtime.

New components are interfaced as named instances of the Widget Server *Client* class and are fully integrated into the base server using a simple text messaging scheme similar to that already used by client and GUI communications in the Widget Server. New GUI elements may be uploaded from the component and created when needed, and additional callback procedures can be defined that subscribe the component to events in the GUI (frame changes, cursor movement, etc.) to which it can react. Unlike the traditional *plug-in* model, the compiled portion of a component executes as an external process, allowing components

---

[1]National Optical Astronomy Observatories, operated by the Association of Universities for Research in Astronomy, Inc. (AURA) under cooperative agreement with the National Science Foundation

[2]Image Reduction and Analysis Facility, distributed by the National Optical Astronomy Observatories

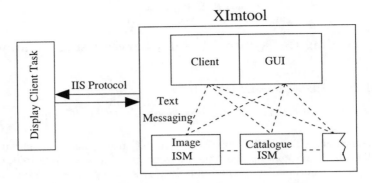

Figure 1. Architecture of the new system.

to be written in any language. The main server and components communicate over an IPC socket layer which is transparent to both. Components attach themselves by requesting a connection on a dedicated socket. The module and XImtool then negotiate a private channel for communications, allowing other modules to later connect as needed.

WCSPIX, the first of these Image Support Module (ISM) components to be implemented using this new facility, provides real-time pixel value and WCS display using direct access to the actual disk image used to load the display. The module is written as an IRAF task to allow easier access to all image formats supported by the IRAF interfaces (including Mosaics), something that could not be easily done using the old architecture. Other ISM components may now be easily written to provide new functionality such as vector graphics display, network catalog access, or other image analysis. With only a minimal understanding of the XImtool GUI, users can write their own components using a standard API provided.

2. Architecture

The basic XImtool architecture (Figure 1) is a single process in which text messages (e.g., to tell the application to change the frame, or the GUI to set the label of a button) are exchanged between the client and GUI to provide the interactivity and core functions of the task. New functionality is supplied by one or more ISM components running as a separate process and connected by text messaging over a socket layer between the component and the base server. ISM modules may send messages to the XImtool client, its GUI or other ISM modules as needed.

Widget Server messages are normally of the form "**send** *object message*". As named *Client* objects, messages sent to an ISM require no special formatting and are simply directed to the channel that was negotiated when the ISM first connected. The syntax and content of the message sent to a component is arbitrary and determined by the needs and abilities of the ISM itself. ISM messages sent to XImtool follow the same rule, however the message itself *does* require a special syntax. Namely, messages must be preceded with keyword

Extending the XImtool Display Server Using Components

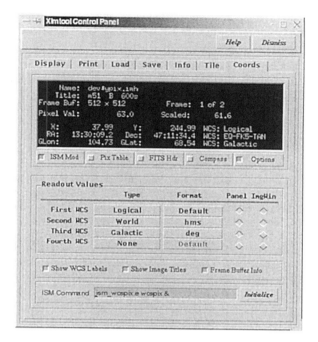

Figure 2. Integrated control panel showing the coordinate readout panel.

source if the message contains GUI Tcl code to be executed (e.g., callbacks needed by the ISM, new GUI objects, etc.), **alert** if the message is some text to be displayed on the XImtool alert panel, or **deliver** if the message text should be delivered to the ISM-specific callback that was previously uploaded in a *source* message.

More complex components (such as the WCSPIX module described here) may also require support code in the compiled portion of XImtool. To deal with this, a list of ISM callbacks is searched when a component first connects and any registered procedures are executed. In the case of the WCSPIX module, such callbacks are executed when it first connects (to initialize the ISM with a list of the currently displayed image regions) and when it disconnects (to clean up). A more general command callback for each ISM can also be registered so that new code specific to the ISM is isolated from the core system, minimizing the impact of an upgrade or a replacement of the component at a later time. Although not implemented in this release, such compiled code could be loaded dynamically as a shared object, further isolating it from the core system.

Formatting of the messages and the details of how they are used by the transport layer are hidden in an interface on each side of the connection. A standard API to connect the component and handle messages is available to support both IRAF and other compiled tasks using a simple IRAF library or an extension to the Client Display Library (Fitzpatrick, 1998).

3. The Image/WCS ISM

The Image/WCS ISM, **WCSPIX**, is written as a general purpose interface to all types of images supported by IRAF and provides a set of generic methods to query an image object for pixel or header data, coordinate systems available for the image (logical, world, or derivatives such as sky coordinate transformations) and translations of coordinates. It is also responsible for caching all images currently displayed in the server. The module is written to allow image objects to be subclassed using new functions to provide the actual image interaction. This means support for new formats such as Mosaics, spectral data or even tables can be added without requiring a new ISM.

The client display task defines a WCS for each region of the display frame buffer, and this information is passed to WCSPIX which initializes the object by opening the image and placing it in the cache. Cursor events in XImtool are translated to coordinates within each image region on the display, and these are then passed to the task to be translated and formatted to any of the WCS values requested by the user in the GUI. The XImtool GUI is responsible only for displaying the strings on the panel (Figure 2) and plays no direct part in the coordinate translation.

4. Other New Features

Along with the new component architecture providing real-time pixel and WCS readout, several other new features have also been added:

- The various panels for Control/Print/Load/Save have now been integrated into a single window selectable with a Tab widget.
- Keystroke commands will cause the cursor to centroid on an object and/or adjust the size of the centering box, providing greater accuracy for cursor commands.
- Frames can be registered with an offset and will maintain the registration of all frames while panning or zooming the image.
- All 16 frames permitted by the IIS protocol may now be used (more are possible). Blinking and registration are supported for all frames.
- A new panel was added to allow more control over the display of tiled images. All 16 frames may be displayed in any user-defined configuration.
- A *compass* indicator for the display orientation on the panner window is now available.
- Pixel table readout of an area around the cursor has also been added.

References

Fitzpatrick, M. 1998, in ASP Conf. Ser., Vol. 145, Astronomical Data Analysis Software and Systems VII, ed. R. Albrecht, R. N. Hook, & H. A. Bushouse (San Francisco: ASP), 200

Tody, D. 1995, in ASP Conf. Ser., Vol. 77, Astronomical Data Analysis Software and Systems IV, ed. R. A. Shaw, H. E. Payne, & J. J. E. Hayes (San Francisco: ASP), 89

Astronomical Data Analysis Software and Systems X
ASP Conference Series, Vol. 238, 2001
F. R. Harnden Jr., F. A. Primini, and H. E. Payne, eds.

Linux BoF

P. N. Daly

National Optical Astronomy Observatories, Tucson, AZ 85726, U S A.

1. Summary

The Linux BoF was attended by $\sim$40 people and included brief talks and an audience Q&A session. Martin Bly reported on Starlink's experiences of running Linux. With Linux on $x86$-compatible hardware in use > 3 years, it is clear that it provides a cost effective solution for reducing data. Linux is entering the mainstream alongside Solaris *etc.* Hardware support issues are reduced to problems with bleeding-edge technology with some (big) vendors issuing Linux drivers/patches or providing information for driver developers. The Linux management tools are very sophisticated with several distributions using the same tools. Software tools are much improved with an integrated C/C++/Fortran compiler set and a number of desktop environments available. Linux integrates well into multi-OS data processing environments.

Phil Daly updated the group on developments in real-time Linux. There are now 130 devices supported by COMEDI and LabVIEW looks set to support the NI cards via this route. The principal advances for *RTLinux* and *RTAI* were ports to other architectures ($x86$, PowerPC, α), 2.4 kernel support and integrated debugging tools.

Luca Fini talked about the effectiveness of Linux in embedded (industrial) applications by presenting two ways to build a protocol converter. He made comparisons between an implementation based on a PLC+SBC running a proprietary RTOS and a PC running Linux. Luca discussed a few details of how to realize a small footprint Linux system and how to cope with special devices used in embedded systems such as flash memory and non-standard interfaces. He stressed some aspects relating to the software development process.

Dave Mills updated the group on recent open source product releases (see http://www.sourceforge.net) and described implementing software guiders and auto-focus tools using Linux, SDL, and frame-grabbers. Dave also provided free CDs of *TurboLinux* and *Linux for Astronomy*.

Peter Teuben gave a talk about Beowulf clusters that covered the basic issues in applications needing such a beast and some tips on building one. It was clear that there was considerable interest within the BoF regarding Beowulf clusters.

The BoF concluded with an 'ask a guru' session, with the audience asking such Linux related questions as:

- How can photo-quality print services be achieved under Linux?
- Which distributions should be avoided?

One audience member also pointed out that the headquarters of the Free Software Foundation was just around the corner from the conference hotel!

Part 5. Science Data Pipelines

The SDSS Imaging Pipelines

Robert Lupton, James E. Gunn, Željko Ivezić, Gillian R. Knapp

Princeton University Observatory, Princeton University, Princeton, NJ 08544

Stephen Kent

Fermi National Accelerator Laboratory, Batavia, IL

Abstract. We summarise the properties of the Sloan Digital Sky Survey (SDSS) project, discuss our software infrastructure, and outline the architecture of the SDSS image processing pipelines. We then discuss two of the algorithms used in the SDSS image processing, the KL-transform based modelling of the spatial variation of the PSF, and the use of galaxy models in star/galaxy separation. We conclude with the first author's personal opinions on the challenges that the astronomical community faces with major software projects.

1. Introduction

The SDSS (York et al. 2000) consists of four major components: a dedicated 2.5m telescope at Apache Point, New Mexico, along with a separate 50cm telescope used to monitor the extinction and to provide calibration patches for the main telescope; a large format imaging camera (Gunn et al. 1998) containing 30 2048 × 2048 (13×13 arcmin) photometric CCDs with $u'g'r'i'z'$ filters and 24 2048 × 400 astrometric and focus CCDs; two 320-fibre-fed double spectrographs, each with two 2048 × 2048 CCDs; and lots and lots of software, with contributions from most of the SDSS institutions (listed in the acknowledgments).

The primary goals of the project are to survey the Northern Galactic Cap ($\approx 10^4$ square degrees) in five bands to (PSF) limits of 22.3, 23.3, 23.1, 22.3, and 20.8, and to carry out a spectroscopic survey of 10^6 galaxies, 10^5 QSOs, and a few $\times 10^4$ stars.

The SDSS is now in operational mode, and as of this writing (late January 2001) has imaged some $1600 \deg^2$ and obtained about 120,000 spectra as part of its commissioning and initial operations phases. These data have allowed dramatic new astronomical discoveries to be made, discoveries that we shall not further discuss here (e.g., Blanton et al. 2001; Fan et al. 2000, 2001; Fischer et al. 2000; Ivezić et al. 2000; Leggett et al. 2000).

2. Software Infrastructure

2.1. Configuration Management and Bug Reporting

The SDSS took an early decision to use public domain software wherever possible; in practice this has largely been applied to our infrastructure rather than scientific codes.

Our software engineering tools are entirely public domain (with the exception of compilers).

We adopted cvs[1] as a source code manager and have been pleased with its performance. We currently have about 1.7 GB in our cvs repository (including at least one version of IRAF). We have found that, after an initial period of distrust, scientists have found cvs to be extremely useful; in at least some cases, people sitting next to each other at the observatory in New Mexico have communicated via a cvs repository in Illinois.

While cvs allows us to control individual pieces of software, it does not provide a means of controlling complete systems. We have used a Fermi National Accelerator Laboratory (FNAL) utility called ups[2] which allows us to associate a set of *dependent products* with a piece of our software. For example, version v5_2_10 of the image processing pipeline depends upon v7_15 of our infrastructure routines. This enables us to guarantee that at any time in the future we can reconstruct an entire system, using exactly the same bits and pieces. The particular versions (e.g., v5_2_10) correspond to tags in the cvs repository. We have adopted a procedure that stable versions of our pipelines correspond to *branch tags* in cvs; this has allowed us to proceed with development while giving us the ability to fix bugs found in the stable, delivered, code.

We have used gnats[3] as our problem report and bug database. Since July 1998 we have acquired 1799 entries in the database; the last thousand have been filed since February 2000.

2.2. Command Interpreter

We use a heavily enhanced version of TCL 7.4 (actually, of TCLX) as our command language. Much of the work developing this system (known as *dervish*, née *shiva*, Sergey et al. 1996) was carried out at FNAL.

In addition to what now appear to be basically cosmetic changes (which we regret), the major enhancements that we made were:

Memory tracing/defragmentation/debugging: A common problem with programs that make heavy use of dynamically allocated memory is that the memory acquired from the operating system becomes *fragmented*, or that the program forgets to free resources. Both of these problems can be resolved by adding a layer above malloc, and we have done so. Figure 1 shows that the total memory used in the steady state by the frames pipeline (see below) is well controlled.

---

[1] http://www.cvshome.org/

[2] http://www.fnal.gov/docs/products/ups/

[3] http://sources.redhat.com/gnats/

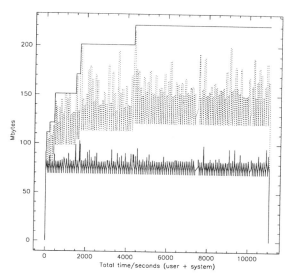

Figure 1. A trace of the memory used while processing 121 fields (3.6 GB) of an SDSS imaging run on a single 800 MHz alpha processor. A total of 165029 objects were detected and charcterised in five bands, giving a rate of 13.4 ms/object/band for processing from raw CCD frame to reduced catalog. The figure has three lines illustrating memory usage versus time. The lower line is the memory actively in use; the middle dotted line shows the memory in heap, and the top line shows the memory allocated from the system. The difference between the upper two lines is *guaranteed* to be in 10 Mb blocks, all except one of which is completely unused, and can safely be assumed to be swapped out to disk.

Support for C datatypes at tcl level: We wrote a processor that scans the C include files ('.h files') and generates a description of the schema of all the types declared therein. This was originally used to implement a primitive persistent store, but proved more useful in making the C data elements available at the TCL prompt; this greatly increased the power and flexibility of our command language, allowing us to build the command-and-control parts of our pipelines in TCL rather than having to use compiled C. For example,
```
assert {[exprGet $c.calib<$i>->filt<0>] == $f}
handleSet $fieldparams.frame<$i>.fullWell<0> $fullWell(0,$f)
```
where a 'handle' is an address and a datatype.

Easy(ish) bindings from C to tcl: We implemented a set of library calls that made it possible to bind C commands to TCL in a way that, if not simple, at least required no thought and could be handled by pasting appropriate boiler-plate code.

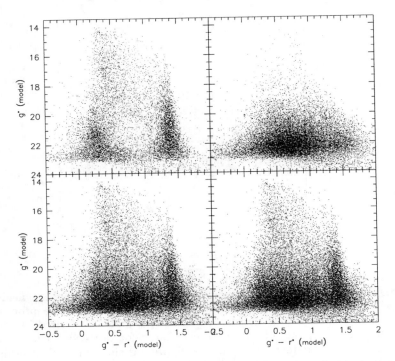

Figure 2. A g' vs. $g' - r'$ colour-magnitude diagram containing 31803 objects from SDSS commissioning data. The bottom two panels show all objects, the top left shows only stars and the top right only galaxies. The disk and halo turnoffs are clearly seen in the stellar diagram. If you are viewing this figure in colour, green points are stars; red points are galaxies classified morphologically as having deVaucouleur-like profiles; cyan points have exponential profiles; and magenta points are unclassified galaxies.

If we were starting this problem today, we would probably not use TCL (maybe python in its PyRAF incarnation?), and we would certainly make greater efforts to use *vanilla*, up-to-date, versions of our chosen system.

3. Imaging Pipelines

The SDSS has quite a large number of pipelines which must be run in order to fully process the data; we shall not discuss the spectroscopic reductions or the operational and scientific databases.

`Astroline`: On the MVE167 processors (running vxWorks) used to archive the raw data on the mountain, we also run a pipeline that processes the pixels before they're written to disk/tape. We generate star cutouts ('Postage Stamps') and column quartiles; this is all that we save from the 22 astrometric CCDs.

MT Pipeline: Process the Photometric Telescope camera data. This consists of a set of staring-mode observations of fields of standard stars, used to define the extinction and photometric zero-points for the 2.5m scans.

Serial Stamp Collecting (SSC) Pipeline: Reorganise the data stream, cut a more complete set of Postage Stamps.

Astrometric Pipeline: Process the centroids of stars from astroline/SSC and generate the astrometric transformations from pixels to $(\alpha, \delta)_{J2000}$ and between bands.

Postage Stamp Pipeline (PSP): Estimate the flat field vectors, bias drift, and sky levels, and characterise the PSF for each field.

Frames Pipeline: Process the full imaging data, producing corrected frames, object catalogues, and atlas images.

Calibration: Take the outputs from MT pipeline and frames, and convert counts to fluxes.

One major gain from splitting responsibilities in this way is that once we get to the frames pipeline, fields ($10' \times 13'$ patches on the sky) may be processed independently and in any order.

4. Interesting Algorithms

The SDSS imaging pipelines employ a number of novel, and even interesting algorithms, which are slowly being written up for publication; for example, the image deblender (Lupton 2001). Here we shall only discuss a couple of features connected to handling the point spread function (PSF) and the related problem of star/galaxy separation.

4.1. PSF Estimation

Even in the absence of atmospheric inhomogeneities the SDSS telescope delivers images whose FWHMs vary by up to 15% from one side of a CCD to the other; the worst effects are seen in the chips furthest from the optical axis.

If the seeing were constant in time one might hope to understand these effects ab initio, but when coupled with time-variable seeing the delivered image quality is a complex two-dimensional function and we chose to model it heuristically using a Karhunen-Loève (KL) transform.

Why We Need to Know the PSF: The description of the PSF (as derived in the next subsection) is critical for accurate PSF photometry, i.e., for all faint object photometry—if the PSF varies, so does the aperture correction.

We also need to accurately know the PSF in order to be able to separate stars from galaxies; after all, the *only* valid discriminant that isn't based on colours or priors is that galaxies don't look like stars.

A good knowledge of the local PSF is also needed for all studies that measure the shapes of non-stellar objects (e.g., weak lensing studies, Fischer et al. 2000).

KL Expansion of the PSF The first step is to identify a set of reasonably bright, reasonably isolated stars from our image. We then use these stars to form a KL basis, retaining the first n terms of the expansion:

$$P_{(i)}(u,v) = \sum_{r=1}^{r=n} a_{(i)}^r B_r(u,v) \qquad (1)$$

where $P_{(i)}$ is the ith PSF star, the B_r are the KL basis functions, and u,v are pixel coordinates relative to the origin of the basis functions. In determining the B_r, the $P_{(i)}$ are normalised to have equal peak value.

Once we know the B_r we can write

$$a_{(i)}^r \approx \sum_{l=m=0}^{l+m \leq N} b_{lm}^r x_{(i)}^l y_{(i)}^m \qquad (2)$$

where x,y are the coordinates of the centre of the ith star, N determines the highest power of x or y to include in the expansion, and the b_{lm}^r are determined by minimising

$$\sum_i \left(P_{(i)}(u,v) - \sum_{r=1}^{r=n} a_{(i)}^r B_r(u,v) \right)^2 ; \qquad (3)$$

note that all stars are given equal weight as we are interested in determining the spatial variation of the PSF, and do not want to tailor our fit to the chance positions of bright stars.

Application to SDSS data: For each CCD, in each band, there are typically 15–25 stars in a frame that we can use to determine the PSF; we usually take $n = 3$ and $N = 2$ (i.e., the PSF spatial variation is quadratic). We need to estimate n KL basis images, and a total of $n(N + 1)(N + 2)/2$ b coefficients, and at first sight the problem might seem underconstrained. Fortunately we have many *pixels* in each of the $P_{(i)}$, and thus only the number of spatial terms $((N+1)(N+2)/2$, i.e., 6 for $N = 2$) need be compared with the number of stars available.

In fact, rather than use only the stars from a single frame to determine that frame's PSF, we include stars from both preceeding and succeeding frames in the fit. This has the advantage that the spatial variation is better constrained at the leading and trailing edges of the frame; that the PSF variation is smoother from frame to frame; and that we have more stars available to determine the PSF.

We have found that optimal results are obtained by using a range of ± 2 frames to determine the KL basis functions B_r and $\pm 1/2$ frame to follow the spatial variation of the PSF. If we try to use a larger window we find that variation of the a^r coefficients is not well described by the polynomials that we have assumed. We have not tried using a different set of expansion functions (e.g., a Fourier series).

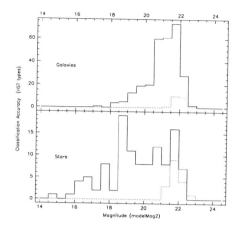

Figure 3. Star-Galaxy separation in the SDSS. The bottom panel shows object that are classified as stars based on their HST morphology; the top panel shows galaxies. The x-axis is the r' model magnitude. The solid line shows the number of objects classified correctly by the SDSS pipeline, the (red) dotted line shows the objects misclassified. It is clear that the performance is quite good, even close to the plate limit at about 22nd magnitude.

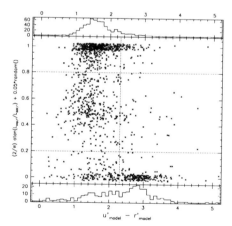

Figure 4. The relationship between morphological classification, based on the ratio of the deVaucouleurs and exponential likelihoods. The x-axis is the $u' - r'$ colour, which divides the galaxies nicely into two classes, presumably early- and late-type. The y-axis shows the likelihood ratio (mapped into the range $[0, 1]$); above and below the plot are shown the marginal distributions of galaxies which lie *outside* the pair of dotted lines. The correlation of colour with morhology is clearly seen.

4.2. Model Fitting and Star/Galaxy Separation

We fit three models to every object, in every band: a PSF, a pure deVaucouleurs profile, and an exponential disk; the galaxy models are convolved with the local PSF (as estimated using the KL expansion of the previous section). This is potentially an expensive operation as it involves a 3-dimensional $(r_e, a/b, \phi)$ non-linear minimisation; each iteration requires the calculation of a 2-D analytical model of a galaxy followed by convolution with the PSF and the calculation of χ^2 by summing over many pixels of the image. We make heavy use of pre-calculated tables of models, and pre-extract the radial profile into a series of annuli, each containing twelve 30° sectors; in consequence fitting a single galaxy model in a single band takes of order 1.5 ms on an 800 MHz alpha.

The primary use of these models is in star/galaxy separation and morphological classification of galaxies. We initially hoped to use the relative likelihoods of the PSF and galaxy fits to separate stars from galaxies, but found that the stellar likelihoods were tiny for bright stars, where the photon noise in the profiles is small, due to the influence of slight errors in modelling the PSF. Instead we found the ratio of the *flux* in the best-fit galaxy model to that in the PSF to be an excellent discriminant.

Figure 2 shows a colour-magnitude diagram from a small area of SDSS imaging data. The top left panel shows only objects classified as stars; note that most objects with colours of $g' - r' \approx 0.9$ are preferentially classified as galaxies. The star/galaxy separation is done independent of the object's colours, so this rejection *must* be a measure of how well the star/galaxy classification is working.

Studies of the performance of the SDSS S/G separation in the Groth strip data (where accurate classification is available from HST imaging) indicates that separation is reliable to at least a r' of 21.5 in data that has a 5σ limit of $r' \approx 22$.

The $u' - r'$ colour of galaxies is a good discriminant of Hubble type (Strateva et al. 2001). Figure 4 shows $u' - r'$ plotted against what is essentially the likelihood ratio for deVaucouleurs and exponential models shows that the galaxy likelihoods provide clear *morphological* clasification to $r' \approx 20$, in data with a PSF 5σ limit of about 22.5.

5. Conclusions and Software Sociology

As far as he knows, this section represents the views only of the primary author and not of his coauthors. Those of you who know him will have heard these opinions before.

The SDSS has been very challenging technically, scientifically, and managerially. In all categories the software stands out: The hardest technical aspect of building the SDSS was probably the software, although building the mosaic camera wasn't easy; some of the software was a major *scientific* challenge; and the software was undoubtedly the hardest part of the project to manage.

Let me expand upon some of these issues. We have found it extremely hard to hire good people to work on astronomical software. There is no career path within the universities for software specialists, despite the fact that there's no logical distinction between building hard- and soft-ware instruments. Smart and sensible graduate students, desirous of a career in astronomy, simply don't choose to specialise in the software required to reduce modern observational datasets.

Hiring computer professionals is not the solution to this problem. Besides being (if competent) too expensive for the average astronomical project, they simply don't possess the skills needed to solve the *scientific* challenges posed by astronomical data. We need *scientists* to resolve scientific problems, albeit with support from people whose job it is to know about optimizers, LALR(1) grammars, and good software engineering practices. We also need our software-scientists to be in rich scientific environments, where they can talk with (say) the quasar-scientists about the data analysis that they are carrying out.

If we, as a community, knew how to reuse software from one project on another some of these problems might be alleviated, but I don't believe that they would go away. The availability of good numerical libraries hasn't made the development of new cosmological codes stop; the impetus for change comes from the desire to do things better, not just from the not-invented-here syndrome.

I believe that part of the problem is that we, as a community have not yet faced the reality that software is *difficult*, and that the dynamic range between the really good and the average programmer is as great as that between Lyman Spitzer and the average graduate student. This makes management difficult; imagine trying to get a collaboration of 100 self-opinionated astronomers to agree about the best way to solve a problem, and tell me why this is any easier than running a large modern collaboration involving large amounts of software. I reluctantly believe that we must learn to run large software projects (and all large projects nowadays are large software projects) as benevolent dictatorships—of course with the implicit hope that I shall be the dictator (but not the manager).

Acknowledgments. The Sloan Digital Sky Survey (SDSS) is a joint project of The University of Chicago, Fermilab, the Institute for Advanced Study, the Japan Participation Group, The Johns Hopkins University, the Max-Planck-Institute for Astronomy, New Mexico State University, Princeton University, the United States Naval Observatory, and the University of Washington. Apache Point Observatory, site of the SDSS telescopes, is operated by the Astrophysical Research Consortium (ARC).

Funding for the project has been provided by the Alfred P. Sloan Foundation, the SDSS member institutions, the National Aeronautics and Space Administration, the National Science Foundation, the U.S. Department of Energy, Monbusho, and the Max Planck Society.

The SDSS Web site is http://www.sdss.org/.

References

Blanton, M., et al. 2001, The Luminosity Function of Galaxies in SDSS Commissioning Data, submitted to AJ

Fan, X., et al. 2000, AJ, 120, 1167

Fan, X., et al. 2001, AJ, 121, 54

Fischer, P., et al. 2000, AJ, 120, 1198

Gunn, J. E., et al. 1988, AJ, 116, 3040

Ivezić Z., et al. 2000, AJ, 120, 963

Leggett, S. K., et al. 2000, ApJ, 536, L35

Lupton, R. H., et al. 2001, SDSS Image Processing I: The Deblender AJ, submitted

Sergey, G., et al. 1996, in ASP Conf. Ser., Vol. 101, Astronomical Data Analysis Software and Systems V, ed. G. H. Jacoby & J. Barnes (San Francisco: ASP), 248

Strateva I., et al. 2001, in preparation

York, D. G., et al. 2000, AJ, 120, 1579

The Networked Telescope: Progress Toward a Grid Architecture for Pipeline Processing

Raymond Plante, Dave Mehringer, Damien Guillaume[1], Richard Crutcher[1]

National Center for Supercomputing Applications

Abstract. Pipeline processing systems for modern telescopes are widely considered critical for addressing the problem of ever increasing data rates; however, routine use of fully automated processing systems may discourage the typical user from exploring processing parameter space or trying out new techniques. This issue might be particularly important with regard to radio interferometer data in which the post-calibration processing required to create an image for scientific analysis is not well defined. We describe our architecture for the BIMA Image Pipeline which attempts to address this issue. The pipeline by default is automated and uses NCSA supercomputers to carry out the processing; however, as we further develop the system, we will progressively add tools that enable the astronomer to guide the automated processing. This same system can also be used by the astronomer to create new processing projects using data from the archive. We report our progress on this system, highlighting the design features that allow for greater user interaction, including a research-oriented data model, web-based portal interfaces, and use of the AIPS++ toolkit. The ultimate goal is to evolve the system into a flexible, computational grid for processing radio interferometer data.

1. Vision: The Data Life Cycle within a Grid-based Ecosystem

The BIMA Data Archive was built to deliver data automatically in real–time from the BIMA interferometer to a repository at NCSA where it can easily be accessed from NCSA supercomputers for high–performance processing or delivered to astronomers via the Web for local processing (Crutcher 1994, Plante & Crutcher 1997). We are now in the process of expanding the archive system to support automated calibration and construction of images from the raw visibility data, a process traditionally done interactively by the investigating astronomer. Pipeline processing of modern astronomical data is a problem well-suited to a Grid environment because of the inherent distributed nature of the hardware, software, data, and people involved. As part of NCSA's efforts to build Grid infrastructure, we have adopted a Grid model for implementing the BIMA Image Pipeline.

---

[1] Astronomy Department, University of Illinois Urbana–Champaign

Our vision of a grid for radio astronomy starts by viewing the flow of data not strictly as a pipeline but rather as a cycle. The cycle begins when an astronomer turns an idea into an observing proposal and plan. When the data are generated by the telescope, they can be transfered through a variety of channels to multiple, distributed archives, and then processed using multiple, distributed compute engines. Further analysis by the astronomer often incorporates information from a variety of network-based services (that need not be tightly integrated with the rest of the Grid). Finally, the results and data are published to feed ideas into new research projects. Much of our previous work with the BIMA Data Archive and the Astronomy Digital Image Library[2] (ADIL; Plante et al. 1996) concentrated on data archiving, publishing, and delivery; we are now turning our attention to problems of distributed computing.

Our pipeline is motivated by the same issues that are driving other pipelines in use and in development today. These issues have been described in previous ADASS papers; however, a few of the motivators are worth highlighting. Ever-increasing data production rates driven by improvements in hardware threatens radio astronomy just as it does other fields. As new millimeter arrays come on-line (CARMA, ALMA), there is a greater need to understand the data as least as fast as they are being produced. Furthermore, a pipeline that operates on new data from the telescope can just as easily be applied to data from the archive; thus, value is added to the data when the pipeline enables archival research. Finally, when the pipeline incorporates high-performance computing resources, not only can we tackle larger observing projects, we can explore more of the processing parameter space. This can be important when processing parameters are not well-defined, as is typically the case in radio astronomy.

Building the pipeline within a grid environment adds value as well. Other papers in this volume describe what constitutes a "grid" and why it might be useful; however, again, a few reasons are worth highlighting. First, a grid-based pipeline can provide users flexible access to high-performance computing. It can provide the infrastructure needed to integrate data from diverse sources. People are also an important component of a grid; thus, it can cultivate a community for developing and disseminating new processing techniques. These motivators extend beyond the BIMA community, which is why we have been collaborating with NRAO to develop a general blueprint for a data grid for radio astronomy.[3]

2. Architecture: A Portal for Guided Automation

One approach to building a pipeline might be to take the tools one uses to process data interactively, wrap them up, and connect them together so that they can run automatically. We see our approach as the exact opposite: we want to build a system that is inherently automated and then extend it to add ever increasing amounts of interactivity, resulting in an architecture that might be described as "guided automation." Here are some ways we want to allow users

---

[2]http://adil.ncsa.uiuc.edu/

[3]http://monet.astro.uiuc.edu/papers/COBRA/NGRA_white_paper.pdf

to interact with the pipeline: (a) **prior to observations:** the astronomer can override default processing parameters to better suit the scientific goals of the project; (b) **during observations:** the astronomer can monitor the telescope and data via the web; (c) **after observations:** the astronomer can browse the archive's holdings using customizable displays; (d) **prior to processing:** the astronomer can create his/her own scripts for reprocessing archival data; (e) **during processing:** optional viewers can be opened up to monitor, and possibly steer, the deconvolving process.

One of the challenges to enabling all these diverse features is delivering interfaces to users over the network. This is a problem for most any type of grid; thus, NCSA has been developing a framework for *scientific portals*. A scientific portal can be thought of as a collection of network-based services and documents integrated into a single, customizable web environment for the purpose of conducting scientific research. NCSA has recently released its first version of such a framework called the *Open Portal Interface Environment* (OPIE).

3. Data Management through Data-centric Programming

In many ways, the hardware and scientific processing software are the simpler parts of the pipeline. The rest of the system is about information management; thus, metadata have an important role: they drive the data through the system. We encode the metadata used to archive, process, and deliver datasets to astronomers in XML. Our schema[4] is based on a research-oriented data model that organizes data into hierarchical collections (i.e., projects, experiments, trials, and datasets). Users can browse these metadata via the web; XSLT is used to turn XML into HTML on-the-fly. The goal of the HTML rendering is to make the relationships between data and collections intuitively clear.

Our use of XML has resulted in software that is very *data-centric* (as opposed to process-centric, see Guillaume & Plante 2001). We found XML to very useful for rapid modeling of our data; object-oriented software built around entities in our model followed naturally. In this regard, our application has two important challenges: first, our model is large (we currently define about 100 objects), and second, we know it will evolve as we add new capabilities. The cost of these challenges on software can be quite high; however, our use of the Quick software package controls this cost. Not only can the package convert XML documents into intelligent Java objects, but (through our contributions to the package) it can also convert the schema definition document into custom Java classes (see Guillaume & Plante 2001).

Our pipeline is "metadata-driven;" this means that the task of processing the data is a matter of converting metadata into processing instructions. This can be handled quite effectively using XSLT: instead of converting XML to HTML, we convert it into high level Glish scripts (see §4.) that select and configure predefined processing recipes. XML is also effective for storing state information for processing that may proceed incrementally over a period of months or years as more data becomes available.

---

[4] http://monet.astro.uiuc.edu/BIP/xml/bimaarch_dtd.txt

4. Data Processing with AIPS++

The processing is carried out using AIPS++, which employs the Glish scripting language (Scheibel 2000) to glue processing objects together. Its event-driven programming model (combined with the toolkit nature of AIPS++) makes it ideal for building automated processing in a distributed environment. An important role for NCSA, as a member of the AIPS++ development consortium, is to enable support for parallel processing on a range of mildly to massively parallel machines, with a particular emphasis on Linux clusters. The Intel Itanium-based supercluster that will be brought on–line at NCSA this year will handle the bulk of the imaging and deconvolution chores for the pipeline, while smaller machines will handle the serial processing.

The pipeline's processing engine presents its own set of interesting challenges. Foremost is making effective use of parallel processing. Common forms of imaging and deconvolution can be broken down into naturally self-contained components that can be parsed out to separate processors. The simplest example is dividing the problem into independent frequency channels. A more interesting application to widefield imaging is described by Golap et al. (2001). MPI is used to orchestrate the parallel processing (Roberts et al. 2000). An important focus of current research is parallel I/O via MPI-2. Another challenge is making remote grid processing appear interactive within the AIPS++ interface. Grid processing in general is typically queue-scheduled, which can introduce latencies and unnatural barriers one would not see in traditional interactive environments; nevertheless, users will want to interact with processing in real-time, and so computing resources must be negotiated in real-time as well. Finally, a powerful pipeline can be an excellent tool for exploring processing parameter space from which we hope to develop diagnostic measures that can not only evaluate various processing strategies but also predict the most appropriate strategy for a given experiment. Such diagnostics would further improve the quality of the data products coming from the pipeline.

References

Crutcher, R. M. 1994, in ASP Conf. Ser., Vol. 61, Astronomical Data Analysis Software and Systems III, ed. D. R. Crabtree, R. J. Hanisch, & J. Barnes (San Francisco: ASP), 409

Golap, K., Kemball, A., Cornwell, T., & Young, W. 2001, this volume, 408

Guillaume, D. & Plante, R. 2001, this volume, 221

Plante, R. L. & Crutcher, R. M. 1997, Proc. SPIE, 3112, 90

Roberts, D. A., Crutcher, R. M., Young, W., & Kemball, A. J. 1999, in ASP Conf. Ser., Vol. 172, Astronomical Data Analysis Software and Systems VIII, ed. David M. Mehringer, Raymond L. Plante, & Douglas A. Roberts (San Francisco: ASP), 15

Schiebel, D. R. 2000, in ASP Conf. Ser., Vol. 216, Astronomical Data Analysis Software and Systems IX, ed. N. Manset, C. Veillet, & D. Crabtree (San Francisco: ASP), 39

The ESO Imaging Survey Project: Status and Pipeline Software

Richard N. Hook[1], Stephane Arnouts, Christophe Benoist, Luiz da Costa, Roberto Mignani, Charles Rité, Michael Schirmer, Remco Slijkhuis, Benoît Vandame, Andreas Wicenec

European Southern Observatory, Karl-Schwarzschild Str.2, D-85748 Garching bei München, Germany

Abstract. The ESO Imaging Survey (EIS) is a major public imaging survey project being conducted by the European Southern Observatory using several different telescopes and detectors at La Silla in Chile. The primary aim is the identification of samples suitable for more detailed study using the ESO Very Large Telescope (VLT). The first part of the project consisted of two parts: EIS-Wide, an optical survey to moderate depth ($I_{AB} \lesssim 24$) covering about 20 square degrees; and EIS-Deep, which covered smaller areas of sky to greater depth ($B_{AB} \lesssim 26.5$) and also included deep near-IR imaging. These surveys are essentially complete and the data are available to users world-wide through the EIS web pages.[2]

Following a successful pilot project with the new Wide Field Imager (WFI) on the 2.2m telescope at La Silla, the EIS team is now engaged in a public imaging survey using the WFI telescope in conjunction with IR imaging using SOFI on the NTT. Three deep fields of one degree square, including the Chandra Deep Field South (CDF-S), are being imaged in the optical along with smaller sub-areas in IR bands. In addition, 160 stellar fields have been imaged as part of the preparations for the use of the FLAMES fibre-feed system for spectrographs on the VLT.

To support the large data volume from this survey and to facilitate its rapid scientific exploitation, a complete end-to-end pipeline system has been developed. Here we give a brief outline of the project so far, and describe the motivation and architecture for the pipeline software.

1. Origins and Historical Evolution

The ESO Imaging Survey (EIS) was established at the European Southern Observatory in 1997 to perform appropriate public surveys to help identify samples of interesting objects for the ESO Very Large Telescope, which was then approaching completion. Existing surveys at that time did not go deep enough to reach the spectroscopic limits of the new generation of large telescopes. A

---

[1]Space Telescope – European Coordinating Facility

[2]http://www.eso.org/eis

working group was established to oversee the surveys, and a visitor programme established to allow experts from the community to be funded to work at ESO as part of the EIS team. EIS aimed to establish a framework for future imaging surveys and to foster collaboration with the European astronomical community.

The first phase of the survey was EIS-Wide, which was devoted to moderately deep imaging ($I_{AB} \lesssim 24$) in BVI in four patches of the southern sky. A total of 17 square degrees in I and smaller areas in B and V were obtained using the EMMI instrument on the ESO NTT. This was complemented by EIS-Deep, which imaged two smaller fields (the Hubble Deep Field South and the Chandra Deep Field South) to greater depth ($B_{AB} \lesssim 26.5$) and also included near infrared imaging with SOFI. Data were made public in 1998 and described in accompanying papers (Nonino et al. 1999; Olsen et al. 1999a; Prandoni et al. 1999; Zaggia et al. 1999; Olsen et al. 1999b; Benoist et al. 1999; Scodeggio et al. 1999).

In 1999 the Wide Field Imager (WFI) on the 2.2m telescope at La Silla was available. This efficient, wide-field mosaic camera system was used in a pilot project to complete EIS-Wide and image many stellar fields. From November 1999 this was extended into a formal Deep Public Survey which is currently in progress. It covers three fields, each 2×0.5 degrees in the UBVRI optical bands and covering 450 arcmin^2 at JKs in the near infrared. This is supplemented by 160 stellar fields in preparation for the FLAMES multi-fibre spectrograph to be installed on the VLT. These and the earlier fields are shown on Figure 1.

2. Science Results and Data Examples

An example of a small piece of an EIS-Wide patch is given as Figure 2, which shows how many overlapping small fields, from different telescopes and cameras, can be mosaiced effectively and aligned from separate astrometric solutions. Some initial science results on distant galaxy clusters have been published (Olsen et al. 1999a, 1999b; Scodeggio et al. 1999) along with preliminary lists of interesting colour-selected point-sources (Prandoni et al. 1999; Zaggia et al. 1999). A much larger area of sky is now available in BVI from the combined pilot and EIS-Wide surveys and is yet to be fully explored. Extensive follow-up observations, with the VLT and other telescopes, are currently in progress.

3. Early Software Approaches

The initial data volume and rate were modest and a pragmatic approach of re-using existing software was adopted (Hook et al. 1998). Tools pressed into service included SExtractor (Bertin & Arnouts 1996) for object catalogue preparation, IRAF for standard frame reduction, Drizzle (in adapted form) for image coaddition, and Eclipse (Devillard 1999) for reducing IR data. The LDAC tools, developed for the DENIS survey (Epchtein et al. 1998) were used for the demanding astrometric and photometric calibration steps. These items were controlled from shell scripts. Unfortunately the simultaneous observing, software development, reduction, data-release, science analysis, and publication of results by a small team, combined with the increase in data volume from the WFI and crit-

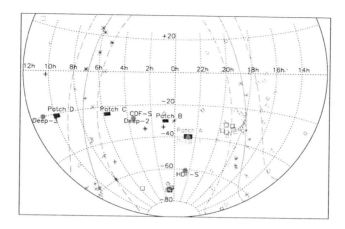

Figure 1. The Southern Sky showing the positions of the EIS fields along with the Milky Way. The main fields are individually labeled. Asterisks are open cluster, diamonds globular cluster, crosses local group galaxies triangles Milky Way bulge/halo fields and squares Sagittarius Dwarf and LMC/SMC fields.

ical staff changes led to an impossible situation and a major pipeline overhaul was conducted in 1999.

4. The Current EIS Pipeline

The current pipeline design aims for robustness and ease of use but accepts the constraint that it would be impossible to create a fully automatic system using the resources available. Extensive use of a Sybase database allows tracking of the processing and storage of some intermediate information. The use of Objectivity was considered but inadequate resources were available for such an initiative. The ESO archive provides both a repository for the raw data and a mechanism for distributing the data products to the community. Python has been adopted as the pipeline glue language, with a GUI (written in Tkinter) layered on top for ease of use and training for a rapidly changing group. The GUI provides both high level pipeline control panels and access to lower-level functions for expert use. It is an end-to-end system with components to address aspects of the dataflow from proposal preparation through to catalogue preparation and science backend tasks such as preparing colour-colour plots and determining photometric redshifts. Some components remain from the earlier work but entirely new pipeline software has been created for the Deep and Pre-FLAMES reductions. The pipeline components and infra-structure are mostly in place at the time of writing but not all are in full operation.

Figure 2.   Example of a small piece of EIS-Wide Patch D. The blue and green channels come from 2.2/WFI B and V data and the red from earlier NTT/EMMI I band imaging.

5. Future ESO Sky Surveys

The current EIS survey is only a small part of the planned European imaging surveys for the near future. Over the next two years the VLT Survey Telescope, a 2.5m dedicated survey telescope to be sited adjacent to the VLT, and VISTA, a UK project for a 4m optical-IR survey telescope, also on the same site, will conduct much larger surveys.

References

Benoist, C., et al. 1999, A&A, 346, 58

Bertin, E. & Arnouts, S. 1996, A&AS, 117, 393

Devillard, N. 1999, in ASP Conf. Ser., Vol. 172, Astronomical Data Analysis Software and Systems VIII, ed. David M. Mehringer, Raymond L. Plante, & Douglas A. Roberts (San Francisco: ASP), 333

Epchtein, N., et al. 1998, ESO Messenger, 87, 27

Hook, R. N., Bertin, E., da Costa, L., Deul, E., Freudling, W., Nonino, M., & Wicenec, A. 1998, in ASP Conf. Ser., Vol. 145, Astronomical Data Analysis Software and Systems VII, ed. R. Albrecht, R. N. Hook, & H. A. Bushouse (San Francisco: ASP), 320

Nonino, M., et al. 1999, A&AS, 137, 51

Olsen, L. F., et al. 1999a, A&A, 345, 363

Olsen, L. F., et al. 1999b, A&A, 345, 681

Prandoni I., et al. 1999, A&A, 345, 448

Scodeggio, M., et al. 1999, A&AS, 137, 83

Zaggia, S., et al. 1999, A&AS, 137, 75

Astronomical Data Analysis Software and Systems X
ASP Conference Series, Vol. 238, 2001
F. R. Harnden Jr., F. A. Primini, and H. E. Payne, eds.

GIGAWULF: Powering the Isaac Newton Group's Data Pipeline

Robert Greimel,[1] Nicholas A. Walton, Don Carlos Abrams

Isaac Newton Group, Apartado de Correos 321, 38700 Santa Cruz de La Palma, Canary Islands, Spain

Mike Irwin, James R. Lewis

Institute of Astronomy, University of Cambridge, Madingley Road, Cambridge, UK

Abstract. The Isaac Newton Group of Telescopes (ING) is currently implementing a unified data flow and processing system to support all its astronomical instrumentation systems. The ING data infrastructure is tightly integrated, encompassing the initial data acquisition, pipeline processing and subsequent archiving and distribution of raw and processed data products.

In this paper we describe *Gigawulf*—a Beowulf cluster based on commodity PCs running Linux—which provides cost effective processing power for the ING's image data pipeline processing. The configuration and performance of Gigawulf is discussed. The operation of the data pipeline on the cluster together with the integration of the *Gigawulf* pipeline processor system with the DVD-R tower archiving system is detailed.

1. Introduction

The Isaac Newton Group (ING) operates three telescopes, including the 4.2-m William Herschel Telescope (WHT) and the 2.5-m Isaac Newton Telescope (INT). In the era of 8-m telescopes, there is increasing pressure to operate 4-m class telescopes in the most economic, efficient, and effective manner possible. The ING is addressing these three 'E's, in part with its implementation of an improved streamlined data flow system encompassing data acquisition to archiving of processed data products.

This paper describes the ING's implementation of the processing unit, 'Gigawulf,'[2] a Beowulf class cluster, which ensures sufficient computational capability to handle the ING data flow.

---

[1]Institut für Geophysik, Astrophysik und Metorologie, Universität Graz, Austria

[2]http://zwolle.ing-slo.iac.es/Beowulf

2. The ING Data Flow System and Data Rates

In order to support new large array detectors and increase on-sky observing efficiencies, the ING has recently introduced a new Data Acquisition System (DAS, Rixon et al. 2000). A key consideration in its implementation was that it would provide the front-end to the ING data flow system (Walton et al. 1998). Thus attention was paid to items such as correct and sufficient FITS (see e.g., Walton & Rixon 2000) header information being made available for each science and calibration data file.

All raw data are archived on DVD-R media, mass on-line availability provided by a dual juke box system with a current capacity of $\sim$ 6TB. The data are accessible via a WWW based front end to a Sybase database (Lewis & Walton 1998).

The advent of large format CCD arrays and large infra-red detectors has led to an explosion in data volumes. For example, the current data rates at the ING are determined by the Wide Field Camera on the INT which typically generates $\sim$8 GB/night and the IR camera on the WHT generating some $\sim$4 GB/night. In total, data flows can amount to 15–20 GB/night from all telescopes.

The current day availability of affordable processing power and storage capability has opened the possibility to provide (semi-)processed data products at the point of data origin to the visiting astronomer. These reduced data products will form the core data resource of new 'Virtual Observatories' (see e.g., http://www.astro.caltech.edu/nvoconf/).

3. The ING Data Reduction System

Details of the ING's data processing pipeline, as implemented for the reduction of imaging data, are described elsewhere (Irwin & Lewis 2001). The basic data reduction steps are: linearity correction, bias subtraction, flat fielding and the application of a basic astrometric solution. Additional steps in the pipeline are de-fringing, an accurate astrometric solution, object detection, classification and catalogue generation.

Currently, the pipeline (*quick look* and *full*) is implemented for the reduction of imaging data, optical and near infra-red, only. A pipeline will be introduced for Echelle and multi-fibre spectroscopic data by the end of 2001.

The *quick look* pipeline delivers a processed image to the observers within five minutes of image acquisition. This enables immediate assessment of the image quality and instrument performance. The *quick look* pipeline differs from the science pipeline in two areas: it applies calibration files from the most recent science pipeline run instead of the calibration frames from the current run, and it implements a subset of the full reduction, terminating with the de-fringing stage.

The science pipeline provides the observer with reduced data shortly after the end of the observing run. To provide the highest quality data a limited amount of human intervention is necessary, mainly in rejecting poor calibration frames. This pipeline has been running on a Sun UltraSparc system servicing ING Wide Field Survey Data since August 1998 (Lewis et al. 1999).

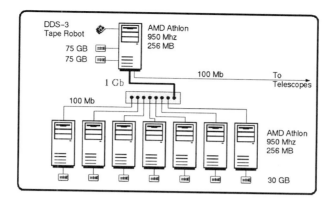

Figure 1.   Schematic overview of the *Gigawulf* cluster.

4. Powering the Pipeline: *Gigawulf*

In order to provide an economic processing unit for the pipeline, it was decided to use commodity PC components. Recent advances in the clustering of PCs, making use of the Linux (see e.g., http://www.linux.com) operating system have made this feasible. Linux based PC farms have been found by a number of groups to offer a powerful and cost effective solution for large computational problems. Indeed, a small number of PC systems have been developed for use in astronomical data processing environments (see e.g., Gravitor[3]).

The ING data pipeline is a coarse grained parallel processing case. To a first approximation, each science data frame is processed in an identical fashion, with no cross reference to any other. Therefore a night's data can be equally distributed between the nodes. PC clusters are ideally suited for this case (see discussion by Brown 1999[4]).

4.1. The *Gigawulf* Structure

Gigawulf is a 'Beowulf' type cluster (see http://www.beowulf.org for a definition and related links) of eight high end PCs. Each node consists of an AMD Athlon 950 MHz processor with 256 MB of main memory. The seven slave nodes have 30 GB EIDE hard disks while one node, subsequently called the head node, has two 75 GB EIDE hard disks. The head node also has a DDS-3 DAT tape robot as well as a second network card which provides the connection to the telescopes and data archives. The network in the cluster is 100 Mbps apart from the head node, which has a gigabit connection. A schematic view of the system is shown in Figure 1.

---

[3]http://obswww.unige.ch/~pfennige/gravitor/gravitor_e.html

[4]http://www.phy.duke.edu/brahma/beowulf_advanced.ps

4.2. 'Gigawulf' Software Issues

To minimise the operational and maintenance overheads, the Scyld Beowulf extension (Scyld Computing Corporation[TM5]) to Linux (currently based on RedHat's[6] 6.2 distribution) has been used as the operating system for *Gigawulf*.

Scyld Beowulf supports standard Linux interfaces and tools. It enhances the Linux kernel with features (provide by bproc[7]) that allow users to start, observe, and control processes on cluster nodes from the cluster's head node (Hendriks 1999[8]). With this arrangement, software needs only to be configured on the head node. The result is that the cluster appears to be more like a traditional multi-processor computer to a user or developer. This reduces the cost of cluster application development, testing, training, and administration.

The existing pipeline software, has been ported to run on *Gigawulf* with only small modifications. Scripts have been developed to handle the input of data from the telescopes and output of reduced data to archiving media (DVD-R and DDS-3 tape).

5. Conclusion

In terms of performance, the *Gigawulf* cluster is currently $\sim$ 10 times more cost effective than using alternative computing hardware, for instance UltraSparc computers. The cost of use has been minimised because only limited changes have been required in order to run the existing data reduction pipeline software on it. For the future, *Gigawulf* will be enlarged with the addition of more slave nodes.

References

Irwin, M. J. & Lewis, J. R. 2001, NewAR. (in press)

Lewis, J. R. Bunclark, P. S., & Walton, N. A. 1999, in ASP Conf. Ser., Vol. 172, Astronomical Data Analysis Software and Systems VIII, ed. David M. Mehringer, Raymond L. Plante, & Douglas A. Roberts (San Francisco: ASP), 179

Lewis, J. R. & Walton, N. A. 1998, in SPIE Proc., Vol. 3349, 263

Rixon, G. T., Walton, N. A., Armstrong, D. B., & Woodhouse, G. 2000, in SPIE Proc., Vol. 4009, 132

Walton, N. A., Bunclark, P. S., Fisher, M. P., Jones, E. L., Ress, P. C. T., & Rixon, G. T. 1998, in SPIE Proc., Vol. 3351, 197

Walton, N. A. & Rixon, G. T. 2000, ING Newsl., 3, 31

---

[5]http://www.scyld.com

[6]http://www.redhat.com

[7]http://www.beowulf.org/software/bproc.html

[8]http://www.beowulf.org/software/bproc.html/bproc-LinuxExpo99.ps.gz

Using OPUS to Perform HST On-The-Fly Re-Processing (OTFR)

Michael S. Swam, Edwin Hopkins, Daryl A. Swade

Space Telescope Science Institute, 3700 San Martin Drive, Baltimore, MD 21218

Abstract. The Hubble Space Telescope (HST) OPUS implementation of On-The-Fly Calibration (OTFC) processing currently provides the benefit of applying the most current calibration algorithms, reference files, and repairs of errant header keyword values (aperture, shutter, etc.) to Level-1b datasets as they are retrieved from the HST archive. While OTFC has performed well, a number of concerns about maintenance and flexibility have resulted in the evolution towards an On-The-Fly Re-Processing (OTFR) System. Also based on the OPUS pipeline architecture, OTFR carries further the notion of creating products for archive users at the time of their request, by completely regenerating calibrated (Level-2) data products for an exposure from the base telemetry files sent from the HST (Level-1a data). By starting processing at this earlier state, and taking advantage of the changes to the data processing software that are made as that software matures, improved, more consistent calibrated (Level-2) data products are produced. Use of OPUS distributed multi-processing, the relatively small size of HST datasets, and the efficiency of the data processing and calibration software results in a very small impact on the overall time it takes to complete an archive retrieval. There could be an impact to archive research, however, since the archive catalog meta-data will not completely reflect the reprocessed products as they would be delivered to the archive user. This problem will be addressed by performing catalog updates for any major discrepancies. This paper will describe the concerns raised about OTFC, the design of the OTFR pipeline system, and the benefits of using the OPUS architecture in this design.

1. The Existing On-The-Fly Calibration (OTFC) System

The Hubble Space Telescope archive at the Space Telescope Science Institute (STScI) has had an on-the-fly calibration system (Swam & Swade 1999) in place for the Space Telescope Imaging Spectrograph (STIS) and the Wide-Field Planetary Camera 2 (WFPC-2) instruments since December 1999. Whenever calibrated STIS or WFPC-2 products are requested from the HST archive, the exposures are automatically recalibrated using the latest reference files, calibration algorithms, and known repairs to the FITS header keyword values. These improved products are then delivered to the user who made the archive request.

2. Problems with OTFC and Reasons for On-The-Fly Re-Processing

While the WFPC-2 instrument support group at STScI has been very pleased with the OTFC system, the STIS instrument group has had to do quite a bit more work to get their OTFC output products in optimal form. Because STIS is a more complex and less mature instrument than WFPC-2, over 200,000 header keyword repairs (most only for aesthetic reasons) are specified in the OTFC keyword repair database table (Lubow & Pollizzi 1999) for STIS exposures, compared with around 700 for WFPC-2. This large set of repairs now has to be checked for layering issues as each new STIS repair is implemented.

More importantly, there is separate pre-archive pipeline software that converts HST raw telemetry (Level-1a data) into calibrated data products for loading the archive catalog (Swade, Hopkins, & Swam 2001). Therefore new repairs must be made in both the keyword repair table for data already in the archive, and in the pre-archive pipeline software for any new exposures taken with HST. There are also some cases where repairs to the data format of STIS exposures in the archive are warranted (e.g., time-tag observations of very bright sources, Sahu 2000), and the OTFC system cannot automatically make these repairs, since it currently only fixes header keyword values. The exposures must be pulled out of the archive and reprocessed through the pre-archive pipeline software, with the resulting products then re-ingested into the archive.

For the Near-Infrared Camera and Multi-Object Spectrometer (NICMOS) and the Advanced Camera for Surveys (ACS), the complication of adding multi-file associated exposures is required. This capability already exists in the pre-archive pipeline code.

These problems all point to using the pre-archive pipeline software in the OTFC system, so that each archive request results in reprocessed products using the most complete, up-to-date data processing software. This is the on-the-fly re-processing system, OTFR.

3. Design of the OTFR System

The OTFR system makes use of the OPUS pipeline architecture (Rose & Miller 2001, Miller 1999, Swade & Rose 1999) to spread the automated re-processing over a number of nodes in a cluster of CPUs. The pre-archive pipeline software is already implemented using OPUS, so it was straightforward to add OPUS pipeline stages to perform OTFR. Stages were added before the pre-archive pipeline code to collect OTFR requests from the archive and feed them in for processing, and stages were added after the pre-archive pipeline code to collect processed products and feed them back to the archive for distribution.

The stages before the pre-archive pipeline have to:
- poll for OTFR requests from the archive,
- convert requested exposure names into the names of the corresponding raw telemetry files (a many to many relationship, held in database tables),
- request the raw telemetry files from the archive,
- segment raw telemetry files into one exposure per file,
- pass the segmented raw telemetry files to the pre-archive pipeline.

The stages after the pre-archive pipeline have to:

- poll for completed products,
- convert all products to FITS format, if not already,
- return the completed products to the archive,
- clean out completed products from the pre-archive pipeline.

4. Using OPUS in OTFR

The OPUS system provides a number of features beneficial to the implementation of OTFR:
- distributed multi-processing keeps throughput high;
- integration of scripts and programs into pipeline stages is straightforward;
- support for file arrival, time trigger, and blackboard events provides the flexibility for integrating the OTFR pipeline with the archive system;
- a single executable or script can be configured multiple ways using ASCII resource files, allowing one set of pre-archive code to meet many needs (Boyer & Choo 1997).

This last point is very important in meeting one of the goals of OTFR; to allow the pre-archive pipeline software to be used to both populate the archive catalog, and to reprocess exposures for archive users. This reduces maintenance and promotes code reuse, allowing fixes and enhancements to one set of software to benefit both tasks.

5. Issues for OTFR

The use of on-the-fly re-processing does have implications for the state of the archive catalog meta-data. This meta-data, consisting of astronomical target information, instrument configuration, and other ancillary parameters, is populated when the exposure is initially received from HST and processed through the pre-archive pipeline. This implies that the meta-data reflects the state of processing at that instant in time. When an archive user later requests this exposure through OTFR, the exposure is again processed through the pre-archive pipeline code, but this code could have evolved since the original population of the catalog meta-data. This means that the catalog meta-data may not accurately reflect the exposure data at retrieval time. This has implications for archive users, as they search the catalog to find observations that meet certain criteria. To address this issue, the HST archive is committed to updating the catalog meta-data when significant differences exist between the original catalog values, and new values coming out of OTFR. The intent is to eventually automate this process, using special processing flags for the OTFR system.

6. The Future of OTFR at STScI

OTFR is currently being tested, and has been scheduled for deployment at STScI in early 2001 for STIS and WFPC-2. The NICMOS will be added shortly thereafter, while the ACS will be an OTFR instrument from the moment it is added to HST during servicing mission 3B, currently scheduled for fall of 2001. It is projected that the future Cosmic Origins Spectrograph (COS) and the Wide Field Camera 3 (WF3) will also be OTFR instruments as soon as they are installed in the HST.

Acknowledgments. Many thanks to those who reviewed the design and code for the OTFR system at STScI: Dorothy Fraquelli, Chris Heller, Warren Miller, Jim Rose, John Scott, Lisa Sherbert, and Steve Slowinski.

References

Boyer, C., & Choo, T. H. 1997, in ASP Conf. Ser., Vol. 145, Astronomical Data Analysis Software and Systems VII, ed. R. Albrecht, R. N. Hook, & H. A. Bushouse (San Francisco: ASP), 42

Lubow, S., & Pollizzi, J. 1999, in ASP Conf. Ser., Vol. 172, Astronomical Data Analysis Software and Systems VIII, ed. David M. Mehringer, Raymond L. Plante, & Douglas A. Roberts (San Francisco: ASP), 187

Miller, W. 1999, in ASP Conf. Ser., Vol. 172, Astronomical Data Analysis Software and Systems VIII, ed. David M. Mehringer, Raymond L. Plante, & Douglas A. Roberts (San Francisco: ASP), 195

Rose, J. F. & Miller, W. 2001, this volume, 325

Sahu, K. 2000, private communication

Swade, D. A. & Rose, J. F. 1999, in ASP Conf. Ser., Vol. 172, Astronomical Data Analysis Software and Systems VIII, ed. David M. Mehringer, Raymond L. Plante, & Douglas A. Roberts (San Francisco: ASP), 111

Swade, D. A., Hopkins, E., & Swam, M. 2001, this volume, 295

Swam, M., & Swade, D. A. 1999, in ASP Conf. Ser., Vol. 172, Astronomical Data Analysis Software and Systems VIII, ed. David M. Mehringer, Raymond L. Plante, & Douglas A. Roberts (San Francisco: ASP), 203

HST Data Flow with On-The-Fly Reprocessing

Daryl A. Swade, Edwin Hopkins, Michael S. Swam

Space Telescope Science Institute, 3700 San Martin Drive, Baltimore, MD 21218

Abstract. This paper documents the flow of data through the HST ground system in the OTFR era, both on initial receipt from the spacecraft and from the point of view of archive retrieval. The goal is to provide an understanding of what processing has been performed on the data and to indicate the state of data at various stages of processing.

1. Introduction

Traditionally, the Hubble Space Telescope (HST) data flow sequentially processed science observations through an OPUS[1] pipeline (Rose 1998) and archived these data with no further processing in the post-observation ground system. The pre-archive pipeline consisted of receipt of science telemetry, packaging exposures into observations, transforming the images or spectra into a standard format, and calibrating the data. Pre-archive calibration used the reference files available in the pipeline at the time of the observation, and, in almost all cases, this calibration was not optimal. Upon issuing an archive request, HST data users would receive the original processed version of the data, which required them to perform re-calibration on their own with Space Telescope Science Institute (STScI) provided software.

A new paradigm and enhancement to On-The-Fly Calibration (Lubow & Pollizzi 1999, Swam & Swade 1999), On-the-Fly Reprocessing (OTFR) is being implemented at STScI for HST data for most current and future science instruments. With OTFR, data retrieved from the HST archive are completely reprocessed from the raw telemetry (pod files) using the latest data processing and calibration software. Users automatically get the advantage of keyword updates, data processing software upgrades and bug fixes, as well as the latest calibration. Further details about the OTFR system can be found in a companion paper in this volume "Using OPUS to Perform HST On-The-Fly Reprocessing" (Swam, Hopkins, & Swade 2000).

Figure 1 shows the data flow within the HST ground system from the spacecraft and upon retrieval from the archive.

---

[1]http://www.stsci.edu/software/OPUS/

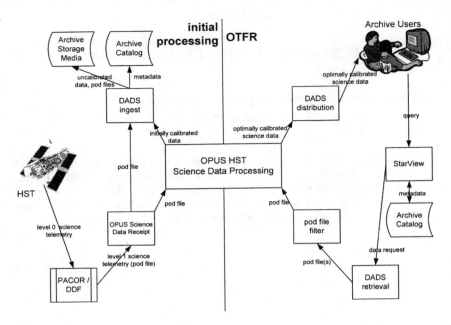

Figure 1. Data flow within the HST ground system from the spacecraft and upon retrieval from the archive.

2. Initial Data Processing

HST data is dumped from the on-board science recorder and transmitted as level 0 telemetry through a TDRSS satellite to the White Sands TDRSS downlink facility. From White Sands the telemetry is sent over domestic communication satellite to the PACOR data capture facility at Goddard Space Flight Center in Greenbelt, MD. At PACOR the data are binned into Reed-Solomon corrected packets to generate telemetry pod files (level 1 data). From PACOR the pod files are transmitted over the GSFC network to STScI via ftp.

At STScI, the pod files are received by the OPUS Science Data Receipt pipeline and passed into the OPUS Science Data Processing pipeline (described in section 4 below). OPUS science data processing converts the telemetry into calibrated FITS format data files.

After processing, the data are transferred to the Data Archive and Distribution System (DADS) for archiving. Keyword value information from calibrated headers is incorporated into metadata in the archive catalog. Pod files and uncalibrated FITS format data are written to the storage media.

3. Data Processing on Archive Retrieval

An archive researcher uses StarView to interface with the archive catalog. Using StarView, the archive researcher can query the archive catalog on a large number of parameters to search for specific observations. Once these observations are found, a data request can be submitted to DADS through StarView.

The OTFR system will then retrieve from DADS the relevant pod files for the requested observations. Since a single pod file may contain more than one exposure, or a single exposure may span more than one pod file, a pod file filtering process has been inserted into the data flow to insure that only the necessary science telemetry enters the OTFR pipeline. At this point the pod files that correspond to the data request enter the OTFR pipeline and are processed with the exact same software and input information as the current run-time (pre-archive) data processing pipeline (described in section 4 below).

4. Details of the Science Data Processing Pipeline

HST data flows sequentially through a set of processes in the OPUS science data processing pipeline. Here we describe a generic pipeline and the primary operations that occur at each step. For HST, each science instrument has its own specific pipeline that deviates to some degree from the process order described. For example, the data collector process, where the exposures are collected for further processing as an associated unit, can occur at any point in the pipeline.

- Data Partitioning - accepts pod file telemetry as input, sorts data into individual observations, checks for bad data segments, sets data quality flags, and inserts fill into telemetry dropouts when necessary. The output of Data Partitioning is an intermediate data product, the EDT dataset, which is a combination of packetized binary and ASCII files.
- Support Schedule - extracts information from proposal database to populate keywords. This information is saved as ASCII file in the EDT dataset.
- Data Validation - dredges keyword values from telemetry and applies telemetry conversions if necessary, as well as compares actual vs. planned observation parameters.
- World Coordinate System - converts pointing information into standard FITS WCS keywords. The pointing information comes from planning databases and telemetry. This information is also saved as an ASCII file in the EDT dataset.
- Generic Conversion - converts unformatted EDT dataset into FITS data arrays or tables, properly orients the image or spectra, and populates keyword values in FITS headers.
- Data Collector - holds processing for individual exposures that require further processing as a single unit (association).
- Calibration - performs instrument specific calibration.

5. System Deployment

Current plans at STScI are to transition STIS and WFPC2 data processing from OTFC to OTFR early in 2001. A NICMOS OTFR system will be deployed soon after, and subsequent upgrades to the NICMOS OTFR pipeline will include temperature dependent darks and automated pedestal correction. OTFR will be developed as a baseline data processing functionality for ACS, COS and WFC3.

Hence, it will be available for these HST science instruments at the time of their launch. There are no plans to develop OTFR systems for the HST legacy science instruments WFPC, FOC, HSP, FOS, and GHRS.

6. Implications for Archive Catalog

The archive catalog metadata is populated from the header keywords of the initial data processing within a few hours of when the observation is downlinked from the HST. In the time between the initial processing and a request for archival data, it is likely that various updates will have been made to either the calibration reference files or algorithms associated with the data. Hence, it is possible that fields in the archive catalog will not agree with keyword values in data processed through OTFR. The OTFR keyword values should be considered the current best data. Attempts will be made to keep all common search fields and fields that determine calibration reference files up to date in the archive catalog. The latter is required for the BESTREF utility that is available through StarView[2] and the web[3].

7. Future Possibilities

OTFR allows for additional processing steps that cannot be realized in the near real time processing. At the time of archive data retrieval additional information, such as data quality information, engineering telemetry, and more exposures from the same field, will probably be available for the observations. OTFR will be able to provide higher quality and enhanced data products such as combining exposures from different visits or incorporating engineering data parameters in the science data. This also leads to the possibility of OTFR generating new data products that could not be realized in initial, near real-time processing.

References

Lubow, S., & Pollizzi, J. 1999, in ASP Conf. Ser., Vol. 172, Astronomical Data Analysis Software and Systems VIII, ed. David M. Mehringer, Raymond L. Plante, & Douglas A. Roberts (San Francisco: ASP), 187

Rose, J., 1998, in ASP Conf. Ser., Vol. 145, Astronomical Data Analysis Software and Systems VII, ed. R. Albrecht, R. N. Hook, & H. A. Bushouse (San Francisco: ASP), 344

Swam, M., & Swade, D. A. 1999, in ASP Conf. Ser., Vol. 172, Astronomical Data Analysis Software and Systems VIII, ed. David M. Mehringer, Raymond L. Plante, & Douglas A. Roberts (San Francisco: ASP), 203

Swam, M., Hopkins, E., & Swade, D. A. 2001, this volume, 291

---

[2]http://archive.stsci.edu/starview.html

[3]http://www.stsci.edu/cgi-bin/cdbs/getref.cgi

Automated Reduction and Calibration of SCUBA Archive Data Using ORAC-DR

Tim Jenness

Joint Astronomy Centre, 660 N. A'ohōkū Place, Hilo, HI 96720

Jason A. Stevens

Mullard Space Science Laboratory, University College London, Holmbury St Mary, Dorking, RH5 6NT, Surrey, United Kingdom

Elese N. Archibald, Frossie Economou, Nick Jessop, Ian Robson, Remo P. J. Tilanus, Wayne S. Holland

Joint Astronomy Centre, 660 N. A'ohōkū Place, Hilo, HI 96720

Abstract. The Submillimetre Common User Bolometer Array (SCUBA) instrument has been operating on the James Clerk Maxwell Telescope (JCMT) since 1997. The data archive is now sufficiently large that it can be used for investigating instrumental properties and the variability of astronomical sources.

This paper describes the automated calibration and reduction scheme used to process the archive data with particular emphasis on the pointing observations. This is made possible by using the ORAC-DR data reduction pipeline, a flexible and extensible data reduction pipeline that is used on UKIRT and the JCMT.

1. Introduction

The Submillimetre Common User Bolometer Array (SCUBA) (Holland et al. 1999) consists of two arrays of bolometers (or pixels); the Long Wave (LW) array has 37 pixels operating in the 750 micron and 850 micron atmospheric transmission windows, while the Short Wave (SW) array has 91 pixels for observations at 350 micron and 450 micron. Each of the pixels has diffraction-limited resolution on the telescope (approximately 14″ and 8″ FWHM respectively), and are arranged in a close-packed hexagon. Both arrays have approximately the same field-of-view on the sky (diameter of 2.3′), and can be used simultaneously by means of a dichroic beamsplitter.

2. The SCUBA Archive

All data from the JCMT are archived in Hilo (Tilanus et al. 1997) and at the Canadian Astronomy Data Centre (CADC). The SCUBA data rate is on average only 100 MB per night (peak of 200 MB/night). This is relatively small, and all

SCUBA data ever taken (since 1996) are available on disk for instant access. The current archive totals approximately 50 GB.

The data headers are stored in a Sybase database with a web front-end to simplify searches. We provide the ability to return the location of the data to users rather than the data itself to minimize the disk requirements for local researchers. To give some idea of the number of observations in the archive, there are approximately 2500 850 micron map observations of Uranus, our primary calibrator.

3. ORAC-DR

ORAC-DR (Economou et al. 1999; Jenness & Economou 1999) is a modular and flexible data reduction system initially developed for the United Kingdom Infrared Telescope (UKIRT) as part of their ORAC project and subsequently used by the James Clerk Maxwell Telescope (JCMT) for automating the reduction of data from SCUBA. The primary design goals of ORAC-DR were to simplify data reduction whilst observing and to provide near-publication quality results. This approach can also be applied to the processing of archive data, allowing time-dependent data such as telescope efficiencies, stability of secondary calibrators, and the flux changes of variable astronomical sources such as blazars to be processed with minimal effort.

4. Calibration

The accuracy of the calibration is critical for automated reduction of large datasets. If the error on the calibration is too large it becomes impossible to determine long-term trends in data since they are likely to be swamped by calibration errors. The calibration of submillimetre continuum data requires two different pieces of information:

- the amount of flux absorbed by the atmosphere (the atmospheric extinction) usually represented by a sky opacity (tau);
- the sensitivity of the telescope and the instrument; this is called the flux conversion factor (FCF) and is expressed in terms of janskys per volt.

4.1. Extinction Correction

Currently, the best method of correcting data for sky opacity is to use a combination of the CSO tau data (generating a 225 GHz data point every 10 minutes) and the SCUBA 850 micron skydips (taken infrequently throughout the night) and to fit a polynomial to the data. This has the advantages of reducing the instrumental scatter in the CSO data and allows for a human to assess the usefulness of the night for calibration; if the tau data are difficult to fit the night is not used for automated reduction. The comparison between CSO tau and skydip data is also important since the tau meter looks at a fixed azimuth whereas skydip data are taken at the current telescope azimuth. When the skydip data and CSO tau data do not agree it is usually because the extinction is dependent on azimuth. More details on the opacity correction can be found in Archibald et al. (2000).

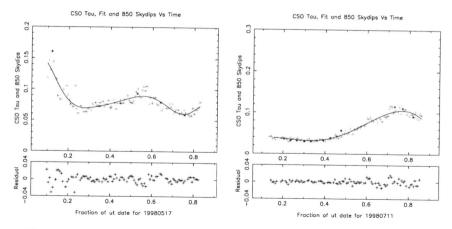

Figure 1. Two nights of CSO tau (crosses) and SCUBA skydip data (stars, scaled to CSO tau values) with corresponding polynomial fits.

The polynomial fitting (e.g Figure 1) has been performed for every single night since SCUBA began operating and provides a crucial quality control to guarantee that data are only processed if the sky is stable enough. This is very important when 1000 observations are to be reduced and they can not all be examined by hand.

4.2. Flux Conversion Factors

The flux conversion factor depends on the dish shape, the filter profiles and other instrumental factors. It is therefore expected to change when filters are changed or the dish is out of shape, although at 850 microns it is approximately constant for most conditions. The FCF has been determined at 850 microns by reducing the calibrators Uranus and CRL 618 using the polynomial fits described above. Figure 2a shows the FCF determined from observations of CRL 618 and reduced using ORAC-DR. The variation is less than 10 per cent and the shift in sensitivity at MJD 51400 is caused by an upgrade to the 850 micron filter. These data indicate that an accuracy of 10 per cent can be achieved when processing 850 micron archive data using this automated system.

5. Initial Results

The calibration techniques described above were applied to the pointing data present in the archive. These data provide a homogeneous dataset since every pointing is observed using the same technique and at the same wavelength. Automatic reduction of these data has allowed us to determine more accurate fluxes of our secondary calibrators and to provide numerous blazar light curves (Robson et al, 2001). An example light curve for 3C 273 is shown in Figure 2b. It should be noted that analysis of the full pointing archive is simply impossible without the use of an automated system.

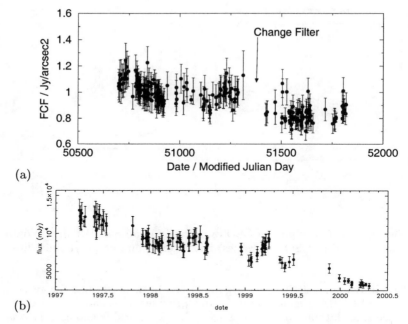

Figure 2. (a) Flux conversion factor determined for CRL 618 at 850 microns. (b) Light curve for 3C 273 at 850 microns.

References

Archibald, E. N., Wagg, J., Jenness T., 2000, SCUBA Calibration Database System Note 2, James Clerk Maxwell Telescope

Economou, F., Bridger, A., Wright, G. S., Jenness, T., Currie, M. J., & Adamson, A. 1999, in ASP Conf. Ser., Vol. 172, Astronomical Data Analysis Software and Systems VIII, ed. David M. Mehringer, Raymond L. Plante, & Douglas A. Roberts (San Francisco: ASP), 11

Holland, W. S. et al. 1999, MNRAS, 303, 659

Jenness, T., & Economou, F. 1999, in ASP Conf. Ser., Vol. 172, Astronomical Data Analysis Software and Systems VIII, ed. David M. Mehringer, Raymond L. Plante, & Douglas A. Roberts (San Francisco: ASP), 171

Robson, E. I., Stevens, J. A., Jenness, T., 2001, MNRAS, in preparation

Tilanus, R. P. J., Jenness, T., Economou, F., Cockayne, S. 1997 in ASP Conf. Ser., Vol. 125, Astronomical Data Analysis Software and Systems VI, ed. G. Hunt & H. E. Payne (San Francisco: ASP), 397

The Chandra Automatic Processing/Archive Interface

Sreelatha Subramanian, David Plummer

Harvard-Smithsonian Center for Astrophysics, 60 Garden St. MS-81, Cambridge, MA 02138, Email: latha@head-cfa.harvard.edu

Abstract. The Chandra Automatic Data Processing System (AP) requires quick access to previously generated data. Potential inefficiencies are avoided by introducing a layer between the pipelines and the archive. This archive interface layer includes an archive request queue, a data archiving server (darch), and an archive "cache". The design and functional operation of each of these components are presented in this paper.

1. Cache

The first step in improving the AP system is the implementation of a cache. In theory, the cache operates as a typical cache—data is retrieved from a slow-access storage facility and is stored to a higher-speed retrieval location, with the expectation that it will be used again.

In our implementation, the cache is actually a directory where the data files reside, as shown in Figure 1. While a pipeline is processing, it will store its data products to the cache directory. These data products can then be stored to the archive at any time, separating data product ingestion from processing. When another pipeline requires those data products, the cache is checked first. If the data products are not in the cache, then the pipeline retrieves the files from the archive database. This method is especially significant when processing data for the first time since the system can access all the data products from cache, avoiding the archive database. If re-processing is required, data will be retrieved from the archive if it is no longer in cache.

1.1. Inactive Cache

The implementation of a cache introduces the issue of memory usage. The cache server loads all the files in the cache directory into its memory, which allows for quick search and retrieval of the data. However, as the size of the cache increases, the cache server requires more memory. To reduce the number of files in cache, an inactive cache directory structure is created. The cache server does not load files that exist in the inactive directory into its memory. It stores a few pieces of information about the file along with the filename, in order to facilitate retrieval. The implementation of the inactive cache eliminates the issue of memory usage, while maintaining the usefulness of the cache.

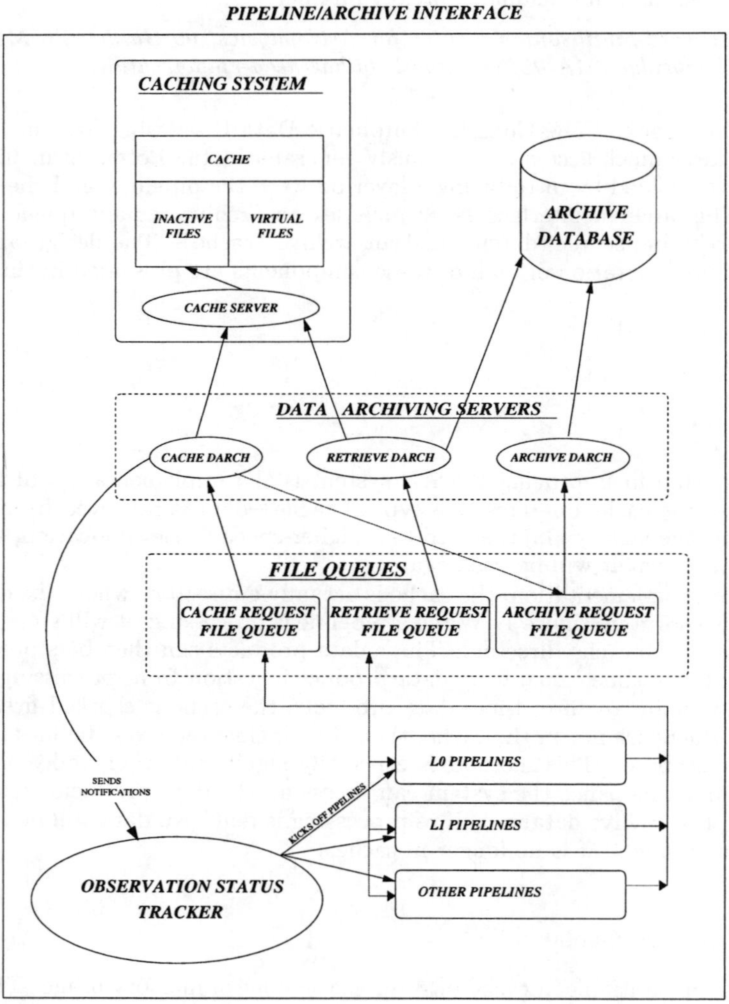

Figure 1. AP/Archive Interface

1.2. Virtual Cache

A significant improvement to the cache system is made through the implementation of virtual files. In our system, virtual files are simply files that do not exist in the cache, but whose filename and metafile still exist. With this information, the cache can be queried to retrieve a file based on time, data product id or other information. The cache returns the filenames of the files it found, and also returns an error code which indicates that the files are virtual. The AP system, recognizing that the retrieved files are virtual, can then use the filename to retrieve the file from the archive database. This modification is significant because retrieving a file from the archive using the filename as the key is more efficient than retrieving using another key (such as data product id, or time).

The other important factor is that the implementation of virtual files allows us to search for file(s) by using cache specific tag information. Tag information, which is stored in the metafile, is not entered into the archive. Therefore it was previously not possible to retrieve files from the archive using tag information. With virtual files, however, this becomes possible. A search by tag function searches the cache, using the tag. As this information is in the metafile, the function will return a corresponding filename, which can then be used to retrieve the file from the archive.

2. File Queues

Another important modification to the AP System is the implementation of file queues. File queues are implemented to eliminate the time a pipeline must wait for its data products to be ingested into the archive database. Instead of waiting for a file to be ingested, the ingestion request is sent to a file queue. The request is then picked up at a later time by the data archiving server, or darch, and processed. The implementation of file queues is simply a file which contains all the requests. The pipeline will not need to wait for ingestion since it has already stored the file to cache. Therefore, ingestion of the file does not need to occur simultaneously, and the file queue allows for another process to handle the requests at any time.

3. Conclusion

The archive interface layer was not part of the original design. However, the necessity for it was identified during the processing of ground calibration data of the system before launch. This allowed us to implement and integrate the archive interface layer into the AP system, to avoid possible bottlenecks during the mission.

4. Acknowledgments

This project is supported by the Chandra X-ray Center under NASA contract NAS8-39073.

A High Throughput Photometric Pipeline

Michael L. Reid[1], Denis J. Sullivan, Richard J. Dodd[1,2]

School of Chemical and Physical Sciences, Victoria University of Wellington, New Zealand

Abstract. The advent of large format CCD detectors in projects measuring time-critical astrophysical phenomena has resulted in an explosion in data generation rates. The current generation of gravitational microlensing and all-sky surveys have, on an ad-hoc basis, developed data processing software which currently meets their needs. However, the next generation of these projects will require faster, more advanced software to manage the data flow. As part of the Japanese/New Zealand microlensing collaboration, MOA (Microlensing Observations in Astrophysics), a software pipeline has been developed with the intention of being highly scalable, automated, portable, flexible, and robust. The core of the system is built around a high performance object database optimised for time series work and flexible reduction software that can use the PSF fitting packages DoPHOT and DAOphot II, as well as the ISIS Optimal Imaging Subtraction software. Evolution of the software has made it suitable for general purpose astronomical photometric reduction. This paper provides an overview of the software system.

1. Introduction

The Microlensing Observations in Astrophysics Project (MOA[3]) is a collaboration between Japanese and New Zealand scientists to search for gravitational microlensing events. Gravitational microlensing occurs when a massive object bends light from a luminous background object resulting in a time-dependent apparent brightness change of the background object. Such events are produced by rare alignments between the luminous source, massive body and the observer and to have a reasonable chance of observing some microlensing events the brightness of millions of stars must be measured over many nights. MOA performs nightly observations from Mount John University Observatory (MJUO) of ten million stars towards the Galactic Bulge and Magellanic Clouds.

The MOA Project has a custom microlensing detection system (Bond 2000) at MJUO but also exports data to member institutions. Lessons learnt from this

---

[1]Carter Observatory, Wellington, New Zealand

[2]Faculty of Science, University of Auckland, New Zealand

[3]http://www.vuw.ac.nz/scps/moa/

system were used to build software suitable for automated digital image (CCD) reduction which could be used with any telescope/detector combination and for any astronomical time-series research. The software is specifically designed for use in large astronomical projects with mosaic CCD detectors which produce too much data for the reductions to be guided by manual intervention.

2. Automated Reduction

A software package called Autophot has been developed to automate reduction of CCD images. High throughput is achieved by reducing the images in parallel using a task scheduler (most useful on multi-processor computers). The reduction can be performed using DAOphot II (Stetson 1987), or DoPHOT 3.1 (Schechter 1993) – ported by the authors to ANSI C. Autophot has been adapted to use the ISIS Optimal Image Subtraction software (Alard & Lupton 1998, Alard 2000) and modifications to ISIS to improve the consistency of the image subtraction are in progress. The system has been designed for robustness: failure to reduce one image does not halt the pipeline and prevent the reduction of other images.

A new task scheduler which uses Java Remote Method Invocation (RMI) is being developed to allow parallel reduction on a cluster of workstations. This will allow improvements in reduction throughput which scale, almost linearly, with the size of the cluster. Implementation of the software with portability in mind should allow heterogeneous computer types to be added to the cluster (including all the Windows PCs which are unused at night).

3. Data Storage

Knowledge of typical data access patterns required for fast time-series photometric analysis allows data storage to be optimised for fast retrieval. An object database implemented in C++ has been built ('StarBase') which has shown itself to provide fast access to approximately one hundred observations of several million-star star fields made by the MOA mosaic SITe CCD. The database files written by StarBase are platform-independent and can be read by the Java java.io.DataInputStream classes too. The database is customised by programming to a C++ API and a set of Java bindings have been tested.

The database implementation uses algorithms which were selected (or developed) to perform well on large data sets. As an example, built into the database is the ability to find non-variable stars and use them to estimate corrections for atmospheric transparency (called 'homogenisation').

4. Homogenisation

The least-squares atmospheric transparency estimation method suggested by Honeycutt (1992) takes about one hour and ~256 MB RAM to perform on ~2000 observations made by Sullivan et al. (2000) on the pulsating white dwarf GW Librae. An iterative method developed by the authors estimates the same transparency corrections to within one milli-magnitude of the least-squares estimates.

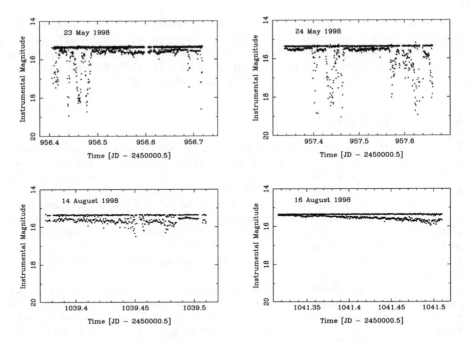

Figure 1. Observations of a reference star near GW Librae. The lighter (red) points are the observed instrumental magnitudes and the darker (blue) points are the instrumental magnitudes corrected for atmospheric transparency variations (presumably due to transient clouds).

The iterative method requires 30 seconds and 10 MB RAM and is able to deal with an ensemble of stars with missing observations. Using an algorithm which scales well with increasing data had an enormous effect in this case, and has guided development of the data pipeline. Reduction of these white dwarf data (which were not obtained by the MOA collaboration) demonstrates that the pipeline can be used in other astronomical projects.

5. Performance - Case Study

This reduction pipeline software has been tested on a number of MOA observation targets including the possible planetary microlensing event MACHO-98-BLG-35 (Rhie et al. 2000) and the finite-source microlensing event MACHO-95-BLG-30 (Alcock et al. 1997), successfully reducing observations made in both poor and good seeing conditions. Observations made on MACHO-95-BLG-30 by the MOA Project were originally reduced manually (by the authors) using IRAF/DAOphot over six man-months. The same images were reduced using Autophot/DoPHOT with the same hardware (a 4–CPU SGI Indigo) but removal of human interaction allowed a reduction time of four hours using parallel processing. Estimation of atmospheric transparency corrections by manual plotting

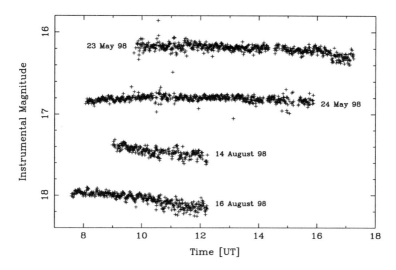

Figure 2. Observations of GW Librae with most atmospheric effects removed (nights earlier than 16 August have been offset for clarity). The brightness pulsations of the white dwarf are now visible.

and examination of star lightcurves took ~weeks, a process now performed by StarBase in 30 seconds.

6. Acknowledgments

The authors would like to thank all members of the MOA Project and acknowledge financial support from Carter Observatory New Zealand, Victoria University Science Faculty Leave and Grants Committee, and the Marsden Fund of New Zealand. Additional thanks go to the Astronomical Data Analysis Software and Systems X (ADASS 2000) Committee for conference financial assistance.

References

Alard, C. 1998, A&AS, 144, 363
Alard, C.& Lupton, R. 1998, ApJ, 503, 325
Alcock, C., et al. 1997, ApJ, 491, 436
Bond, I. 2000, "Microlensing 2000", ASP Conf. Series (in press)
Honeycutt, R. 1992, PASP, 104, 435
Rhie, S., et al. 2000 ApJ, 533, 378
Schechter, P. 1993, PASP, 105, 1342
Stetson, P. 1987, PASP, 99, 191
Sullivan, D., et al. 2000, Baltic Astronomy, 9, 223

Interactive Analysis and Scripting in CIAO 2.0

S. Doe, M. Noble, R. Smith

Harvard-Smithsonian Center for Astrophysics, MS 81, 60 Garden Street, Cambridge, MA 02138 Email: sdoe@head-cfa.harvard.edu

Abstract. Interpreted scripting languages are now recognized as essential components in the programmer's (and user's) tool chest, and, as amply demonstrated at ADASS 1999, have infiltrated the scientific community with great effect.

In this paper we discuss the utilization of the S-Lang interpreted language within the Chandra Data Analysis System (CIAO, or Chandra Interactive Analysis of Observations). In only a few months, with substantial reuse and comparatively little manpower and code bloat, this effort has increased by an order of magnitude the analytical power and extensibility of CIAO.

We summarize our design and implementation, and show brief fitting, modeling, and visualization threads that demonstrate capabilities roughly comparable with those of commercial packages. Finally, we present a beta version of the CIAO spectroscopic analysis module, GUIDE – largely a collection of S-Lang scripts, glued with C++ enhancements to Sherpa and ChIPS – to illustrate in more depth the range of new functionality and the rapid prototyping now available in CIAO.

1. Extending CIAO Applications with S-Lang

S-Lang[1] is an interpreted language created by John Davis of the Center for Space Research at MIT. S-Lang is a popular language, with several hundred users, and several applications that use it as an extension language. Embedding such a language into an application can enhance its flexibility and power by allowing users to more easily extend its capabilities. We have added S-Lang to CIAO, where it is currently used most heavily in ChIPS (Chandra Imaging and Plotting Software) and Sherpa (the CIAO fitting application).

2. Reasons for Choosing S-Lang

Several candidate languages were evaluated before S-Lang was chosen, among them Python[2], Tcl and Glish[3]. Some of the reasons we did not choose one of these alternatives are:

---

[1] http://space.mit.edu/~davis/slang/

[2] http://www.python.org/

[3] http://aips2.nrao.edu/docs/glish/glish.html

- Tcl, a string based language, historically under-performs for numerical work;
- Python does not support multi-dimensional arrays, so other packages (such as NumericalPython) also need to be installed;
- Python is typically not used as an embedded language; it has a larger footprint, and would be more complicated to integrate with CIAO;
- Glish at heart is an event dispatcher, and CIAO has already adopted XPA for event/message communication between CIAO processes; also, the Glish parser is not meant to be embedded within another parser.

What makes S–Lang attractive for CIAO are:

- built-in multi-dimensional array handling,
- well-defined grammar,
- ease of installation and integration with CIAO,
- lighter footprint, and fast execution,
- intrinsic arithmetic and mathematical functions.

3. VARMM – Interface Between S–Lang and C++

In addition to embedding S–Lang into CIAO applications, we have created a new library that acts as an interface between C++ code and S–Lang. This is the VARMM (Variables, Math and Macros) library. As shown in Figure 1, the VARMM library is a thin layer, providing a clean, simple API, and useful classes for representing data. The VARMM library provides several benefits:

- application code is shielded from explicit knowledge of the interpreted language;
- a clean, focused interface simplifies application development;
- OO abstractions naturally model the data usage patterns;
- less developer effort is required to add S–Lang to new CIAO applications.

The VARMM library provides two classes for accessing data stored in S–Lang variables. S–Lang variables may store scalars, arrays, or structures that encapsulate several scalars and/or arrays. The Varmm class provides access to the contents of an S–Lang variable; for uniformity, Varmm treats all S–Lang variables as arrays. A Varmm object implicitly synchronizes with its S–Lang variable, so that changes to variables in S–Lang scope are automatically reflected in the C++ scoped object when S–Lang variable data members are accessed.

The library also provides the VarmmDSet class, which is an associated array of Varmms. The VarmmDSet class can be used to represent data read from file; the data from each column of a table are stored in Varmms, and additional information about the file (e.g., the file name, the path, etc.) is stored in other VarmmDSet data members. Data from particular file types, for which we need to store additional information (images, PHA and RMF files), can be stored in objects of classes which derive from the VarmmDSet class. S–Lang variables can be accessed by an application via the Varmm and VarmmDSet public functions.

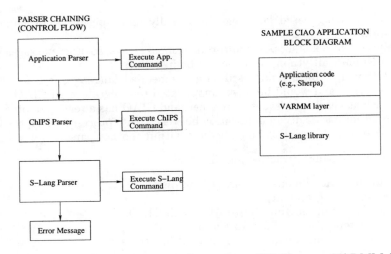

Figure 1. CIAO Application Structure. CIAO's new VARMM library is a thin layer between S–Lang and an application. The application also allows ChIPS and S–Lang commands to be parsed.

4. Other VARMM Functions

VARMM provides the programmer with additional functions, to a) create, copy and destroy Varmms and VarmmDSets from within an application; b) get a reference to the data to which a Varmm points; c) get references to Varmms and VarmmDSets given the names of the corresponding S–Lang variables; and d) pass commands to the S–Lang parser (see Figure 1).

The VARMM library also provides a number of user functions for reading data from ASCII and FITS files, and assigning the data to S–Lang variables. These functions are: `readfile, readascii, readbintab, readpha, readarf, readrmf, readimage, writeascii` (to read from and write to various file formats) and `print, printarr` (to print the contents of variables).

5. An Example

Here is a simple example, in which we use a S–Lang variable in ChIPS:

```
chips> pldata = readfile("phas.dat")
chips> print(pldata)
filename        = phas.dat
path            = /home/sdoe/
filter          = NULL
ncols           = 2
nrows           = 124
col1            = Float_Type[124]
col2            = Float_Type[124]
chips> plot x pldata.col1 y pldata.col2
```

```
chips> pldata.col2 = log(pldata.col2 + 1.0) + 10
chips> redraw
```

An S–Lang variable called `pldata` is created, and data are read from a file named "phas.dat". The ChIPS `plot` command creates a plot, using data from `pldata`. The data in `pldata` can then be manipulated via S–Lang commands at the ChIPS command line. When the plot is drawn again, the updated y-values are immediately apparent.

In addition, the user can customize ChIPS and Sherpa with user–defined scripts. For example, the user could ignore points below a certain y-value by creating the following script:

```
define ignoreLowPoints(data, floor) {
   data.col1 = where(data.col2 > floor);
   data.col2 = where(data.col2 > floor);
   return data;
}
```

The script could then be used in ChIPS:

```
chips> evalfile("ignoreLowPoints.sl")
chips> ignoreLowCounts(pldata, 10)
chips> redraw
```

The new plot would contain only those points where the y-value exceeded 10.

6. GUIDE

Embedding S–Lang in ChIPS and Sherpa not only opens up the data for the user, but also allows the user to create and use scripts to further extend CIAO capabilities. Such scripts may consist of Sherpa, ChIPS and S–Lang commands combined.

For CIAO 2.0, we present a package of such scripts, to facilitate analysis of Chandra high–resolution spectra. This package is called GUIDE (Grating User Interactive Data Extension) and is a modular extension to Sherpa.

GUIDE extends Sherpa in several ways. First, we provide a new file format for storing the state of a Sherpa session. More importantly, GUIDE provides many functions for analysis of high–resolution Chandra spectra. These include functions to identify emission features; to provide access to collisional plasma models, such as the APEC or Raymond & Smith models (to get predicted line emissivities); and to use the modeled emissivities to fit a differential emission measure model to the line fluxes.

While GUIDE is beta software, we already see that embedding S–Lang in ChIPS and Sherpa has enabled us to provide tools for spectroscopic analysis, to do so much more rapidly than we would have otherwise, and to involve scientists more directly in their creation. We believe this approach will be useful for future CIAO releases and for users who develop their own S–Lang packages for use with CIAO software.

Acknowledgments. This project is supported by the Chandra X-ray Center under NASA contract NAS8-39073.

Infrared Spectroscopy Data Reduction with ORAC-DR

Frossie Economou, Tim Jenness, Brad Cavanagh[1]

Joint Astronomy Centre, 660 N. A'ohōkū Place, Hilo, HI 96720

Gillian S. Wright, Alan Bridger

UK Astronomy Technology Centre, Blackford Hill, Edinburgh, EH9 3HJ, Scotland

Tom H. Kerr, Paul Hirst, Andy J. Adamson

Joint Astronomy Centre, 660 N. A'ohōkū Place, Hilo, HI 96720

Abstract. ORAC-DR is a flexible and extensible data reduction pipeline suitable for both on-line and off-line use. Since its development it has been in use on-line at UKIRT for data from the infrared cameras UFTI and IRCAM and at JCMT for data from the sub-millimetre bolometer array SCUBA.

We have now added a suite of on-line reduction recipes that produces publication quality (or nearly so) data from the CGS4 near-infrared spectrometer and the MICHELLE mid-infrared Echelle spectrometer. As an example, this paper briefly describes some pipeline features for one of the more commonly used observing modes.

1. Background on ORAC-DR

As part of the Observatory Reduction and Acquisition Control project (ORAC[2], Wright et al. 2001) for the United Kingdom Infrared Telescope (UKIRT), a flexible and extensible data reduction (DR) pipeline has been developed. The ORAC-DR pipeline design was presented previously by Economou et al (1998) and Jenness & Economou (1999) and makes extensive reuse of existing data reduction packages (Economou et al. 1999). ORAC-DR is a data-driven pipeline with simple ASCII files called recipes that contain data reduction instructions. ORAC-DR has been in use at UKIRT for on-line reduction of data from the UFTI and IRCAM/TUFTI infrared cameras, as well as at the James Clerk Maxwell Telescope for the on-line reduction of data from the Submillimeter Common-User Bolometer Array (SCUBA, Holland et al. 1999).

---

[1] University of Victoria, Victoria, BC V8W 2Y2, Canada

[2] ORAC is a joint project of the Astronomy Technology Centre, Edinburgh and the Joint Astronomy Centre, Hawaii.

This paper describes the recent addition to the pipeline of recipes and primitives for the reduction of data from the UKIRT infrared spectrometer CGS4. The support for infrared spectroscopy added for the CGS4 instrument will also form the basis for the support of the Mid Infra-Red Echelle Spectrometer (MICHELLE) when it is in use at UKIRT[3].

2. Requirements for IR Data Reduction at UKIRT

Since the primary purpose of on-line data reduction is to give observers the ability to assess the quality of their data in near real-time and modify their observing strategy as necessary, this is a primary requirement for the IR spectroscopy component of ORAC-DR. However, since experience with other supported instruments has shown it is often possible to obtain publication quality results in the on-line data reduction process, this too is a goal whenever practical.

Online data reduction becomes even more important with the advent of multi-wavelength observers (who may not be experienced infrared observers) and flexible scheduling (where the observer is not necessarily acquainted with the scientific aims of the observations that he or she may be carrying out). There is an additional concern for the potential future user who may be accessing the data through a public archive. Online reduction at the telescope to produce near-publication quality data ensures that all suitable calibration frames are taken and that the data can even be automatically reducible at retrieval from the archive (see e.g., Jenness et al. 2001).

3. Feature Highlights

The UKIRT 1–5 micron spectrometer CGS4 is typically used in the following observing mode: the usual array calibration frames (for example bias, flat, dark and arc) are taken, then a star of known magnitude and spectral type at an airmass near that of the astronomical target is observed. This is used later for removal of the atmospheric response from the data. Both standard stars and astronomical targets are usually observed with a nodding technique to enable accurate sky removal. The calibration frames, standard star frames, and target frames all have to be processed and retrieved at appropriate points during the reduction in order to produce meaningful scientific output for observers.

The flexible architecture of ORAC-DR has allowed us to provide "added-quality" features with very little programming effort. For example, a data reduction step requiring user intervention is processing the spectrum of a standard star into a normalised calibration frame subsequently used to remove atmospheric features from the astronomical target. Prior to the advent of ORAC-DR, this process requiring knowledge of the standard star's temperature and infrared magnitude was not performed automatically in real time. The ORAC-DR pipeline can now use either a local catalogue or an http connection to an astronomical database, together with the standard star name or the RA and Declination coordinates from the FITS header, to determine the spectral type

---

[3]The MICHELLE instrument is a shared between the UKIRT and Gemini projects

and magnitude. From these the stellar temperature and color and hence the appropriate infrared magnitudes are derived. All this takes place in the time it takes to acquire the next frame.

Future plans call for the addition of such features as automatic wavelength calibration from observation of an arc lamp.

4. Implementation Details

For spectroscopy ORAC-DR currently makes use of the following external applications: Figaro (Shortridge et al 1999; Shortridge 1993), Kappa (Currie & Berry 2000) and CCDPACK (Draper, Taylor & Allan 2000), all supported by Starlink and driven by the ADAM messaging system (Allan 1992). ORAC-DR itself is written in Perl and is described by Economou et al. (2000).

Acknowledgments. We would like to thank STARLINK for distributing ORAC-DR and supporting many of the packages it uses. We would also like to thank our users for their input and suggestions.

References

Allan, P. M. 1992, in ASP Conf. Ser., Vol. 25, Astronomical Data Analysis Software and Systems I, ed. D. M. Worrall, C. Biemesderfer, & J. Barnes (San Francisco: ASP), 126

Currie, M. J., Berry, D. S. 2000, Starlink User Note 95, Starlink Project, CCLRC

Draper, P. W., Taylor, M., & Allan, A. 2000, Starlink User Note 139, Starlink Project, CCLRC

Economou, F., Bridger, A., Wright, G. S., Rees, N. P., & Jenness, T. 1998, in ASP Conf. Ser., Vol. 145, Astronomical Data Analysis Software and Systems VII, ed. R. Albrecht, R. N. Hook, & H. A. Bushouse (San Francisco: ASP), 196

Economou, F., Bridger, A., Wright, G. S., Jenness, T., Currie, M. J., & Adamson, A. 1999, in ASP Conf. Ser., Vol. 172, Astronomical Data Analysis Software and Systems VIII, ed. David M. Mehringer, Raymond L. Plante, & Douglas A. Roberts (San Francisco: ASP), 11

Economou, F., Jenness, T., Currie, M. J., & Adamson, A. 2000, Starlink User Note 230, Starlink Project, CCLRC

Holland, W. S., et al. 1999, MNRAS, 303, 659

Jenness, T. & Economou, F. 1999, in ASP Conf. Ser., Vol. 172, Astronomical Data Analysis Software and Systems VIII, ed. David M. Mehringer, Raymond L. Plante, & Douglas A. Roberts (San Francisco: ASP), 171

Jenness, T., et al. 2001, this volume, 299

Shortridge, K. 1993, in ASP Conf. Ser., Vol. 52, Astronomical Data Analysis Software and Systems II, ed. R. J. Hanisch, R. J. V. Brissenden, & J. Barnes (San Francisco: ASP), 219

Shortridge, K., et al. 1999, Starlink User Note 86, Starlink Project, CCLRC

Wright, G. S., et al. 2001, this volume, 137

Infrared Imaging Data Reduction Software and Techniques

C. N. Sabbey, R. G. McMahon, J. R. Lewis, M. J. Irwin

Institute of Astronomy, Madingley Road, Cambridge, CB3 0HA, UK

Abstract. Developed to satisfy certain design requirements not met in existing packages (e.g., full weight map handling) and to optimize the software for large data sets (non-interactive tasks that are CPU and disk efficient), the InfraRed Data Reduction[1] software package is a small ANSI C library of fast image processing routines for automated pipeline reduction of infrared (dithered) observations. The software includes stand-alone C programs for tasks such as running sky frame subtraction with object masking, image registration and co-addition with weight maps, dither offset measurement using cross-correlation, and object mask dilation. Although currently used for near-IR mosaic images, the modular software is concise and readily adaptable for reuse in other work.

1. Introduction

The Cambridge Infrared Survey Instrument (CIRSI), a near-IR mosaic imager containing a 2×2 array of Rockwell Hawaii I 1024×1024 detectors (Beckett et al. 1996; Mackay et al. 2000), has been in operation for about two years, obtaining almost 1 TB of imaging data. The uniquely wide field accessible by CIRSI on a 2–4 m class telescope makes it ideal for moderate depth large-area surveys. Preliminary results from two such current surveys include the measurement of galaxy clustering at intermediate redshift (McCarthy et al. 2001), and the demonstration of a reddening-independent quasar selection technique based on combined deep optical and near-IR color diagrams (Sabbey et al. 2001).

However, the CIRSI data reduction poses several challenges. With the large data rate (5–10 GB of data taken per night, ∼100 nights per year, and a significant data backlog currently) the software has to be very efficient and completely automated. Also, the software should handle diverse data sets, from Galactic center observations to very sparse fields at high Galactic latitude. From thousands of individual images taken over many nights, wide-field, deep mosaic images are generated. Since the gaps between detectors comparable to detector size, filled mosaic images are made using the co-added dither sets from different chips and telescope pointings. Thus weight maps and accurate astrometry are crucial and the common simplification of clipping the dither sets to their intersection region is not appropriate.

---

[1] IRDR, available via anonymous ftp at ftp.ast.cam.ac.uk in pub/sabbey

Although existing software packages are used when possible (see below), the decision was made to write core image processing routines to satisfy certain design requirements. For example, a two pass reduction provides: object masks derived from first pass co-added dither sets, subpixel image registration/co-addition using full-weight maps without image clipping, optimizations for CPU and disk efficiency, customized artifact cleaning (destriping and defringing), and reusable tools (from having stand-alone tasks to be glued together with a high level scripting language, to making C library calls, or even extracting portions of source code). The basic processing steps, described below, are: flatfield correction, running sky frame subtraction, dither offsets measurement, dither set co-addition, and mosaic image creation.

2. Data Reduction

2.1. Flatfield Correction

Flatfield images are produced by subtracting stack medians of lamp-off from lamp-on dome flats. The flatfield images are divided by the mode of the chip 1 flatfield to produce a gain map per chip. Bad pixels are automatically identified in the gain maps (and set to 0.0) by looking for outliers ($> 5\sigma$ from the median in 15x15-pixel blocks) or pixels with extremely low or high sensitivity ($> 30\%$ from the median gain). Because bad pixels often occur in clumps, these are eliminated during image co-addition rather than by interpolation in an initial cleaning pass. The data frames are multiplied by the inverse of the gain map.

The image stacking is done using `cubemean.c`, which calculates the median, robust standard deviation, or robust mean plane with a choice of weights (none, scalar, or maps) and image scaling or zero offsets using the image modes. This is not a general purpose tool like IRAF's `imcombine`, but for the specific task of calculating the median plane was found to be 2.5 times faster (for a stack of 50 data frames). The flatfield image is converted to a gain map and bad pixels identified using `gainmap.c`. The data are flatfield corrected using `flat.c`.

2.2. Running Sky Frame Subtraction

With `skyfilter.c`, a sky image is constructed from the robust mean of the eight nearest frames in the observation sequence and subtracted from each data frame. Objects detected in the co-added dither sets from the first pass reduction are masked out during sky frame creation in the second pass. The object masks are produced using the check image OBJECTS option to SExtractor (Bertin and Arnouts 1996), which produces a FITS image with non-object pixels set to 0. This is simpler and more effective than building masks from a catalog of object coordinates and shape parameters. The object regions (detection isophotes) generated by SExtractor are then expanded by a multiplicative factor of 1.5 (using `dilate.c`), thereby growing the mask regions for large objects more than small objects. The object masks for individual frames are obtained on the fly using pixel offsets (i.e., the dither offsets) into the master dither set masks.

Running sky frame subtraction is normally a significant bottleneck in processing infrared imaging, so optimizations are important. To do running sky frame subtraction for a stack of N images, the program `skyfilter.c` uses a sliding window (circular buffer of image pointers) to require only N image reads and N image mode calculations. In contrast, putting this logic into a script normally involves N × M image reads and mode calculations, where M is the width of the sky filter in frames. Also, the disk I/O (and storage) for non-co-added data uses short integers (2 bytes deep), even though most calculations are done in floating point (4 bytes). Some calculations work with short integer data however to allow optimizations. For example, the almost trivial distribution sort can be used to obtain an image histogram, sorted image array, and median value in $O(n)$ time (∼5 times faster than running an optimized median routine on typical data images).

2.3. Dither Offsets Measurement

A typical dither sequence consists of nine observations in a 3 × 3 grid with offsets in each direction of ∼10″. The approximate dither offsets stored in the FITS header WCS information are refined using cross-correlation analysis (offsets.c). The non-zero (object) pixels of the reference frame object mask (SExtractor OBJECTS image) are stored in a pixel list (x, y, brightness), and this list is cross-correlated against the object mask images of the following frames in the dither set. The SExtractor OBJECTS image conveniently removes the background (important for cross-correlation methods) and identifies the object pixels more reliably than a simple thresholding algorithm (e.g., especially in images with a non-flat background, large noise, and cosmic rays). Using an object list in the cross-correlation focuses on the pixels that contribute to the cross-correlation signal and is faster than cross-correlating two images.

The cross-correlation technique uses coordinate, magnitude, and shape information and is more reliable than matching object coordinate lists (improvement was noted in extreme cases, e.g., Galactic center images and nearly empty fields with an extended galaxy). Subpixel offset measurement accuracy of ∼ 0.1 pixel is obtained by fitting a parabola to the peak of the cross-correlation image. This cross-correlation method was found to be ∼ 10 times faster (for typical survey data and a relatively large search box of 100 pixels) than IRAF STSDAS crosscor. Although the success rate is ∼100%, offset measurements corresponding exactly to the search area border, or a small fraction of object pixels overlapping in the aligned data images, would indicate failure.

2.4. Dither Set Co-addition

A weight map is generated on the fly for each data frame with the weight for pixel p_i given by: $w_i = g_i \cdot t/V$, where g_i is the gain for pixel p_i (0.0 for bad pixels), t is the exposure time, and V is the image variance. The data frames and corresponding weight maps in the dither set are registered using bi-linear interpolation modified to account for bad pixels and image weights. Each output (interpolated) pixel value P is calculated from the weighted average of the four overlapping input pixels p_i of the input image:

$$P = \frac{1}{W} \sum_{i=1}^{4} a_i w_i p_i, \quad W = \sum_{i=1}^{4} a_i w_i$$

where a_i are pre-calculated fractional areas of overlap of P with p_i, and w_i are image weights for p_i. The weight maps are registered similarly, but weight W

for pixel P is calculated via a weighted sum. An alternate registration method sometimes recommended is to replicate each image pixel into N x N pixels and do an integer shift in units of these new pixels. Although this will approximate the above bi-linear interpolation scheme as N becomes large, it requires more work for less precision. Since the default approach is fast and reasonable, especially given the low signal-to-noise ratio of the individual data frames, higher order interpolators (see e.g., Devillard 2000) have not been tested.

With `dithercubemean.c`, dither frames are combined by calculating the weighted mean pixel value at each (x, y) position of the dither stack, with pixel values $> 5\sigma$ from the median at each position rejected. Images borders are added during registration to avoid clipping the data to the intersection of the dither frames. The standard deviation (σ) at each position is calculated from $\sigma = $ MAD $/ 0.6745$, where MAD is the median absolute deviation from the median (simpler methods such as minmax rejection are inappropriate when taking averages of small numbers of values, e.g., 5–9 frames per dither set). The weight maps are combined by calculating the sum at each (x, y) position of the stack of weight maps (for pixels not clipped during co-addition).

2.5. Mosaic Creation

The current astrometry pipeline, a small SExtractor Perl script that produces an object catalog for each co-added dither set, runs APMCAT (a stand-alone C program)[2] to download over the network the APM sky coordinates of objects in each field of view and then runs IMWCS from WCSTools (Mink 1999) to calculate the astrometry fit and update FITS header WCS information. Co-added dither sets and weight maps from different chips, telescope pointings, and nights are then drizzled onto a wide-field mosaic image using EIS Drizzle[3]. Astrometry residuals between the mosaic image and the APM catalogue show a random error of $\sigma \approx 0.3''$ without significant systematic effects.

References

Beckett, M. G., et al. 1996, SPIE, 2871, 1152

Bertin, E., & Arnouts, S. 1996, A&AS, 117, 393

Devillard, N. 2000, The Messenger, 100, 48

Mackay, C. D., et al. 2000, SPIE, 4008, 1317

McCarthy, P., et al. 2001, in ESO "Deep Field" Workshop, held in Garching, Germany, October 2000, astro-ph/0011499

Mink, D., 1999, in ASP Conf. Ser., Vol. 172, Astronomical Data Analysis Software and Systems VIII, ed. David M. Mehringer, Raymond L. Plante, & Douglas A. Roberts (San Francisco: ASP), 498

Sabbey, C. N., et al. 2001, in New Era of Wide Field Astronomy, held in Preston, England, August 2000, astro-ph/0012294

---

[2] http://www.ast.cam.ac.uk/~apmcat

[3] http://www.eso.org/eis

ADS's Dexter Data Extraction Applet

Markus Demleitner, Alberto Accomazzi, Günther Eichhorn, Carolyn S. Grant, Michael J. Kurtz, Stephen S. Murray

Harvard-Smithsonian Center for Astrophysics, 60 Garden Street, Cambridge, MA 02138

Abstract. The NASA Astrophysics Data System (ADS) now holds 1.3 million scanned pages, containing numerous plots and figures for which the original data sets are lost or inaccessible. The availability of scans of the figures can significantly ease the regeneration of the data sets. For this purpose, the ADS has developed Dexter, a Java applet that supports the user in this process. Dexter's basic functionality is to let the user manually digitize a plot by marking points and defining the coordinate transformation from the logical to the physical coordinate system. Advanced features include automatic identification of axes, tracing lines and finding points matching a template. This contribution both describes the operation of Dexter from a user's point of view and discusses some of the architectural issues we faced during implementation.

1. Dexter's Operation

The ADS provides access to the full-text of over 200,000 scientific papers published in astronomical journals, conference proceedings, newsletters, bulletins and books, for a total of over 1.3 million pages. The ADS article service allows users to view individual pages using any web browser with graphical capabilities. When viewing a scanned page, the Dexter applet can be started by following the link available below the image. As described in its help page[1], Dexter can easily be used to extract data; a relatively simple case is depicted in Figure 1.

After starting Dexter, the user can select the portion of the page containing a plot or figure to be analyzed. (It is advisable to keep this portion as small as possible, both to reduce the Java Virtual machine memory requirements and to facilitate the automatic feature extraction algorithms.) When Dexter's main window has popped up, it is generally worthwhile to attempt an automatic detection of the axes (in the Recognize menu). In the example of Fig. 1, this has worked well; in other cases it may be necessary to mark the axes manually or correct Dexter's axes by click-and-dragging the ends of the axes. Next one fills in the text fields for the start and end values of the axes (marked with a large "1" in Fig. 1), which completes the information needed by the applet to display

---

[1] http://adsabs.harvard.edu/Dexter/Dexterhelp.html

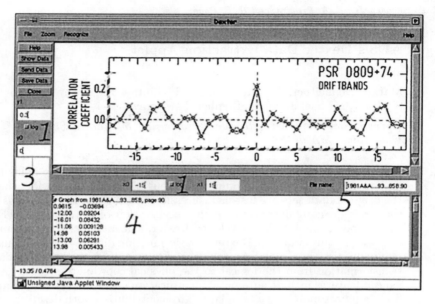

Figure 1. An illustration of Dexter's main window after the data from a graph was extracted. Marked points and axis markings have been emphasized to enhance their visibility in this grey-scale rendering.

physical (graph) instead of logical (pixel) coordinates in the status line ("2" in Fig. 1).

To actually mark points (the crosses in the figure), one uses the "Find Points" function from the Recognize menu and clicks on a sample point. Dexter then marks all similar points, where the similarity threshold can be adjusted in "Recognizer Settings". Occasionally Dexter will miss overlapping points and encounter similar mishaps, or it may have a hard time following the correct path when tracing lines. In these cases, it is necessary to mark the points manually, which is done by clicking on their positions in the graph. Error bars can be added by clicking on a point and dragging the bar (error bars are only available after the coordinate system has been set up). The magnifying glass ("3" in Fig. 1) may help here and can be activated by clicking on it. To work around bugs in some Java virtual machines, it is inactive by default.

When all the points are marked, any method listed in the File menu can be used to obtain the resulting data set – "Show Data" outputs it in the text field labeled "4" in Fig. 1, "Send Data" usually opens a browser window, and "Save Data" will save it in a text file on the user's machine (provided the browser is correctly configured). The name of the text file can be set in the "File name:" field ("5" in Fig. 1) and defaults to "bibcode.page" (bibcode being the article's bibliographic code and page the page's sequential number within the article).

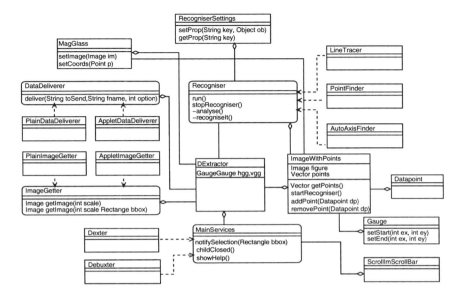

Figure 2. An abridged representation of Dexter's architecture.

2. Dexter's Architecture

Fig. 2 shows a raw sketch of Dexter's architecture in a graph inspired by the Unified Modeling Language (UML). Rectangles symbolise classes, rounded rectangles stand for interfaces, with dashed arrows from their implementation. Connecting lines indicate that two classes are talking to each other, with the diamond pointing from an embedding class.

In the central position there is the DExtractor class that imple. ients most of the user interface and contains the logic responsible for the transformation from screen to graph coordinates through the GaugeGauge class that in turn controls the text fields for entering the start and end values for the graphical gauges. The DExtractor class is derived from the Abstract Windowing Toolkit's Frame class.

The panel containing the image of the figure is displayed by a different class, ImageWithPoints, that also embeds the classes handling the graphical Gauges and the Datapoints. These latter classes handle their events themselves. For example, the handling of error bars is completely encapsulated within the Datapoint class.

The MainServices interface hides the startup from DExtractor. Two implementations of MainServices exist: Dexter, the applet interface, and the standalone Debuxter useful for debugging. It should not be difficult, however, to implement MainServices that, for example, would use ghostscript to provide Dexter's capabilities for PostScript articles. To build such a PostScript-Dexter, one would also need to adjust the implementations of the ImageGetter and DataDeliverer interface, the first accepting requests for scaled and cropped versions of the image, the second delivering the extracted data after the user has requested to save the data or get it sent. Among the implementations of these interfaces, the AppletDataDeliverer is somewhat involved since an (unsigned) applet cannot access the client's disk or portably send data directly to the browser due to Java's security model. In order to save data or display it in a browser window, it is relayed back to the host and tunneled to the client through a pipe on the host.

The Recogniser interface, derived from Java's Thread class, defines how classes that try to do automatic feature extraction interact with both DExtractor and ImageWithPoints. It also contains some utility functions, for example to acquire the pixels from the graph image. Images are stored as arrays of signed bytes by the Recognisers; the dynamic range from 0 to 127 is more than adequate for the images Dexter deals with, and the sign is rather useful for flagging purposes, e.g., in the flood filler used in the PointFinder Recogniser.

Recognisers need to communicate with ImageWithPoints to access the image and to set points or gauges they may have found, as well as with DExtractor, giving prompts in the status line and telling it when they are finished. DExtractor needs to know this to re-enable some critical operations that are disabled while a Recogniser is running (changing the resolution of the graph image, sending data). Most Recognisers will need some sort of user input, e.g., to get a start point for line tracing or a template for point matching. Since this is done under control of the Recogniser thread, (almost) all that has to be done from Dexter's main thread is to make a call to the Recogniser's start Method.

Since Recognisers do not register themselves automatically with DExtractor, some source code changes in both DExtractor (that controls the menu bar in which the Recognisers are registered) and ImageWithPoints (that actually starts Recognisers) are necessary when a new Recogniser is written. Given the current scope of the project (about 5000 lines of source code), a more elaborate plug-in scheme did not seem necessary. If Recognisers have adjustable parameters, they can use the RecogniserSettings class, containing both a Hashtable to store the property values and the logic to display a dialog to change them.

All three Recognisers currently implemented (AutoAxisFinder to locate the axes, PointFinder for point matching, LineTracer for automatically digitising lines) use naive syntactic algorithms. In a tool like Dexter intended to be interactive, the computational effort for performing Fourier or Hough transforms on entire images makes these approaches currently unattractive, given the poor performance of the Java virtual machines on some architectures.

3. Conclusions

Dexter has shown itself to be a very useful tool for those interested in obtaining numerical data from ADS articles. The applet could be easily adapted by other data archives providing scanned publications, and even extended to work with postscript or PDF files. Dexter's source code is available under the GNU General Public License on Sourceforge.net[2]

---

[2] http://Dexter.sourceforge.net

The OPUS CORBA Blackboards and the New OPUS Java Managers

Walter Warren Miller III[1], James F. Rose[2]

Space Telescope Science Institute

Abstract. OPUS is a generic pipeline environment for running multiple instances of multiple processes in multiple paths on multiple nodes. This is the platform which has been used successfully at the Space Telescope Science Institute for over five years to process the HST telemetry. This paper presents the basic concepts of the new OPUS caching blackboards and illustrates these concepts with the new OPUS Java Managers.

1. Background

OPUS[3] is a generic pipeline environment that has been used successfully at the Space Telescope Science Institute for over five years to process HST telemetry. It is generic enough to be applicable to other pipeline requirements and has been distributed on CD-ROM to other NASA/ESA observatories free of charge. As a consequence it is now being used by NASA's great observatories (*Chandra, Hubble*, and SIRTF), and by other observatories throughout the world.

The original OPUS architecture (Rose et al. 1994) was very simple and evolved to meet a variety of needs such as changing requirements and operating systems. A recent review of the original design revealed both the growing complexity of the system and some tremendous opportunities that could enhance the base system to make it more maintainable, extensible, and powerful.

As a result, the OPUS Application Programming Interface OAPI; Miller 1999[4] was born: an object oriented rework of the original OPUS architecture and a complete and published interface allowing client access to the internal workings of the OPUS blackboards.

2. Caching Blackboards

Riding on the generalization of the OPUS architecture introduced with the OAPI, blackboards can reside anyplace: on the file system, in a database, or in

---

[1] AURA: Association of Universities for Research in Astronomy

[2] CSC: Computer Sciences Corporation

[3] http://www.stsci.edu/software/OPUS

[4] http://www.stsci.edu/software/OPUS/OAPI

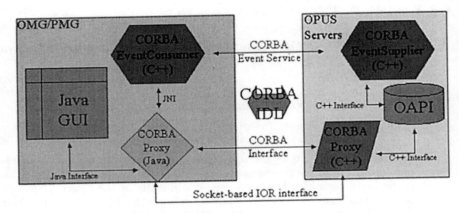

Figure 1. OMG/PGM, CORBA server, and OAPI OPUS interfaces

memory. Instead of using an overworked file server to provide the OPUS cache, a specially designed OPUS server has been developed. The OPUS caching blackboard is an in-memory blackboard (which nevertheless uses the file system as a backing store) where all interprocess communication is provided through a CORBA (Common Object Request Broker Architecture) interface.

Using the new facility of the OPUS caching blackboards to push events, the new OPUS Java Managers (the Observation Manager and the Process Manager) no longer are required to initiate polling. Instead they receive notification from the OPUS Server whenever the blackboard is modified. In consonance with the caching philosophy, a new channel was developed to serve the processes that monitor the OPUS pipelines.

Although a variety of approaches might be taken to incorporate these features in the new managers, CORBA can deliver a homogeneous solution since it promotes both programming language and locational transparency. It offers an object-oriented interface architecturally similar to the OAPI, and includes the desired push event mechanism. CORBA is an industry-standard specification for an object request broker[5] that acts as a message bus for the transmission of invocation requests and their results to distributed CORBA objects.

The diagram in Figure 1 illustrates relevant interfaces between components in the new managers, between the new managers and the CORBA servers, and between the CORBA servers and the OAPI.

3. Java Managers

Two pipeline managers come with the OPUS system. These are full Java applications that assist the user in monitoring the system. The Process Manager (PMG) not only assists with the task of configuring the system, but monitors what processes are running on which nodes, and what they are currently doing.

---

[5] http://www.omg.org

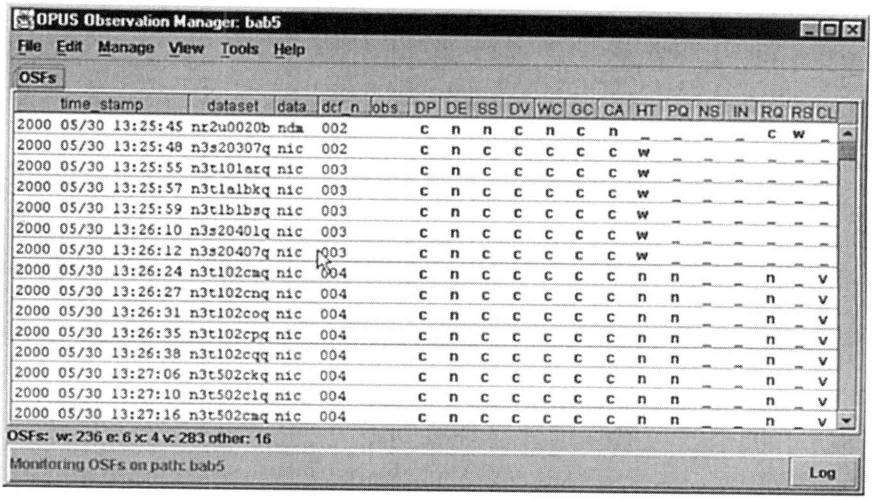

OPUS CORBA Blackboards and OPUS Java Managers 327

Figure 2. OPUS Observation Manager

The Observation Manager (OMG; see Figure 2) takes a second view of the pipeline activities, monitoring what datasets are where in the pipeline and alerting the operator when observations are unable to complete pipeline processing.

Because the interprocess communication mechanism is CORBA based, the Managers have been written as portable Java applications that communicate with the OPUS blackboards using the standard TCP/IP protocols. This, in turn, frees the Managers from having to operate within the confines of the pipelines they are managing: so long as they have a secure link with the OPUS server, the pipeline monitors can be served anywhere.

Freeing the graphical user interface from the responsibility of tracking events directly resulted in much more responsive tools in the hands of the OPUS user. The OPUS Java managers have taken all the capabilities of the Motif managers they replaced and added important new functionality. The locational transparency afforded by the new OPUS CORBA server implies we can run any number of Managers over the network on our workstations.

The new OPUS Observation Manager continues to support the flexibility built into the original Motif tools, e.g., configurable from external ASCII files, able to handle thousands of entries without loss of responsiveness, and easy modification of the status of any selection of entries. Consistent with the use of the managers as a view into the OPUS blackboards, the GUI provides its own user-defined views into the event lists.

Thus the user can create a new view that displays only the observations associated with a particular instrument, started on a particular day, or having completed a particular step in the pipeline. All views are accessible by clicking on tabs at the top of the display. Full use is made of "tooltips" to expand the use of abbreviations and mnemonics.

The new OPUS Process Manager (see Figure 3) maintains the functionality of the original tools and can monitor many pipelines simultaneously. Like

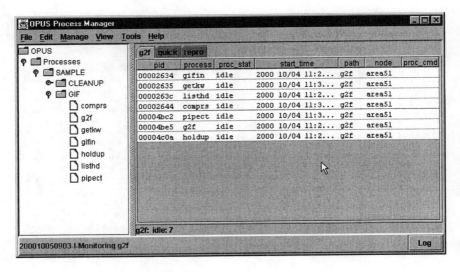

Figure 3. OPUS Process Manager

the new Observation manager that uses the full functionality of the new Java Swingset, the Process manager's display can be easily configured on the fly by each user. Thus all paths defined by the user will be displayed in a separate tab. All OPUS processes are made accessible, and can be dragged to a path to define an operational pipeline.

References

Rose, J., et al. 1994, in ASP Conf. Ser., Vol. 77, Astronomical Data Analysis Software and Systems IV, ed. R. A. Shaw, H. E. Payne, & J. J. E. Hayes (San Francisco: ASP), 429

Miller, W. W. III 1999, in ASP Conf. Ser., Vol. 172, Astronomical Data Analysis Software and Systems VIII, ed. David M. Mehringer, Raymond L. Plante, & Douglas A. Roberts (San Francisco: ASP), 195

An Extragalactic Point Source Simulator for SIRTF Pipeline Testing

Sonali Kolhatkar, Fan Fang, Mehrdad Moshir

SIRTF Science Center, California Institute of Technology, Pasadena, CA 91125

Abstract. We designed and developed a package which simulates extragalactic point source data within a given field of view and SIRTF band. Our method uses input empirical relations, such as known luminosity functions and galaxy spectral energy distribution templates. It also uses instrument filter information to produce a list of sources, their redshifts, positions, and fluxes. The output fluxes in the specified band are based on the user-specified luminosity and redshift evolution modes. We designed the simulator so that the empirical input and filter information can be easily modified. The simulator is modular and extensible to enable multi-wavelength simulations, has been optimized by utilizing parallel processing capabilities, and will be used to test pipeline software.

1. Design

The extragalactic simulator has a modular, linear design and consists of six stand-alone C modules. These modules are represented in Figure 1 as (red) boxes labeled: in (1), "GENERATE...", "EVOLVE", and "DISTRIBUTION...", in (2), "GENERATE FLUXES", and in (3), "POSITION..." . There are several additional input tables of data and parameters which are bound to the modules via a single C-shell script. The design of the simulator is well-suited to future extensions. Each module can be run separately at the command line for testing and debugging purposes and also produces its own output and log files. All user-specified parameters reside in a separate file that can easily be up-to-dated with new inputs (e.g., SED templates, filter responses, etc.).

The simulator has been streamlined for high-speed processing and is capable of generating tens of thousands of extragalactic sources in minutes. This is enabled by splitting the most time consuming process over several processors (currently up to 20) and running them in parallel. Users can simply list available machines and run processes in the background of each machine.

2. Algorithm

In the first stage of the simulator, a set of extragalactic sources is generated for a given set of input parameters. The simulated sources fall within a user-specified area on the sky and have three basic attributes—see part (1) of Figure 1:

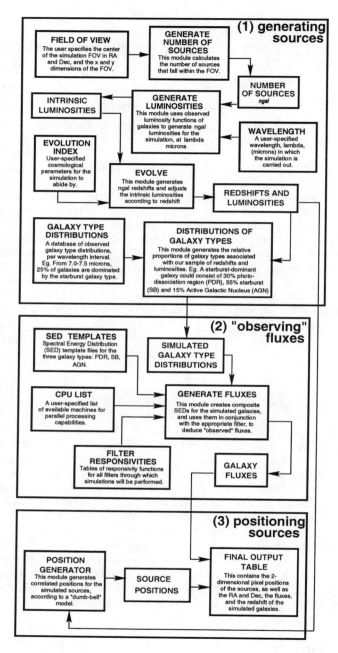

Figure 1. Schematic diagram of the extragalactic simulator illustrating its three main stages; see text. (In the on-line version of this article, blue boxes represent data input modules; red, data processing modules; and green, output modules.)

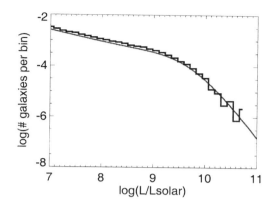

Figure 2. Distribution of simulated $8\,\mu m$ luminosities. Overlaid smooth line is the parent $12\,\mu m$ function from which luminosities are drawn.

- **Intrinsic luminosities** are chosen on the basis of an appropriate luminosity function (see Figure 2).
- **Redshifts** are derived independently for each galaxy using a $(1+z)^{-\beta}$ distribution, appropriate for a co-moving density evolution.
- **Galaxy types** are determined from intrinsic luminosities and the band of interest and emission is modeled as a linear combination of the average emission from "Photo-dissociation Regions (PDR)", "Star Bursts (SB)" and "Active Galactic Nuclei (AGN)", determined from empirical data.

In the second stage of the simulator, available Spectral Energy Distribution (SED) templates for PDR, SB, and AGN galaxy types are used—see part (2) of Figure 1—to construct individual composite SEDs based upon the proportions of types calculated in the previous stage. With the filter response function of the specified SIRTF wavelength, the "integrated luminosities" of the galaxies are calculated by multiplying by the normalized SEDs and integrating to determine flux densities from the following relation: $Flux_{Band} = L_{int}/4\pi D^2$ where L_{int} is the integrated luminosity, and D is the luminosity-distance of the galaxy.

In the last stage of the extragalactic simulator, galaxies are placed on the 2-dimensional sky—see part (3) of Figure 1—with correlated positions calculated according to a dumb-bell model of fractal structure (Soneira & Peebles 1978). The final output of the simulator contains luminosities, redshifts, fluxes, and positions in Right Ascension and Declination of the simulated galaxies.

3. Discussion

The extragalactic simulator has been tested for four filters on the SIRTF IRAC instrument: $3.6\,\mu m$, $4.5\,\mu m$, $5.8\,\mu m$, and $8.0\,\mu m$. The table look-up method used to generate the luminosities of the simulated galaxies works well at all four wavelengths (see Figure 2).

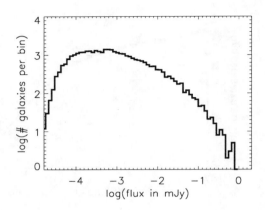

Figure 3. Log plot of number counts of 34,557 extragalactic sources vs. flux.

The extragalactic simulator is capable of generating large numbers of galaxies within a given patch of sky for SIRTF pipeline testing. Tens of thousands of galaxies can be simulated within minutes and further techniques to speed up the process are under consideration. This includes "table look-up" methods in the flux calculation module that could replace calculation of individual fluxes. When available, luminosity functions can be replaced by newer ones recorded at wavelengths close to SIRTF instrument wavelengths. SED templates can also be replaced by newer, more accurate ones when available. In order to simulate an observation of the real sky, this simulator will soon work in conjunction with the SKY model galactic simulator (Cohen 1993) to produce data sets that contain external galaxies as well as objects within our own.

A common test of cosmological models is a $logN$ - $logS$ plot of galaxies. An effective test requires a complete sample of galaxies out to relatively high redshifts, and Figure 3 shows a plot of $logN$ versus $logS$ for 34,557 simulated extra-galactic sources with redshifts in the range $z=0$ to 5, "observed" using the IRAC $8.0\,\mu m$ filter. The extragalactic simulator produces galaxies with fluxes that are at the limit of IRAC sensitivity ($\approx 1\,\mu Jy$).

References

Cohen, M. 1993, AJ, 105, 1860

Soneira, R. M. & Peebles, P. J. E. 1978, AJ, 83, 845

Interfacing Real-time Linux and LabVIEW

P. N. Daly

National Optical Astronomy Observatories, 950 N. Cherry Avenue, P. O. Box 26732, Tucson, AZ 85726-6732, USA

Abstract. Real-time Linux is a set of extensions to the kernel that provides hard real-time functionality with low, bounded latencies and deterministic response. The main methods for communicating between kernel space and user space are fifos and shared memory. LabVIEW is the well-known commercial product for developing control systems and engineering applications. This paper, presents the fifos and shared memory virtual interface (VIs) that allow LabVIEW to communicate and share (bulk) data with the real-time core.

1. Introduction

The cardinal programming rule of real-time Linux is to do as much as possible in user space with only critical sections of code in the real-time core. This demarcation requires methods of exchanging simple, structured and/or bulk data between the real-time side and the application side. Two such methods exist: fifos, for simple or moderately structured data, and shared memory, for bulk data. For the work described in this paper, the interested reader will require the common mbuff 0.7.1[1] and fifos 0.5[2] packages built under either the RTLinux 2.2[3] or RTAI 1.3[4] kernel patches to Linux 2.2.14[5].

2. Code Samples for Command Line Fifos and Shared Memory

To understand what is intended, a quick tour of using fifos and shared memory from command line applications follows.

---

[1] ftp://crds.chemie.unibas.ch/PCI-MIO-E/mbuff-0.7.1.tar.gz

[2] http://www.realtimelinux.org/CRAN/software/rtai_rtl_fifos-05.tar.gz

[3] ftp://rtlinux.com/rtlinux/v2/rtlinux-2.2.tar.gz

[4] ftp://www.aero.polimi.it/RTAI/rtai-1.3.tgz

[5] ftp://ftp.kernel.org/pub/linux/kernel/v2.2/linux-2.2.14.tar.gz

2.1. Fifos

Fifos are simple, character-based devices which must be created in the real-time core before being used by an application. To enable /dev/rtf0, for example, with a buffer 1024-bytes long, the code in the init_module entry point would be:

```
int err = 0;
if ( (err=rtf_create(0,1024)) < 0 ) return -EINVAL;
```

A periodic real-time task might then write data in the handler to the fifo using a code segment such as:

```
unsigned int sent = 0;
while ( 1 ) {
#ifdef RTL       // RTLinux 2.2 re-scheduling point
  (void) pthread_wait_np();
#else            // RTAI 1.3 re-scheduling point
  (void) rt_task_wait_period();
#endif
  (void) rtf_put(0,&sent,sizeof(sent));
  sent++;
}
```

The application then reads the fifo:

```
unsigned int recv = 0;
if ( (fd=open("/dev/rtf0",O_RDONLY)) < 0 ) return (fd);
while ( read(fd,&recv,sizeof(recv)) ) {
  (void) printf("
}
(void) close(fd);
```

It is easy to reverse this process to put data into the core and/or make the operation pass structured data. This fifo is destroyed in the cleanup_module using:

```
(void) rtf_destroy(0);
```

2.2. Shared Memory

Shared memory *must* be declared in the init_module and cast to the desired data type or structure. The user space application uses symmetric calls. For example, to obtain a 1 Mb array of signed integers, one could use:

```
static int *mptr = (int *) NULL;
mptr = (int *) mbuff_alloc("myints",1024*1024);
#ifdef __KERNEL__      // code was called by init_module
  if ( mptr == (int *) NULL ) return -ENOMEM;
#else                  // code was called by user application
  if ( mptr == (int *) NULL ) return (EXIT_FAILURE);
#endif
```

The memory can then be accessed using straightforward de-referencing and pointer arithmetic to traverse the array. The memory is released using:

```
mbuff_free("myints",(void *)mptr);
```

Note that the name of the memory section must be common between the real-time core and application. Several applications may also references the same memory segment. The only restriction on the size of a memory section is the amount of system memory available. In the WIYN Top-Tilt Module project (Daly 2000), for example, 128 Mb are mapped using a single call to *mbuff_alloc*.

3. The *lvrtl* Package

The functionality described in the previous section is implemented in LabVIEW using the *lvrtl* package developed by the author. Two versions exist, one for LabVIEW 5.1[6] and one for LabVIEW 6i[7]. Only the LabVIEW 6i code will be developed further as needs arise. Documentation on the design and implementation of *lvrtl* is available in *lvrtl.pdf*[8] as included in either tarball.

This package provides a shared library (/usr/lib/liblvrtl.so) and 78 VIs that furnish access to fundamental data types. It also includes extensive test code for either hard real-time Linux variant. The following VIs are implemented in this release where *dtype* is one of the set {int8, int16, int32, uint8, uint16, uint32, float32, float64, string}:

1. rtf_open, rtf_close, mbuff_open and mbuff_close.
2. rtf_get_*dtype* and mbuff_get_*dtype*.
3. rtf_put_*dtype* and mbuff_put_*dtype*.
4. rtf_read_*dtype* and mbuff_read_*dtype*.
5. rtf_write_*dtype* and mbuff_write_*dtype*.

Note that *rtf_put_string* and *mbuff_put_string* are the only two routines that add a NULL byte to the data before transfer. The real-time core must be set up to accept this extra byte. If the number of elements to put is 0, the code will size the data array and transfer the whole structure (rather than return an error).

The 'get' and 'put' VIs both accept a number of elements of the given data type to read/write. On error, -1 is returned. On success, the number of *bytes* read or written is returned and *not* the requested number of elements. Since the *mbuff_get*_dtype and *mbuff_put*_dtype VIs also require an offset from the start of the memory array, poor G-code programming could result in a request that exceeds the bounds of the memory segment. Such an error is not trapped by this version of the code and the result will likely be an illegal pointer operation in the kernel resulting in a system crash. The programmer is urged to make sure that all inputs to these VIs are valid.

The fifo may be accessed using a blocking read, non-blocking read, read-write or write-only operation.

As an example, Figures 1 and 2 show the front panel and G-code diagram for a *rtf_read_float32* operation. This example reads five floating point numbers

---

[6]ftp://orion.tuc.noao.edu/pub/pnd/lvrtl.1.1.51.tgz

[7]ftp://orion.tuc.noao.edu/pub/pnd/lvrtl.1.1.60.tgz

[8]http://www.realtimelinux.org/documentation/lvrtl.pdf

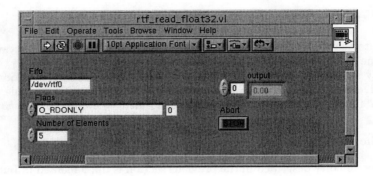

Figure 1.   rtf_read_float32

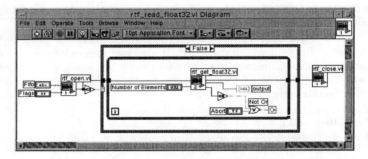

Figure 2.   rtf_read_float32

from /dev/rtf0. Note how any output error from the embedded *rtf_get_float32.vi* is trapped within the G-code environment and terminates the loop. The data from each read appears in the 'output' array.

4. Structured Data

Clearly, *lvrtl* can handle basic data types but what of structured data encapsulating heterogeneous data types? There are two approaches. First, one could write one's own suite of VIs and add code to the shared library. There should be enough detail in the documentation and the examples to permit coders to do this. Alternatively, one could wire together VIs handling the data types of the individual elements of the structure. In this latter case, more than one read is required to completely obtain the structured data.

References

Daly, P. N. 2000, in ASP Conf. Ser., Vol. 216, Astronomical Data Analysis Software and Systems IX, ed. N. Manset, C. Veillet, & D. Crabtree (San Francisco: ASP), 388

The Basic Calibrated Data Processing Pipeline for SIRTF IRS

Fan Fang, Bob Narron, Clare Waterson, Jing Li, Mehrdad M. Moshir

SIRTF Science Center, Caltech

Abstract. The design and implementation of the data processing pipelines for the Infrared Spectrograph (IRS) onboard Space Infrared Telescope Facility (SIRTF) are presented. This includes pipeline for generating Basic Calibrated Data (BCD) in which instrument artifacts are removed, and calibration pipelines generating calibration data that allows reduction of the science data to the BCD level. Further reductions of BCD to generate Post-BCD products is also briefly discussed.

1. Introduction

The Infrared Spectrograph (IRS) will be one of the three instruments onborad NASA's Space Infrared Telescope Facility (SIRTF). Four instrument modules of IRS are designed and built to observe the mid-infrared ($5\,\mu m$ to $40\,\mu m$) spectra of astronomical sources in four overlapping wavelength channels with low- and medium-resolution dispersion optics and As:Si and As:Sb BIB detectors. The IRS Basic Calibrated Data (BCD) pipelines are designed to remove instrument artifacts introduced by a combination of optics and detector effects. The end-products are two-dimensional images containing spectra. Post-BCD processing pipelines will enable further reduction of the BCD frames to extract the one-dimensional spectra of observed sources.

2. Instrument Signatures

The spectra of astronomical sources are taken by the two-dimensional BIB arrays operated at a temperature of $\sim 5\,K$. There are four readout channels, and the analog-to-digital convertor is saturated at a level below the pixel full-well. The nominal data-taking mode for observing spectra is the so-called sample-up-the-ramp, in which pixels are read non-destructively during one Data Collection Event (DCE), and all readings are downlinked to the ground. So the pipeline typically deals with a data cube with multiple sampled layers in one DCE. A two-dimensional IRS spectral image has the following instrument artifacts:

1. dark current that is always present at the operating temperature, independent of exposure to outside the detector.
2. nonlinearity when a detector pixel contains more than a certain number of electrons.
3. pixel response variations across the array.
4. readout-channel dependent gain variations.

5. detector transient effects causing "jail-bar" patterns along the slow-read direction.
6. a "droop" effect in which the readout value of one pixel is affected by the presence of electrons in all other pixel wells in the array.
7. a short-timescale (< 32 sec) nonlinearity behaving differently from pixel to pixel.
8. radiation hits in space environment.
9. amplifier drift during consecutive DCEs.
10. a muxbleed effect in which a bright pixel trails in the fast-read direction every several pixels.
11. a blaze-function introduced by dispersion optics.
12. fringing caused by multiple reflections from instrument and optical interfaces.

Since the IRS has no shutter it will direct its slit to blank sky areas to take reference dark frames for calibration of dark current. The non-linearity effect is not expected to be significant since the A/D converter saturates at a level below full pixel well, but will be corrected. Although the detector has pixel-dependent ramp nonlinearity at short time-scale, the same behavior is persistent in different integration ramps with the same sample duration, so this can be corrected by layer-by-layer subtraction of a reference data cube such as the reference dark cube.

Radhits can be identified and removed by examining the discontinuities in the charge ramp. A segmented fit of the ramp is performed for each pixel and the probability and strength of the radhits are estimated using a Bayesian approach. The "droop" effect can be caused by both illumination and radhits so it should be accounted for before radhits are removed. Saturation at the A/D converter needs to be corrected since electrons are still accumulating in pixel wells and contribute to the "droop" level. Any remaining "droop", such as that caused by radhits in saturated pixels, can be removed by examining and subtracting the remaining median levels in unilluminated regions. Such unilluminated regions, where spectra orders are well-separated and order cross-talk effects are small, exist in all IRS arrays. Any channel-dependent effects can be similarly removed by examining and correcting remaining cross-channel median levels.

Flatfielding corrects for both optical dispersion function and pixel response variation. To estimate flatfield, a number of calibration sources with known continuum and spectral lines will be observed. The spectral lines in different sources will be masked, and a slit profile will be incorporated into the continuum spectrum which is removed from data.

3. The IRS Processing Pipelines

Figure 1 shows the current design of the BCD science and two calibration pipeline threads. They are shown as flowcharts illustrating the data reduction sequence and the software components in boxes that perform specific tasks. In the reference dark-current calibration, the amplifier drift is removed for consec-

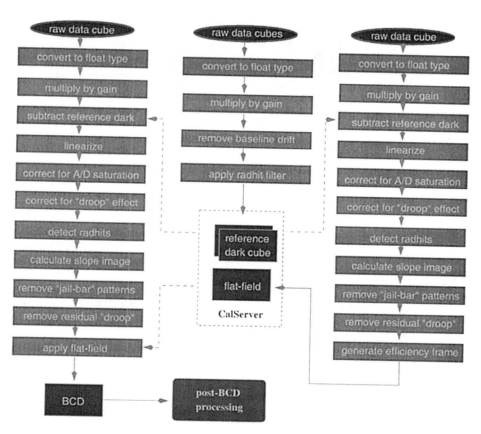

Figure 1. The current IRS BCD science data reduction (*left*), dark-current (*middle*) and flatfield (*right*) calibration pipelines shown as flow charts. The muxbleed effect correction and fringing removal will be included in these pipelines in their subsequent deliveries.

Figure 2. Reduction of the IRS BCD science pipeline using stimflash data. The image at left shows the highest sample in a ramp in the data cube as the pipeline input; the image at right shows the reduced result. So far we do not have ground spectra data in this data-taking mode and flatfield calibration is not applied to this reduction.

utive dark exposures. This is for correctly identifing small radhits during the subsequent median-filtering of the same data layers of many such exposures. The flatfield calibration follows nearly the same reduction steps as in the science pipeline, since flatfielding is performed at the end of the science pipeline. A result from the science pipeline reduction using stimulator data is shown in Figure 2.

Further reduction of the BCD is needed to generate spectra. The purpose is to make it straightforward to extract one-dimensional spectra from the two-dimensional BCD. This basically involves "straightening" of the spectra in both two-dimensional pixel space and one-dimensional wavelength space. We have implemented post-BCD pipelines that take into account the curvature of spectra and wavelength resolution elements to perform the "straightening". The pipeline uses a calibration file containing elements optimally sampled in wavelength space and their locations in pixel space in order to estimate the average profile of spectra orders and perform spectra extraction based on the profile.

The components in a pipeline are all stand-alone modules. Operationally the calling of a software module is communicated via wrapper scripts of the modules. A wrapper script can have extra capabilities such as communicating with calibration servers, checking input files, directing output files, setting database flags, etc., in addition to executing the corresponding module.

Part 6. Software History

Astronomical Software—A Review

Keith Shortridge

Anglo-Australian Observatory, P.O. Box 296, Epping, NSW 1710, Australia

Abstract. It is now impossible to imagine 'doing astronomy' without using software. Sometimes it is hard to remember that it has not always been like this.

Over a timescale now measured in decades, the art (or science) of astronomical programming has evolved. Once it involved the squeezing of hand-crafted assembler routines into insufficient memory. Now it includes the design of ambitiously large frameworks for data acquisition and reduction. The organisation required for the production of such software has had to grow to match these new ambitions.

This review looks back on the path taken by this fascinating evolutionary process, in the hope that it can provide a background that may let us imagine where the next years will lead.

1. Introduction

What follows is essentially a verbatim transcript of the rather informal talk given at ADASS. A more formal treatment of some of the ideas presented here can be found in Shortridge (2001).

2. Early History

I'm going to look back into the murky history of astronomical computing. There are usually two reasons for looking back like this. One is that it's fun. This is what memories are made of—and once upon a time memories were made of little bits of core with wires threaded through them... The other reason is that history provides a context for understanding the present and guessing the future.

This talk has been billed as being about astronomical software since the '60s, but it wouldn't hurt to remember that computation in astronomy goes back much earlier than that. I don't want to go into the speculation that structures such as Stonehenge were computers... But debugging them must have been fun. 'Comment out this stone and we'll try again next mid-summer's day...' And some smart Alec will say, 'Of course, when we go from 1000BC to 999BC you'll wish you'd used more than two stones for the year...'

Even without going that far back, astronomy—or at least, astrometry—has always depended on calculations. I was once told of a report written by an early Astronomer Royal who complained that 'the computers were a disorderly, drunken, rabble.' These computers were, of course, what computers are now: a resource astronomers use to do tedious mathematical calculations. Except that these were people—some of them a bit the worse for drink.

You still find the term 'computer' being used to mean a person into the early 20th century. But technology advances all the time.

In 1951 the then Astronomer Royal was still having trouble with 'computers'. His report included: 'The model 602A calculator, which is the only machine that can multiply, is the key machine as far as general computing work is concerned. Unfortunately, the calculator was delivered from the USA without the relays for division, so there is still a considerable lack of flexibility for complicated calculations.'

You know just how he felt, don't you?

The next year, it still couldn't divide. But in the next year, 1953, they finally delivered the division relays. Let's hear it for field service!

Now a 602A calculating punch was—well, it's a computer, Jim, but not as we know it. It could be programmed. It had a pluggable patch panel with up to 60 program steps. It couldn't just divide (eventually), it could loop. It didn't store the program in memory. It didn't treat program and data interchangeably; a program couldn't really be the output from another program.

But it needed programmers.

I'd like to emphasise, by the way, that all this was *before my time*. I'm not doing this review because I was there through it all. I'm not that old. I was 30 this year. Unfortunately, I was twenty-F last year. Come on, how many people here can still work that out in their heads?

3. Mainframes and Mini-computers

Let's move on. Round about 1958, the term 'software' was coined. In 1959 came the IBM 7090; a recognisable mainframe computer with—gasp—a FORTRAN compiler. You could do ephemeris calculations, you could run model atmosphere codes. You had card readers, you had paper tape punches...

Back then, you knew how big a program module had to be. There was a rule of thumb. If you couldn't hold all the cards in one hand without dropping them, it was too big. About 400 lines. Or you'd be scrambling around on the floor muttering, "Next time, I'll leave out the comments."

Have you noticed how programmers go on about how it used to be? "When I first used PCs, they couldn't address more than 640K." "640K?—I once got an operating system, a compiler, an editor, an assembler and useful applications into 2K!" "2K? We had to make do with single bits—we had to sign forms for each one we used!" "Real memory—luxury! We used graduate students; they stood up for one, sat down for zero; the floating point unit took up a whole football field."

Come on. Who here's played that game in the last few days?

A problem with mainframes is that you can't control instruments with them. Obviously, one huge change came in when mainframes appeared in computer centers. But—for us—an equally big change was when mini-computers came into the labs. (And then, of course, microprocessors into the instruments. And networks, which is a different kind of change.) Mini-computers allowed the use of computers for control as well as calculation. Most of the programs I write don't actually 'compute' very much—they control. And that happened in the late 60s and early 70s.

That was when the PDP-8 came out, followed a bit later by the 16-bit machines that were the mainstay of control systems for over ten years. The DEC PDP-11 was probably the most ubiquitous, but there were also the Data General Nova and Eclipse, and the machine that introduced me to all this: the Interdata 70.

These were the machines that controlled the 4-metre class optical telescopes of the 70s. When the Anglo-Australian Telescope went into operation in 1974 it had a marvelous control system (which was nothing to do with me—the credit goes to Pat Wallace et al.) This modeled the deformation of the telescope and allowed it to set to better than 1.5 arc-seconds, which was astonishing. Just to reminisce for a moment, these machines used at the AAT had 64 kilobytes of memory, a 4 MHz system clock—although the current infatuation with Megahertz was unknown then—and 5 MB of disk space.

I gave a talk a bit ago in Sydney where I passed around an 8 KB core memory board from one of these machines. This big (14 inches square) and 8 KB (read my lips, *Kilo*bytes) of memory. I'd like to do that now, but I wasn't allowed to bring it with me. You see, the current AAT control computer is ... yes, it's that same Interdata 70 and that board is one of the vital set of spares we keep for it. You shouldn't assume your control software won't still be being used in twenty five years time.

Let's really talk about software. These machines were programmed in assembler. You add two numbers by 'add register one to register two'. Or in FORTRAN, where you do it by A = A + B. (In capital letters, because card punches don't do lower case.) You knew—and cared—exactly how long each instruction took, and they had a good real-time operating system. And you didn't have to worry about whether your driver was at a lower priority than the Ethernet driver, because there wasn't one.

The main programming methodology was 'whatever works fast enough and will fit in memory will do,' and the only thing even resembling a standard library was the built-in FORTRAN I/O package—and I usually didn't use that because it used up all of 8 KB.

4. Memory Constraints

Back then memory usage was one of the biggest constraints. Generally, a 16-bit machine can address 64 KB. If you like splitting time up into eras, then the predominant number of address bits is one criterion. This was the 16-bit era. We're now arguably at the end of the 32-bit era. Once you can physically buy all the memory you can address, you clearly need more address bits. And you can now afford to buy memory in Gigabytes.

Partly because of these memory considerations, the FORTH language enjoyed a vogue then. FORTH defined a FORTH machine, with an extensible set of operations (words). Words were defined in terms of other words, and because they were very lightweight you reused them a lot and got very dense code. I really did once get an operating system, compiler, editor and applications into

2 KB, and it was done in FORTH. It sacrificed clarity for conciseness: A = A + B is now A @ B @ + A ! which is only arguably an improvement.

FORTH originated in astronomy, and was used a lot then fell out of favour. Eventually, its compactness was not so important and its disadvantages became more apparent. One thing it missed, and I think this is still important though I've never heard it discussed much, was 'locality of code'. You couldn't look at just one part of a program listing and understand it. You never were familiar with the words used and they were defined elsewhere—in terms of other words defined elsewhere. It's an aspect of complex code that's introduced by modularity. You can reduce it by designing your components as intuitively as possible, and by encapsulation, but it's still a big issue for code maintainability.

What removed the memory constraints?—More memory. 32-bit machines.

Perkin-Elmer released a 32-bit version of the Interdata 70 they called the 'Megamini'. You could order one with a megabyte of memory. I remember a meeting that ended up specifying one for RGO with 512 KB, because *nobody* could think of any possible reason a control computer could need a Megabyte of memory. And the next year I think they ordered the other 512 KB ...

But the machine most will remember from the 80s was the VAX. Out in 1977, the VAX 11/780 had 32-bit addressing *and* virtual memory. VAX—Virtual Address Extension, because we all know 'Extend' is spelled with an 'X'. It looked like a wardrobe, but it was a wardrobe that could address 4 gigabytes.

5. Software Frameworks

Moore's Law has been given a good airing at this meeting, but nobody has actually stated it. It says that the processor power needed to run Microsoft Word doubles roughly every 18 months, but fortunately the hardware keeps up. But it isn't just processor speed that drives the changes we've seen. Speed lets you do things faster, but memory—disk as well as core—lets you do more complex things.

With unimaginable amounts of memory now available, programmers could start to build the sort of new systems they now realised they'd always wanted to write. This is where we start to see the emergence of the big systems. You know these, you use them now. IRAF, ADAM, AIPS, MIDAS.

Looking back, these are component software frameworks. You define the way a program gets run, the way it gets its parameters, how it handles disk files, and an application becomes an easily-written component that fits into the framework, providing a facility that wasn't previously available under that framework.

Persuade people that such components are easy to write (once you've mastered the framework) and more components will be written and the framework becomes richer and richer. The people who write the programs don't get richer and richer, because we don't work in that sort of environment. Fame has to be the spur. And it is. Don't you get a kick from getting bug reports from all around the world?

AIPS came out around 1978, IRAF in the early eighties. ESO's early IHAP system, running on 16-bit HPs, was replaced by MIDAS. These are data reduction systems. Data acquisition frameworks are harder, because they're real-time systems, but you can do it if you're rash enough. ADAM emerged as a data acquisition framework, originally for that Perkin-Elmer 'Megamini', and was used a lot, particularly by places that had UK connections.

There was another, more or less parallel trend. The emergence of standards. What format do you use for data interchange? FITS. FITS has been around a long time. Who uses 2880 as a PIN number?—it's one of those numbers that rolls trippingly off the tongue. FITS has been a great triumph. So have standard subroutine libraries—SLALIB for astrometry—and standard components like SAO image that can be massaged into a number of frameworks.

Well, things change. The VAX lasted a long time, but the combination of RISC chips and UNIX was unstoppable. Most of the frameworks moved over to UNIX. Some had been there all the time. With UNIX came C, and $A = A + B$ is now in lower case and has a semicolon after it—so you can squeeze a lot of statements onto a line in the interests of readability...

Then C++, and you don't just add numbers together anymore; now you need to know what sort of thing they represent so you can encapsulate them into a class and define how the 'plus' operator works on them. And the productivity gains are amazing!

It's easy to poke fun, but actually I've found I enjoy writing C++ and Java. I think that's because there's no obvious real-life metaphor for procedural programming, but we're all used to working with things—particularly people—with different skills and specialities, and getting them to work together is something we understand. And Java is a framework all to itself... And then there's the Web, and the Grid...

Back a step. As UNIX started to dominate the data reduction world, it also took over the top levels of data acquisition systems. But it generally doesn't go all the way down to the sharp end—the instrumentation hardware. You find the same RISC processors there now—SPARCS, PowerPCs. (PowerPCs—you have to love a chip with an instruction called EIEIO. Enforce In-order Execution of I/O. It does something to the cache, but I'm glad to say I don't know exactly what. You should always take home one fact from any talk—but maybe not that one.) These hardware control processors can have as much or more memory as the workstations, and Ethernet connections, but they aren't running the same software, generally.

The data acquisition frameworks are now controlling highly complex networked systems. Both ESO and Gemini, for example, have systems that use a database paradigm, where hardware components map to database entries and changing the database entry is supposed to have a direct effect on the hardware. In both cases there's a distinct boundary between the real-time parts and the top levels.

In Gemini, the low levels are supposed to look like EPICS databases. In the ESO VLT system you have to think carefully about whether you put software items into the workstations or the low level LCU systems. Both systems also have elements of the more conventional 'send a command, wait for a response' systems. The AAO's DRAMA system—an ADAM descendant—has the same API at all levels, but is a pure command/response system and misses out on some advantages of the database approach.

Interestingly, all these use the VxWorks real-time kernel at the low level, but they all hide it so much you'd not know it was there. Knowing VxWorks—or UNIX—doesn't help much when it comes to learning these systems.

And learning matters.

6. Summary

Looking back, there's been a steady progression, not just in the speed of systems, but also in the increasing amounts of memory available, which allows not just the handling of more data but also the production of more complex software systems. And complexity has many ramifications. One can tackle complexity through the use of packages, components, objects, but these increase the learning curve and tend to reduce the code 'localisation'—the ability to understand a piece of code just by looking at a page on a screen and knowing the language used.

And, perversely, as we move to more complex astronomical frameworks, some of our organisations are moving to organisational structures where the components written for these structures are written by outside contractors. Both ESO and Gemini outsource their instrumentation. But you can't just advertise in the IT section of the local paper for someone with expertise in the VLT software environment.

These were thoughts that came to me as I tried to look back a bit. Everything you could possibly know about FORTRAN IV fitted into a large typeface IBM manual maybe a third of an inch thick. Who here has 'Java in a Nutshell' on their desks? It's a series of three thick volumes in a specially condensed font.

One last thing. A point that was made earlier in the SETI talks, and which I heard first from Ron Ekers at ATNF. Some time back, costs crossed over. Computer hardware is now a consumable that the software uses. The software represents the capital investment.

So, next time the organisation's bean counters want to stick an asset number on your workstation, tell them it's just a consumable, like a box of printer paper. Tell them to stick their asset number—wait for it—tell them to stick it on your program code. They won't, but it might give you a warm inner glow, and that's what writing software should be all about.

References

Shortridge, K. 2001, Software in Astronomy, in The Encyclopedia of Astronomy and Astrophysics, ed. Paul Murdin (London: Institute of Physics Publishing)

Reflections on a Decade of ADASS

Richard A. Shaw
Space Telescope Science Institute

Elizabeth Stobie, Irene Barg
University of Arizona

Abstract. Since its inception ten years ago, the ADASS conference series has been the premiere forum for software development related to the acquisition, calibration, analysis, and dissemination of astronomical data. We review the history of the conference itself, and highlight a few of the most important advances in the software and systems, and ponder what future advances might bring. We conclude with a description of the new adass.org Web site, and the services it offers.

1. Introduction

The past ten years have witnessed extraordinary changes in the way astronomers pursue their science. These changes in the way they build and operate the national observatories, the way they obtain, reduce, and analyze their data, and the way they communicate their results to their peers and the public, have all been profoundly changed, and changed primarily, through the revolution in software, networks, and information technology. The Astronomical Data Analysis and Software Systems (ADASS) Conference Series was established ten years ago as a forum on the development of software and systems for the acquisition, calibration, analysis, and dissemination of astronomical data. This annual conference has given those of us on the forefront of the astronomical information age a chance each year to pause and reflect on the trends and techniques of our craft, and on the application of the myriad of advances in the computer industries to our profession.

2. ADASS Past and Future

The landscape of astronomical software development prior to the 1980s was typified by small, heroic, but generally isolated groups that built highly specialized software for the support of specific missions or observatories. Later, as major observatories took on the role of supporting the data reduction and analysis needs of broader and growing communities of observers, more general reduction and analysis systems such as AIPS, IRAF, and MIDAS came into wide use. Yet it was generally difficult for non-experts to add major applications and otherwise extend these systems. As a result, software development for many projects continued its divergent course.

The ADASS conference series began in 1991 as somewhat of an experiment: to see whether astronomers and the developers of software systems for data reduction and analysis, archiving, etc. could be brought together (even once) to share ideas, experiences, and techniques for the collective good of astronomy. The hope was that by raising the awareness of the various approaches to common problems across systems and wavelength domains, a willingness to cooperate and even collaborate would result. This first conference was hosted in the fall of that year by the National Optical Astronomy Observatories (NOAO), and was attended by more than 300 people. Shortly after its inception, ADASS became the premiere, world-wide conference on astronomical software; it has been hosted by a number of major astronomical institutions in the US, Canada, and Europe (see Table 1). ADASS conferences have generally been well attended, although the attendance is somewhat lower when the venue is outside the US (see Figure 1).

Table 1. Past and Future ADASS Conferences

Year	Location[a]	Year	Location[a]
1991	Tucson, AZ	1998	Champaign, IL
1992	Boston, MA	1999	Waikoloa, HI
1993	Victoria, BC, Canada	2000	Boston, MA
1994	Baltimore, MD	2001	Victoria, BC[b]
1995	Tucson, AZ	2002	Baltimore, MD[b]
1996	Charlottesville, VA	2003	Strasbourg, France[b]
1997	Sonthofen, Germany	2004	Pasadena, CA[b]

[a]All locations in USA unless otherwise indicated.

[b]Future venue.

The typical three and one-half day ADASS program consists of a few dozen oral presentations (including invited talks), numerous display presentations, software demonstrations, "Birds of a Feather" (BoF) sessions on special topics, and one or more tutorials on important new technologies. The legacy of ADASS is captured in the proceedings volumes, which have been published each year by the Astronomical Society of the Pacific in hardcopy and (since 1993) electronic form.[1] The combined volumes comprise nearly 1200 papers, which include the oral and display presentations and summaries of some demonstration sessions. Perhaps not surprisingly, many of these papers have been cited elsewhere in the astronomical literature (some with enviable citation rates).

---

[1]http://www.adass.org/proceedings/

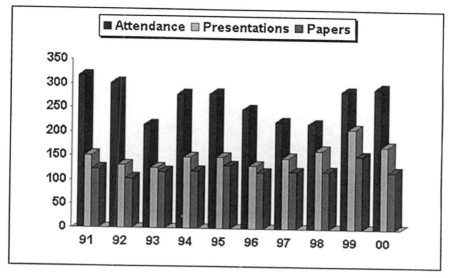

Figure 1. Annual totals of ADASS attendance (*left bar*), presentations (*middle bar*), and papers that appeared in the proceedings (*right bar*).

2.1. Demographics

The ADASS Conference series was conceived with astronomers and software developers (who are often one in the same) in mind. So, too, were technical leads and managers, though the program content did not explicitly target this demographic group in the early years. The largest segment of ADASS attendees received formal, post-graduate education in astronomy; few attendees have a formal background in computer science or software engineering. Surveys of the ADASS attendees in 1993, 1998, and 1999 yield an interesting insight: Figure 2 shows a breakdown by formal professional education or training, and Figure 3 by primary job responsibility. Evidently, most developers (and managers) of astronomical software came in to their current careers as astronomers.

3. Projects that Changed Astronomy

Looking back on a decade of ADASS, it is interesting to search for the best or most compelling papers, or those projects that were closest to the cutting edge of technology. We have created a database[2] of authors, subjects, and paper titles from the first decade of ADASS to enable such searches, which will be kept up to date as a resource for the community. We selected for this review those papers that describe projects which arguably have had the most profound impact on astronomy and astronomical data analysis during the last ten years.

---

[2]http://www.adass.org:8080/ADASS/Database

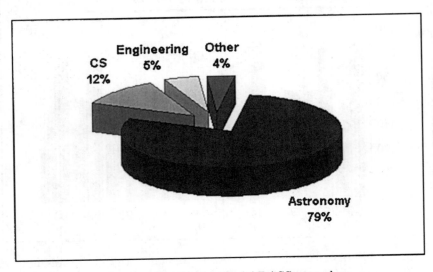

Figure 2.   Professional background of ADASS attendees.

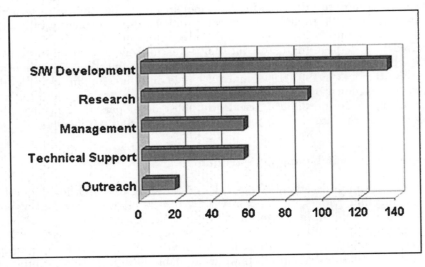

Figure 3.   ADASS attendees by professional duties.

Regrettably, time and space limitations prohibit us from highlighting more than a few papers.

The 1980s and 90s witnessed the introduction and maturation of the major data reduction and analysis systems, with which most astronomical data are processed. ADASS has always been an excellent forum for developers to hear of the latest enhancements and plans for these systems, e.g., AIPS++ (Croes 1992) and IRAF (Tody 1992). These and other systems are routinely featured in ADASS BoF sessions. The increasing importance of space-based observatories during that time frame, particularly in 1990 with the launch of the *Hubble Space Telescope*, drove the development of sophisticated software in a number of areas. These included the application of AI for astronomical scheduling problems (Johnston 1992), and observation planning tools (Koratkar & Grosvenor 1999).

The archiving of digital data has also been a major focus of space missions for a few decades now, but it is fair to say that the supporting technologies only reached maturation in the 1990s. More than 100 papers have been presented at ADASS on the technologies behind digital archives and archive science. In the past few years, larger institutions have been leveraging their archiving expertise and infrastructure to build multi-mission archives. This is not to say that the generic problem of archiving ground-based data has been ignored, as discussed by Crabtree (1996). The SDSS planned perhaps the most ambitious electronic archive of ground-based data for its era (Brunner et al. 1996).

If space missions and electronic archives have ushered in a golden age in astronomy, then the extensive surveys that populate the archives are a significant portion of the precious ores. The science goals and the technologies employed in many of these surveys have been presented at ADASS, including FIRST (Becker 1994), the Sloan Digital Sky Survey (Kent 1994; Szalay et al. 2000), the Hubble Deep Fields (Ferguson 2000), and 2MASS (presentation by Cutri). What has become abundantly clear, though, is that it is now easier to populate archives with vast quantities of survey data than it is to mine and refine the best science from them. This very point was addressed in the recent Decadal Survey of Astronomy[3] (NRC 2000), which advocated the construction of a National Virtual Observatory. The NVO will be based on a distributed model of data repositories and of computational resources, and will provide the necessary infrastructure for astronomers anywhere on the Internet to access any public archive to resolve sophisticated queries (Szalay 2001).

The "killer app" of the 1990s, NCSA Mosaic (the precursor of Netscape), was presented at ADASS III by Joseph Hardin, before most of the world was even aware of the Internet. This single application revolutionized the way in which individual astronomers interact with observatories, professional journals, and each other. The metamorphosis of the Astrophysics Data System (ADS) into the virtual repository of astrophysical literature is a prime example of the profound effect of Web browser technology: compare, for example, the initial presentation on ADS by Good (1992) to that by Accomazzi et al. (1995). Web browsers also made possible the sharing of data, information, and the results of research more rapidly and effectively than ever before.

---

[3]http://www.nap.edu/books/0309070317/html/

The technology behind some of the most profound science results of the past decade have also been presented at ADASS. For example, perhaps the most widely used (and cited) tool for photo-ionization modeling, CLOUDY, was described in detail by its author and curator, Gary Ferland at ADASS VI (Ferland, Korista & Verner 1997). Also that year Paul Butler reviewed the Lick Observatory planet search, which was one of the most successful techniques behind the then-new discovery of Jupiter-sized planets around nearby stars. Results such as these, and the supporting technologies, have helped to shape the direction of the NASA "Origins" program as well as components of the Decadal Survey (NRC 2000). Looking to the future, ADASS X attendees learned of the objectives and techniques of the Search for Extra-Terrestrial Intelligence (SETI) (presentation by Tarter), including the familiar distributed computing technologies used to process the accumulating data (presentation by Werthimer) on an amazing number of idle workstations and PCs around the world.

4. A Look Ahead

What can we expect to be presented at future ADASS conferences? Given the extraordinary and largely unanticipated advances in the last decade, offering a serious answer to this question invites, at worst, comparisons to salesmen, politicians, and fools; more likely, it will provide punch-lines for the ADASS XX review speaker.

Some developments are not too hard to foresee: the continuing development of XML as a de-facto standard of astronomical data and meta-data exchange, for example would be quite complementary to its use in commercial applications. The design and construction of NVO (though we are confident it will be a "GVO," or Global Virtual Observatory) will also almost certainly come to pass. What is less clear is whether NVO will be successful at first (given the cost and public profile, we hope so), or whether the fundamental GRIDS technology involved will be stable for the duration of the project (we're not taking any bets here). Concomitant with NVO will come data mining software, and other middle-ware that will provide generic methods to access data, and to map it to/from one of a few "standard" models. We are already seeing applications that can access data from multiple archives and display images, catalogs, etc. on areas of the sky. We suspect that this is only the first wave of "net-savvy" applications that will be built or modified from other analysis software. This trend certainly has precedent: consider the evolution of simple word-processing software from simple text-entry to include WYSIWYG presentation, spell-checking, graphics and spread-sheet importation, Web-page generation, etc. Simple file browsers for local file selection will not be able to compete with VO-aware applications.

NVO is an example of a system where alternative computing models are critical to success. But we have already seen examples of Beowulf and other parallel or at least distributed systems for specialized applications in astronomy. We expect this trend to continue in the next decade. We also suspect that wireless applications will find interesting uses in astronomy—beyond the portable computer: telescope control, data taking, and instrument control are potential applications. Beyond that, we expect to see more (semi-) autonomous systems used in observatory operations. Studies of automated scheduling and data acquisition and calibration are underway for the Next Generation Space Telescope and other missions. Can self-calibrating detectors, self-constructing data pipelines, and automated data analysis be far behind?

4.1. Implications of the Virtual Observatory

The Virtual Observatory has the potential to change astronomy as profoundly as the major, space-based archives have done individually. But as proposers for observing time at major facilities know, programs are unlikely to survive the peer-review process unless the proposals demonstrate that archival data has been used to its fullest extent. In this sense, observational astronomy is changing in a very profound way: proposers are being held responsible for *every photon* in *every public archive!* We suspect this point is under-appreciated in the community, but we believe it will drive the development of observation planning and data analysis applications that can access archives transparently.

5. ADASS On-Line

In view of the continuing role of ADASS to the professional community of astronomers and software developers, we have begun developing an ADASS Web site with the rather ambitious goal of "one-stop shopping for astronomy software." The electronic version of the conference proceedings and the database of authors and papers were described above. Beyond that, we plan to provide resources such as centralized conference registration services, presentation and publications policies, and templates and style files for proceedings authors. We also plan resources that will benefit a somewhat broader community, such as master author and subject indexes for the combined proceedings, links to other conferences of potential interest to ADASS attendees, and career services for professionals in astronomical software. Links to these resources will be found on the ADASS home page at http://www.adass.org/.

Acknowledgments. The ADASS conference series would not have been possible without substantial support from the sponsoring institutions, whose material, financial and in many cases staff support have helped make ADASS a success each year. Table 2 lists the past sponsors to whom we are all indebted.

References

Accomazzi, A., Grant, C. S., Eichhorn, G., Kurtz, M. J., & Murray, S. S. 1995, in ASP Conf. Ser., Vol. 77, Astronomical Data Analysis Software and Systems IV, ed. R. A. Shaw, H. E. Payne, & J. J. E. Hayes (San Francisco: ASP), 36

Becker, R. 1994, in ASP Conf. Ser., Vol. 61, Astronomical Data Analysis Software and Systems III, ed. D. R. Crabtree, R. J. Hanisch, & J. Barnes (San Francisco: ASP), 165

Brunner, R. J., Csabai, I., Szalay, A., Connolly, A. J., & Szokoly, G. P. 1996, in ASP Conf. Ser., Vol. 101, Astronomical Data Analysis Software and Systems V, ed. G. H. Jacoby & J. Barnes (San Francisco: ASP), 493

Crabtree, D. 1996, in ASP Conf. Ser., Vol. 101, Astronomical Data Analysis Software and Systems V, ed. G. H. Jacoby & J. Barnes (San Francisco: ASP), 473

Croes, G. A. 1993, in ASP Conf. Ser., Vol. 52, Astronomical Data Analysis Software and Systems II, ed. R. J. Hanisch, R. J. V. Brissenden, & J. Barnes (San Francisco: ASP), 156

Ferguson, H. 2000, in ASP Conf. Ser., Vol. 216, Astronomical Data Analysis Software and Systems IX, ed. N. Manset, C. Veillet, & D. Crabtree (San Francisco: ASP), 395

Ferland, G., Korista, K. T., & Verner, D. A. 1997, in ASP Conf. Ser., Vol. 125, Astronomical Data Analysis Software and Systems VI, ed. G. Hunt & H. E. Payne (San Francisco: ASP), 213

Good, J. C. 1992, in ASP Conf. Ser., Vol. 25, Astronomical Data Analysis Software and Systems I, ed. D. M. Worrall, C. Biemesderfer, & J. Barnes (San Francisco: ASP), 35

Johnston, M. 1993, in ASP Conf. Ser., Vol. 52, Astronomical Data Analysis Software and Systems II, ed. R. J. Hanisch, R. J. V. Brissenden, & J. Barnes (San Francisco: ASP), 329

Kent, S. 1994, in ASP Conf. Ser., Vol. 61, Astronomical Data Analysis Software and Systems III, ed. D. R. Crabtree, R. J. Hanisch, & J. Barnes (San Francisco: ASP), 205

Koratkar, A., & Grosvenor, S. 1999, in ASP Conf. Ser., Vol. 172, Astronomical Data Analysis Software and Systems VIII, ed. David M. Mehringer, Raymond L. Plante, & Douglas A. Roberts (San Francisco: ASP), 57

NRC 2000, *Astronomy and Astrophysics in the New Millennium*, Astronomy and Astrophysics Survey Committee, Board on Physics and Astronomy—Space Studies Board, Commission on Physical Sciences, Mathematics, and Applications, National Research Council (Washington, D. C.: National Academy Press)

Szalay, A. S., Kunszt, P., Thakar, A., Gray, J., & Slutz, D. 2000, in ASP Conf. Ser., Vol. 216, Astronomical Data Analysis Software and Systems IX, ed. N. Manset, C. Veillet, & D. Crabtree (San Francisco: ASP), 405

Szalay, A. S. 2001, this volume, 3

Tody, D. 1993, in ASP Conf. Ser., Vol. 52, Astronomical Data Analysis Software and Systems II, ed. R. J. Hanisch, R. J. V. Brissenden, & J. Barnes (San Francisco: ASP), 173

Table 2. Past Sponsors of the ADASS Conference Series

Institutions	Corporations
Associated Universities, Inc.	APUNIX
Canada-France-Hawaii Telescope	Barrodale Computing Services, Ltd.
Canadian Astronomical Data Centre	Co Comp, Inc.
European Southern Observatory	CREASO
European Space Agency	Digidyne, Inc.
Gemini 8m Telescopes Project	Digital Equipment Corp.
Infrared Processing and Analysis Center	GE Fanuc Automation
National Aeronautics and Space Administration	Hughes STX
National Center for Supercomputer Applications	Network Computing Devices
National Optical Astronomy Observatory	Open Concepts, Inc.
National Radio Astronomy Observatory	Pink Aviation Services
National Research Council of Canada	Research Systems, Inc.
Smithsonian Astrophysical Observatory	Resource One
Space Infrared Telescope Facility Science Center	Silicon Graphics, Inc.
Space Telescope Science Institute	Sprint Communications, Inc.
Space Telescope European Coordinating Facility	Sun Microsystems, Inc.
The Vatican Observatory	Sybase, Inc.
University of Arizona, Steward Observatory	ZED Data
University of Illinois Astronomy Dept.	
University of Victoria	
University of Virginia	

Astronomical Data Analysis Software and Systems X
ASP Conference Series, Vol. 238, 2001
F. R. Harnden Jr., F. A. Primini, and H. E. Payne, eds.

The Evolution of GIPSY—or the Survival of an Image Processing System

M. G. R. Vogelaar and J. P. Terlouw

Kapteyn Astronomical Institute, Postbus 800, 9700 AV Groningen, The Netherlands

Abstract. Since its introduction in the early seventies, GIPSY has constantly evolved. We present an overview of the developments over the last few years. These include the introduction of event-driven user interaction and the addition of a set of highly interactive graphical user interface (GUI) components. The GUI has been built on top of the existing user interface with which it is completely compatible. We also present examples of applications based on these developments.

1. Introduction

GIPSY[1], the Groningen Image Processing SYstem (Allen et al. 1985; Van der Hulst et al. 1992), has its roots in the early seventies. Since then it has constantly evolved. Stimulated by close contact with its users, the exploration of many new ideas in image analysis and user interaction has been possible. GIPSY's user group is modest in size but is still growing and has extended beyond those interested in the analysis of spectral line synthesis data.

The system is organized as a set of independent application programs, 'tasks', which are controlled by the user interface program Hermes (Allen & Terlouw 1981).

2. Event-driven Tasks and Graphical User Interfaces

Originally, all tasks within GIPSY were programmed in a procedural fashion. For many tasks this is still the case. Operation is enhanced by the user interface program Hermes. One of Hermes' most important functions is maintaining the user inputs for every active task. These are stored in the form of keyword-value pairs. The user can supply input to Hermes at any time. When a task needs input, it presents the keyword to Hermes, which then returns the associated value. If no value is available, the user may be prompted.

2.1. The Event Mechanism

Event-driven tasks are fundamentally different from procedural tasks. Procedural tasks are in control of the order in which their different parts are executed

---

[1] http://www.astro.rug.nl/~gipsy

and consequently of the required order of their inputs. An event-driven task, by contrast, is programmed so that it ideally can handle any input at any moment. In this way the user, not the task, is in control.

To support the event-driven operation of tasks, some features had to be added, but the original scheme was kept intact. Hermes was modified so that it sends a message to the task whenever there is a change in its set of inputs. The task, in turn, must be prepared to receive this message. For this purpose an event-dispatching routine was written. When a task is to be notified of changes related to particular keywords, it can register one or more functions with the event-dispatcher for every such keyword. Whenever a change occurs for any of these keywords, the dispatcher will call all functions registered for that keyword. Each function can then obtain the associated value in the usual way.

2.2. The Graphical User Interface

GIPSY's graphical user interface has been implemented as a collection of 'elements', each consisting of an X Toolkit widget, such as a button, a menu, a text input field, a plot canvas, etc., and code which connects the widget to the event mechanism. Most elements are associated with a user input keyword. The widget always reflects the keyword's value, regardless of how it was set. When the widget is manipulated by the user, e.g., by typing text into an input field, the keyword will be updated. This update then causes an event that can be received and handled by the application code. Implemented in this way, there is a strict separation between form and meaning. A benefit of this is that the application programmer need not know anything about the details of the GUI, but only has to deal with a uniform, abstract, event mechanism. It also allows the separate development of application code and user interface.

2.3. Plot Windows and PostScript Driver

In earlier versions of GIPSY, the PGPLOT library (by T. J. Pearson, California Institute of Technology) was only used for vector graphics. Image display was done by GIPSY's image display server GIDS. In the current version PGPLOT's image capability has become an important utility.

GIPSY's graphical user interface contains a PGPLOT device driver supporting up to 16 independent plot windows. The color maps for these windows can be partially or completely shared, or can be separate. Because PGPLOT's procedural cursor interaction cannot be used in the event-driven environment, a special mouse interface routine is provided instead.

Associated with each window, off-screen images can be stored which can quickly be brought back to the screen. These can be used for backup purposes, e.g., when it is too time consuming to regenerate the window while the application regularly needs to overwrite parts of it. They can also be used to implement movie loops, etc.

For images with a small number of pixels, which usually have a 'blocky' appearance when displayed, interpolation between pixels has been implemented. This feature is available both in the display windows described above and in GIPSY's PostScript driver. To prevent the PostScript output from becoming excessively large, in this case the interpolation was implemented as an algorithm written in the PostScript language.

3. New Applications

3.1. Applying the New Input Facilities

The developments described above allowed us to write more user-friendly programs. The graphical interface is used to facilitate input for applications, especially those where it is important to understand the relations between variables used for fine tuning. Examples are the fitting parameters in the GIPSY tasks GAUFIT2D and XGAUFIT. The first fits the parameters of a two-dimensional Gaussian distribution to data in an image and uses a robust moments calculation to find initial estimates for a least-squares fit. The second task fits the parameters of a one-dimensional Gaussian distribution to profiles, usually used to derive a velocity field from data in an H I data cube. It also measures deviations from the Gaussian shape using higher order terms of the so called Gauss-Hermite series. (Van der Marel & Franx 1993)

Another example is program RENDER which renders 2-D and 3-D GIPSY data and displays 2-D slices at any angle. It is based on the subroutine library PGXTAL[2] by D. S. Sivia. The sky coordinate transformation task SKYTOOL is an example of matured functionality with a new interface. It is based on intensively used and tested transformation routines. One can enter a position in any of the GUI fields, either formatted or in degrees, and the program will immediately update the other fields with the corresponding coordinates.

3.2. Enhanced Task Interactivity

We also used the new techniques to create highly interactive applications. Examples are the interactive profile fitter XGAUPROF, based on the same algorithms as used in XGAUFIT, and ROTMAS, which fits the mass components contributing to a rotation curve to an 'observed' rotation curve. All components can be interactively varied to explore parameter space. ROTMAS uses the (modified) values of the variable parameters as initial estimates for a least squares fit. The composition of the curve can be either a function selected from a menu or a user defined expression. ROTMAS also includes special functions such as a Hernquist or Isotherm halo, exponential functions or expressions entered by the user.

3.3. New Procedures for Image Analysis

Inspired by the new applications, users proposed new functionality which could not be implemented before. One example is INSPECTOR. This task displays slices through a three-dimensional GIPSY data set (channel maps) and overlays radial H I velocities (i.e., 'tilted ring' circular velocities) on velocities in position-velocity diagrams extracted from the data and allows the user to adjust the tilted ring parameters interactively in order to get a better fit. A full description of how to derive rotation curves with INSPECTOR and its advantages and disadvantages compared to other methods is given in Swaters (1999). A second example is SLICEVIEW. This is a multi-purpose inspection tool for GIPSY data sets. It has an easy to use movie facility for the display of subsets, e.g., channel maps.

---

[2]http://www.isis.rl.ac.uk/dataanalysis/dsplot/pgxtal.htm

It also has a flexible way of creating images of slices through a 3-D data set. Slice types are a straight line, an ellipse or a cubic spline with control points positioned by the mouse.

3.4. Tools for Education

Some new GIPSY applications are also suitable for education. The task FUNPLOT is an example. It is a utility to explore mathematical expressions. The user enters an expression containing functions and variables. For each variable a slider is created which enables the user to interactively change the value of that variable. The function, as well as its derivative and Fourier transform, can also be plotted. The interactive change of the variables' values can reveal unexpected behaviour of the Fourier transform which can be very instructive.

3.5. Re-use of Procedural Tasks

The majority of tasks within GIPSY is still of the procedural kind. Of course these tasks can still be used the way they were originally designed, but with GIPSY's standard facility that allows one task to run another, some of these tasks can be re-used in novel ways. For example the task ROTMAS can delegate the calculation of a rotation curve to the old task ROTMOD. It is worth mentioning that some tasks are quite old (e.g., ROTMOD dates from 1984), but this is an advantage rather than a disadvantage. Many tasks have matured over time and are in excellent condition.

4. Conclusion

Evolution, rather than redesign, has been the main mechanism by which GIPSY has been able to keep up with modern requirements. This has enabled our focus to remain on the development of applications that implement new ideas in data analysis.

References

Allen, R. J. & Terlouw, J. P. 1981, in Proceedings of the Workshop on IUE Data Reduction, ed. W. Weis. (Vienna Observatory), 193

Allen, R. J., Ekers, R. D. & Terlouw, J. P. 1985, in Data Analysis in Astronomy, ed. V. di Gesù, L. Scarsi, P. Crane, J. H. Friedman & S. Levialdi, (Plenum Press, N.Y.), 271

Swaters, R. A. 1999, Ph.D. thesis Rijksuniversiteit Groningen

Van der Hulst, J. M., Terlouw, J. P., Begeman, K., Zwitser, W. & Roelfsema, P. R. 1992, in ASP Conf. Ser., Vol. 25, Astronomical Data Analysis Software and Systems I, ed. D. M. Worrall, C. Biemesderfer, & J. Barnes (San Francisco: ASP), 131

Van der Marel, R. P. & Franx, M. 1993, ApJ, 407, 525

Part 7. Software Applications

Astronomical Data Analysis Software and Systems X
ASP Conference Series, Vol. 238, 2001
F. R. Harnden Jr., F. A. Primini, and H. E. Payne, eds.

The LWS Interactive Analysis (LIA) Package

S. J. Chan, S. D. Sidher,[1] B. M. Swinyard, M. G. Hutchinson

Space Science Department, Rutherford Appleton Laboratory, U.K.

S. Lord, S. Molinari, S. J. Unger[1]

Infrared Processing & Analysis Center, California Institute of Technology, U.S.A.

S. J. Leeks

Department of Physics, Queen Mary, University of London, U.K.

Abstract. We report on our on-going software project *the LWS Interactive Analysis (LIA)*, which has been developed under the IDL environment. This is a processing and analysis software package for the Long Wavelength Spectrometer (LWS) aboard the Infrared Space Observatory (*ISO*). There are three classes of routines are available under LIA: Inspection Routines, Recalibration Routines, and Interactive Routines.

1. LWS: Long Wavelength Spectrometer

The European Space Agency (ESA) Infrared Space Observatory (*ISO*) was launched on 17 November 1995. The Long Wavelength Spectrometer (LWS) is one of two complementary spectrometers aboard *ISO*. The LWS contains ten dectectors to cover the wavelength range 43–196.9 μm. It operates in two observing modes, corresponding to medium and high spectral resolution. In medium resolution mode, the LWS contains a reflection grating to give a spectral resolving power ($\lambda/\Delta\lambda$) from $\sim$ 150 to 200. In high resolution mode, a Fabry-Perot (FP) Interferometer is inserted before the grating to give a spectral resolution ranging from 6800–9700 across the entire wavelength range (Clegg et al. 1996; Swinyard et al. 1999).

2. LIA: LWS Interactive Analysis

2.1. Introduction

The LWS Interactive Analysis (LIA) is a software package designed to allow users to inspect, reprocess, and recalibrate their LWS data with the possibility

---

[1]Department of Physics, Queen Mary, University of London, U.K.

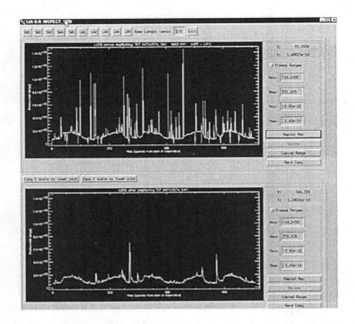

Figure 1. Inspection Routine: inspect_spd – An interactive tool displays SPD results before and after the application of the standard deglitching algorithm.

of interactively customizing the various data reduction stages to their particular set of data. It is a joint development of the ISO-LWS Instrument Team at Rutherford Appleton Laboratories (RAL, UK—the PI institute) and the Infrared Processing & Analysis Center (IPAC, USA). LIA development started in 1997 (Sidher et al. 1997). The current version is LIA 8.1. The final version will be LIA 10, which will operate on products from OLP 10—the final pipeline processing software to be released. The current version of LIA can be obtained via LIA/UKIDC web site[1] and LIA/IPAC web site.[2]

2.2. LIA and ISAP

LIA is written under the Interactive Data Language (IDL) environment. The programming language used in this package is also IDL. IDL provides the ability to plot and visualize data easily and immediately. The *ISO* Spectral Analysis Package (ISAP), which is required to run LIA, is also written in IDL.

ISAP is a software package for the reduction and scientific analysis of the *ISO* SWS (Short Wavelength Spectrometer) and LWS Auto Analysis Results (AARs). AARs are the end product of the official automatic pipeline processing,

---

[1] http://jackal.bnsc.rl.ac.uk/isouk/lws/software/lia/lia.html

[2] http://www.ipac.caltech.edu/iso/lws/lia/download_lia80.html

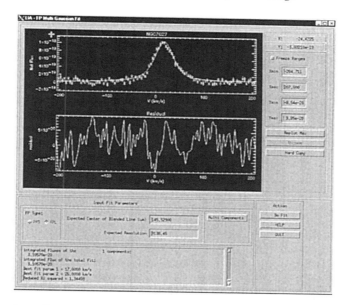

Figure 2. Interactive Routine: gui_fpmf – This tool is written for Interactive FP Multi Gaussian Fitting for a single "blended" line.

which processes the raw data as they are received from the satellite, via a number of intermediate stages.

During the early post-*ISO* mission stage, the calibration accuracy of these products was generally good to within 20% on average. One of the advantages of LIA is that it allows users to check the quality of their scientific and calibrated data, and also to recalibrate them if necessary.

2.3. The LIA routines

There are three classes of routines are available under LIA. They are Inspection Routines, Recalibration Routines, and Interactive Routines.

Inspection Routines allow users to inspect their pipeline products as produced through the Standard Processing Stage and the Auto Analysis Stage (e.g., inspect_spd routine, see Figure 1).

Interactive Routines allow users to customize the data reduction process and to tune the algorithms used in the data reduction (e.g., gui_fpmf routine, see Figure 2).

Recalibration Routines allow users to recalibrate the data reprocessed using the interactive routines, or to make a complete non-interactive reprocessing using defaults other than those used by the automatic Off-Line Processing (OLP; e.g., fp_proc routine, see Figure 3).

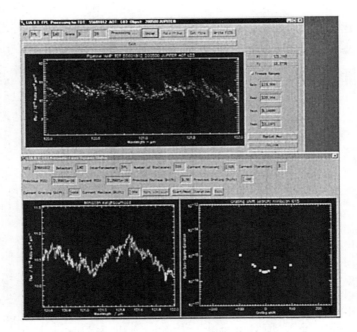

Figure 3. Recalibration Routine: fp_proc – An interactive tool produces calibrated LSAN files for FP observations.

2.4. Routines Planned for LIA 10

Automatic routines: One of the advantages of LIA 10 will be that it process large amounts of data without manual intervention.

- auto_avg – to average spectra automatically
- irasphot – to calculate synthetic photometry (using iras filter profiles)
- nirleak – to apply near-infrared leaked correction automatically.

Planned recalibration routines

- non-linear correction of a strong source.
- aperture correction for an extended source.
- transient correction

Acknowledgments. S. J. Chan thanks the Program Organizing Committee of the Tenth Annual Conference on Astronomical Data Analysis Software and Systems for offering her financial support to attend the conference.

References

Clegg, P. E., et al. 1996, A&A, 315, L38
Sidher, S. D. 1997, ESA SP-419, 297
Swinyard, B. M., et al. 1998, in SPIE Proc., Vol. 3354, 888

Prototypical Operations Support Tools for NASA Interferometer Missions: Applications to Studies of Binary Stars Using the Palomar Testbed Interferometer

Raymond J. Bambery and Charles Backus

Jet Propulsion Laboratory, Caltech, Pasadena CA 91109

Abstract. The Palomar Testbed Interferometer (PTI) is a 110-meter baseline K-Band infrared interferometer located at Palomar Mountain, California. In 1999 an effort was started to provide more observer-friendly observation planning and monitoring tools, such as might be used in NASA interferometer missions. This session illustrates how these prototype tools aid in the observation of spectroscopic binary stars. Animations, using IDL, show how the measured visibilities relate to the positions of the secondary star during its orbital period.

1. Introduction

The Palomar Testbed Interferometer (PTI) is a two-element infrared interferometer located at Palomar Mountain, San Diego County, California (Colavita 1999). Although it has 3 telescopes, only two of them (1-baseline) can be used at one time. PTI was developed as a proof-of-concept for three NASA optical and infrared interferometers: the Keck Interferometer (a ground-based visible and infrared interferometer), the StarLight Mission (a space-based formation-flying visible wavelength interferometer) and the Space Interferometry Mission (a space-based visible astrometric interferometer). In spite of its testbed origins, PTI is a fully-functional astronomical interferometer capable of performing scientific investigations. From 1997–98 an off-line suite of routines performed analyses on spectroscopic binaries, stellar diameters and atmospheric modeling. In 1999 an effort was undertaken to prototype mission operations support tools at PTI. This session illustrates how these prototype tools aid in the observation of spectroscopic binary stars.

2. PTI Data

2.1. Astrometric Precision

PTI has a fringe spacing of $\sim$5.0 milliarcseconds (mas) in the K-Band infrared (2.0–2.4 microns) for its 110-meter North-South baseline. By comparison, the Hubble Space Telescope WFPC-2 camera has a pixel size of $\sim$46 mas and a point spread function at Full-Width Half-Maximum of 50 mas (Figure 1).

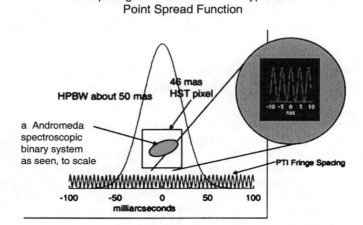

Figure 1. PTI resolution compared to HST.

2.2. PTI Observables

In its simplest operating mode PTI yields unphased visibility (actually visibility squared), which is the fringe contrast of an observed brightness distribution on the sky normalized to [0:1]. As described by Boden (1999) the Visibility modulus, V, for a single star in a uniform disk model is:

$$V = 2J_1(\pi B\theta/\lambda)/(\pi B\theta/\lambda) \qquad (1)$$

where J_1 is the first order Bessel function, B is the Projected baseline vector magnitude at the star position, in meters, θ is the apparent angular diameter of star, in radians, and λ is the center-band wavelength, in meters. Visibility is related to the angular size of the star. When the size is point-like relative to the fringe spacing then the visibility approaches unity, but as the size approaches the fringe spacing ($\sim$5.0 mas) the visibility approaches zero.

Double star visibility squared in a narrow pass-band is:

$$V^2 = V_1^2 + V_2^2 r^2 + 2rV_1V_2 \cos(2\pi \mathbf{B} \cdot \mathbf{s}/\lambda)/(1+r)^2 \qquad (2)$$

where V_1 and V_2 are visibility moduli for each component, r is the brightness ratio between primary and secondary, **B** is the projected baseline vector at the star position, and **s** is the primary-secondary angular separation vector on plane of sky. Since the angular separation of the two stars in the spectroscopic binary is greater than the fringe spacing, visibility goes to a maximum when the centers of the two stars lie on fringe maxima (multiples of the fringe spacing). Visibility goes to a minimum when the center of the secondary star lies on a fringe minimum. (The center of the primary star is always on a fringe maximum.)

Delay line jitter, the measured movement of the delay position of the central fringe (measured in nanometers) over the duration of the integration time, is

Prototypical Support Tools for NASA Interferometer Missions 371

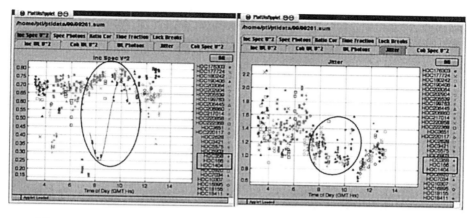

Figure 2. PTI Night 261: (a) visibility vs. time, (b) jitter vs. time.

converted to phase error in radians. Jitter provides a measure of both the instrument and atmospheric stability for each observation and yields the formal errors in the visibility calculation. Observations on each target are accompanied by measurement of calibrator targets, whose angular sizes are computed from astrophysical models. Visibilities from the calibrators are used to determine the system (instrumental and atmospheric) response for a nightly run.

2.3. Observation Program

During 1999 α-Andromedae, a B8 IVmnp spectroscopic binary, was observed on 7 nights over 75 days of α-Andromedae's 96.7-day period. Using the visibilities and baseline orientation from observations at PTI, combined with the radial velocity semi-amplitudes, K1 and K2 from Ryabchicova (1999), allowed monitoring of the orbital motion of the secondary star.

3. Tools

For the years 1997–98, PTI science observers used software developed by the instrument engineers to monitor their nightly runs. However, this software only manages the current target observation and has no ability to recall earlier observations. The night observer would run an off-line batch process to display the earlier observations to note trends in the instrument performance or seeing conditions. In 1999, an effort was started to provide more observer-friendly observation planning and monitoring tools. For example, a science analysis routine was modified to provide a real time monitoring tool, *rtvis*. The output of *rtvis* is read by a Java tool that continually monitors instrument output as the data is collected.

Figures 2a and 2b show two of the diagnostics that summarize the results of an observation. Figure 2a shows a typical night (Night 261 in 1999) for the variation of visibility of α-Andromedae over $\sim$2 hours. Figure 2b shows the corresponding jitter value for the same period. Note that PTI is queue-scheduled

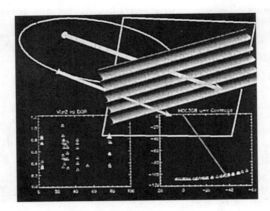

Figure 3. PTI Observations vs. Orbit of α-Andromedae.

for a number of different observers and the figures show these other observations. The values for α-Andromedae (HDC358) and its calibrators HDC1404 and HDC166 are circled in both figures. Note the variation of visibility as the secondary star passes through a fringe maximum and minimum. An IDL animation was created from this data, radial velocity data and orbital parameters. Figure 3 is one frame from that animation and shows the orbit of the secondary around the primary. The width of the arrows extending downward from each star through the fringe pattern indicates the relative contribution of each star to the total visibility. The two lower panels show the visibility values on the left and the baseline orientation on the right. During the animation the values for each observation are highlighted on these panels.

Acknowledgments. The research described in this article was carried out by the Jet Propulsion Laboratory, California, California Institute of Technology, under a contract with the National Aeronautics and Space Administration. A special thanks to Dr. Andrew F. Boden of Caltech's Interferometry Science Center for numerous and valuable discussions, comments, and clarifications of interferometric observations of binary stars at PTI.

References

Boden, A., et al. 1999, ApJ, 515, 356
Colavita, M., et al. 1999, ApJ, 510, 505
Ryabchicova, T., Malanuschenko, V., & Adelman, S. 1999, A&A, 351, 963

Correction of Systematic Errors in Differential Photometry

J. Manfroid, P. Royer, G. Rauw, E. Gosset

Institut d'astrophysique et de géophysique, Université de Liège, Belgium

Abstract. A common cause of errors in CCD differential photometry is an improper calibration of the array. The importance of these errors is evaluated for different cameras with fields between 3 and 30 arcminutes. The usual superflat illumination corrections based on night sky exposures are often found to be unsatisfactory. "Photometric superflats" based on stellar measurements are more reliable and should be used instead.

1. Introduction

Differential photometry performed within a single CCD field can lead to very good accuracy. However, the imperfect nature of the flat-field calibrations results in spatially dependent errors which show up as zeropoint shifts, or time variations if the stars are not centered on the same pixels in every frame. Depending on the centering procedure, the resulting errors may thus seem to be either random or systematic, space or time dependent.

The same problems arise in non-differential photometry but they are more difficult to detect because the exposures are generally isolated, and there are additional sources of error. Moreover, the calibration errors partly cancel out when computing colour indices, so colour diagrams are largely unaffected.

2. Superflat Correction of Data

An example of space-dependent calibration errors combined with poor centering is shown in Figure 1, where the differential V photometry for three stars in a single field is plotted as a function of the position on the detector. The observations were made with the University of Bochum 0.6 m telescope at ESO providing a field of $3\rlap{.}'2 \times 4\rlap{.}'8$.

Clearly, the stars suffer from large space-dependent errors that can be traced to the flat-field calibration. As a consequence, some stars exhibit large time variations (see the light curve of star b in Figure 2). These variations are quite different from star to star. It is difficult for the observer to detect a systematic pattern and, hence, to suspect an instrumental effect.

Dithered observations bring the opportunity to correct the errors. Given a sufficiently large number of stars, diagrams similar to the upper ones of Figure 1 can yield a rough estimate of the 2D illumination correction. A more efficient procedure has been described by Manfroid (1995, 1996). Several improvements have been brought, making the method more robust. Independent fields, hence

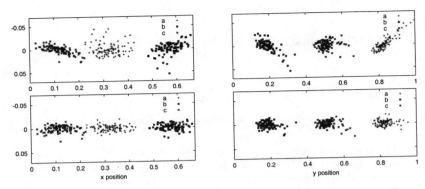

Figure 1. Differential V photometry as a function of the position on the CCD for three stars in a field observed with the University of Bochum 0.6m telescope at the ESO La Silla observatory. The reference is computed from the median of the other stars in the frame. The upper panel shows uncorrected data, i.e., only submitted to the usual flat-field calibration. The stars are brightest towards the edges of the CCD. The lower panel shows data corrected with a proper photometric calibration.

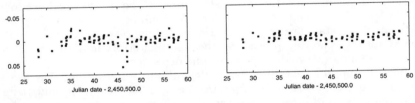

Figure 2. Time variations of star b of Figure 1 before and after calibration.

groups of nights, can be treated at once. A first-order correction of the extinction is included. This is necessary at high air masses and/or in the case of wide field cameras.

The procedure yields a purely photometric calibration which shall be called hereafter the "photometric superflat". It has exactly the same purpose as the superflat obtained from a median of night sky frames.

The photometric superflat has been computed for the Bochum data, by analyzing 67 stars and 106 frames. The data have then been submitted to this illumination correction. The new light curve of star b no longer shows spurious large-amplitude variations (Figure 2, right).

Profiles of the photometric and night-sky superflats for the Bochum telescope data are shown in the first panel of Figure 3. They give opposite results. Obviously, the night-sky superflat technique degrades the data. Without superflat correction, the stars at the edge of the CCD are too bright by about 0.05 mag relative to the stars close to the center, while the sky background shows an opposite pattern. With the night-sky superflat, the sky background is forced to

Correction of Systematic Errors in Differential Photometry 375

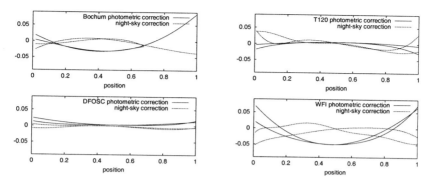

Figure 3. Central profiles (in RA and declination) of both superflat corrections relative to the observations obtained at various telescopes (in magnitude units). The photometric superflats were computed as 4th-order 2D polynomials. The two corrections yield contradictory results, i.e., the sky superflat increases the calibration errors.

be uniform, but the brightness of the outer stars is further enhanced relative to that of the central stars. The correct photometric calibration of the Bochum data gives an *uneven background,* which is brighter by about 0.10 mag at the center. The quality of the calibration can not be judged from the uniformity of the sky background.

Similar conclusions are obtained for other telescopes (Figure 3): the Haute Provence Observatory T120 telescope (field 12′), the ESO 1.54m telescope with the DFOSC camera (13′) and the ESO 2.2m with the WFI mosaic (32′). DFOSC shows the smallest errors, followed by the T120 telescope. The night sky superflat correction mainly corrupts the northern and southern edges of the field at the T120 telescope.

Figure 4 shows the residuals for hundreds of stars observed in I and V with the WFI, plotted along the CCD columns. Data reduced with (i) dome flats only, (ii) dome + night sky superflats and (iii) dome + photometric superflats are compared. Again, the night-sky superflats are seen to degrade the results, in accordance with the calibrations shown in Figure 3. Differential data computed between widely separated stars can show errors of up to about 0.1 mag when a night-sky superflat correction is applied. In the same conditions, CM diagrams over wide fields would show an instrumental effect (increased dispersion) of the same order of magnitude.

Despite its wide field, most of the area of the WFI mosaic can be calibrated within 0.01 mag using a relatively smooth correction (the very edges (40″) must be excluded because of sharp variations). However, when a better accuracy is needed, a higher-order analytic function is needed (see Figure 4, rightmost lower panel in V).

Larger errors were noticed at the 4.2m WHT telescope at La Palma, with a CCD camera using narrow band filters (see Royer et al. 1998). The central stars were too faint by more than 0.20 mag. Such large variations probably had their origin in the design of the camera: unfiltered *white*light reached the detector. In

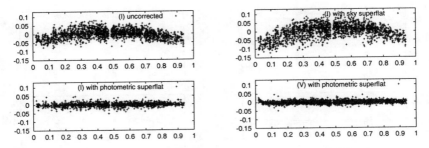

Figure 4. Observed deviations from the average I magnitude as a function of the (relative) row number for the WFI mosaic. The exposures have been calibrated using dome flat fields. Without further corrections (upper left panel) the rms scatter is 0.024 mag. In the upper right panel, a night-sky superflat correction has been included—obviously degrading the data (0.036 mag rms). In the lower left panel, a photometric superflat is used instead, and the deviations are much lower (0.009 mag rms). The plots combine data relative to 2000 I band observations of 400 stars measured on 34 exposures, in 7 distinct fields. Photometrically corrected data for the V band are shown in the lower right panel. The rms is lower than in I (0.007) and it can still be improved by using higher-order functions.

such a case, the high-frequency component of the original flat-field calibration is wrong. The photometric superflat calibration can only correct the large-scale variations.

3. Discussion

The large magnitude of the errors discussed here should be a cause of concern, in differential and non-differential photometry. The background of astronomical images includes an important, non-uniform, component of unfocused (scattered or reflected) light and should not be used to provide the long-wavelength component of the flat-field calibration. In particular, the illumination correction based on median sky frames appears to be quite unreliable. Comparison (Figure 4) of photometrically corrected data and data processed in a more conventional manner shows that a much better calibration can be achieved simply through the use of appropriate software. Additional effects such as geometric distortion are simultaneously corrected. Moreover, the use of this method relaxes the constraint on the flat field acquisition.

References

Manfroid J. 1995, A&AS, 113, 587
Manfroid J. 1996, A&AS, 118, 391
Royer, P., Vreux, J.-M., & Manfroid, J. 2000, A&AS, 130, 407

Projective Transform Techniques to Reconstruct the 3-D Structure and the Temporal Evolution of Solar Polar Plumes

A. Llebaria, A. Thernisien, P. Lamy

Laboratoire d'Astronomie Spatiale (CNRS) Marseille

Abstract. A sequence of 400 images obtained over three days with the C2-LASCO/SOHO coronagraph was used to disentangle the complex evolution of the structures observed on the corona of the Sun's North pole. Projective transforms were used to find and delimit the elusive linear structures on each image ($< 1 : 1$ of SNR). From frame to frame, these structures show strong brightness variations as well as lateral shifts which are linked to rotation of the Sun. Taking advantage of solar corona rotation as a rigid body (of $\sim$28 days period), we are able to extract short sinograms to obtain a 3-D reconstruction with few hypotheses. The whole procedure is described, emphasizing the role of the bilinear transform as a new tool in this process.

1. Introduction

Polar plumes appear in images of the Sun's corona as faint linear enhancements in radial directions emanating from the North and South polar regions during the period of minimal solar activity. For each pole, the plumes seem to diverge from a virtual point located on the polar axis not far from the solar surface. They overlay the smooth and very bright background due to the F-corona.

What are polar plumes? Solar light is scattered by the coronal electrons moving outward from the polar regions at about 500 km/s (solar K-corona). Local enhancements of electron density, outlining the 3-D polar magnetic field, appear to the observer as polar plumes by projection onto the sky. Due to the exponential decrease of electron density with increasing radial distance from the Sun and to projection effects, the plumes parallel to the sky plane outshine the tilted ones, thus bringing up a selection effect. Recent studies (Lamy et al. 1997; Deforest 1998; Llebaria et al. 1998) have shown that plumes are enduring recurrent structures showing a transient activity. We present here the image processing methods used to obtain these results.

Even if plumes seems to be in rigid body rotation with the solar magnetic field, classic inversion with Radon transform is unable to reconstruct such 3-D structure due mainly to transient phenomena. In this paper, the 3-D distribution must be deduced from the temporal analysis using a variant of sinograms called TID (Time Intensity Diagram) introduced in previous works (Lamy et al. 1997; Llebaria & Lamy 1999). The TID is a variation of the classical sinogram of the Radon transform, where angle has been replaced by time and sections are

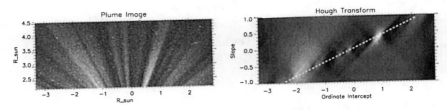

Figure 1. Plume images (left) and correlative picks in the dual plane(right). The dotted line shows a fit to maxima positions. Fit parameters reveal the position of divergence point in image plane.

replaced by the accumulated flux of the plumes along radial directions. We will describe some properties of TID and of its derivatives for the first time.

To improve the TID accuracy, we have extended the concepts developed in these works using the *bilinear dual transform* or BDT. This transform is a member of the projective transforms family, as Radon and Hough transforms are. These points are discussed in detail by Ballester (1994), who introduced the BDT as a variant of Hough transform. We have applied such techniques to an outstanding sequence of high rate C2-LASCO frames obtained in the interval from 1997/03/21 at 22:10:35 UTC to 1997/03/24 at 16:22:41 UTC. This entails a sequence of 402 frames for 66 hours of uninterrupted observations, i.e., a frame every 10 minutes. For 3-D reconstruction, the problems so far have been:

1. Inaccurate exposure times.
2. All plumes have a very low SNR, so the divergence point is poorly defined.
3. Combined motion and transient activity leading to a very confusing effects.

We address points 2) and 3), solved with projective techniques.

2. Bilinear Dual Transform

Plumes appear in coronal images as linear features with few or no structures on the radial direction. As a result of this, and at least for this preliminary approach, we assume that the shape of the radial profiles of solar plumes is constant from one image to the next. Some of these changes are caused by lateral displacements due to coronal rotation.

The increase of SNR is obtained by summing up intensities along radial directions centered on the virtual divergence point. Finding this point in the original images is the role of the *bilinear dual transform* (BDT), defined below.

The BDT is a dual transform which projects lines on points, like the standard Hough transform, and points in lines, unlike the Hough transform (Jain 1989). A straight line $y = x/q + p/q$, defined by slope $1/p$ and shift point p/q in the original plane of (x, y) coordinates, is converted to a point (q, p) in the transformed (or dual) space, also called the plane of parameters. Conversely, a line $q = p/d + c/d$ on the plane of parameters corresponds to a point (c, d) on the image plane.

A unique equation defines this symmetric conversion:

$$qy = x + p$$

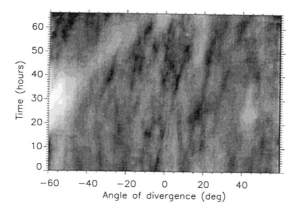

Figure 2. The TID generated from 402 images (~ 66 hours). The bright feature at left is a high latitude solar jet, moving quickly to the East (right), and crossing normal plume paths. Plume traces of variable thickness with different lifetimes appear as obvious features.

where (q, p) is a pair of parameters defining a line in the image plane and a single point in the dual plane. Conversely, (x, y) refers to a point in the original image and they are parameters of a line on the dual plane. The main advantage of BDT is this symmetry. It is clear that deducing the divergence point is easier in the dual plane than in the image plane. Once transformed, the plumes appear as a set of aligned peaks on the parameter plane. We obtain the divergence point fitting a straight line to the set of positions of these peaks. The straight line coefficients define the position of the divergence point on the image plane. Moreover the clustering of positions on BDT reflects interesting geometrical properties of the plumes which are not addressed here.

3. TID Construction and Properties

We recall here the main steps of the TID construction algorithm:
- For each frame, from a center located at the divergence point, and for a regular set of angular positions, plumes are radially integrated.
- A profile of integrated brightness values as a function of angle is generated for each frame.
- The sequence of profiles compose a Flux=F(angle,time) diagram, here called the time intensity diagram (TID)

The TID reveals how plumes behave. Bright pixels delineate continuous paths in the time direction indicating the presence of enduring plumes over this time interval. Interrupted paths show intermittent plumes and local brief brightenings are clearly visible as distinct from crossings of plume paths.

The derivative of TID with respect to angular position ($d\ TID/d\alpha$) shows strong correlations in the time direction. The $d\ TID/d\alpha = 0$ condition corresponds to ridges and valleys of intensity extrema. Paths of positive extrema define the sites with lateral shift. Projecting paths following a family of parametric sinusoids determine the best parameters (phase and amplitude) defining the 3-D position of each plume (we follow a Hough-like transform procedure).

The derivative of TID with respect to time emphasizes the intrinsic intensity behavior. In particular, sudden intensity increases followed by quick decreases reveal "bursts" of plume activity (corresponding to polar "jets" of the solar literature). The plumes rise and fall appear diffuse over the image and correspond in fact to relatively slow changes in activity.

4. Plume Paths and Temporal Activity

Once paths have been defined as described above, it is possible to return to the original image sequence in order to isolate a plume for all the observing time and isolate the corresponding temporal profile. The latter is obtained as an image of integrated brightness depending on time and radial distance to the Sun surface. This temporal profile allows the finest analysis of how plumes evolve. Using projective techniques, once again, with this temporal profile, the apparent radial speed has been found for a significant set of plumes.

5. Conclusions

The bilinear dual transformation, a member of the Radon-Hough family, has been successfully applied to detect converging linear structures in a low SNR environment. An outstanding feature is its linearity in both *a quod* and *ad quem* spaces. Points are converted into straight lines and reciprocally. The latter has successfully been used to find linear convergent structures in presence of strong noise. This algorithm, as well as the modified Radon transform, has been crucial in determining several solar plume characteristics like their mean radial flow speed or the frequent outburst of short-lived "polar jets" (1–2 hours) over a long-term background activity (Llebaria et al. 2000).

References

Ballester, P. 1994, A&A, 286, 1011

Deforest, C. E. 1998, in ESA ESP SP-421 Conf. Ser. ed. T. D. Guyenne, 63

Jain, A. K. 1989, Fundamentals of Image Processing (Englewood Cliffs, N.J.: Prentice Hall)

Lamy, P., Llebaria, A., Koutchmy, S., Reynet, P., Molodensky, M., Howard, R., Schwenn, R., & Simnett, G. 1997 in ESA ESP-404 Conf. Ser. ed. A. Wilson, 487

Llebaria, A., Lamy, P., Deforest, C. E., Koutchmy, S. 1998, in ESA ESP SP-421 Conf. Ser. ed. T. D. Guyenne, 87

Llebaria, A. & Lamy, P. 1999, in ASP Conf. Ser., Vol. 172, Astronomical Data Analysis Software and Systems VIII, ed. David M. Mehringer, Raymond L. Plante, & Douglas A. Roberts (San Francisco: ASP), 46

Llebaria, A., Thernissien, A., & Lamy, P. 2000, in COSPAR Conf. Ser. (in press)

Astronomical Data Analysis Software and Systems X
ASP Conference Series, Vol. 238, 2001
F. R. Harnden Jr., F. A. Primini, and H. E. Payne, eds.

Pile-up on X-ray CCD Instruments

J. Ballet

DSM/DAPNIA/SAp, CEA Saclay, 91191 Gif sur Yvette Cedex, France

Abstract. Pile-up occurs on X-ray CCDs when several photons hit the detector at the same place between two read-outs. In that case they are counted as one (or not at all) and their energies are summed. Pile-up thus affects both flux measurements and spectral characterisation of bright sources. Here I extend a previous work on flux loss to a form directly applicable when the pixel size is comparable to the telescope's point spread function, as in *AXAF/ACIS* or *XMM/EPIC-PN*.

1. Introduction

After the *ASCA/SIS*, the current generation of X-ray astronomy satellites all incorporate CCD detectors (*AXAF/ACIS*, *XMM/EPIC*). The reason is that they are very versatile, offering good quantum efficiency, spatial resolution and reasonable energy resolution (Lumb et al. 1991).

The principle of an X-ray CCD detector is that the whole detector is read out at regular intervals. The contents of the detector as it is read out is called a frame. A local excess in the charge contents is called an event, and is usually associated with a single X-ray impact. The total charge contents of the event is directly proportional to the X-ray energy (with some dispersion of course), and is used as the energy measure.

The underlying hypothesis is that an event is due to only one X-ray. If an event is actually due to several superposed X-rays, then both the count rate and the energy measurements are wrong. This phenomenon (X-ray superposition) is known as pile-up, and is always present for bright sources.

In a previous paper (Ballet 1999), I presented a detailed statistical analysis of the pile-up phenomenon, including formulae quantifying the flux loss, applicable when the pixel size is much smaller than the telescope's point spread function, as in *ASCA/SIS* or *XMM/EPIC-MOS*. I also showed that the perturbations on the spectrum were relatively minor (not more than a few %) for single events (not even diagonally touching another one), so that at first order one could use the standard energy response function after selecting single events only.

In a later paper (Ballet 2000) I presented a second-order approximation, reaching a better precision on the spectrum, applicable both directly (using the auto-convolution of the spectrum) and iteratively in a linear formulation (for XSPEC).

Here I generalise the formulae of Ballet (1999) to the case when the pixel size is comparable to the telescope's point spread function. The formulae of

Ballet (2000) may be generalised in a similar way. Section 3 is directly relevant to *XMM/EPIC-PN*, Section 4 to *AXAF/ACIS*.

The figures are available from the author on request.

2. Framework

The way to identify an event in a frame is to look for a local maximum above some statistical threshold, and then to perform a proximity analysis around this maximum. The charge from an X-ray may either concentrate in one pixel (single events) or appear over several adjacent pixels (split events) forming a specific pattern or grade.

To generalise Ballet (1999), I call $g_{x,y}^{(m)}$ the proportion of X-rays creating the charge pattern defined by m, with $N(m)$ pixels above threshold and a certain orientation, falling at position x, y, for a fixed orientation of the telescope. This encompasses both the sky image $g_{x,y}$ (summing over m) and the pattern distribution $\alpha^{(m)}$ (summing over space). $g_{x,y}^{(m)}$ is also the PSF/pattern distribution of measured events in the ideal situation of a single point source with no pile-up. $g_{x,y}^{(m)}$ is integrated on energy, and normalised such that

$$\sum_{x,y,m} g_{x,y}^{(m)} = 1 \qquad (1)$$

The reason to write it that way is that the PSF may depend on the pattern, for example if events split toward one direction occur preferentially when the X-ray interacted close to the pixel's edge in that direction. For example, in *XMM/EPIC-MOS* index 5 denotes charge patterns with 3 pixels, extending above and to the right of the central pixel (with maximum charge), as in Figure 1 of Ballet (1999), bottom center. Zero denotes strict single events (same figure, top left), 30 generalised single events, possibly touching another pixel above threshold by a corner (same figure, top right).

I call Λ the total incoming X-ray flux/frame and $M^{(n)}$ the expected (expectation value of the) count rate/frame in pattern n. I call $M^{(n)t}$ the expected count rate/frame in pattern n of clean (not piled-up) events (not something you can measure). Local quantities (per pixel) are denoted as lower case so that

$$M^{(n)} = \sum_{x,y} \mu_{x,y}^{(n)} \qquad (2)$$

$1 - M^{(n)}/(\alpha^{(n)}\Lambda)$ is called the flux loss. It is the loss in detection efficiency due to pattern overlap.

$1 - M^{(n)t}/M^{(n)}$ is called the pile-up fraction. It is the fraction of measured events whose energy will be wrong.

Note that all pile-up formulae are naturally written *per pattern* (known in the data). Global quantities are obtained by summing over the patterns.

I have shown (Ballet 1999, Equations 1–3 and A5) that for strict (0) and generalised (30) single events (using the above notations), assuming a (locally) uniform X-ray flux, i.e., a pixel size small with respect to the PSF:

$$\gamma_{x,y}^{(0)} = 9 + 3 \sum_{N(m)=2} g_{x,y}^{(m)} + 6 \sum_{N(m)=3} g_{x,y}^{(m)} + 7 \sum_{N(m)=4} g_{x,y}^{(m)} \quad (3)$$

$$\gamma_{x,y}^{(30)} = 5 + 3 \sum_{N(m)=2} g_{x,y}^{(m)} + 5 \sum_{N(m)=3} g_{x,y}^{(m)} + 7 \sum_{N(m)=4} g_{x,y}^{(m)} \quad (4)$$

$$\mu_{x,y}^{(0)t} = g_{x,y}^{(0)} \exp\left[-\gamma_{x,y}^{(0)} \Lambda\right] \quad (5)$$

$$\mu_{x,y}^{(0)} = \left(\exp\left[g_{x,y}^{(0)} \Lambda\right] - 1\right) \exp\left[-\gamma_{x,y}^{(0)} \Lambda\right] \quad (6)$$

$$\mu_{x,y}^{(30)t} = g_{x,y}^{(0)} \exp\left[-\gamma_{x,y}^{(30)} \Lambda\right] \quad (7)$$

$$\mu_{x,y}^{(30)} = \left(\exp\left[g_{x,y}^{(0)} \Lambda\right] - 1\right) \exp\left[-\gamma_{x,y}^{(30)} \Lambda\right] \quad (8)$$

3. Flux loss

I will here write explicitly the formulae generalising (5–6) or (7–8) to the case when it is not true that the X-ray flux incoming to nearby pixels can be assumed to be identical. This case was already mentioned in section 4.4 of Ballet (1999).

The pattern shape is represented by $P_{i,j}^{(n)} = 1$ wherever the corresponding pixel is above threshold and 0 elsewhere, with $i, j = 0, 0$ at the center of the pattern. For example $n = 5$ (Figure 1 of Ballet 1999, bottom center) is represented by $P_{i,j}^{(5)} = 1$ in $i, j = 0, 0; 0, 1; 1, 0$.

An associated exclusion zone $E_{i,j}^{(n)}$ is then obtained by setting to 1 all pixels above 0 in the convolution with a mask $M_{i,j}$

$$E_{i,j}^{(n)} = \min\left(\sum_{k,l} M_{k,l} P_{i-k,j-l}^{(n)}, 1\right) \quad (9)$$

where the mask is 1 where $max(|i|, |j|) \leq 1$ (strict events, for Equations 5–6) or where $|i| + |j| \leq 1$ (generalised events, possibly touching another pixel above threshold by a corner, for Equations 7–8). Note that this definition corresponds to the white + grey + black pixels in Figure 1 of Ballet (1999), and differs from that in Figure 2 of Ballet (1999), which defined exclusion zones with respect to a particular incoming pattern m and a particular target pattern n. The exclusion zones defined here relate only to the target pattern (n).

The exclusion term $\gamma_{x,y}^{(n)}$ to be injected into Equations (5–6) or (7–8) may then be obtained by summing the terms in the exponential (multiplying probabilities) for all pixels in the exclusion zone.

$$C_{i,j}^{(m)} = \sum_{k,l} P_{k,l}^{(m)} g_{i-k,j-l}^{(m)} \quad (10)$$

$$\gamma_{x,y}^{(n)} = \sum_{i,j,m} E_{i,j}^{(n)} C_{x+i,y+j}^{(m)} = \sum_{i,j,k,l,m} E_{i,j}^{(n)} P_{k,l}^{(m)} g_{x+i-k,y+j-l}^{(m)} \quad (11)$$

where $C_{i,j}^{(m)}$ is the distribution of illuminated pixels for incoming pattern m, obtained by convolving the sky image with the pattern itself. Note that Equation (11) is the result of the correlation (not the convolution) between that distribution and the target's exclusion zone, summed over all incoming pattern types.

This equation is related to Equation (9) of Ballet (1999) by setting $u = k - i, v = l - j$ and

$$\gamma_{x,y}^{(n)} = \sum_{u,v,m} Z_{u,v}^{(n,m)} g_{x-u,y-v}^{(m)} \qquad (12)$$

$$Z_{u,v}^{(n,m)} = \sum_{i,j} E_{i,j}^{(n)} P_{u+i,v+j}^{(m)} \qquad (13)$$

Note that Ballet's (1999) Equation (9) was actually incompatible with the representation of $Z_{u,v}^{(1,m)}$ in its Figure 2 (the diagrams should be reversed, i.e., L in place of R and D in place of U, or the convolution replaced with a correlation).

Figures 1 and 2 illustrate the comparison of Equation (11) with Equation (3) or (4). At moderate count rate ($\sim$ 1 cts/frame) flux loss is less in the narrow PSF case (dashed curve) than in the broad PSF one (full curve) because the neighbouring pixels (at much lower level) do not act to change singles into double events. At higher count rate ($\sim$ 10 cts/frame) flux loss is larger in the narrow PSF case because the high central pixel prevents singles from being detected around it.

The large hump in the pile-up rate in the narrow PSF case (near 3 cts/frame) corresponds to pile-up in the central pixel. It depends sensitively on where the PSF is centered within a pixel, as shown by the comparison between the dashed and dotted curves. Contrary to the broad PSF case, the pile-up rate is not much larger when single events touching others by a corner are included (Figure 2).

4. Attitude Drifts

On *Chandra* the telescope's pointing direction is permanently varied (dithering) to avoid giving too much weight to local irregularities of the detector (CCD edges, bad pixels). This dithering is slow on the time scale of one CCD frame, so that within one frame the telescope's attitude is essentially constant.

As mentioned in Ballet (1999, Section 4.3), pile-up depends on what happens within a frame so the PSF to apply is the original (instantaneous) one. In the continuous case (pixel size much smaller than the PSF), dithering would have no effect at all on pile-up. In the discrete case (pixel size larger than the PSF) such as on *Chandra* it has the effect of averaging the humps in Figures 1 and 2 because at each frame the PSF will be centered differently within a pixel.

References

Ballet, J. 1999, A&AS, 135, 371
Ballet, J. 2000, E1.1-0022 at COSPAR 2000 (Warsaw), Adv. Sp. Res. (in press)
Lumb, D. H., Berthiaume, G. D., Burrows, D. N., Garmire, G. P., & Nousek, J. A. 1991, Exptl. Astron., 2.3, 179

ATV: An Image-Display Tool for IDL

Aaron J. Barth

Harvard-Smithsonian Center for Astrophysics

Abstract. The IDL language offers a powerful environment for reduction and analysis of astronomical data. While there are numerous libraries of publicly available IDL routines, one major drawback has been the lack of an image-display program optimized for viewing astronomical images. ATV is a display program written entirely in IDL and designed to emulate the SAOimage and DS9 programs. It works equally well under the Unix, Linux, Windows, and MacOS versions of IDL, and includes features such as interactive control of color stretch, zoom, and image scaling; image blinking; and creation of publication-quality PostScript output. It also includes a simple photometry package suitable for quick-look reductions. The program is freely available via the Internet.

1. Introduction

Interactive Data Language (IDL) is an increasingly popular tool for astronomical data reduction and analysis. One long-standing drawback of IDL, for astronomical applications, has been the lack of a high-level display routine optimized for viewing astronomical images. IDL's built-in image display routines are primitive, and it is difficult to integrate stand-alone programs such as SAOimage and DS9 into an IDL session. ATV is an image-display program designed to solve this problem. It is written entirely in IDL and offers a range of features similar to the stand-alone image viewers.

ATV takes advantage of the IDL widget interface, which provides a simple mechanism for creating graphical user interfaces, as well as a number of preexisting routines in the IDL Astronomy User's Library. Thus, the ATV code itself is fairly compact, and in some respects it simply acts as a graphical front-end for various library routines. The initial release of the program in 1998 contained basic features such as control of image scaling and color stretch, blinking, and zooming, and since then the program has grown to include other features such as photometry, coordinate tracking, and creation of PostScript output. For IDL users, the major benefit of ATV is that it works within an IDL session, so the user can pass data or FITS filenames directly to ATV from the IDL command line. ATV can be customized to pass data to other IDL routines, and its internal data are stored in common blocks so that they can be accessed or modified by the user.

Since ATV is distributed as IDL source code, it can easily be modified for specialized use. For example, it has been adapted for displaying data from the *FUSE* satellite (http://fuse.pha.jhu.edu/analysis/sw), and a modified

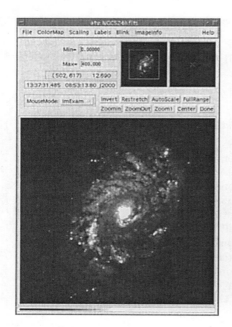
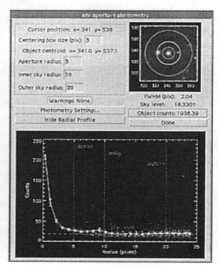

Figure 1. The main ATV display window and the ATV photometry window.

version of ATV is being used as part of the CEDAR software suite for the *HST* Cosmic Origins Spectrograph (Béland & Penton, this volume).

2. Features

The main ATV window is designed to look similar to SAOimage and DS9, with a panning window, a zoom window that tracks the cursor motion, and information boxes that track the pixel coordinates, pixel value, and world coordinates (Figure 1). ATV's features include:

- Ability to read in FITS images, including FITS extensions, and nonstandard data types such as *HST* WFPC2 arrays.
- Image input via pull-down menus or from the IDL command line. Data arrays in memory can be passed directly to ATV from the IDL command line.
- Interactive control of zoom, color table, brightness and contrast, and data range mapped into the color table. As in other display programs such as SAOimage and DS9, brightness and contrast stretching are done by dragging the cursor over the main display window.
- Choice of linear, log, or histogram-equalized scaling.
- Tracking of the cursor position in pixel units and coordinate units if the image header specifies a coordinate system. User can choose RA/Dec, Galactic, or Ecliptic coordinates. For 2-D spectroscopic images with wavelength

information in the header (such as *HST* STIS images), the wavelength at the cursor position is displayed.
- Ability to overplot text, contours, or arbitrary line graphics over the image.
- Ability to plot a compass and scale bar for images with coordinate information in the header.
- Ability to create publication-quality PostScript output, and output in Tiff image format.
- Image blinking with up to three saved images.
- Row, column, surface, and contour plots.
- Point-and-click aperture photometry, including radial profile plots. In the photometry window, the user can control the sizes of the photometric and sky apertures, the centering box size, and photometric zeropoint. Output is in counts or magnitudes.
- Ability to view the image header in a separate window.
- ATV works in 8-bit and 24-bit color. (IDL currently does not support 16-bit color for the X window system, but may do so in the future.)
- ATV works equally well under the Unix, Linux, Windows, or MacOS versions of IDL. Some features require the use of a 3-button mouse (for example, storing 3 blink images), but the program has been designed so that almost all features can be accessed with a 1-button or 2-button mouse.

3. Future Development Goals

Development of ATV continues with new versions released every few months. Some areas for future development include:
- Improvements in speed and memory management.
- A more complete photometry package, including saved results in a log file.
- Ability to read and display multi-CCD mosaic images from various telescopes.
- Ability to handle extremely large images.
- Better user control over pixel ranges and plot details for line plots, and PostScript output for line plots.

4. Credits

Some important sections of the ATV code were written and contributed by David Schlegel, Douglas Finkbeiner, Michael Liu, and Wesley Colley. The FITS I/O, astrometry, and photometry routines used by ATV are part of the IDL Astronomy User's Library, maintained by Wayne Landsman at NASA/GSFC[1]. ATV also uses a PostScript configuration tool written by Craig Markwardt[2]. Instructions and complete documentation can be found at the ATV web page: http://cfa-www.harvard.edu/~abarth/atv.

---

[1] http://idlastro.gsfc.nasa.gov/homepage.html

[2] http://cow.physics.wisc.edu/~craigm/idl

A New Field-Matching Method for Astronomical Images

C. Thiebaut, M. Boër

Centre d'Etude Spatiale des Rayonnements (CESR-CNRS), BP 4346, 31028 Toulouse Cedex 4, France

Abstract. We propose a new Field-Matching algorithm for astronomical images. This new method is based on a multiresolution analysis. We tried two cases: first, we compared the test image with a synthetic image built from a point source catalog; second, we used a reference image of the relevant portion of the sky. Structures of images are obtained at different scales by applying the wavelet transform. An appropriate thresholding of the wavelet coefficients gives the significant pixels in the Wavelet Transform Space. In order to compare the selected coefficients between the test and the reference images we used a genetic algorithm. We applied this method on images taken by the automatic TAROT (Rapid Action Telescope for Transient Object) telescope. The reference data are taken from the USNO-A2.0 Catalog and from the Digital Sky Survey. The results are more robust and reliable than those obtained with the FOCAS algorithm. Moreover, the new algorithm is faster than FOCAS.

1. Introduction

The field-matching consists in recognizing the field of an image with unknown coordinates in a reference image taken from a catalog. The existing field-matching methods, like FOCAS, require good knowledge of the approximated centroïd position. We have developed a new method of field-matching which takes into account the geometrical characteristics of the test image and tries to compare these characteristics with those of a reference image. To do this, we use a multiresolution analysis which will give us the details of the image at the different scales. First, we use Mallat's analysis, which is anisotropic. Because the images we study are quite isotropic (astronomical objects: stars, galaxies etc.) we try to use an isotropic analysis too: the "à trous" algorithm. After having obtained the different wavelet plans, we have to threshold the coefficients in order to keep the most significant ones. The pixels we keep represent the image structure.

We obtain two structures that we match using a genetic algorithm. It will give us the vertical and horizontal offsets between the two structures and then, the offset between the two original images.

2. The Developed Method

2.1. The Multiresolution Analysis

Thanks to a multiresolution analysis, we obtain the details of an image at different scales by applying a wavelet transform. Mallat introduced this concept in 1989, which led to the discrete wavelet transform (Mallat, 1989).

Mallat's Analysis This analysis is a non-redundant one because the amount of data is divided by two at each scale: this is called a dyadic analysis. In two dimensions, Mallat's analysis uses three wavelets which leads to an anisotropic analysis. We obtain the horizontal, diagonal and vertical details of the image at each scale. We used the Daubechies wavelet of degree four. However, the astronomical objects are usually isotropic and without privileged directions. That is why we use the so called "à trous" algorithm (Starck et al. 1995).

The "à trous" Algorithm This analysis is isotropic but redundant. The image is smoothed on the different scales. Because of the redundancy, this algorithm is not as fast as Mallat's algorithm.

2.2. Threshold of the Coefficients

When we have obtained the structures at the different scales, we have to keep only the significant wavelet coefficients. Because of the wavelet form, the best coefficients of Mallat's analysis are the most negative and positive ones, whereas the best coefficients of the "à trous" analysis are the most positive ones. As a consequence, we take the absolute value of the wavelet plans from the first analysis and we keep the 20 best coefficients of these images. For the "à trous" analysis, we keep the 20 best coefficients of the original wavelet plans.

At each scale and for each details image, we have 20 significant pixels. The set we obtain is called the structure of the studied image. We apply one of the algorithms on the test image and on the reference image and we obtain two structures. Finally, we have to match both structures and find the original offset between the two images.

2.3. Matching Both Structures: the Genetic Algorithm

To match both obtained structures we use a genetic algorithm (Houck et al. 1995). These algorithms are inspired by natural evolution theory: they maintain and manipulate a family or a population of solutions and implement a "survival of the fittest" strategy in their search for better solution. Those algorithms have been shown to solve linear and nonlinear problems by exploring all regions of the state space and exponentially exploiting promising areas through mutation, crossover, and selection functions applied to individuals in the population.

Here we want to find the vertical and horizontal offsets between the two original images. The population of our algorithm is a vertical and horizontal offset pair to apply to one of the structures. The algorithm will converge to the best offset pair. Taking into account the scale of the studied structure, we can find the original offset by multiplying by the relevant factor.

3. Application and Results

We want to find the offset position between the astronomical images taken by the automatic TAROT telescope (Boër et al. 2001) and a reference image. The reference image of a relevant portion of the sky is an image taken from the Digitized Sky Survey. The second reference image is built from the point source catalog USNO-A2.0. The pixel image is convolved with a Gaussian which represents the Point Spread Function of the TAROT telescope.

We made a Matlab implementation of the algorithms. For the genetic algorithm, we took a population of 100 offset pairs, the selection function was a tournament. We only used two crossover and two mutation functions. With a convergence time of 40 s, the Mallat's analysis is faster than the "à trous" algorithm (120 s). We then decided to use the anisotropic method. We matched the horizontal and vertical details images of the third scale.

Then, we compare the matching with the DSS images and the one with the USNO-A2.0 catalog images. In Table 1, we give the original vertical and horizontal offset, and those found after the convergence. Finally, we show the results of the FOCAS method (matching with the USNO catalog): we give the number of matched stars and the number of stars found on the image.

For the DSS images, 30 images of 31 are matched. For the only non-matched image, we took the structures of the second scale. The new found offset is $(-40, -64)$, and the matching is then done. For the USNO images, only 13 of 31 images are matched.

4. Conclusion

The matching with the USNO images is not as good as the one with DSS images, which is very good. In fact, a multiresolution analysis is perhaps not well adapted to such constructed images, which present no structure. We could apply the genetic algorithm directly to the brightest objects of both images.

Nevertheless, the new method is faster and more robust than other methods. It does not require a good knowledge of the centroïd coordinates.

References

Boër, M. et al. 2001, this volume, 111

Valdes, F., Campusano, L., Velasquez, J., & Stetson, P. 1995, PASP, 107, 1119

Starck, J.-L., Murtagh, F., & Bijaoui, A. 1995, in ASP Conf. Ser., Vol. 77, Astronomical Data Analysis Software and Systems IV, ed. R. A. Shaw, H. E. Payne, & J. J. E. Hayes (San Francisco: ASP), 279

Lega, E., Bijaoui, A., Alimi, J. M., & Scholl, H. 1996, A&A, 309, 23

Houck, C., Joines, J., & Kay, M. 1995, NCSU-IE TR 95-09

Mallat, S. 1989, IEEE Trans on Pattern Anal. and Math. Intel., 11, 7

Table 1. Summary of the results.

		New method			FOCAS	
Image	Initial offset	DSS Images	USNO Images	Matched	Objects	Matching
1a	(7,44)	(8,48)	(8,−232)	23	593	OK
1b	(19,53)	(16,48)	(−40,−102)	624	1295	OK
1c	(16,46)	(24,48)	(−24,−104)	28	141	OK
1d	(20,52)	(16,48)	(8,56)	1408	1668	OK
2a	(−8,−101)	(−8,−96)	(−8,−88)	382	484	OK
2b	(−1,−101)	(−8,−88)	(0,−72)	4	109	NO
3a	(−15,−71)	(−8,−64)	(−48,16)	3	364	NO
3b	(−35,−63)	(−24,−48)	(−64,40)	205	254	OK
3c	(−38,−65)	(−32,−56)	(0,64)	2	37	NO
3d	(−38,−65)	(−32,−56)	(−8,64)	1	44	NO
4a	(−10,−94)	(−8,−88)	(−16,−88)	198	231	OK
4b	(−20,−98)	(−24,−88)	(−16,−88)	117	155	OK
4c	(−20,−98)	(−24,−88)	(−16,−88)	159	198	OK
5a	(−1,−76)	(8,−72)	(−8,−88)	1	29	NO
5b	(−18,−78)	(−8,−72)	(0,−80)	286	347	OK
5c	(−19,−80)	(−8,−80)	(−32,−120)	1	36	NO
5d	(−20,−82)	(−8,−80)	(−40,−136)	20	69	OK
6a	(−8,−73)	(−8,−72)	(64,136)	1	55	NO
6b	(−32,−79)	(−24,−80)	(80,152)	69	74	OK
6c	(−33,−78)	(−24,−72)	(24, 152)	46	51	OK
7a	(18,−3)	(16,−8)	(32,−64)	1	97	NO
10a	(−40,−60)	(80,−178)	(−136,−72)	68	77	OK
11a	(10,−83)	(16,−72)	(8,−64)	267	311	OK
11b	(0,−88)	(0,−72)	(0,−72)	278	342	OK
11c	(−4,−90)	(0,−80)	(0,−88)	2	62	NO
11d	(−5,−90)	(−24,−80)	(−16,−80)	2	75	NO
12a	(10,−14)	(16,0)	(136,−160)	127	149	OK
12b	(−2,−14)	(0,−8)	(120,−136)	157	175	OK
12c	(−14,−12)	(−16,8)	(32,−48)	2	36	NO
13a	(10,19)	(8,24)	(8,16)	7	202	NO
13b	(9,18)	(8,24)	(−120,−96)	22	57	OK

A System for Web-based Access to the HSOS Database

Ganghua Lin

National Astronomical Observatories, Chinese Academy of Sciences, Beijing, China

Abstract. Huairou Solar Observing Station's (HSOS) magnetogram and dopplergram are world-class instruments. Access to their data has opened to the world. Web-based access to the data will provide a powerful, convenient tool for data searching and solar physics. It is necessary that our data be provided to users via the Web when it is opened to the world. In this presentation, the author describes general design and programming construction of the system. The system will be generated by PHP and MySQL. The author also introduces basic feature of PHP and MySQL.

1. Introduction

Huairou Solar Observing Station's (HSOS) magnetograph can simultaneously measure the solar 2-D magnetic field and velocity field with different spectral lines. The photospheric vector magnetograms and dopplergrams, chromospheric longitudinal magnetograms and dopplergrams, and corresponding filtergrams can be obtained by this system. So far, 13 years of observations have been obtained.

Because of the international character of astronomy, we are planning to develop software that is independent of the archive architecture and which provides users with Web access capabilities.

Given the large amount of data, and that the data is not only plain text but also image file, visual file, we select a relational database—MySQL—to build the database management system, and use PHP to build the database engine and user query interface.

The final system is expected to allow users to get good response time and easy browsing of the HSOS on-line database by the means of a Web-based user interface, display results graphically, and save them locally for further analysis.

In Figure 1, we give an overview of the architecture of the system:

- content of the database;
- the database structure being developed;
- the extensions we are developing to the database engine to support HSOS data and queries;
- access methods;
- access interfaces currently under development.

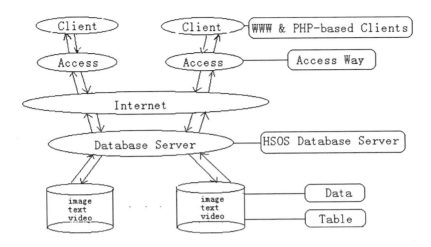

Figure 1. Architecture of the HSOS database on-line system.

In the following sections we briefly describe the items shown as rectangles in Figure 1.

2. Data

At present, only the data which are currently being used or built in the HSOS database project are foreseen to be included in the database. They are: photospheric longitudinal magnetic field data, photospheric filtergram image data and photospheric transverse magnetic field. Other data, for example, chromospheric longitudinal magnetic field, filtergram image, longitudinal velocity field data, etc., may be added in the future depending on users' requirements. Users can learn more about HSOS data through reading the the on-line README file.[1]

3. Table

All data in the collection are described in a metadata structure we call the metadata table. All applications will access the database through the table, so that they will work independently of any data-specific structure. Using a table will also simplify the addition of new data to the collection. Metadata access has been developed in the PHP language.

---

[1] http://sun.bao.ac.cn/observation/datadir.html

4. HSOS Database Server

All the data in the collection are managed by means of the MySQL[2] relational DBMS. The database engine can be extended by means of software modules that plug in to MySQL Dynamic Server and extend its capabilities in order to store, retrieve, and manipulate new data types, besides the primitive ones provided with the server. These modules are transaction manager, retrieve server, database manager, save server, and print server.

5. Access Methods

We are following the recent efforts directed at giving users direct access to HSOS database through the HSOS Web site. We will give users integrated access to world-wide distributed astronomical data collections. Our efforts will be directed to providing open access to the system by means of standard interfaces. In our view, this is the most effective way to maximize our system's utilization.

6. WWW & PHP-based Clients

We are currently prototyping simple WWW-based clients able to submit queries to the on-line system. These clients are implemented in PHP[3]. In this way, users are not required to install special client software and more interactivity is possible at the user workstation level. After this prototyping phase, we plan to design and build more sophisticated, intuitive interfaces to allow users to submit complete queries to the database, possibly in a graphical way.

7. PHP and MySQL

PHP has the following useful features:
- PHP is a server-side, cross-platform, HTML embedded scripting language;
- Much of its syntax is borrowed from C, Java, and Perl, with a couple of unique PHP-specific features thrown in. It has powerful database access functions. It is easy to learn;
- It has nearly all the features needed for building a Web site: designing objects, database access, network protocols, and security.

Its quick response time, multi-threaded nature, and transparency to the end user make it ideal for developing dynamic Web sites and applications.

The critical factors for a database management system are:
- Performance. A lot of clients can connect to the database server at the same time. Clients can use many databases simultaneously, and submit many queries;
- It should support SQL and ODBC;

---

[2]http://www.mysql.com/information/index.html

[3]http://www.zend.com/zend/art/intro.php

- Connectivity and security. Its database should can be accessible and shared by anyone on the Internet.

MySQL offers a rich and very useful set of functions. The performance, connectivity, speed and security make MySQL highly suited for accessing databases on the Internet. MySQL is a client/server system that consists of a multi-threaded SQL server that supports different backends, several different client programs and libraries, administrative tools, and a programming interface.

PHP and MySQL work with each other very well. Both PHP and MySQL are open source software. In combination with our Linux operating system, they form a ideal network database environment.

Uniform Data Sampling: Noise Reduction & Cosmic Rays

J. D. Offenberg[1], D. J. Fixsen[1], M. A. Nieto-Santisteban[2],
R. Sengupta[1], J. C. Mather[3], H. S. Stockman[2]

Abstract. As computers and scientific instruments become more complicated and more powerful (Moore's Law), we can perform astronomical observations never before contemplated. As larger data volumes are acquired, as more complex instruments are designed, and as observatories are placed in distant space locations with constrained downlink capacity, the need for automated, robust image processing tools will increase.

We present a robust, optimized algorithm to perform automated processing of array image data obtained with a non-destructive read-out. We present the derivation of the noise effects of this algorithm and compare alternative strategies.

1. Introduction

The effects of radiation and cosmic rays can be a formidable source of data loss for a space-based observatory. Several solutions to the problem of identifying and removing cosmic rays exist. We evaluate Up-the-Ramp sampling with on-the-fly cosmic ray identification and mitigation, which is described in detail by Fixsen et al. (2000) and compared to Fowler Sampling (Fowler & Gatley 1990). We concentrate on Up-the-Ramp sampling for study because it provides better signal-to-noise in what is probably the most difficult-to-measure regime, the read-noise limit. In the absence of cosmic rays, Up-the-Ramp sampling provides modestly ($\sim 6\%$) higher signal-to-noise than does Fowler Sampling (Garnett & Forrest 1993). The fact that an Up-the-Ramp sequence can be screened for cosmic rays and other glitches improves this result. Furthermore, on-the-fly cosmic ray rejection allows longer integration times which also improves the signal-to-noise in the faint limit (Offenberg et al. 2001).

The following discussion is largely an excerpt from Offenberg et al. 2001.

2. Fowler Sampling

Fowler sampling reduces the effect of read noise to $\sigma'_r = \sigma_r \sqrt{4/N}$ (for an observation sequence consisting of N samples, $N/2$ Fowler-pairs). However,

---

[1] Raytheon ITSS, 4500 Forbes Blvd, Lanham MD 20706

[2] Space Telescope Science Institute, 3700 San Martin Dr., Baltimore MD 21818

[3] Code 685, NASA's Goddard Space Flight Center, Greenbelt MD 20771

when a pixel is impacted by a cosmic ray during an observation, the cosmic ray essentially injects infinite variance and reduces the signal to noise to zero at that location. If we start with the Fowler sampling signal-to-noise function in the read-noise limit, from Garnett & Forrest (1993; Eqn. 6),

$$SN_F = \frac{FT}{\sqrt{2}\sigma_r}\sqrt{\frac{\eta T}{2\delta t}}\left(1 - \frac{\eta}{2}\right) = \frac{FT}{\sqrt{V_0}} \quad (1)$$

where F is the flux of the target, T is the observation time, σ_r is the read noise, η is the Fowler duty cycle, and δt is the time between sample intervals (determined by engineering or scientific constraints on the system). We note that this formula breaks down for relatively small numbers of samples (i.e., δt large with respect to T). FT is the signal, so the remaining terms are the noise, which is the square-root of the variance, V_0. If we consider two cases, "no-cosmic-ray" and "hit-by-cosmic-ray," and combine the variances according to

$$V_{comb} = \frac{V_0 P_0 W_0^2 + V_1 P_1 W_1^2}{(W_0 P_0 + W_1 P_1)^2} \quad (2)$$

we can rewrite Equation 1 as

$$SN_{FC} = \frac{FT}{\sqrt{V_{comb}}} = \frac{FT(W_0 P_0 + W_1 P_1)}{\sqrt{V_0 P_0 W_0^2 + V_1 P_1 W_1^2}} \quad (3)$$

As the weight is the inverse of the variance ($W_i = 1/V_i$), Equation 3 can be rewritten as

$$SN_{FC} = \frac{FT(\frac{P_0}{V_0} + \frac{P_1}{V_1})}{\sqrt{P_0/V_0 + P_1/V_1}} = FT\sqrt{\frac{P_0}{V_0} + \frac{P_1}{V_1}} \quad (4)$$

V_0 is the variance in the no-cosmic-ray case, taken from Equation 1, and P_0 is the probability of a pixel surviving without a cosmic ray hit. For simplicity, we define $1 - P$ to be the probability of a pixel being hit by a cosmic ray per time unit δt, so P is the probability of "survival" and $P_0 = P^{T/\delta t}$. As a cosmic ray hit injects infinite uncertainty, the variance in the cosmic ray case is $V_1 = \infty$. Plugging in to Equation 4, we get the signal-to-noise for Fowler sampling in the read-noise limited case with cosmic rays, Equation 5:

$$SN_{FC} = FT\sqrt{\frac{P_0}{V_0} + 0} = \frac{FT}{\sqrt{2}\sigma_r}\sqrt{\frac{\eta T}{2\delta t}}\left(1 - \frac{\eta}{2}\right) P^{T/(2\delta t)} \quad (5)$$

For a given integration time T and minimum read time δt, the maximum SN_{FC} occurs with duty cycle $\eta = 2/3$. If we plug this back into Equation 5, we get

$$SN_F = \frac{2}{3}\frac{FT}{\sqrt{2}\sigma_r}\sqrt{\frac{T}{3\delta t}} P^{T/(2\delta t)} \quad (6)$$

From here, it is possible to find the value of T which gives the best signal-to-noise for a single observation; it occurs at $T = -3\delta t/\ln(P)$. If, however, we consider the observation as a series of M equal observations with a specific total observation time, T_{obs}, the signal-to-noise for the series is

$$SN'_{FC} = \frac{FT\sqrt{M}}{3\sigma_r}\sqrt{\frac{2T}{3\delta t}}P^{T/(2\delta t)} = \frac{FT\sqrt{T_{obs}}}{3\sigma_r\sqrt{T}}\sqrt{\frac{2T}{3\delta t}}P^{T/2\delta t} \quad (7)$$

If we hold T_{obs} constant and find the optimum T, we find it at $-2\delta t/\ln(P)$. In either case, it is important to note that there is an optimal value for T, and extending the observation beyond that time will ruin the data.

It is worth noting that the result assumes that all cosmic ray events can be identified *a posteriori*. This is not necessarily the case, particularly when it is considered that, in the one-image case, the fraction of pixels surviving without a cosmic ray impact is $P^{-3/\ln(P)} = e^{-3} \approx 0.05$; for the multi-image case, the fraction of survivors is $P^{-2/\ln(P)} = e^{-2} \approx 0.14$. In both cases, the number of "good" pixels is so low that separating them from the impacted pixels will not be a trivial task. For example, the median operation would not be able to identify a good samples, as more than half of the samples would be impacted by cosmic rays. In practice, the detector will often saturate before this limit is reached, but this shorter integration time means that less-than-optimal signal-to-noise will be obtained.

3. Up-the-Ramp Sampling

Up-the-Ramp sampling reduces the effect of read noise to $\sigma'_r = \sigma_r\sqrt{12/N}$, for N uniformly-spaced samples with equal weighting (which is the optimal weighting for the read-noise limited case). When a pixel is impacted by a cosmic ray, the Up-the-Ramp algorithm preserves the "good" data for that pixel. The exact quality of the preserved data depends on the number of cosmic ray hits and their timing within the observation. For example, a cosmic ray hit which just trims off the last sample in the sequence has minimal impact compared to a cosmic ray hit that occurs in the middle of the observation sequence. The variance of a Uniformly-sampled sequence with N_i samples is proportional to $1/N_i(N_i+1)(N_i-1)$. If an Up-the-Ramp sequence is broken into i chunks by a cosmic ray, the variance becomes

$$V_i = V_U \frac{N(N+1)(N-1)}{\sum_{j=0}^{i}(N_j)(N_j+1)(N_j-1)} \quad (8)$$

When there are zero cosmic ray events, of course, $V_0 = V_U$. If there is one cosmic ray event during the sequence, the variance becomes

$$V_1 = V_U \frac{N(N+1)(N-1)}{N_i(N_i+1)(N_i-1) + (N-N_i)(N-N_i+1)(N-N_i-1)} \quad (9)$$

If we assume (as is reasonable) that the cosmic ray events are randomly distributed over time and find the expectation value for all values of $0...N_i...N$, we find that the typical $V_1 \approx V_U * 2$ (plus a small term in N^{-1}, which we will ignore for simplicity). If we perform a similar computation for two cosmic ray events, we find that $V_2 \approx V_U * 10/3$ (again, plus lower-order terms which we ignore). In general, we find that it is possible to find a valid result with a finite variance for any sequence broken up by cosmic ray events provided we have at least two consecutive "good" samples (for all practical purposes, we can ignore the situation where this is not the case). To simplify the following, we consider only three cases: The no-cosmic-ray case $V_0 = V_U$, the one-cosmic-ray case $V_1 = 2 * V_U$ and all multiple-cosmic-ray cases combined as one, $V_{2+} = V_U/\epsilon^2$, where ϵ^2 is a small but non-zero number, roughly 0.3.

The Up-the-Ramp signal-to-noise function for the read-noise limited case (Garnett & Forrest 1993; Eqn. 20) is

$$SN_U = \frac{FT}{\sqrt{2}\sigma_r}\sqrt{\frac{N^2-1}{6N}} = \frac{FT}{\sqrt{V_U}} \quad (10)$$

We combine the variances in the three possible cases with the three-case equivalent to Equation 2, and thus arrive at

$$SN_{UC} = FT\left(\frac{P_0}{V_0} + \frac{P_1}{V_1} + \frac{P_{2+}}{V_{2+}}\right)^{1/2} = \frac{FT}{\sqrt{V_U}}\left(P_0 + \frac{P_1}{2} + \epsilon P_{2+}\right)^{1/2} \quad (11)$$

where P_i is the probability of a pixel being impacted by i cosmic rays during the integration. We note, as did Garnett & Forrest, that there would be no reason to limit the number of samples to anything less than the maximum possible number, so we can set $N = T/\delta t$. Using the definition of P described earlier, $P_0 = P^{T/\delta t}$, $P_1 = (T/\delta t)(1-P)P^{T/\delta t - 1}$ and $P_{2+} = 1 - (P_0 + P_1)$. Putting these values back into Equation 11, we get:

$$SN_{UC} = \frac{FT}{\sqrt{2}\sigma_r}\sqrt{\frac{T^2 - \delta t^2}{6T\delta t}}\left[(1-\epsilon)P^{T/\delta t} + (1-\epsilon)\frac{T}{\delta t}(1-P)P^{T/\delta t - 1} + \epsilon\right]^{1/2} \quad (12)$$

If we seek the maximum value of SN_{UC} with respect to T, we find that SN_{UC} is strictly increasing if $T \geq \delta t$ (otherwise we would have an integration shorter than one sample time, which would be useless), $P \neq 0$ and $0 < \epsilon < 1$ (both of which are true by construction). This result applies whether we are considering one independent integration or a series of observations to be combined later. As the derivative is strictly positive, the signal-to-noise continues to increase with the sample time, although as $T \to \infty$, the gain in signal-to-noise asymptotically approaches zero. So, extending the observing time while using Up-the-Ramp sampling with cosmic ray rejection does not damage the data (although we might be spending time with little or no gain). As noted earlier for the Fowler-sampling case, there is an optimal observing time, beyond which further observation reduces the overall signal-to-noise.

References

Fixsen, D. J., et al. 2000, PASP, 112, 1350
Fowler, A. M. & Gatley, I. 1990, ApJ, 353, L33
Garnett, J. D. & Forrest, W. J. 1993, Proc. SPIE, 1946, 395
Offenberg, J. D., et al. 2001, PASP, 113, in press

New Tools for the Analysis of ISOPHOT P32 Mapping Data in PIA

C. Gabriel

ISO Data Centre, ESA Astrophysics Division, Villafranca del Castillo, P.O. Box 50727, 28080 Madrid, Spain

R. Tuffs

Max-Planck Institut für Kernphysik, Heidelberg, Germany

Abstract. The Infrared Space Observatory AOT P32 is a dedicated mapping mode combining the raster capability of *ISO* with the chopping capability of the ISOPHOT instrument for obtaining images in the far infrared, maximising the spatial resolution. We present diffraction limited maps of the Crab Nebula at 60 and 100 μm, as an illustration of a new tool for the P32 data analysis, integrated within the ISOPHOT Interactive Analysis (PIA).

1. Introduction

The Astronomical Observation Template (AOT) P32 was one of the observational modes (Heinrichsen, Gabriel, Richards, & Klaas 1997) defined for the instrument ISOPHOT (Lemke et al. 1996) on board the Infrared Space Observatory (*ISO*; Kessler et al. 1996). It tried to solve the question of observing extended structures with large detector pixels in the far infrared by a high sky registration and redundancy. The non-linear response of the Ge:Ga detectors affects the calibration of such a mode to a large extent, making the correction of those transient effects one of the most difficult tasks in the ISOPHOT data analysis.

2. The Challenge: Mapping in the Far Infrared

Obtaining photometric maps of extended sources in the far infrared (50–200 μm) is challenging because of the typical large sizes of the detector pixels involved in addition to the diffraction effects in this wavelength regime. The ISOPHOT detector arrays used in the far infrared were PHT-C100 (a 3×3 array of 46″×46″) and PHT-C200 (a 2×2 array of 92″×92″). Mapping in a finite time with a good (Nyquist) sampling therefore requires special observation techniques.

3. The Way: ISOPHOT's AOT P32

In addition to using *ISO*s capability of performing raster observations and the array structure of ISOPHOT's long wavelength detectors, the Astronomical Observation Template P32 used the focal plane chopper of ISOPHOT to rapidly modulate the satellite effective pointing, in steps of a third of the detector pixel pitch, on timescales ranging down to ~0.15 sec. The result is very good sky registration, with the oversampling and redundancy necessary for the best achievable spatial resolution.

The chopper deflection is up to ±90 arcsec in the Y-spacecraft direction, in 15 and 30 arcsec steps respectively for the C100 and C200 detectors. The arrays are on the Y-Z spacecraft plane and the total number of positions seen by a chopper sweep is therefore 13 for C100 and 7 for C200. Every sky position is therefore registered three times in the Y-spacecraft direction within a chopper cycle by a detector pixel and its neighbour pixels. The raster steps in Y and Z directions (with different user defined oversampling factors) also ensure a uniform coverage and high redundancy.

4. The Main Problem: Detector Transients

The ISOPHOT C200 (Ge:Ga stressed) and especially the C100 (Ge:Ga) photoconductor detectors have a complex non-linear response as a function of illumination history on timescales of ~0.1–100 sec, depending on the absolute flux level as well as the flux changes involved (Acosta, Gabriel, & Castañeda 2000). The P32 observation mode, with its high frequency flux modulation, as described above, is in principle always in a non-stabilized state. Under- and overshooting signal effects, caused by a short term "hook" response, complicate the calibration of observations performed in this mode.

5. The Solution: P32_Tools Package

This package, originally developed at Max-Planck Institut für Kernphysik by one of us, is a collection of IDL routines, which:
- solves the non-linear optimisation problem for the set of sky brightnesses illuminating the detector on the grid of sky sampling (arbitrary source morphology),
- optionally solves for detector starting state,
- optionally solves for detector model parameters (through self-calibration), which by default are predetermined.

A complete set of diagnostic plots, images, and text information is produced, which enables the user to assess the quality and reliability of the data treatment.

5.1. Detector Model

The signal to every time is given by the sum of two components: $S = S_1 + S_2$ with a slow S_1 and a fast S_2 part:

$$S_1 = (1 - \beta_1)S_\infty(1 - e^{-t/\tau_1}) + S_{01}e^{-t/\tau_1} \tag{1}$$

$$S_2 = \beta_2 S_\infty (1 - e^{-t/\tau_2}) + S_{02} e^{-t/\tau_2} \qquad (2)$$

The prediction of a change in photocurrent after a flux change is known as the "jump condition":

$$S_{01} = \beta_1 (S_\infty - S_{\infty p}) + S_{1p} S_{02} = S_{2p} \qquad (3)$$

with p as the index for the previous flux level.

5.2. Determination of Parameters

The "default" parameters determination was performed using
- starting exposures of internal calibrators for slow components
- standard celestial calibration sources for fast components, exploiting the redundancy
- illumination dependency in parametrisation of βs and τs.

6. The Results

P32_Tools is already giving very good results, both for point sources and extended objects, although it is still in a testing and enhancement phase.

The nature of the problem, together with the fact that disturbances from cosmic ray hits are difficult to handle in an automatic way in this observation mode, require highly interactive work on the data.

7. The Integration within PIA

P32_Tools has been fully integrated within PIA (Gabriel, Acosta-Pulido, Heinrichsen, Morris, & Tai 1997; Gabriel & Acosta-Pulido 1999) for a better and easier access to all the capabilities already given. This also allowed us to make use of PIA's graphical data handling. Graphical menus for driving processing and parameter handling, display of results and information were specially developed, but wide use is made of already existent PIA tools.

8. An Example, FIR images of the Crab Nebula

P32 observations of the Crab Nebula have been preliminarily reduced with the P32_Tools package. These observations were performed with the aim of trying to understand the origin of the "InfraRed Bump" discovered by IRAS. A comparison of the obtained images at 60 μm and 100 μm using the default PIA processing (Figure 1), and the ones obtained using the P32_Tools package (Figure 2) show a remarkable improvement in angular resolution, indicating that the "Infrared Bump" probably arises from compact line emitting structure superposed on the smooth synchrotron emission.

There is a good correlation between the drift corrected P32 maps and an [O III]λ5007 emission map of the Crab Nebula taken with the Goddard Fabry-Perot Imager (Lawrence et al. 1995). The P32 maps are largely tracing oxygen. Prominent oxygen fine structure emission lines are present in the nebula at 52 μm ([O III]), 63 μm ([O I]) and 88 μm ([O III]), as observed by the spectra

New Tools for ISOPHOT P32 Mapping Data in PIA   403

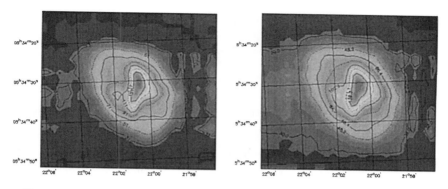

Figure 1. Crab Nebula maps obtained by PIA plain processing

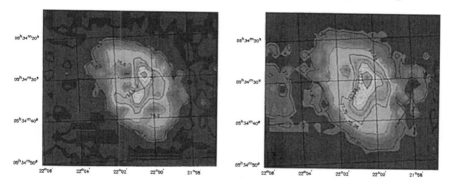

Figure 2. Crab Nebula maps obtained by P32_Tools processing

taken with the ISO-LWS spectrometer in both wavelength regions covered by the ISOPHOT C100 60 μm and 100 μm broadband filters.

References

Acosta-Pulido, J. A., Gabriel, C., & Castañeda, H. 2000, Experimental Astronomy 10, Kluwer Academic Publishers, 333–346

Gabriel, C., Acosta-Pulido, J., Heinrichsen, I., Morris, H., & Tai, W.-M. 1997, in ASP Conf. Ser., Vol. 125, Astronomical Data Analysis Software and Systems VI, ed. G. Hunt & H. E. Payne (San Francisco: ASP), 108

Gabriel, C. & Acosta-Pulido, J., 1999, ESA SP-427, 73

Heinrichsen, I., Gabriel, C., Richards, P., & Klaas, U. 1997, ESA SP-401

Kessler, M. F., et al. 1996, A&A, 315, L27

Lawrence, S. S., et al. 1995, AJ, 109, 2635

Lemke, D., et al. 1996, A&A, 315, L64

AIRY: Astronomical Image Restoration in interferometrY

Serge Correia, Marcel Carbillet, Luca Fini

Arcetri Astrophysical Observatory, Largo E. Fermi 5, 50125 Firenze, Italy

Patrizia Boccacci, Mario Bertero

INFM and Department of Computer and Information Sciences, University of Genova, Via Dodecaneso 35, 16146 Genova, Italy

Antonella Vallenari

Padova Astronomical Observatory, vicolo dell' Osservatorio 5, 35122 Padova, Italy

Andrea Richichi

European Southern Observatory, Karl-Schwarzschildstr. 2, 85748 Garching b.M., Germany

Massimo Barbati

Astronomical Observatory of Torino, Strada Osservatorio 20, 10025 Pino T.se (TO), Italy

Abstract. AIRY is a modular software package designed to simulate optical and near-infrared interferometric observations and/or to perform subsequent image restoration/deconvolution. It is written in IDL and has been designed to be used together with the CAOS Application Builder, version 2.0 or higher. AIRY can be applied to a wide range of imaging problems. We will present in particular an application to the case of interferometric imaging with the Large Binocular Telescope, in which we simulate the observation and scientific interpretation of a synthetic star cluster in the near-infrared. (Related Web page[1])

1. Introduction

AIRY is the acronym describing the activity of a group of astronomers and mathematicians from various Italian institutions (see the Web page indicated above). The aim of the collaboration is to develop methods and software for the restoration of interferometric images, with application to the Large Binocular Telescope (LBT). One of the first results is the package AIRY, IDL-based and

---

[1] http://dirac.disi.unige.it

CAOS-compatible (Fini et al. 2001). AIRY is designed to simulate optical and near-infrared interferometric observations and/or to perform subsequent image restoration/deconvolution. It consists of a set of specific modules which are listed and briefly presented in Section 2. The package also includes a library of ideal and Adaptive Optics (AO) corrected LBT point-spread functions (PSFs). Details can be found in Carbillet et al. (2001). An interesting feature of AIRY is its multiple deconvolution capability, well suited for the LBT case. In the current version the method implemented is the so-called *Ordered Subsets - Expectation Maximization* (OS-EM) algorithm (Bertero & Boccacci 2000a). As an example of application we present in Section 3 a simulated LBT observation together with a scientific interpretation of the results.

2. The Modules of the AIRY Simulation Package

Table 1 shows a complete list, together with a very brief description, of the modules of the current version of the AIRY Simulation Package.

Table 1. Modules of the AIRY Simulation Package.

Module	Purpose
Data simulation modules	
OBJ - OBJect definition	to define the object characteristics among several object types (binary object, open cluster, planetary nebulae, SN remnant, spiral galaxy, YSOs, stellar surface, user-defined)
CNV - object-PSF CoNVolution	to perform convolution
ADN - ADd Noise to image	to add the noise contributions
Data processing modules	
PRE - PRE-processing	to perform image pre-processing
DEC - DEConvolution process	to perform deconvolution (*Ordered Subsets-Expectation Maximization* method)
Data analysis modules	
ANB - ANalysis of Binary	to analyse reconstructed images of binary objects
FSM - Find Star Module	to detect stars in the reconstructed images
Other modules and utilities	
RFT - Read FiTs file format	to read FITS images
WFT - Write FiTs file format	to write FITS images
RSC - Restore im. Struct. Cubes	to restore image structure cubes (XDR format or FITS format)
SIM - Save IMage struct.	to save image structure cubes (XDR format or FITS format)
DIS - DISplay image	to display images

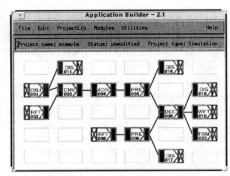

Figure 1. Example of worksheet of a typical simulation.

Figure 1 shows an example of a simulation that can be built with the AIRY Simulation Package. This simulation is essentially composed of three parts. The first part models the observed data by convolving an object map (here a stellar cluster) of given characteristics with a set of PSFs (object-PSFs), extracted, for this example, from the library. The different noise contributions are then added. The second part is the restoration of the observed data set by multiple deconvolution with another set of PSFs (reference-PSFs), after a pre-processing stage. The third part permits the analysis and saving of the deconvolved image.

Note that the modular structure of AIRY also allows using the package for improving real AO data by removing part of the AO-correction residual, and/or to produce images from real interferometric data.

3. Example Application: Scientific Analysis of a Simulated Star Cluster Observed with the Large Binocular Telescope

The goal is to simulate high-resolution interferometric observations of a scientific object of interest with LBT, and to retrieve the scientific parameters of this object after the image restoration process. We have considered a star cluster composed of 1898 stars with the following characteristics: age 4.0 Gyr, metallicity Z=0.008, distance modulus=19($\simeq$ 63 kpc), reddening=0, extension field= $10.''24 \times 10.''24$.

Three object maps (2048×2048 pixels) were modeled in J, H, and K bands. The resulting magnitude ranges were respectively 14.01–24.22, 13.25–23.63 and 12.89–23.56. The worksheet of this simulation is similar to that presented in Figure 1 for each band. We have simulated observations at three parallactic angles (0°, 60°, and 120°) for each band, and with 2000 s integration time for each parallactic angle. PSFs were assumed ideal (coherence, cophasage, no aberrations) for both the reference and the object-PSFs. Multiple deconvolution was carried out for each band using 100 iterations of the *OS-EM algorithm* (see Bertero & Boccacci 2000a, Bertero & Boccacci 2000b, Bertero et al. 2000). Detection and photometry on the restored frames were performed using DAOPHOT, with a 25-sigma detection threshold and a 3 pixels (15 mas) aperture photometry diameter.

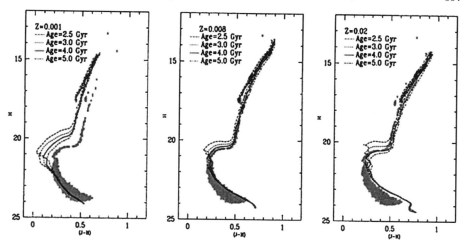

Figure 2. CMD of the star cluster in the plane H vs. (J−H). Detected stars are marked with squares and isochrones corresponding to different ages and for metallicity Z= 0.001 (left), Z= 0.008 (center), and Z= 0.02 (right) are superimposed.

The Color-Magnitude Diagram (CMD) of the star cluster in the plane H vs. (J−H) is shown in Figure 2. Detected stars are marked with squares and isochrones corresponding to different ages and metallicity values are superimposed. From a first visual inspection, we can derive that Z$\simeq$ 0.008 and the age is 4.0–5.0 Gyr, in good agreement with the input parameters.

References

Bertero, M., & Boccacci, P. 2000a, A&AS, 144, 181
Bertero, M., & Boccacci, P. 2000b, A&AS, 147, 323
Bertero, M., Boccacci, P., Correia, S., & Richichi, A. 2000, in *Interferometry in Optical Astronomy*, ed. P. J Lena & A. Quirrenbach, SPIE 4006, 514
Carbillet, M., Fini, L., Femenía, B., Esposito, S., Riccardi, A., Viard, E., Delplancke, F., & Hubin, N. 2001, this volume, 249
Fini, L., Carbillet, M., & Riccardi, A. 2001, this volume, 253

Parallelization of Widefield Imaging in AIPS++

K. Golap, A. Kemball, T. Cornwell, W. Young
National Radio Astronomy Observatory

Abstract. At low frequencies, large synthesis arrays in Radio Astronomy, such as the Very Large Array (VLA), effectively require that a 3-D Fourier transform be used in imaging, rather than the conventional 2-D transform. Given the large data volumes associated with observations of this type, this ensures that these problems are amongst the most computationally demanding in radio astronomy. Typical image sizes are of the order of a few million pixels.

The wide-field imaging problem can be made more tractable by using parallelization. In this paper, we discuss the general wide-field imaging algorithm used in AIPS++, and the techniques used for its parallelization.

1. Overview of the Wide-Field Imaging Problem

A problem occurs when imaging large fields of view with relatively long baselines and non-coplanar arrays. Imaging using synthesis arrays involves inverting the 3-D integral

$$V(u,v,w) = \int I(l,m) \exp j2\pi(ul+vm+wn) \frac{dldm}{\sqrt{1-l^2-m^2}}$$

to obtain the brightness distribution $I(l,m)$ on the sky, from a measured set of visibilities, $V(u,v,w)$, in the uv-plane. In most practical cases the non-coplanar term w can be neglected and the inversion is a direct 2-D Fourier transform. However, for wide-field imaging if the w term is not taken into account there is usually a substantial loss of dynamic range, and it is also impossible to faithfully image regions far from the field center.

2. The Widefield Imaging Algorithm Used in AIPS++

Several algorithms exist to solve the full 3-D problem listed above (Cornwell & Perley 1992). In AIPS++ a multi-faceted transform approach has been chosen for its efficiency. This covers the image plane by a series of facets, in each of which a 2-D transform holds.

We can decompose the visibilities into a summation of re-phased faceted visibilities:

$$V(u,v,w) = \sum_k V_k(u,v) \frac{\exp j2\pi(ul_k+vm_k+w\sqrt{1-l_k^2-m_k^2})}{\sqrt{1-l_k^2-m_k^2}}$$

where :

$$V_k(u,v) = \int I_k(l-l_k, m-m_k) \exp j2\pi(u(l-l_k) + v(m-m_k))dldm$$

The iterative multi-stage algorithm implemented in AIPS++ proceeds as follows:

- Calculate residual images for all facets (using 2-D transforms).
- Partially deconvolve individual facets and update the image model for each facet.
- Reconcile different facets by subtracting the model visibility for all facet models from the visibility data.
- Recalculate residual images and repeat. In the process of making residual images, a uv-plane coordinate system is chosen so that the final image from all facets is projected on a common tangent plane (Sault et al. 1996).

2.1. An Example

Wide-field imaging is computationally expensive. The image in Figure 1 was made using the AIPS++ widefield algorithm with 225 facets. The data are a VLA observation at 74 MHz in the B and C configurations. This image took close to 20 days to process on a desktop workstation (SGI octane). A similar observation in the A-array of the VLA would require some ten times more computer resources to process. Along with other overheads, like better deconvolution algorithms for larger baselines, we are facing computation of 200 to 300 days on a typical desktop. This problem strongly justifies the need for parallelization of this algorithm. The problem will be more pronounced with future arrays such as the Expanded VLA (EVLA).

3. The Parallelization Effort and Progress

For the first level of parallelization we are aiming at parallelizing the nearly embarrassingly parallel sections of the widefield algorithms. There are three distinct sections which we have identified in the widefield algorithms which fall under this category:

- The point spread function (PSF) formation. The PSF for each facet is needed in deconvolution. These can be estimated totally independently of each other, requiring only the uv-coverage seen from each facet.
- The model visibility estimation from the source model components. As the visibilities from different sources (or different facets) are additive, they can be estimated independently for each facet model and cumulatively added into the final model visibility. This has parallel I/O implications.
- The residual image estimation. In calculating the residual image the residual visibility re-projection for the different facets can be estimated independently.

3.1. Progress Made So Far

We have made progress in parallel I/O development and evaluation of different access methods for the visibility data. This includes measuring the efficacy of

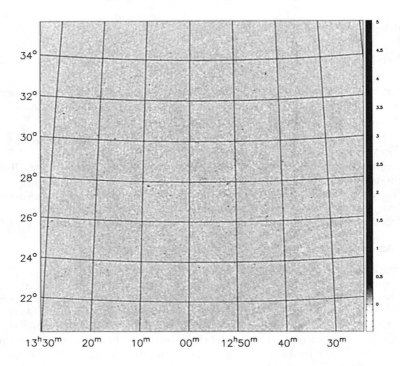

Figure 1. Image of Coma cluster at 74 MHz reduced in AIPS++ using 225 facets

parallelization with multiple processes accessing the same visibility data. We have verified that parallelization of the function to form the PSFs speeds up almost linearly for a few processors, and have parallelized the function to predict the model visibilities.

4. Ongoing and Future Work

The parallel 3-D imaging approach is close to full operational use. Areas of ongoing work include: migration to larger machines or clusters and fully measuring the speed up of each algorithm computed, parallelizing the residual image formation for each facet. further work in Parallel I/O using MPI-2, and investigation of statement level parallelization using OPEN-MP.

References

Cornwell, T. J. & Perley, R. A. 1992 Astron. & Astrophys, 261, 353
Sault, R., Staveley-Smith, L., & Brouw, W. N. 1996 Astron. & Astrophys. Suppl., 120, 375

Ionospheric Corrections in AIPS++

Oleg Smirnov

Netherlands Foundation for Research in Astronomy (ASTRON)
P.O. Box 2
7990 AA Dwingeloo, The Netherlands

Abstract. The ionosphere induces significant distortion in radio astronomical measurements at lower frequencies. For polarimetric work, the most significant effect is Faraday rotation, reaching up to several turns at, e.g., meter wavelengths. Over the course of an observation, variations in the Faraday rotation can completely wash out any polarization in the signal. The effect can be corrected for only if an accurate enough estimate of the electron density distribution along the line-of-sight is somehow obtained.

Campbell (1999) has experimented both with an *a priori* theoretical model of the ionosphere, and with GPS observations, to perform ionospheric corrections in VLBI observations. I present an extension of this approach into the AIPS++ system. AIPS++ now includes an Ionosphere module, encompassing the Parametrized Ionospheric Model (Daniell et al. 1995), and a calibration component that, given a dataset, will automatically estimate and correct for the Faraday rotation. This paper also describes some other applications of the AIPS++ Ionosphere module, such as modelling and simulation.

1. Introduction

The ionosphere induces significant distortion to radio astronomical measurements at lower frequencies. These effects are:

- phase delay,
- Faraday rotation
- refraction, and
- absorption.

For low-frequency polarimetric work, by far the most significant effect is Faraday rotation (FR). Ionospheric FR can reach up to several cycles at, e.g., meter wavelengths. Over the course of an observation, variations in the FR can completely wash out any polarization in the signal. The effect can be corrected for only if an accurate enough estimate of the electron density distribution along the line-of-sight is somehow obtained.

Here I describe ionospheric calibration as being implemented in AIPS++. AIPS++[1] is a data processing system under development by an international consortium of observatories. The latest public release is AIPS++ 1.4. AIPS++ includes extensive support for calibration of radio astronomical data, and correction for FR fits nicely into the framework. Note that the implementation discussed here is currently only available in the development branch of AIPS++, but should make its way into the next public release in 2001.

2. Possible Approaches

Two approaches to the problem of estimating the ionospheric electron density have been pursued. We can use an *a priori* "climatological" model of the ionosphere. One such model is PIM (Parametrized Ionospheric Model), developed at the USAF Phillips Lab (Daniell et al. 1995). Given a time and date, PIM can compute a predicted electron density distribution for any region of the ionosphere. An inherent limitation of *a priori* models is that they can't predict small-scale structure, such as traveling ionospheric disturbances (TIDs), which are, in effect, "clouds" of higher electron content.

Another approach is to use direct observations of the ionosphere. Of these, GPS satellites offer perhaps the most interesting opportunity.

The Global Positioning System (GPS) provides an inexpensive and accurate way to continuously probe the ionosphere. GPS satellites transmit on two L-band carrier frequencies. By measuring the time delay between modulation on the two carriers, the total ionospheric delay (and thus the total electron content) along the line-of-sight to the satellite can be estimated. Relatively inexpensive GPS receivers will readily provide delay data. With at least six GPS satellites overhead at any time, we can derive measurements of the ionosphere along at least six different moving lines-of-sight. The International GPS Service (IGS[2]) continuously collects data from many GPS receivers around the world; it is available via the Internet from various IGS data centers. Besides, most radio observatories have a co-located GPS receiver.

It should be noted that GPS measurements provide only an integral measurement of the electron density (the TEC) along a line-of-sight to the satellite (which, most of the time, is not the direction that we are really interested in). Faraday rotation cannot be derived from TEC directly, as it is also dependent on the magnetic field. To estimate FR, we need to know the electron distribution along the line-of-sight. In effect, the problem becomes one of tomography, with the GPS satellites providing moving slices through the ionosphere. Some sort of fitting of an *a priori* model is still required in order to derive profiles. Erickson et al. (1996) have experimented with a GPS receiver at the VLA, and have had some success even with a relatively simple ionosphere model.

Campbell (1999) has tried a combination of the two approaches. He has used PIM as a base model, and fitted GPS data to build up a field of "corrections" to PIM profiles. The AIPS++ implementation is loosely based on this approach

---

[1] http://aips2.nrao.edu

[2] http://igscb.jpl.nasa.gov

(although secondary GPS-based corrections are not actually implemented at this point in time).

3. PIM in AIPS++

AIPS++ now includes a version of PIM. "Classic" PIM is a huge bulk of FORTRAN code, not especially user- or programmer-friendly. To make life even more difficult, it requires large amounts of diverse external data, such as:

- Model databases, which are distributed in source form with PIM, but must be built into native binary format before use.
- Solar flux ($F_{10.7}$), geomagnetic indices (K_p/A_p), interplanetary magnetic field (IMF) observations for the dates of interest. These observations are available from various data centers on the Internet (NOAA NGDC, NASA GSFC, etc.), but must be downloaded and fed into PIM in a suitable format.

AIPS++ PIM hides all of this complexity from the end-user (or application developer).

The Ionosphere Module. The Ionosphere module of AIPS++ totally encapsulates PIM, and hides the difficulties of running it behind a rather simple class interface. To C++ programmers, Ionosphere provides functions for computing profiles of the ionosphere, TEC, and Faraday Rotation for any specified time, location, and line-of-sight.

The FJones Module. The AIPS++ calibration package is based around the Measurement Equation (ME; Hamaker et al. 1996, Sault et al. 1996). Within the ME, the term responsible for FR is just a 2×2 *Jones* matrix. The FJones class of the calibration package implements this matrix. Time, location, and direction are extracted from the measurement set and passed to Ionosphere. The resulting FR is used to compute the F-Jones term used in the calibration process. Thus, to perform corrections for FR, the user need do nothing more than enable the F-Jones term.

Data handling. All of the data required to run PIM are loaded by AIPS++ automatically. Maintenance scripts on the central AIPS++ site regularly download the latest $F_{10.7}$, K_p/A_p, and IMF observations from appropriate data centers. These data are placed into AIPS++ tables and automatically distributed via the AIPS++ Global Data Repository.

4. GPS Data in AIPS++

Two types of GPS data are required for ionospheric work:

RINEX is the standard output format for GPS receivers. A RINEX file will contain a list of times, satellite IDs, and observed delays, usually for a single day. RINEX files from IGS-participating receivers all over the world can be downloaded via FTP from several data centers (e.g., IGN, GSFC).

Ephemeris data is required to tie RINEX observations to a satellite's position on the sky. Ephemeris files are available by FTP from JPL.

Two modules in AIPS++, RINEX and Ephemeris, are responsible for maintaining this data. Both will automatically download files from relevant FTP sites, and convert them into AIPS++ tables and data structures. They can also combine the data into a single location-direction-TEC table. RINEX and ephemeris services are available to C++ programs, Glish scripts, and interactive users.

5. Other Applications

Glish is the high-level scripting language of AIPS++. Most of the data processing modules, though implemented in C++, have corresponding Glish bindings, so their functionality is available to Glish scripts, or interactively, via the command line. The Ionosphere module is no exception.

Glish programmers can easily use PIM to obtain ionosphere profiles for any given time, location, and line-of-sight. This makes it quite easy to develop various modelling and simulation tasks. Thus, Ionosphere has been used at ASTRON to evaluate requirements for the future Low Frequency Array (LOFAR) instrument.

Acknowledgments. Bob Campbell laid the groundwork for this development, and provided lots of valuable assistance.

References

Campbell, R. M. 1999, New Astr. Rev., 43, 8–10, 617
Daniell, R. E., et al. 1995, Radio Sci., 30, 1499
Erickson, W. C., Perley, R. A., Kassim, N. E., Payne, J. A., Flatters, C. 1996, AAS Meeting, 189, 107.06
Hamaker, J. P., Bregman, J. D., & Sault, R. 1996, A&AS, 117, 137
Sault, R. J., Hamaker, J. P., & Bregman, J. D. 1996, A&AS, 117, 149

Redesign and Reimplementation of XSPEC

Ben Dorman and Keith Arnaud

Laboratory for High Energy Astrophysics, Code 664, NASA/Goddard Space Flight Center, Greenbelt, MD 20771 [1,2]

Abstract. We present a progress report on the redesign of XSPEC. Work has been underway since late 1998 to produce a new version of XSPEC with a code base written in ANSI C++ and with modern design techniques. The new version (XSPEC 12) is expected to be ready for release in 2002.

The new version of the program will implement most of the features of the current release, XSPEC 11. The design allows for both the current Tcl command line interface as well as an optional GUI. The existing scheme whereby users can write theoretical model components in FORTRAN 77 will, however, continue to be fully supported.

Scientifically, the major change will be to support simultaneous fitting to spectral data containing multiple sources, required for coded mask instruments (such as Integral/SPI that are able, in principle, to separate superposed X-ray sources). The focus of the project is, however, to update the program to take advantages of the progress in programming technology that has been made in the field of Computer Science over the last 15 years since XSPEC was originally released.

The main software engineering benefits of the redesign project are an extensible implementation that allows for (a) loosely coupled modules [providing ease of maintenance], including CCfits, a new ANSI C++ interface to the `cfitsio` library developed as a by-product; (b) extension of the supported data formats by 'add-in' user-loadable modules; (c) an independent user interface implementation; and (d) robustness guarantees such as exception safety.

1. Motivation

XSPEC, originally developed by Rick Shafer at Cambridge University in the early 1980s (see Arnaud 1996 and references therein), was designed to be a mission-independent general purpose analysis program for X-ray spectral data. XSPECs general mission support requires the adoption by X-ray missions of

---

[1] Raytheon Information Technology & Scientific Services and Emergent Information Technology, Inc.

[2] Department of Astronomy, University of Maryland, College Park, MD 20742

a set of standard formats for representing X-ray data that are based on the instruments and requirements of the last fifteen years of science spacecraft.

XSPEC (current version is 11.0.1) manipulates theoretical models, X-ray source and background data, and calibration data. Each of these form natural objects for the Object-Oriented programming paradigm, as they have state, behavior, and identity. Encapsulating the abstract behavior of data and models allows the extension of XSPEC capabilities to provide general mission support well into the future. As well, XSPECs internal memory model is somewhat limited by its origins as a statically allocated FORTRAN program, which makes extension to some future needs (e.g., simultaneous fitting of composite data sets to multiple theoretical models) technically difficult and error-prone.

It was decided in late 1998 to re-engineer XSPECs data structure handling in ANSI C++. C++ supports Object Oriented programming, integration with legacy FORTRAN code, and support for generic algorithms and support for numeric computation (the `valarray` and related classes). We expect, with improved compiler implementations expected in the future, that these will become at least as efficient, if not more so, than FORTRAN implementations.

Modernization of XSPECs user interface started with XSPEC 10's adoption of the `tcl` scripting language as its command interpreter. This allows a natural interface to a GUI written with `tk` widgets (as used for the AIPS++ package). ANSI C++'s stream IO facilities allow a user interface which is largely decoupled from internal workings of the program, allowing GUI development side-by-side with the existing command line interface.

This article describes some of the techniques incorporated in the design for the re-engineered XSPEC 12, which is expected to be released in 2002.

2. Methods and Tools

We have adopted Solaris 7 as our development platform because of the availability of a near ANSI-compliant C++ development environment (Workshop 6.0/C++ 5.1), tools such as Rational Purify for efficient resource and access checking, and Rational Rose for design. Rose has the advantage of making class hierarchies and interclass interactions simple to develop and maintain where needed, which we have found to have a strong positive effect on code design. Disadvantages include expense and a degree of unreliability, at least in the current Unix implementations.

3. Code Features

3.1. Data Package

The XSPEC data library design makes use of three design patterns (Gamma et al. 1995) in order to allow a flexible upgradeable capability to read and process X-ray data. Design patterns are described as "elements of reusable object-oriented software" and represent general solutions to common design needs.

The needs XSPEC has in data ingress are:

- Execute a discrete set of steps in processing a data file: read, perform grouping and quality selection operations, select and read background and

response data, and compute statistical quantities (variance, systematic errors) for the analysis.
- Read a set of files with conforming data format.
- Allow for the possibility of adding new data formats without the necessity of modifying the existing code.

These goals are achieved by using respectively the Template, Abstract Factory, and Prototype patterns. Each format is represented by a set of concrete classes, one for each of the files within a given data format (e.g., the widely-used OGIP format has files of type PHA for source and background files, RSP for response files, and ARF for effective areas). These classes have the responsibility of performing the data processing steps. The Abstract Factory technique provides the functionality for minting a set of class instances (PHA, RSP, ARF) with conforming formats. Finally the Prototype pattern is used to store prototypes for formats that the program recognizes. Each file that the user requests to be read is checked against this list, and class instances of the correct type are created and processed. The key advantage of this pattern scheme is that the list of file formats that can be recognized is dynamic. Code can be loaded in shared library modules at run-time, adding to the list of recognized formats without modifying the existing code base. The Template Method specifies and orders the set of operations that must be implemented by the format-specific subclasses.

3.2. Theoretical Models Packages

The theoretical model processing packages comprise class hierarchies dealing with model components and parameters, and parsing classes.

An XSPEC theoretical model is composed of a number of 'additive' Component Groups that represent X-ray sources or combinations of sources whose radiative spectrum is modified by other physical processes (e.g., absorption, represented by a 'multiplicative' model). Component Groups are composed in turn of objects representing sources and absorption (etc.) processes. In the code, these are implemented by a set of classes derived from an abstract Component class. These are instantiated using a parameterized Factory Method, whereby the type of the model component, as specified in the input file, determines the class of the component object created.

The Component Class on construction selects a pointer to a function that generates the modeled flux array from an input set of energies: this function is typically written in FORTRAN 77. Thus, XSPEC12 will continue to support users' ability to extend XSPEC according to their own needs without their needing to learn C++.

The parser class implements the reading of XSPECs model expressions, and a subclass processes expressions linking parameters. The implementation extends XSPECs current capabilities in a natural way to allow much more complex nested model expressions and parameter linking capabilities (for example, a parameter may be set as the product of two existing parameters).

3.3. User Interface

The user interface in XSPEC12 is implemented by deriving from the C++ IOstream library. The class XSstream is derived from `std::iostream` and contains an XSstreambuf object derived from `std::streambuf`. These classes add a pointer to an I/O channel class, XSchannel, to the standard streams. This implementation allows the code to communicate with its user interface through the usual shift operators. The XSchannel pointer is an interface class that mandates the implementation of any I/O device. We are implementing `tcl` and `tk` output channels, so that depending only on a flag set by the user, I/O may be performed through a command line or a widget interface.

4. FITS I/O: The CCfits Library

XSPEC makes extensive use of FITS format I/O. Each individual file format that XSPEC recognizes may use files of five different types, all of which currently require separate code to be maintained, and also duplicated, if new formats are added.

As part of the project we have designed and (partially) implemented a class library which wraps calls to the widely-used `cfitsio` library developed in our Laboratory. This code is independent of XSPEC and can be built as a shared library. It is implemented with the Standard Library containers and algorithms (the "Standard Template Library" or STL) and is in principle portable to any platform that supports the Standard up to member template functions and template partial specializations.

In the CCfits model, a FITS object is created when a FITS file is opened, and data read from the file is loaded into the object which serves as a memory image of the file. Extensions (Images or Tables) are accessed either by index, extension name, or by matching a set of scalar header keys. User code reading FITS extension consists only of supplying a list of keyword and column names to be read.

As simple examples, take the code lines

```
CCfits::FITS dataSource("dSource.fits","HEADER1");
CCfits::FITS dataSource("dSource.fits","HEADER1",true);
```

The first of these will read the following from the file `dSource.fits`: the mandatory primary header keys; the mandatory header keys of the extension `HEADER1` and the column specification for that header. Subsequent calls can be used to obtain the data from the file. The second example will additionally read the primary image and the column data in that extension on initialization.

We expect to be able to release the CCfits library in 2001.

References

Arnaud, K. A. 1996 in ASP Conf. Ser., Vol. 101, Astronomical Data Analysis Software and Systems V, ed. G. H. Jacoby & J. Barnes (San Francisco: ASP), 17

Gamma, E., Helm, R., Johnson, R., & Vlissides, J. 1995 Design Patterns (Reading: Addison-Wesley)

& Astronomical Data Analysis Software and Systems X
ASP Conference Series, Vol. 238, 2001
F. R. Harnden Jr., F. A. Primini, and H. E. Payne, eds.

A Flexible Object Oriented Design for Page Formating

Nancy R. Adams-Wolk

Harvard-Smithsonian Center for Astrophysics, Cambridge MA, 02138

Abstract. The *Chandra* standard data processing now includes a group of summary pages that offer a synopsis of the observation. *Chandra's* instrument and grating combinations form many different spacecraft configurations. For each configuration, a specific summary of the observation is required. We need a flexible and expandable page formatter to handle this situation. One result of this development is the *sum_format_page* tool. This C++ tool is built on object oriented design principles and constrain the flexibility to produce multiple output file formats. Here we discuss the motivations for the tool, the design and implementation, and future enhancements that need to be considered.

1. Introduction

The *Chandra* X-ray Observatory provides two science instruments, a transmission grating, and multiple spacecraft configurations for observers to explore the X-ray universe. The cost of the convenience to the observers is the challenge of developing software that can process these data with minimal human intervention. *Chandra's* Standard Data Processing now creates a data product that summarizes the observation for the principal investigator. The software written for this task needs to be configurable, and fairly simple to update when there are changes in spacecraft operations or enhancement requests. One component of this software is the *sum_format_page* tool. This C++ tool is designed to work with the configurable nature of the summary package.

2. Design Considerations

Since this tool is a portion of a larger suite, we need to look at the specifications for the entire package. The specification process for an observation summary product can be subjective. Each scientist has their own view of what data are important and how these data should be presented. The specification process for the summary products involves polling several scientists for their opinion of what the summary of an observation should contain. The design is based on the items that most scientists want; images, details of the instrument and observing setup, sources, and a quick extracted spectrum if the observation included a grating. Finally, all tools we write need to conform to the *Chandra* X-ray Center Data System (CXCDS) standards.

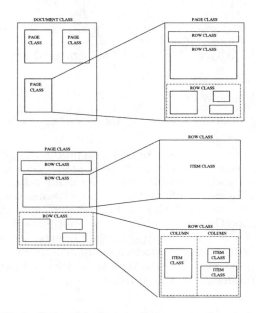

Figure 1. The relationship between the main document classes

2.1. Resulting Design

The result of our discussions with scientists set up the initial requirements. To control the arrangement of the items, we employ an ASCII layout template. The main document classes are designed around this template. Figure 1 displays the relationships between the major document classes. Each row of the layout template represents an item. Items are the input information from which a document is created. An item contains information about the specific input formats and can be written in all valid output formats.

The row class contains items. We allow multiple items in a row instead of implementing a column class. Since the user may want separate pages, rows are stored in pages and the pages are stored in the document. The main function of the program reads the template then populates the document and writes the document in the requested format.

3. Classes

Each specific section of the design is programmed as a class. Here we discuss the basic design of the major classes.

3.1. Template Reading

Two classes work to parse and store the ASCII layout template. The first class, *sumTemplateReader*, reads each line in the template, passing then to a new instance of the *sumTemplateLine*. This class parses the line and stores the item size, type, title, and file information within the class. Each *sumTemplateLine* is returned and stored in standard template Vector in the *sumTemplateReader* class.

The ASCII template format allows the user to customize the document arrangement. Each line describes the item type, size, position within the row, an optional title, and the file that contains the item. The *sumTemplateReader* is passed back to the main to be used in the document class.

3.2. Document Formation

The document is built from the *sumTemplateReader* class. The document class only contains methods to populate itself, check the physical sizes to ensure it will fit on a hardcopy, and write itself to a file. In populating the document, the layout template is read and each page in the document is filled by creating the rows and items to be used in the document.

3.3. Rows and Items

Each row of the document can contain multiple items which allows for columns, with the caveat that the row size cannot exceed the page size for hardcopy formats. The row is responsible for creating the individual item classes and stores the resulting objects in a vector.

Items are the specific inputs to the document. Some are predefined in the source code, such as a horizontal line. Other items are stored in files that are inserted into the document. A figure is an example of this type of item. Figures are expected to be in the correct format for the requested output. For the LATEX format, this is PostScript or Encapsulated PostScript. In the case of an HTML output, the format can be any graphic type allowed in HTML. The currently available input types are lines, new rows, new pages, figures, embedded files, links and table of contents (TOC). Embedded types are files that are in the format of the output file. These are copied verbatim from the source to the output file. The links and TOC items are only used in the HTML format to allow links to other documents and to create local links between the pages and items.

The item classes utilize the polymorphic nature of C++. The item classes are all derived from a singular base class. This class contains virtual functions for writing the outputs in the different formats. The goal of this design is to have the row iterate over its container, writing each item without having to know its type. This design has worked well so far, but recent design discussions have suggested another method of handling item classes as detailed below.

4. Enhancements and Upgrades

One of the design goals of the *sum_format_page* is to make enhancements and upgrades fairly simple. The input formats and file types may change in the future as well as the output formats. For these reasons, the design is flexible enough to simply handle changes to the input and output formats if there are new requirements.

4.1. Input File Type Changes

If a new input type or file is needed, the code can be modified to accept new formats. The changes are fairly well contained. A new item class will need to be defined that is derived from the sumPageItem base class. This class will need all of the virtual functions that are defined in the base class. Once the class is written, the *sumTemplateLine* will need a new identifier in the enumeration of the item types, the ASCII template will need a new identifier to specify the type, and the row class will need to be updated with the new item class to be filled.

At the ADASS X meeting, Ben Dorman (private communication) suggested using a registry to store the input types. Any future upgrades would involve a change in the registry and overloading of functions instead of creating a new class. This is incorporated in the redesign of XSPEC (Dorman et al. 2001) and merits future exploration.

4.2. Output Format Updates

The current output formats may become obsolete in time and a new set of output formats will be needed. In this case, the changes involve adding new functions to write the output format to each of the item classes and the base item class. The parameter file will need to be updated to accept the new format as a valid parameter.

5. Conclusions

The summary package is a powerful set of tools used in the standard data processing to create the summary data product distributed to *Chandra* observers. It could not have been possible without the flexibility of the *sum_format_page* tool. While discussions have shown that we can make the tool even easier to upgrade with future input types, the current design is robust and works well with the multiple templates used in standard data processing.

We expect this tool, and the other tools in the summary package suite, to be used for the remainder of the mission with only minor updates. New configurations of the telescope can be handled with changes in building the ascii layout templates, while new data formats can be added by updating the *sum_format_page* tool.

6. Acknowledgments

The author would like to thank Douglas McElroy and Kenny Glotfelty for assistance and discussions in the design phase of this project. This work was funded by the *Chandra X-Ray Center* NASA contract NAS8-39073.

References

Dorman, B. & Arnaud, K. 2001, this volume, 415

The Stellar Spectra Acquisition, Reduction, and Archiving Systems at the Ondřejov Observatory 2-meter Telescope

Petr Škoda, Jaroslav Honsa, Miroslav Šlechta

Astronomical Institute of the Academy of Sciences of the Czech Republic, 251 65 Ondřejov, Czech Republic

Abstract. The 2-meter telescope of the Ondřejov Observatory near Prague is a middle-class instrument fully devoted to high dispersion stellar spectroscopy. The data are produced by two acquisition systems of different generations. The older linear detector Reticon 1872 AF is controlled by a DOS program providing quick-look display capability as well. The second is a CCD driven by a stand-alone Linux program. The quick-look is provided by SAOimage or XImtool through data pipes. An easy access to both the raw and reduced data is provided by a simple WWW-based archiving system.

We give a basic overview of these systems and software methods used for their communication with the spectrograph control system.

1. Scientific Instruments

The 2-meter telescope built by Carl Zeiss, Jena, was put into operation in 1967 and since the beginning it was devoted to high dispersion spectroscopy of early-type (mainly Be) stars in its 64-m coudé focus. The plate archive of the 2-m telescope contains almost 6000 spectral plates exposed until the beginning of year 1993, when the Reticon AF1872 detector was commissioned. Using it, more than 4000 stellar spectra were secured, giving with calibrations a total of 16600 frames until the end of the year 2000. The Reticon was then replaced by a new CCD camera with ISA 2000 × 800 chip.

2. The Spectrograph Control System Overview

All detectors are installed in the coudé spectrograph which can operate either with the 700 mm or the larger 1400 mm camera. In the summer of 1998 the old-fashioned manual control was replaced by a new microcontroller-based system which takes care of a number of controlled devices:
Flat field lamp - to switch on the projector flat.
Flat field mirror - to tilt the mirror into the lightpath.
Comparison arc lamp - to switch on the ThAr lamp.
Comparison arc mirror - to tilt the ThAr lamp mirror into the lightpath.
Grating - to set desired angle by AC motor, measuring by incremental angle encoder, two limit sensors.
Slit - to set desired slit height by stepping motor, two limit sensors, the position

must be reset by driving to the limit sensor.
Collimator mask - to set desired mask (open, close, open-left, open-right) by rotating mask wheel by AC motor, optical encoders are used to recognize masks.
Dichroic mirrors - to set desired dichroic mirror by rotating wheel with mirrors by stepping motor, optical encoders are used to recognize actual position.
Spectral filters - to set desired filter by rotating filter wheel by stepping motor, optical encoders are used to recognize actual position.
Focusing the 700 mm and 1400 mm camera - to set desired position by stepping motor, two limit sensors.
Correction plate for 700 or 1400 mm camera - read-only signal to distinguish the plate position (in beam, out of beam, intermediate).
Exposure-meter - to start/stop counting pulses from photo-multiplier, to measure frequency (per second) and sum of pulses during exposure, to zero this sum.

The spectrograph control system is based on these ideas:
- Each controlled device in the spectrograph has its own driver unit based on a single-board microcontroller. All units are connected to a common serial line bus.
- The main unit periodically reads the status of controlled devices and sends commands to driver units over this line. This unit is connected to the serial port of the control computer.
- The control computer runs a daemon which is used like a server for clients running on an arbitrary computer in the local net.

2.1. Server for Communication with Spectrograph Driver Units

The server runs as a daemon on the control computer and is started at system startup. After startup the server forks into two processes:
- Process-1 listens on one port for a UDP connection from a client and replies with status data from controlled spectrograph devices on the same port. It also listens on second port, reads a command from it, converts to a special format, and sends to a serial port of the local host (to main control unit).
- Process-2 periodically reads the serial port of the local host and saves the data to memory shared with Process-1.

This configuration has several advantages. The server itself knows at every moment the status of controlled devices and can send the answer immediately to a client; the spectrograph may be controlled from any computer which is connected to Ethernet and is allowed to do so; the client can be linked to various programs and so it can extend their functions; and several clients can run at the same time, some of them only monitoring the status of the spectrograph. The only disadvantage is that it does not work for DOS machines.

2.2. Client-1 - Monitor and Control Program of Spectrograph

This program shows the status of all controlled spectrograph devices and allows the user to control them. The program is written in Tcl/Tk. The client part for communication with the server is written in C as a set of functions compiled as shared libraries by SWIG. This organization radically simplifies the design of new monitor and control programs since the routine work of making a TCP/IP connection to a computer, reading data from it, sending data to it, or closing the connection is reduced to only four Tcl-like functions. Using these functions is enabled by a `load my_library.so` command at the beginning of a Tcl program.

2.3. Client Part of a Program Controlling CCD

The client for communication with the server is included in a large program which controls the CCD. A function which establishes a UDP connection to the server is called at the start of the program and at the end, after quitting other functions, and closes the connection. The function for getting data from the server is periodically called in a loop during the execution of the program. It displays the status of the CCD. The function for sending commands to the spectrograph is called whenever the CCD needs a change of status of a spectrograph device. The program then waits for the finish of the change—it is sometimes unpleasantly time demanding.

The function for reading data from the server must be designed to anticipate possible failure of the UDP connection. The wait for a reply after sending a request for data is limited by a timeout. The timeout is made by a system call of the `alarm()` function.

2.4. Spectrograph DOS Driver

The reticon 1872 AF detector is controlled by a home-made ISA card (containing PIO 8255 and timer 8253) and driven by the program `creticon`, which was written in Turbo Pascal 5.5 and contains time-critical code for the synchronous Reticon readout. This was achieved by disabling interrupts, including waiting loops and direct reading I/O port with embedded assembler commands. Most critical and time-consuming was the tuning of the loops to fit the particular computer's (*alkaid*) speed and compiler options.

As this program cannot easily be replaced due to its complexity (it is also taking care of the target star catalogues, maintaining the observing logs, providing the simple debiasing and flat-fielding, as well as the quick-look display and many more services). We decided to change only several small routines that make the very calls to hardware. As the new spectrograph controlling computer *mizar* is a stand-alone Linux machine connected to the ethernet network, we needed an easy way of reading the status of devices and sending them commands over the network without the need of fundamental changes in the code. The solution we have developed is quite tricky:

- We have installed the TCP/IP stack called PCTCP by FTP Inc. on the DOS machine *alkaid*.
- At the top of this stack was installed the NFS client called Interdrive.
- The small partition /home/coude on the Linux computer *mizar* is NFS exported using the `pcnfsd` and mounted as the logical drive `Q:` on *alkaid*.
- When the `creticon` program is run the special daemon `couded` is started on *mizar* using a `rsh` command. The daemon periodically reads the status of the devices from the serial bus and writes the results in a short ASCII file /home/coude/data1 each second.
- The daemon waits 200 msec to give the client enough time to read the data by simple Pascal `readln` from `Q:data1` into an memory array. After that the file is deleted and new one written with updated status.

- Each status file also includes unique counter which is incremented after new data from the serial bus units is read. So the client only checks this number, and when it is the same, it does nothing.
- At the same time the presence of the file command is checked by the daemon expc on the *mizar*. When the DOS program sends a command to the particular device it will create the file Q:command with simple commands like FF=1 exit.
- If the file /home/coude/command exists, it is opened by expc and the commands for the driving units sent through the serial line. After finishing all hardware changes the file command is deleted by the daemon.
- Before the opening of files on the client side the existence of them should by checked by the IOResult routine in loop.

The quite interesting result of this setup is the speed of status updates though NFS—more than 250 readouts per second. The unique counter is very good for timeout control in case the communication is broken.

3. Data Reduction and Analysis Software

At present the most important data for the current research at Ondřejov are the Reticon electronic spectra. They are small (about 4 kB) and quite simple to handle. For the basic reduction and analysis of these data a proprietary program SPEFO[1] written by J. Horn has been used for years in the stellar department.

The new spectra from the CCD detector (about 3 MB per image) are processed by standard packages in IRAF. All the necessary FITS headers are provided by the data acquisition software.

4. Data archiving

In order to be able quickly find the detailed information about the exposure of a particular spectrum (plate or Reticon file) a SQL-based archive of observing logs has been under development in the Stellar department. Its preliminary version is based on the free RDBMS PostgreSQL and a Web interface, WDBI, is accessible on the 2-m telescope home page[2].

The raw data will not be available on-line (due to its specific nature as a result of long-term monitoring observation) but those interested may contact the staff of the stellar department by e-mail to stelveda@sunstel.asu.cas.cz.

Acknowledgments. This work was partially supported by the Ministry of Education, Youth and Sports of the Czech Republic (grant LB98251).

---

[1] http://iraf.noao.edu/ADASS/adass_proc/adass_95/skodap/skodap.html

[2] http://stelweb.asu.cas.cz/

HST NICMOS Residual Bias Removal Techniques

Howard Bushouse

Space Telescope Science Institute, 3700 San Martin Drive, Baltimore, MD 21218

Abstract. *HST* NICMOS detectors suffer from temporal bias drifts, which lead to incomplete bias subtraction during image calibration. The residual bias signal is spatially constant, so that application of flatfield images during calibration gives rise to an imprint of the flatfield structure. In images of sparse fields the flatfield imprint is easily identifiable and removable through the use of iterative techniques to minimize the imprint. Fields with complicated source structure, however, require more sophisticated approaches, utilizing spatial filtering techniques to isolate and separate source and residual bias signals.

1. Introduction

The Hubble Space Telescope (*HST*) Near Infrared Camera and Multi-Object Spectrometer (NICMOS) uses detectors that are read out non-destructively. An image is formed by subtracting readouts at the beginning of an exposure from those at the end. It is often the case that many readouts are taken throughout the duration of an individual exposure. The initial readout serves as a record of the initial state of the detector pixels and is analogous to a bias image. However, systematic bias signals can only be completely removed from subsequent readouts if the bias is stable during the exposure.

The NICMOS detectors suffer from temporal bias drifts, which are thought to be due to temperature instablilites. Subtraction of the initial or bias readout from all subsequent readouts of an exposure can often leave a residual bias signal. Furthermore, each quadrant of the NICMOS detectors is readout by a separate amplifier, so that the amount of residual bias can vary from quadrant to quadrant within an image. Within a given quadrant the residual bias is usually spatially constant, such that subsequent division by a flatfield image leaves an imprint of the flatfield structure. The existence of this imprint sets an upper-limit to the achievable sensitivity and signal-to-noise ratio. In addition, the detector bias can vary non-linearly during the multiple readouts of an exposure, which causes source signals to appear to accumulate non-linearly.

Several software tools have been developed, which use a variety of techniques to correct NICMOS data for these problems. The tasks run within the IRAF/STSDAS environment, are written in ANSI C, and use the IRAF CVOS interface to perform parameter and data I/O. These tasks are all available in the STSDAS `nicmos` package.

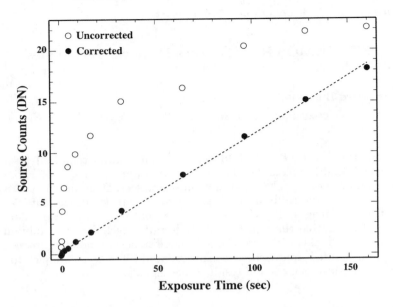

Figure 1. Application of biaseq.

2. The BIASEQ Task

The biaseq task is designed to correct for non-linear drifts in the bias level from readout to readout within an exposure. The basic approach is:
- Form a model or average image of the scene from a user-selectable subset of the exposure's readouts. Typically the later readouts are used, which usually have the longest exposure times and are the least affected by bias drifts. The image combination process uses a simple mean with min/max rejection to remove cosmic-ray hits and otherwise bad pixels.
- The averaged image is subtracted from each readout. Changes in bias between readouts are left as residual signal in the subtracted images.
- The median residual signal in each quadrant of each readout is subtracted from the readout, thus forcing the median signal to accumulate linearly with exposure time.

The subtraction of the average image before measuring the residual bias signal in each readout effectively removes signal from real sources and therefore makes this task insensitive to image source content. Figure 1 shows an example of the uncorrected (non-linear) and corrected accumulating counts for a source in an image. The dashed line is a linear fit to the corrected data.

While biaseq is able to force signals to accumulate linearly, an uncertainty in the overall slope of the signal vs. time relation remains present in the corrected data. This uncertainty is due to net *linear* drifts in bias between readouts that still need to be removed. This remaining bias signal manifests itself as a constant zeropoint offset in each quadrant of the corrected image. The tasks described next must be used to apply this final correction.

3. The PEDSKY Task

The pedsky task is designed to measure and remove any remaining net bias residual (often referred to as "pedestal") from the final image that is formed after combining the corrected readouts of an exposure. The task derives its name from the fact that it measures and removes both pedestal and sky background signals in an image. The sky signal is assumed to be constant across the entire image, while the pedestal level is measured and removed individually for each image quadrant.

The basic assumption in the pedsky algorithm is that signal that is near the sky background level is composed of the true sky, which is modulated by the spatially-varying quantum efficiency of the detector, plus the residual bias or pedestal signal, which is assumed to be constant in each quadrant. In mathematical terms, the task assumes for each pixel that contains only sky and pedestal

$$I_{xy} = sky \times Q_{xy} + bias, \qquad (1)$$

where I_{xy} is the total signal in the pixel at coordinates (x, y), sky is the sky background signal, Q_{xy} is the relative quantum effeciency (flatfield) value of pixel (x, y), and $bias$ is the bias or pedestal signal for the quadrant.

The pedsky task can solve this relation for sky and bias by one of several user-selectable methods. First, it can perform a direct least-squares solution to Equation 1 using the values for all pixels that are near the sky level. Second, it can solve Equation 1 by iteratively subtracting trial sky and bias values from the image, seeking the optimal combination that produces an image with the minimum remaining rms deviation in pixel values. Both of these methods essentially seek to minimize the expression

$$\sigma^2 = \sum_{xy} (I_{xy} - sky \times Q_{xy} - bias)^2, \qquad (2)$$

where σ is the standard deviation of corrected pixel values. A third method allows the user to specify a known sky value, in which case the task subtracts that sky value and then only measures and removes the remaining bias signal from each quadrant. Figure 2 shows an example of the application of pedsky to an image. Note that the flatfield imprint has been removed, as well as the DC offsets between quadrants.

Because pedsky relies on the use of image pixels that only contain sky and bias signal, it can only work effectively on images that are sparsely populated with real sources. The pedsub task, described next, is designed to work with images of any source content.

4. The PEDSUB Task

The pedsub task is similar to pedsky, but it does not rely on being able to measure sky values in order to determine the residual bias. Like pedsky, it relies on the fact that the application of the flatfield to an uncorrected image leaves an imprint, which artificially increases the rms pixel spread. The task solves for the pedestal value in each quadrant by seeking to minimize the spread

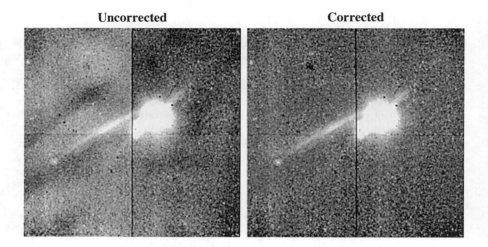

Figure 2.   Application of pedsky.

in pedestal-subtracted pixel values. It assumes that the flatfielded signal in each pixel can be expressed as

$$C_{xy} = I_{xy} + bias \times Q_{xy}, \qquad (3)$$

where C_{xy} is the calibrated value of pixel (x, y), I_{xy} is the intrinsic signal from sky and sources, $bias$ is the residual bias signal in a given quadrant, and Q_{xy} is again the relative quantum efficiency (flatfield) value.

Pedsub loops over a range of trial pedestal values for each quadrant, generating a trial image that has the $bias \times flatfield$ values subtracted from each pixel, and then measuring the remaining spread in pixel values in the trial image. It iterates this process until the pixel spread is minimized.

Spatial filters can optionally be applied to each trial image in order to remove unwanted features or spatial frequencies that might artificially bias the pixel spread. A choice of filter types is available. A ring-median filter can be used to remove small sources, which when applied essentially confines the pixel spread computation to low Fourier frequencies. Alternatively, an unsharp mask filter option (median filtered image subtracted from the unfiltered image) can be used to remove low frequency information, such as signal from extended sources. This option confines the pixel spread computation to only high-frequency image information.

The availability of these options makes the pedsub task useful for images with almost any type of source content, while pedsky can often give incorrect results when applied to images with large amounts of source signal.

Acknowledgments. Many of the algorithms in these tasks were developed from concepts originally devised by Mark Dickinson and Roeland Van der Marel (STScI), and Brian McLeod (Harvard CfA).

Astronomical Data Analysis Software and Systems X
ASP Conference Series, Vol. 238, 2001
F. R. Harnden Jr., F. A. Primini, and H. E. Payne, eds.

Deconvolution of "Drizzled" and Rotated HST/WFPC2 Images: Faint-object Photometry in Crowded Fields

Raymond Butler, Aaron Golden, Andrew Shearer

Department of Information Technology, National University of Ireland Galway, University Road, Galway, Ireland

Abstract. We have further adapted our deconvolution-based HST/WFPC2 reduction techniques, which our simulations have shown to be very effective in crowded fields (Butler 2000), to take advantage of the extra resolution afforded by "dithered" observations. This appears to be the first attempt at *non-linear* restorations of data which had already been sub-sampled and combined in a *linear* manner using the *STSDAS/drizzle* software, thus attaining a net subsampling factor of 4×4. The motivation was a search for the optical counterpart of the millisecond pulsar PSR B1821-24 in the globular cluster M28; the technique is illustrated by our photometric search for candidates in the radio-derived error circle, in archival HST/WFPC2 images in both the F555W & F814W bands of the core field of M28.

1. Introduction: HST/WFPC2 Photometry Challenges

Although difficult, ground-based photometry of faint objects in crowded fields does not really compare with crowded-field photometry with a space-based telescope/camera, e.g., HST/WFPC2, which features an undersampled and spatially varying point-spread function (PSF), with high-frequency structure in the PSF "wings". These factors, plus the sheer number of stars which may require measurement as a single group, cause a variety of practical problems for star detection and photometry, which were detailed in Butler (2000).

2. An Optical Counterpart to PSR B1821-24 in M28?

In this paper we examine solutions to the above issues by taking an example, the post-core-collapse globular cluster M28/NGC 6626. We obtained two sets of archival HST/WFPC2 observations of M28 as follows: (1) 424 sec in F555W & 423 sec in F814W taken on 8/08/97 (P. I. Gebhardt); (2) 1128 sec in F555W & 1508 sec in F814W taken on 12/09/97 (P. I. Buonanno). Both datasets were taken in dithered mode, with steps of 3.666 pixels (epoch 1) or 2.745 pixels (epoch 2) in X and Y on the PC1 chip. There is a relative rotation of $12°.875$ between the two datasets. Finally, the cluster core is centred on the PC1 chip. There is an object of great interest to us in this core field. PSR B1821-24 is a millisecond radio pulsar (period $P \sim 3$ ms; see e.g., Cognard et al. 1996) in M28. The fact that the magnetic field & spin coupling is of a similar magnitude to that of the Crab pulsar in the vicinity of the light cylinder has suggested that the millisecond pulsar may well be an efficient nonthermal emitter. The confirmation by the ASCA satellite of a strong synchrotron dominated hard X-ray pulse fraction (Saito et al. 1997) encourages such a viewpoint. Using phenomenological models of pulsar magnetospheric emission (Pacini & Salvati 1987; Shearer & Golden 2000), the predicted optical luminosity is estimated at $V_0 \sim 21.5$–23.5, which would be reduced to an observable $V \sim 23$–25 by the interstellar extinction towards M28 of $A_V \sim 1.4$ mag (Davidge et al. 1996). This would yield the first optical millisecond pulsar, indeed the first optical pulsar in a globular cluster or any such old stellar population. But PSR B1821-24 lies only $\sim 12''$ from the center of M28, so a targeted high-resolution search of the entire radio error circle to $V \sim 25$ is a challenge—even with HST.

3. Image Processing & Photometry

There are many advantages to "drizzling" HST/WFPC2 images (Fruchter & Hook 1997). It genuinely restores some of the resolution lost to undersampling; it resamples onto a regular astrometric grid, with the option of subsampling the images by 2×2; and it is a *linear* reconstruction method, so the resulting stellar profile shapes are not dependent on signal/noise and the noise statistics remain "physical." This makes it an excellent starting point for our *non-linear* subsampled MEM (Maximum Entropy Method) deconvolution approach to star detection and crowded-field photometry on HST/WFPC2 images. The latter also has many advantages. Star detection is improved because the actual HST PSF shape, in all its complexity, is used (via deconvolution) to "Gaussianise" the PSF while also better separating the stars from each other. Photometry is also slightly improved; our simulations (Butler 2000, Butler & Shearer 2001) have shown that aperture photometry on the subsampled MEM-deconvolved images is superior to all of the following conventional reductions of the original data: aperture photometry, profile-fitting photometry, and the hybrid method of aperture photometry on neighbour-subtracted images (e.g., Yanny et al. 1994).

We combined both these processes as follows. The images were "drizzled" and cleaned in the normal way, subsampling by a factor of 2. A series of "dithered" synthetic PSF grids were also "drizzled:" the "stars" were a uniformly distributed 6×6 grid of normal-sampled Tiny Tim (Krist & Hook 1996) synthetic PSFs for each WFPC2 chip & filter combination, computed at high spatial subsampling, shifted to reproduce the dither offset, rebinned to normal sampling and convolved with the pixel scattering kernel. Instances of the PSF were obtained for deconvolution, at any desired position and with a further 2× subsampling factor, after combining them using a quadratically variable DAOPHOT-II (Stetson 1994) model. Highly overlapping subimages of the field were deconvolved with these PSFs; the deconvolved subimages were reassembled into a whole sharpened, subsampled image for each filter and epoch. The four deconvolved images (F555W and F814W each at two epochs) were combined with a moderate rejection threshold: this eliminated nearly all artifacts, because (a) the radial structure of the PSF changed (due to the waveband dependence of the PSF shape), and (b) the position-angle of the PSF structure on the sky changed (due to the rotation changes). We used this deep, clean, sharp

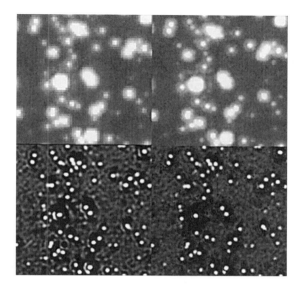

Figure 1. Section of HST/WFPC2 image of the globular cluster M28, $2\rlap{.}''91$ on a side (corresponding to 64 PC1-chip pixels). Clockwise from upper left: Original (F814W epoch 2), Drizzled (F814W epoch 2), Deconvolved (F814W epoch 2), and Coadded after Deconvolution (F814W & F555W, both epochs: used for star detection).

coadded image for star detection. We then performed PSF-fitting photometry on the original "drizzled" images with this starlist and the existing PSF models. All but ≈ 120 selected bright "PSF stars" were subtracted from each image and a spatially-varying empirical PSF model was computed. The deconvolution & photometry steps were repeated with the refined PSFs. The final photometry was aperture photometry on these improved deconvolved images—both fast and accurate.

4. Results & Conclusions

Examination of the radio-derived error circle yielded several potential candidates, down to a magnitude of $V_0 \sim 23.0$; but both in the context of the CMD of M28, and with regard to phenomenological models of pulsar magnetospheric emission, none of them exhibited emission expected from a magnetospherically active pulsar (Golden, Butler, & Shearer 2000). The *key point*, however, is that the starlist in the field of PSR B1821-24, obtained via our drizzling plus deconvolution technique, is more reliable than that obtained by Sutaria (2000) with a subset of this same data (i.e., F555W epoch 2), which appears to be contaminated with several faint spurious detections. We therefore believe that deconvolution of "drizzled" and rotated images (≥ 2 spacecraft roll angles) is the optimal way to detect and measure faint objects in crowded fields imaged with HST, and we recommend such an observing strategy.

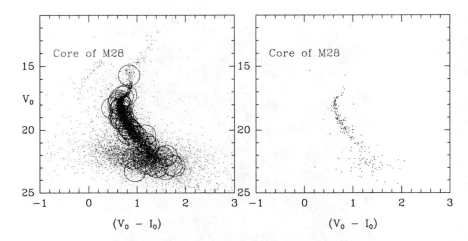

Figure 2. Left: The M28 CMD - de-extincted and re-reddened - for all stars on the PC1 chip of HST/WFPC2. The stars within the radio-derived error circle of PSR B1821-24 are shown by large circles. Right: The M28 CMD for the 4″.55×4″.55 region centered on PSR B1821-24.

Acknowledgments. We gratefully acknowledge financial support from Enterprise Ireland (Basic Research Programme) and the European Commission (TMR Fellowship ERBFMBICT972185 funded much of this work, performed by RB at the University of Edinburgh, UK [TMR host: Prof. Douglas Heggie]). This work was based upon HST data obtained from the ST-ECF (ESO) archive.

References

Butler, R. F. 2000, in ASP Conf. Ser., Vol. 216, Astronomical Data Analysis Software and Systems IX, ed. N. Manset, C. Veillet, & D. Crabtree (San Francisco: ASP), 595

Butler, R. F. & Shearer, A. 2001, in preparation for PASP

Cognard, I., et al. 1996, A&A, 311, 179

Davidge, T. J. 1996, ApJ, 468, 641

Fruchter, A. & Hook, R. N. 1997, Proc. SPIE, 3164, 120

Golden, A., Butler, R. F., & Shearer, A. 2000, submitted to A&A

Krist, J. & Hook, R. 1996, Tiny Tim User's Manual, V4.2 (Baltimore: STScI)

Pacini, F. & Salvati, M. 1987, ApJ, 321, 447

Saito, Y., et al. 1996, ApJ, 477, L37

Shearer, A. & Golden, A. 2000, accepted for publication in ApJ

Stetson, P. B. 1994, PASP, 106, 250

Sutaria, F. K. 2000, in Pulsar Astronomy—2000 and Beyond, Proc. IAU Coll. 177, ed. M. Kramer, N. Wex, N. Wielebinski (San Francisco: ASP), 313

Yanny, B., Guhathakurta, P., Schneider, D., & Bahcall, J. 1994, ApJ, 435, L59

The Chandra X-ray Observatory PSF Library

M. Karovska, S. J. Beikman, M. S. Elvis, J. M. Flanagan, T. Gaetz,
K. J. Glotfelty, D. Jerius, J. C. McDowell, A. H. Rots

Harvard-Smithsonian Center for Astrophysics

Abstract. Pre-flight and on-orbit calibration of the Chandra X-Ray Observatory provided a unique base for developing detailed models of the optics and detectors. Using these models we have produced a set of simulations of the Chandra point spread function (PSF) which is available to the users via PSF library files. We describe here how the PSF models are generated and the design and content of the Chandra PSF library files.

1. Introduction

The Chandra X-Ray Observatory (Chandra) produces sharper images then any other X-ray telescope to date (less then 1″ on-axis), and therefore provides a unique opportunity for high-angular resolution studies of cosmic X-ray sources. Crucial to these studies is the understanding of the characteristics of the Chandra Point Spread Function (PSF).

The unprecedented Chandra resolution is mainly due to the innovative design of this observatory, including the guidance systems, the mirror assembly (High Resolution Mirror Assembly, or HRMA), and the science instruments. HRMA consists of four pairs of nested mirrors (a Wolter Type I design), support structures, and additional thermal and optical baffles system. Four science instruments are located in the telescope focal plane (HRC-I, HRC-S, ACIS-I and ACIS-S). The instruments' resolution is well matched to capture the sharp images formed by the mirrors and to provide information about the incoming X-rays: their number, position, energy, and time of arrival.

A set of simulated PSFs is available to the users for data analysis via the standard PSF library files. In addition to these standard PSF libraries, the user may construct their own library files as long as the FITS HDUs and the PSF images conform to the general Chandra PSF library format. The user can extract the desired PSF model image from PSF library files by using the Chandra Interactive Analysis of Observations package (CIAO) tool **mkpsf**.

2. Chandra PSFs

The HRMA PSFs vary significantly with source location in the telescope field of view (FOV), as well as with the spectral energy distribution of the source. Because of the Wolter Type I design, the image quality is best in a small area centered about the optical axis. The mirrors were designed to concentrate better than 85% of the energy at 0.277 keV within a 1″ diameter. This is why a substantial pre- and post-launch effort was directed at creating a faithful model of the HRMA's mechanical and optical systems. The detailed modeling and metrology of the optics followed by extensive testing at the X-Ray Calibration Facility at the Marshall Space Flight Center in Huntsville and the on-orbit calibration of the point spread function showed that we are close to reaching the goal of calibrating the optics' performance to 1% (Jerius et al. 2000, Proc. SPIE 4012).

The simulated Chandra PSFs used in the standard PSF library files are generated in two steps:

(1) ray files are generated using SAOsac, a ray-trace code which models the interaction of photons (rays) passing through the HRMA (Jerius et al. 2000). The initial number of rays for these simulations was approximately 10^5.

(2) PSF model images are made by projecting these rays to the detector surface and then creating images with pixel sizes smaller than the pixel sizes of the detectors.

We produced a large set of PSFs at many off-axis angles covering the field of view of the detectors and at several energies ranging from 0.277 keV to 8.6 keV. Figure 1 shows examples of the HRC-I PSFs at several off-axis angles.

3. Chandra PSF Libraries

3.1. General PSF Library Definition and Format

The Chandra PSF library consists of two dimensional PSF model image "postage stamps" stored in multi-dimensional FITS images (hypercubes). In the following we summarize the PSF library general format:

HDU type 1 – the PSF image: These are n-dimensional images, hypercubes primary array) that extend along a minimum of five coordinates. The known coordinate axes are:

l - spatial x-direction of the PSF image (PSFX)

m - spatial y-direction of the PSF image (PSFY)

X - spatial x-direction offset coordinate (DETX)

Y - spatial y-direction offset coordinate (DETY)

E - energy (ENERG)

f - defocus (DEFOCUS)

Every image is required to have the following axes: (l, m, E, X, Y, f). Each coordinate may be regularly sampled, in which case the sample points are defined by the usual CTYPEi, etc., keywords; or irregularly, in which case the sample points are defined in a table extension (in the same file).

Each coordinate has to have one or more pixels, but one is expressly allowed. If there is only one point along any of the required axes, the axis still needs to be present and its coordinate value is defined in the usual way (CTYPEi, etc.). The

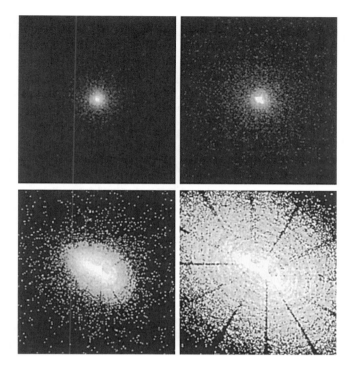

Figure 1. Model PSFs for the HRC-I instrument at 1.4967 keV as a function of off-axis angles (log display); clockwise from the top, off-axis angles 0 (on axis), 1.'5, 6', and 12'. The size of the FOV is about 0.'5.

coordinate axes (most notably the spatial ones) have several aliases defined in the header. The headers of these images contain the required Caldb keywords. These images have SUMRCTS=1.0.

HDU type 2 – the irregularly sampled coordinate definition tables: These are binary tables which allow an unambiguous translation of "bins" or "pixels" to physical coordinates (e.g., energy, defocus).

HDU type 3 – SUMRCTS image: The SUMRCTS images contain the information on how many photons there are per individual PSF data in the PSF hypercubes. These images match the PSF hypercubes exactly, except that the l and m axes are missing. The image pixels indicate the number of counts used for each PSF image. The SUMRCTS images are kept in IMAGE extensions.

4. Standard PSF Library Structure and Content

There are four standard PSF libraries provided per instrument. The standard library files are comprehensive, covering the entire instrument FOV, or a section of the FOV in a regular grid. The HDU for each of those files contains the self-contained PSF data, as a single hypercube (6-D; NAXIS=6) with image extension containing information on the number of photons (weights) needed for normalization of the PSF models, and one or more binary tables containing irregular coordinate definitions.

The PSF models incorporated in the PSF library hypercubes are single size arrays (images) on a fixed, regular grid. They do not contain spreading due to aspect. The PSF model images are made using the nominal aim-point and the nominal SIM-Z position for each detector. Currently, the standard libraries contain PSF models calculated for one defocus position (defocus=0) and 5 energies (0.277 keV, 1.4967 keV, 4.51 keV, 6.4 keV, and 8.6 keV).

When designing the standard library we had to make a compromise between the need for fine spatial and spectral resolution in the PSF grids, and the need for a reasonable file size of each individual library hypercube delivered to the user. In fact, the large size of the individual PSF "postage stamp" image files was a limiting factor on how many can be incorporated in the standard PSF library files, and still provide a useful off-axis angle and energy sampling of the PSFs, and at the same time cover a reasonable FOV. In the following we summarize the current structure of the library files:

1. High resolution library (I) with a step of 1 arcminute between images: contains 1 μm pixel images covering a -1 to $+1$ arcminutes grid (3×3) about the optical axis. [NB: ACIS pixel size is 24 μm; HRC pixel size is 6.4294 μm). The actual resolution ("effective pixel size") for HRC can be obtained by convolving with a Gaussian with σ of 1.5 HRC pixels.

2. High resolution library (II) with a step of 1 arcminute between images: contains 2 μm pixel images covering a -6 to $+6$ arcminutes grid (11×11) in azimuth and elevation (in a telescope fixed system) about the optical axis.

3. Medium resolution library with a step of 1 arcminute between images: contains 6 μm pixel images covering a -10 to $+10$ arcminutes square grid for ACIS-I and HRC-I (21×21), and a -10 to $+10$ arcminutes in elevation and -5 to $+5$ arcminutes in azimuth grid (21×11) for ACIS-S and HRC-S about the optical axis.

4. Low resolution library with a step of 5 arcminutes between images: contains 12 μm pixel images covering a -25 to $+10$ arcminutes in elevation and -10 to $+10$ arcminutes in azimuth (8×5) grid for ACIS-I, a -25 to $+25$ arcminutes in elevation and -5 to $+5$ arcminutes in azimuth (11×3) grid for ACIS-S, a -25 to $+25$ arcminutes in elevation and -25 to $+25$ arcminutes in azimuth (11×11) grid for HRC-I, and a -30 to $+30$ arcminutes in elevation and -5 to $+5$ arcminutes in azimuth (13×3) grid for HRC-S about the optical axis.

The PSF library grids are very coarse (azimuth and elevation angular offsets of 1' or 5', only 5 energies, and only one defocus position). Therefore, the user needs to interpolate in these grids to get a PSF for the off-axis angle and the energies (spectrum) of the observed source. The standard libraries can be used to view the general distortions and structure of the HRMA PSFs, as well as the PSF variations as a function of off-axis angle and energy. Since the variation in the PSF can be significant even for small off-axis angles, the current PSF library should be used with caution when performing a detailed spatial/spectral analysis or for deconvolution.

Acknowledgments. This work was supported by NASA contracts NAS8-39073.

On the Fly Bad Pixel Detection for the Chandra X-ray Observatory's Aspect Camera

Mark Cresitello-Dittmar, Thomas L. Aldcroft, David Morris

Harvard-Smithsonian Center for Astrophysics

Abstract. The Chandra X-ray Observatory uses an optical CCD in its aspect camera. As with all space-based CCD detectors, radiation damage will accrue with time and substantially increase the dark current of individual pixels, resulting in "warm pixels." In order to obtain the most accurate aspect solution possible, it is necessary to identify and compensate for these regions when processing the guide star images. If a warm pixel is included in a guide star image, it will bias the centroid location for that image. As the spacecraft dithers, this bias will introduce a wobble to the star location that translates to a wobble in the aspect solution. Special dark current calibration observations can be taken to provide a full-frame dark current map, however, it is not operationally feasible to obtain a new map for each observation.

The CXC data systems group has developed software to analyze the star image data and identify warm pixels as part of standard processing. This "on the fly" determination allows us to adjust for variations in CCD conditions between dark current calibration observations and provides useful information for identifying bad regions on the Aspect camera CCD.

1. Motivation

In order to achieve the unprecedented accuracy of Chandra's aspect solution, it is necessary to get the most accurate star centroid locations possible. As the aspect camera CCD is exposed to radiation, warm pixels will develop. These warm pixels can affect the centroid locations of the stars by creating a bias in that direction. If the pixel is very warm, this bias can be quite pronounced and create relatively large centroid errors. It is possible to correct for these warm regions during processing by applying a background subtraction using a dark current map. These dark current maps show the expected number of counts to be registered by each CCD pixel when no source photons fall on it. They are generated through special dark current calibration observations. Since it is not feasible to conduct a dark current calibration for each observation, a mechanism is needed to identify and correct for new warm pixels as they evolve. This can be accomplished by analyzing data from the guide star images themselves.

2. Accumulate Pixel Data

Chandra's Aspect camera detector is a 1024 × 1024 pixel CCD. To save bandwidth, only a small subset of pixels centered on each star is telemetered to the ground.

To determine dark current, we want to collect data from pixels with few or no counts from the source. The typical star image will have a FWHM of 1.8 pixels. This means that the outer rim of pixels in a 6 × 6 pixel image are largely unaffected by star light, and can be used in the analysis.

If the spacecraft did not dither, the same set of pixels would be seen in each image and we would have only a few pixels to analyze. However, spacecraft dither will cause the star image to move along the CCD. The aspect camera tracks this motion and adjusts the set of pixels used so that the star remains centered in the field. As a result, a much larger sampling of pixels can be obtained.

Background subtracted image values are accumulated for each pixel matching the above criteria. For each pixel, we also determine the average total image counts of all images containing that pixel.

3. Determine Dark Current Level and Threshold

To determine if a pixel is warm, it must consistently show a number of counts above some threshold. Each pixel will show random fluctuations in counts from background radiation. They may also show higher count levels from extended source emissions or from elevated background levels in the vicinity of the star. Stars located near the outer ends of the CCD will have elongated PSFs which could cause source photons to land in the outer rim pixels. Since there are several factors that affect the number of counts seen in these 'background' pixels, we cannot simply apply a static threshold to all pixel data to determine if it is warm. We use a two-tiered method for calculating the dark current threshold level to apply. The dark current threshold is defined to be the greater of:

1. An absolute threshold level (default = 200 counts/sec).
2. A fraction of the average total image counts (default = 0.005).

The dark current value for each pixel is determined by a percentile method.

$$Dark\_current_p = pixel\_value[N * numvals] \qquad (1)$$

where N is the Percentile level, typically 0.10, $numvals$ is the number of pixel values accumulated, and $pixel\_value$ is the sorted array of pixel values.

Any pixel whose dark current level is above the dark current threshold is considered 'WARM.' Its location and dark current level are stored.

4. Application to Data

Once the warm pixels have been identified, this new information must be applied to the star images in order to remove the effects these pixels will have on the image centroids.

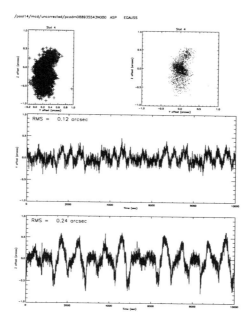

Figure 1. Reconstructed star location without bad pixel correction.

The dark current map is updated to reflect the elevated dark current levels for all bad pixels found. Since the image data we use has already been background subtracted, this correction is additive:

$$Dark\_current[pixel\_row, pixel\_col] + = Dark\_current_p \qquad (2)$$

The raw image data is then re-run through the background subtraction process. With the proper dark current subtracted, the centroids will not be biased by the elevated counts, and a more accurate centroid can be obtained.

5. Results

When a warm pixel contaminates a star image, it produces an offset to the image centroid in the direction of that pixel. As spacecraft dither moves the image along the CCD, the direction of this offset changes, creating a periodic wobble in the star locations. This wobble is apparent in the aspect solution. The effect is reduced by the use of multiple guide stars and smoothing techniques, but it can still have a noticeable impact on pointing accuracy.

Figures 1 and 2 show a series of plots that characterize the accuracy of a star's centroids. The plots show the difference between the star centroid and that star's 'expected' location. Since a guide star's actual position is well known, one can use the spacecraft motion described by the aspect solution to predict where that star should fall on the CCD as a function of time. By comparing these values with the locations described by the star centroids, one can gain a sense of the accuracy of these centroids.

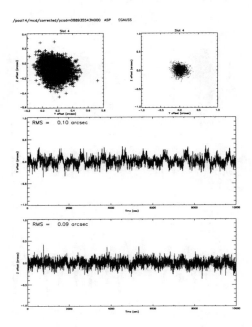

Figure 2. Reconstructed star location with bad pixel correction.

The two spatial plots at top show the same data at different scales. These show the distribution of the centroid offsets in the spacecraft Y and Z axes from expectations. With good centroids and a good solution, this distribution should be centered on 0.0 with a small random spread. The other plots show the offsets for each axis as a function of time. These allow the periodic nature of the effect to be seen.

Figure 1 shows the results of a run containing two bad pixels. The warm pixel detection was turned off during processing. The spatial plots show a significant elongation in the Z direction. This is a result of the warm pixels moving in and out of the image field as the star dithers on the CCD. The plots of offset vs. time show the periodic nature of the effect. The RMS of the offsets is indicated on these plots.

Figure 2 shows the same data when warm pixel detection is applied. Notice the spatial plot shows significant improvement in the distribution. The plots of offset vs. time also show significant improvement, especially in the Z axis. The RMS has dropped from 0.24 to 0.09 arcsec. The remaining periodicity is most likely due to pixels that are warm, but not yet above the threshold level.

Acknowledgments. This project is supported by the Chandra X-ray Center under NASA contract NAS8-39073.

The Sliding-Cell Detection Program for Chandra X-ray Data

T. Calderwood, A. Dobrzycki, H. Jessop, and D. E. Harris

Harvard-Smithsonian Center for Astrophysics, 60 Garden St. Cambridge, MA 02138

Abstract. The Chandra X-ray Observatory provides unprecedented resolution over a large field of view with large collecting area. With these advancements, different and/or improved detection algorithms are a necessity for Chandra data analysis. We here present an overview of *Celldetect*, a source detection program for Chandra. *Celldetect* is descendent from *Einstein* and *ROSAT* data analysis programs (Harnden et al. 1984; DePonte & Primini 1993). It is part of the Chandra Interactive Analysis of Observations (CIAO) software package and is also used in automated processing of Chandra data.

1. Introduction

Celldetect identifies point sources in the presence of background. It can do so by three methods: estimating background locally, accepting a background map, or accepting a background value. The first approach is to examine the counts in a candidate source region and estimate the number of the counts in that region that are due to background. The candidate region is called the "detect cell" (Figure 1). The size of the cell is chosen to be some large fraction of the Point Spread Function (PSF) size. It is surrounded by a "background frame" of approximately equal area. For an isolated point source centered in the detect cell, the cell would contain source and background counts, while the frame would have only background counts. Using the counts in the frame, an estimate is computed for the number of counts in the cell which are due to background. These are subtracted from the cell counts. If the remaining detect cell counts are significantly higher than the estimated background counts, a source is detected.

When a background map or value is supplied to the tool, the background is not estimated and *celldetect* does not use a background frame.

Regardless of method, the cell is started in one corner of the dataset. It then "slides" repeatedly by one third its size, with a new source test made at each position. At the far end of the dataset, the cell slides down by one step and is repositioned horizontally, on the starting side.

This whole approach is essentially the "Local Detect" procedure described in DePonte & Primini (1993). The following sections outline advances built on that technique.

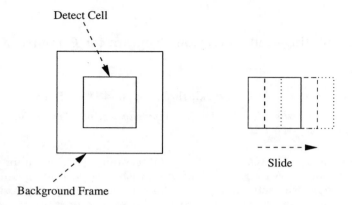

Figure 1.   The sliding Detect Cell.

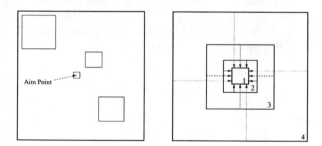

Figure 2.   Varying cell sizes *(left)* and recursive blocking *(right)*.

2. Varying Detect Cell Size

As the size of the PSF varies with off-axis angle, the size of the detect cell should also vary (Figure 2). At each point where a detection is attempted, *celldetect* uses an appropriately sized cell. This size will typically be smallest at the aim point, and largest at the periphery of the dataset. *Celldetect* implements the cell size variation in an inelegant but practical manner. The tool does not attempt to switch between cell sizes as it slides the cell. It does a complete scan of the data at each cell size found in the dataset. A detection calculation at any cell placement point is only made when the PSF size at that point is appropriate for the size of the cell that is currently sliding. Correctly varying the cell size and placement during one scan is nontrivial.

3. Recursive Blocking

Modern X-ray detectors are large, and a spatially complete dataset can overwhelm the memory of a typical user's computer. However, the PSF usually grows with off-axis angle, and full resolution is not needed throughout the dataset. *Celldetect* starts analysis with a 2048 × 2048 unblocked "window" at the cen-

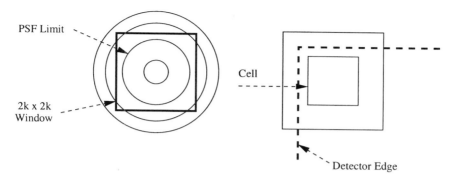

Figure 3. PSF limits *(left)* and detector edge effect *(right)*.

ter (aim point) of the dataset, then blocking down successively larger regions of the data into the same window (Figure 2). The analysis at each successive pass excludes the spatial extent of the previous pass. In each pass, the window is scanned at each of the cell sizes appropriate for that window. For a given instrument, the size of the window must be chosen large enough so that the blocked pixels still oversample the PSF size in later passes.

4. Consistent Analysis at Each PSF Size

When the pixels from a blocking pass are examined, there will be a range of PSF sizes in the analysis window. Some contours of constant PSF size will lie entirely within the window, others will lie only partially in the window (Figure 3). In each pass except the last, *celldetect* limits its analysis to the region of the largest PSF size which is completely contained within the window. This ensures that the data at any given PSF size are all examined at the same blocking factor.

5. Exposure Variation

When the background frame extends over an area of low exposure (relative to the detect cell), the background counts will be artificially reduced (Figure 3). This will make the detect counts appear artificially significant. The consequence is that *celldetect* is prone to declaring false sources near detector edges and other regions of low exposure. To assist the user, *celldetect* provides an "exposure ratio" calculation for each detection. This is a ratio of the average exposure (in either seconds or effective collecting area) for pixels in the background frame to the average exposure for pixels in the detect cell. Under-illuminated backgrounds will cause the ratio to be less than unity. The source list can then be filtered on this value to exclude questionable detections.

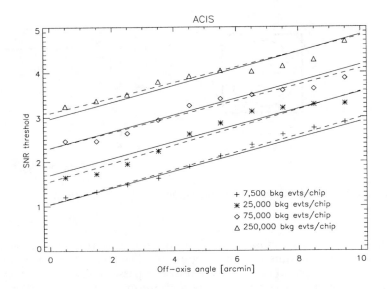

Figure 4. Graph of good SNR values for ACIS off-axis angles for 4 different densities of background events. Based on *celldetect* runs on MARX simulations containing only background events (in which *celldetect* should find no sources) (Dobrzycki et al. 2000).

6. Future Work

The background statistics in *celldetect* change as the cell size increases. This means that a fixed value of the SNR threshold is not appropriate for best source detection. A SNR varying with cell size (and hence off-axis angle) gives better results (Figure 4). A varying SNR is under consideration for a future release of *celldetect*.

Acknowledgments. This project is supported by the Chandra X-ray Center under NASA contract NAS8-39073.

References

Harnden, F. R., Fabricant, D. G., Harris, D. E., & Schwartz, J. 1984, SAO Report No. 393

DePonte, J. & Primini, F. 1993, in ASP Conf. Ser., Vol. 52, Astronomical Data Analysis Software and Systems II, ed. R. J. Hanisch, R. J. V. Brissenden, & J. Barnes (San Francisco: ASP), 425

Dobrzycki, A., Jessop, H., Calderwood, T., & Harris, D. E. 2000, AAS/High Energy Astrophysics Division, 32, 2708

SLIM: A Program to Simulate the ACS Spectroscopic Modes

N. Pirzkal, A. Pasquali, R. N. Hook, J. R. Walsh, R. A. E. Fosbury, W. Freudling, R. Albrecht

ST-ECF, Karl-Schwarzschild Str-2, Garching bei Muenchen D-85748, Germany

Abstract. SLIM is a program developed at ST-ECF to simulate direct and slitless spectroscopic images. It is written in Python to keep development time short and features all of the spectral elements available in the Advanced Camera for Surveys (ACS) which is to be installed on-board the *Hubble Space Telescope*. SLIM produces slitless spectroscopy simulations which include the effects of geometric distortions, fringing, and the field-dependence of the spectral dispersion. We illustrate the use of SLIM with simulations of the HDF-N with the G800L grism using both the Wide Field Channel and with the High Resolution Channel.

1. A Slitless Spectroscopy Simulator

A slitless spectroscopy simulator was required to create realistic data of the upcoming Advanced Camera for Surveys (ACS) slitless spectroscopic modes. ACS is scheduled to be launched in 2001, and simulated data are needed to test future slitless spectroscopy extraction software. Our basic requirement was to have a simulator that could, based on our current knowledge of the instrument, generate both geometrically and photometrically realistic images. While this simulator would in principle be used mainly to generate ACS images, an effort was made to keep it general enough to be used to simulate other instruments. This ensured that the simulator would remain usable to generate ACS simulations even if the performances and/or the characteristics of the instrument were to change in the future.

2. Why Python?

Creating such a simulator can be facilitated greatly if one chooses a language which can easily handle complex data types such as arrays, and interpolated functions. In addition to these, Python has the advantage of being a free, well documented, and well supported language, which can be used on almost all known combinations of hardware and operating systems. While a scripting language, Python has also been shown to be very flexible and appropriate for both small and large projects. One thing that made Python attractive for this project was the existence of many Python extension packages which provide useful high-level data types. For example, we have made extensive use of the

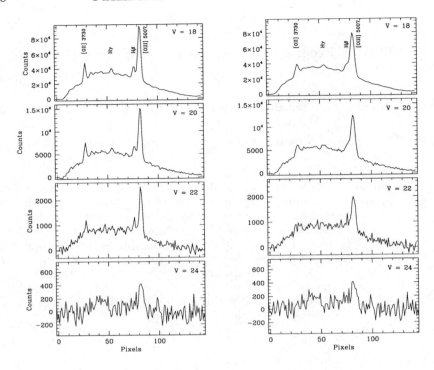

Figure 1. Simulated and extracted spectra of starburst galaxies at z=0.7 (1000 s SLIM simulations). The object is an ellipse of size 2 × 5 pixels, with PA of 90° (left) and 45° (right) with respect to the dispersion direction. The change in the final spectroscopic resolution is clearly visible.

Python Numeric package, which allowed us to store images as sets of arbitrarily large arrays. Furthermore, input spectra and throughput curves were loaded as Interpolated functions using the Python Scientific module. This allowed us to treat them as if they were continuous functions, and not discrete datasets, allowing us to avoid all the problems associated with non-uniform sampling of the input spectra and throughput curves. The speed of an interpreted scripting language such as Python is somewhat lower than that of a low level C program. This drawback is, however, offset by the lowered development time that is achieved using a modern, feature rich, object oriented language which requires no compilation phase. Furthermore, in reality one can reach close to C performance with extensive use of Python's high level data types and functions, which operate at C language speed. For example, using Numeric array types, one can add two 1000 × 1000 arrays with a simple Python declaration like c = a + b. The actual work of adding up all the elements of the two arrays is done in the underlying, optimized C Numeric module, and performance becomes very close to that achieved by some stand-alone C code.

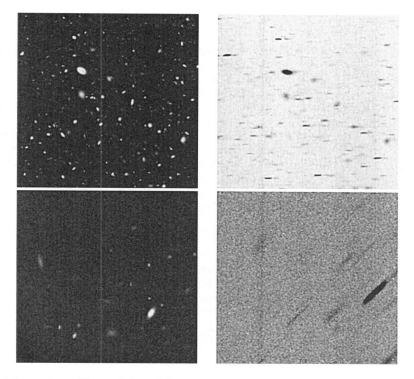

Figure 2. Upper-left: Color composite (F435W, F606W, F814W) using SLIM simulations of ACS/WFC images of an HDF-N like field and (upper-right) the same field when simulated as seen with the WFC grism GL800. The true HDF-N distribution of objects was preserved near the center of this field. Integration time is 1000 seconds in each filter. The field of view is $202'' \times 202''$. Lower panels: SLIM simulations of the same HDF-N like field in 10 hour HRC direct and grism observations. The field of view is $27'' \times 27''$.

3. SLIM Interface

The program interface was kept as simple as possible. Every aspect of the simulator is controlled though the use of three simple text files containing the various simulation parameters. The first one contains a list of input objects (positions, spectral types, red-shifts, and reference magnitudes). The second one contains general simulation parameters such as mirror size, gain, pixel size, and a list of throughput files. Finally, the last configuration file contains a field-dependent polynomial description of the polynomial dispersion relation to apply to the input spectra.

4. SLIM Main Features

The main feature of SLIM are (a) configuration using simple text files, (b) creating slitless simulations by dispersing a 1-D input spectrum and, at each wavelength, convolving it with a PSF of the appropriate shape and size, ¡c) an arbitrary number of throughput files can be used to simulate any combination of mirrors, windows, filters, dispersive elements, and detectors, (d) photometric zero points are computed on-the-fly using the input list of filters and throughputs and a spectrum of Vega, (e) arbitrary n^{th} order polynomial descriptions of the dispersion relations can be used, allowing both near-linear (grism) and highly non-linear (prism) simulations to be generated, and field dependence to be accounted for, (f) no requirement for any a-priori sampling of the input spectrum or throughput files, and (g) producing ACS simulations that agree photometrically within 1 percent with the ACS ETC.

5. Examples

Several examples of spectra which were extracted from simulated ACS WFC grism images are shown in Figure 1. Figure 2 contains simulations of the HDF-N as seen in a three color direct image composite and with the GL800 grism when using the WFC. The same field as seen with the High Resolution Camera (HRC) is also shown. While providing a smaller field of view, the HRC has a higher resolution than WFC in both the direct and grism modes.

6. Conclusion

The use of Python allowed us to efficiently produce a slitless spectroscopic simulator which has been successfully used to produce geometrically and photometrically correct images. More information about SLIM can be found at http://www.stecf.org/software/.

Automated Spectral Extraction for High Multiplexing MOS and IFU Observations

Marco Scodeggio[1], Alessandra Zanichelli, Bianca Garilli

IFC–CNR, Milano, Italy

Olivier Le Fèvre

LAM–CNRS, Marseille, France

Giampaolo Vettolani

IRA–CNR, Bologna, Italy

Abstract. The distinguishing characteristic of VIMOS is its very high multiplex capability: in MOS mode up to 800 spectra can be acquired simultaneously, while the Integral Field Unit produces 6400 spectra to obtain integral field spectroscopy of an area approximately 1×1 arcmin in size. To successfully exploit the capabilities of such an instrument, it is necessary to expedite as much as possible the analysis of the very large volume of data that it will produce, automating almost completely the basic data reduction and the related bookkeeping process. The VIMOS Data Reduction Software (DRS) has been designed specifically to satisfy these two requirements. A complete automation is achieved using a series of auxiliary tables that store all the input information needed by the data reduction procedures, and all the output information that they produce. We expect to achieve a satisfactory data reduction for more than 90% of the input spectra, while some level of human intervention might be required for a small fraction of them to complete the data reduction. The DRS procedures can be used as a stand-alone package, but are also being incorporated within the VIMOS pipeline under development at the European Southern Observatory.

1. Introduction

VIMOS is the first of a pair of imaging spectrographs that are being built for the unit telescopes of the European Southern Observatory Very Large Telescope. The instrument field of view is split into four separate quadrants, each one covering approximately 7 × 8 arcmin. On one side of the instrument is anchored the head of the Integral Field Unit (IFU), consisting of a lenslet array of 6400 lenslets organized in an 80 × 80 array that covers a field of view of 54 × 54 arcsec.

---

[1] On behalf of the VIRMOS Consortium

The light collected by the IFU head is fed to the main spectrograph via optical fibers.

This instrument was designed specifically to carry out survey work, and to have a very high multiplexing capability. Because of this, it will also require a rather new approach to the process of data reduction: it will simply be impossible to reduce by hand the very large amount of data that VIMOS will produce, as the following example demonstrates. Working "by hand" using IRAF (or a similar package), even with a number of *ad hoc* scripts designed to carry out repetitive tasks, an astronomer could reduce one spectrum in 3 to 5 minutes. Since VIMOS in Multi Object Spectrograph (MOS) mode will produce between 150 and 200 spectra per quadrant for each exposure, and it has 4 quadrants, it would then take between 30 and 70 hours of work to reduce a single exposure. Therefore, reducing just one night of observations "by hand" would require between 300 and 700 hours of work, which is 40 to 150 working days (8 hours a day, doing nothing else)!! With IFU observations, each one producing 6400 spectra, things would obviously be even worse.

For this reason it was decided to implement a completely automatic data reduction pipeline for the processing of VIMOS data. As part of the agreement for the construction of the instrument, the VIRMOS consortium is developing the core components of this pipeline, which we call here the VIMOS Data Reduction Software (DRS). The European Southern Observatory is in charge of putting this core component into a completely automatic pipeline within the framework of their Data Flow System. Although the DRS has been designed specifically for VIMOS, its conceptual lay-out is general enough that it could be easily adapted to any kind of instrument with capabilities comparable to those of VIMOS.

2. Instrument and Individual Mask Calibrations

It is very difficult to design completely automated tasks to reduce complex datasets like those produced by VIMOS. It was decided that the DRS will *always and only* work from a "reasonable" first guess about all calibration parameters of the instrument. This will require the setup of an instrument calibration database, and the periodic execution of calibration procedures that will produce the necessary calibration data. When reduced using DRS procedures, these data will produce: (a) a mapping between positions on the slitlets mask and positions on the CCD frame where the slitlets images are recorded; (b) a mapping of the distortions that affect each individual spectrum; (c) a mapping of the wavelength dispersion solution that associates a wavelength to each pixel coordinate of a spectrum image; and (d) two mappings between celestial coordinates and coordinates in the plane of the slitlets mask and in the plane of the CCD frame. These mappings, generally in the form of second to fourth order spatial polynomials, are stored in calibration matrix coefficients inside FITS binary tables.

Starting from the general instrument calibrations, specific calibrations are obtained for each specific mask used for an observation (see Figure 1): (a) the Mask Preparation Software used by the astronomer to define the slitlets positions for MOS observations (see Bottini, Garilli, & Tresse 2001) produces an Aperture

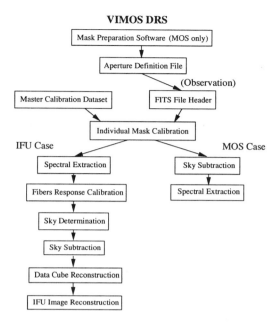

Figure 1. A block diagram scheme of the DRS spectral extraction process

Definition File (ADF), containing the positions of all slitlets in the mask plane; (b) the ADF is inserted into the FITS file header at the end of the observation (for IFU observations, where the ADF is always the same, this is stored in a separate table); (c) using the slit positions derived from the ADF, and the general calibration matrices, the approximate location of each 2-D spectrum on the CCD frame is computed; (d) a flat field exposure is used to measure the exact location of each 2-D spectrum on the CCD frame; and (e) a calibration lamp exposure is used to measure the position of a number of spectral lines and refine the approximate wavelength solution provided by the general calibration matrix.

3. Spectral Extraction

The final result of the calibration phase is the complete definition of a set of apertures to be used for the final spectral extraction. The extraction itself is carried out following two different procedures for MOS and IFU observations.

For observations carried out in MOS mode, it is assumed that slitlets always contain a region of "pure sky," where the contribution of the sky background to the composite spectrum of sky plus target astronomical object can be estimated. Each 2-D spectrum is collapsed along the wavelength dispersion axis, to produce a 1-D intensity profile along the entrance slit. Groups of at least three adjacent pixels, all with intensity above a certain threshold (derived from the rms noise in the profile itself) are considered as "object," while all other pixels are considered as "sky." 1-D spectra for each object so identified are obtained using the Horne (1986) optimal extraction method.

For observations carried out in IFU mode, 1-D spectra for each lenslet are obtained by adding together the flux collected by the corresponding optical fiber. Each one of these spectra can belong to one of two categories: either it is the superposition of sky and astronomical object contributions, or it is a pure sky spectrum. It is therefore necessary to identify those spectra in this second category to build an estimate of the sky background intensity, before subtracting it from all the spectra. This identification is done on the basis of the distribution of total light intensities registered in the various spectra, considering as pure sky ones those that have an intensity below the mode of the distribution. For extremely crowded fields (or very deep exposures), the number of pure sky spectra is expected to be a very small fraction of the total number of spectra. In this case an interactive tool will be provided, to allow the astronomer to specify "by hand" the spectra to be used to build the sky background estimate.

After the pure sky spectra are identified, they are median-averaged together, to build an estimate of the sky spectrum, which is subtracted from all the 6400 1-D spectra. Since the optical fibers redistribute the light collected by the lenslet array over the four VIMOS quadrants following a rather complex pattern, the 1-D sky-subtracted spectra are then re-assembled in a data cube to provide a complete and spatially coherent reconstruction of the given data-set. Also, all spectra can be collapsed along the wavelength axis to provide a 2-D image of the area covered by the IFU field of view.

4. Final Remarks

The VIMOS DRS is written entirely using ANSI C code. Highly specialized tasks are carried out by external libraries, including the CFITSIO and WCS libraries, and the SExtractor object detection software. All input and output files are standard FITS files (image and binary table format). Heavy usage is made of ESO hierarchical keywords within headers (this means IRAF would have problems handling those headers).

The final DRS output will be: for MOS observations a 2-D FITS image, containing one extracted spectrum for each row of the image, plus a table containing the identification parameters for each one of those spectra; for IFU observations a 3-D FITS image containing the IFU data cube, plus a table containing the identification parameters for each spectrum.

References

Bottini, D., Garilli, B., & Tresse, L. 2001, this volume, 455
Horne, K. 1986, PASP, 98, 609

The VIMOS Mask Preparation Software

Dario Bottini[1], Bianca Garilli

Istituto di Fisica Cosmica G. Occhialini, CNR, Milano

Laurence Tresse

Laboratoire d'Astrophysique de Marseille, France

Abstract. The main scientific characteristic of VIMOS (VIsible Multi Object Spectrograph, to be placed on ESO Very Large Telescope) is its high multiplexing capability, allowing astronomers to obtain up to 800 spectra per exposure. To fully exploit such a potential a dedicated tool, the VIMOS Mask Preparation Software (MPS), has been designed. The MPS provides the astronomer with tools for the selection of the objects to be spectroscopically observed, including interactive object selection, handling of curved slits, and an algorithm for automatic slit positioning that derives the most effective solution in terms of the number of objects selected per field. The slit positioning algorithm must take into account both the initial list of preferred objects, and the constraints imposed either by the instrument characteristics or by the requirement of having easily reduceable data. The number of possible slit combinations in a field is in any case very high (10^{73}), and the task of slits maximization cannot be solved through a purely combinatorial approach. We have introduced an innovative approach, based on the analysis of the function (Number of slits)/(slit length) vs. (slit length). The algorithm has been fully tested with good results, and it will be distributed to the astronomical community for the observation preparation phase.

1. Introduction

The VIRMOS (Visible and Infrared Multi-Object Spectrograph) project consists of two spectrographs with enhanced survey capabilities to be installed on two unit telescopes of ESO Very Large Telescope (Chile): VIMOS (0.37–1 μm) and NIRMOS (0.9–1.85 μm), each one having a large field of view ($\sim 14' \times 16'$) split into four quadrants and a high multiplexing factor (up to approximately 800 spectra per exposure).

To exploit such potential, a dedicated tool, the VIRMOS Mask Preparation Software (MPS), has been implemented. It provides the astronomer with tools for selecting objects to be observed spectroscopically, and for automatic slit

---

[1]On behalf of the VIRMOS Consortium

positioning. The output of MPS is used to build the slit masks to be mounted in the instrument for the spectroscopic observations.

2. Requirements

At a limiting magnitude$_I$ < 24, the density of objects in the sky is such that more than 1000 galaxies are visible in a VIMOS quadrant. Of course, not all these objects can be observed spectroscopically, as some requirements imposed by data quality have to be taken into account when placing slits: the minimum slit length will depend on the object size, since the slit must contain some area of "pure sky" to allow for a reliable sky subtraction; spectra must not overlap either along the dispersion or the spatial direction; as each first order spectrum is coupled with a second order spectrum which will contaminate the first order spectrum of the slit above, a good sky subtraction can be performed only if slits are aligned in columns (same spatial coordinate) and, within the same column, have the same length. All these factors lead to a theoretical maximum number of spectra per quadrant of approximately 200.

Another requirement for MPS is set by the very good VLT seeing which allows the use of slits widths of $0.''3$–$0.''4$. Such narrow slits imply an extremely precise slit positioning, with maximum uncertainties of order $0.''1$. Thus the need for some (1–2 per quadrant) manually selected reference objects (possibly bright and point-like) to be used for mask alignment. Moreover, the user must have the possibility to choose manually some particularly interesting sources to be included (Compulsory objects) and some others to be excluded (Forbidden objects) from the spectroscopic sample. A tool for manual definition of curved or tilted slits, to better follow the shape of particularly interesting objects, must also be provided.

The MPS starts from a VIMOS image, to which a catalogue of objects is associated. The catalogue can be derived from the image itself or from some other astronomical dataset. In this second case, a way to correlate the celestial coordinates of the objects in the catalogue with image coordinates is to be provided. Some catalog handling capabilities, to allow for the selection of classes of sources among which to operate the choice of spectroscopic targets, some image display and catalogue overlay capabilities have to be provided by the package.

3. MPS Graphical User's Interface

As MPS will be distributed to the astronomical community, it should be based on some already known package (Not Yet Another System). It was therefore decided to base the MPS GUI on the SKYCAT[1] tool distributed by ESO. This tool allows astronomers to couple VIMOS images and catalogues on which to operate selections of objects over which to place slits.

---

[1] http://archive.eso.org:8080/skycat/

A new panel for catalogue display and object selection (Reference, Compulsory, Forbidden object) has been implemented. For each type of catalogued object, a different overlay symbol has been defined.

3.1. Curved/Tilted Slits Definition

A dedicated zoom panel allows the definition of curved/tilted slits. Curved slits are defined by fitting a Bezier curve to a set of points chosen by clicking on the zoom display. The fitted curve is then automatically plotted. The slit width is chosen through a scale widget.

Tilted slits can be defined as curved slits and then straightened. If the astronomer wants to have slits of a width different from the one chosen for the automatic slit placements, he can define them as tilted slits and then align them to the other automatically placed slits.

4. Slit Positioning Optimization Code

The core of Mask Preparation Software is the Slit Positioning Optimization Code (SPOC). Given a catalog of objects, SPOC maximizes the number of observable objects in a single exposure and computes the corresponding slit positions.

SPOC places slits on the field of view taking into account: special objects (reference, compulsory, forbidden), special slits (curved, tilted or user's dimension defined), spectral first order superposition, spectral higher order superposition and sky region parameter (the minimum amount of sky to be added to an object size when defining a slit).

4.1. The Optimization

The issue to be solved is a combinatorial computational problem. Because of the constraint of slits aligned in the dispersion direction, the problem can be simplified slightly: the quadrant area can be considered as a sum of strips which are not necessarily of the same width in the spatial direction. Slits within the same strip have the same length and the alignment of orders is fully ensured. The problem is thus reduced to be one dimension. It is easy to show that the number of combination is roughly given by: $N_{combination} = (N_{possible\ strip\ widths})^{(average\ number\ of\ strips)}$. The slit length (or strip width) can vary from a minimum of 4 arcsec (20 pixels, i.e., twice the minimum sky region required for the sky subtraction) to a maximum of 30 arcsec (150 pixels, limit imposed by the slit laser cutting machine). The average number of strips can be estimated as the spatial direction size of the FOV divided by the most probable slit length: assuming the latter to be 50 pixels (10 arcsec), we would have 2048/50 = 41 strips. The number of combinations would then be: $N_{combinations} = 130^{41} \simeq 4.7 \times 10^{86}$. Computing these many combinations would correspond to 10^{60} years of CPU work! The problem is similar to the well known traveling salesman problem: in the standard approach, this is solved by randomly extracting a "reasonable" number of combinations and maximizing over this subsample. In our case, due to computational time, the "reasonable" number of combinations cannot be higher than $10^8 - 10^9$, so small with respect to the total number of combination that the result is not guaranteed to be near the real maximum. Our approach has been to consider only the most "probable" combinations, i.e., the ones that have the highest probability of maximizing the solution.

Step 1: For each spatial coordinate, we can vary the strip width from the given minimum to the given maximum, count how many objects we can place in the strip, and build the diagram of the number of slits in a strip divided by the strip width as a function of the strip width. For each spatial coordinate, only the strip widths corresponding to peaks in this histogram are worth considering, as they correspond to local maxima of the number of slits per strip. The exact positioning of the peaks varies for each spatial coordinate, but the shape of the function remains the same. The position of the peaks can be easily found in no more than 6–7 trials (using a partition exchange method).

Step 2: For each spatial coordinate we have K (where K is the number of peaks) possible strips, each with its own length and number of slits. Although the number of combinations to be tested is decreased, it is still too big in terms of computational time.

Step 3: A further reduction can be obtained if, instead of considering all the strips simultaneously, we sequentially consider M subsets of N consecutive strips, which together cover the whole FOV. At this point, we should vary N (and consequently M) to find the best solution. In practice, when N is higher than 8–10, nothing changes in terms of number of observable objects. For N=10, thus M=4 (i.e., 2048/(4 × 50)), the number of combinations is reduced to only $4 \times 4^{10} - 4 \times 10^6$, which means a few seconds of CPU work. Unfortunately, as a consequence of the optimization process, small size objects are favored against the big ones.

4.2. An Alternative Optimization

A second, less optimized algorithm has been implemented within SPOC. This alternative algorithm does not optimize all strips simultaneously but builds the $N_{Slit}/Strip$ function strip by strip without considering object sizes, and takes only the maximum of the distribution. Then it enlarges each strip width by taking into account object sizes. In this way the number of placed slits decreases by a few percent but the object dimension bias disappears.

5. SPOC Graphical User Interface

A dedicated panel for SPOC setup has been implemented within SKYCAT. Through this panel, users can select the grism, the slit width, the sky region parameter, the number of masks to be obtained for the given field, and the type of SPOC maximization.

The number of input slits and placed slits for all kinds of objects (Reference, Compulsory, etc...) is printed in a text box.

The slit catalogue produced by SPOC can be loaded as a normal SKY-CAT catalog with overlay symbols defined for all kinds of objects, and it is also possible to plot the slit and spectrum overlay for all SPOC catalog objects.

… *Astronomical Data Analysis Software and Systems X*
ASP Conference Series, Vol. 238, 2001
F. R. Harnden Jr., F. A. Primini, and H. E. Payne, eds.

IFU Data Products and Reduction Software

Alasdair Allan

School of Physics, University of Exeter, Stocker Road, Exeter, EX4 4QL, U.K.

Jeremy Allington-Smith, James Turner

Department of Physics, University of Durham, Science Labs., South Road, Durham, DH1 3LE, U.K.

Rachel Johnson

Institute for Astronomy, University of Cambridge, Madingley Road, Cambridge, CB3 0HA, U.K.

Bryan Miller

Gemini Observatory, Operations Center, c/o AURA Inc., Cassilla 603, La Serena, Chile

Frank Valdes

NOAO, IRAF Group, 950 N. Cherry Ave., P.O. Box 26732, Tuscon, Arizona, 85726-6732, U.S.A.

Abstract. We present a summary of the current status of Starlink and UK data reduction and science product manipulation software for the next generation of IFUs, and discuss the implications of the currently available analysis software with respect to the scientific output of these new instruments. The possibilities of utilising existing software for science product analysis is examined. We also examine the competing science product data formats, and discuss the conventions for representing the data in a multi-extension FITS format.

1. Introduction

Integral Field Spectroscopy (IFS) is a technique to produce spectra over a contiguous 2-D field, producing as a final data product a 3-D data cube of the two spatial coordinate axes plus an additional axis in wavelength. Although existing techniques, such as stepping a longslit spectrograph or scanning a Fabry-Perot device, can produce such a data cube the IFS technique collects the data simultaneously with obvious savings in observing efficiency. However, IFS has only recently approached maturity as a hardware technique (e.g., Haynes et al. 1998; Haynes et al. 1999; Allington-Smith et al. 2000).

2. Reduction Software

Initial data reduction to remove instrumental effects such as flat fielding and cosmic ray removal, and mapping between the 2-D detector coordinates and the data cube, is highly instrument dependent.

There are two paradigms for IFS data reduction. First, the "traditional" method, adapted from multi-object spectroscopy (MOS), where the output from each fibre is extracted by tracing the spectrum and accounting for wavelength dependent distortion (normally referred to as the *MOS paradigm*). More recently, with the arrival of TEIFU, where the fibre outputs are under-sampled by the detector, an alternative paradigm has arisen (usually referred to as the *longslit paradigm*). Although the independence of the spatial samples is lost due to the under-sampling of the PSF by the detector, it can be shown that this is irrelevant so long as the target is critically sampled by the IFU (Allington-Smith & Content 1998). Here the methods adapted from MOS cannot be used and the resulting dataset bears more resemblance to traditional longslit spectroscopy than to MOS data.

While data reduction software is available for the currently operating IFUs, e.g., SMIRFS (Haynes et al. 1999), software to deal with data from the next generation of instruments, such as GMOS (Allington-Smith et al. 2000) or GNIRS, is either still in development or it is unclear who is tasked with providing the software. This is worrying, as it seems unlikely that (with currently available resources) a comparison between the two data reduction paradigms will be made for the upcoming generation of IFUs, many of which fall between the two reduction paradigms (e.g., GMOS).

3. Analysis Software

While the initial data reduction software for IFUs is highly instrument dependent, the data analysis of the final science data product for all these instruments should be fairly generic. The end product of the data reduction for IFS is, almost naturally, an (x,y,λ) data cube. Once assembled, with associated variance and quality arrays, scientifically interesting information can be extracted from the cube.

While not every possible operation can be anticipated there are several standard processes that most observers will want to carry out during the data analysis stage:

Mosaicing Mosaicing data cubes obtained from different observations, offset in both position and wavelength, with appropriately chosen re-sampling algorithms.

Visualisation Extraction of individual spectra, and image planes corresponding to spectral features or chosen passbands.

Mapping Construction of radial velocity, line strength, and ratio maps from the data cube (see Figure 1).

A lot of these required tasks can be carried out using pre-existing Starlink software with only minor or no modifications necessary to the code. This sit-

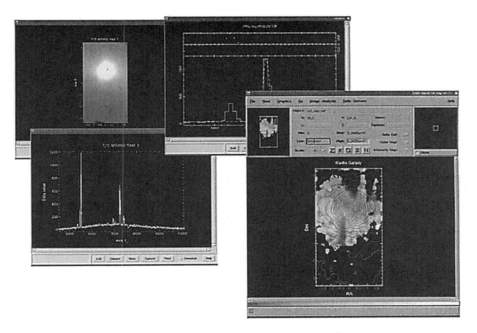

Figure 1. A reconstructed velocity field of the 5007Å OIII line (lower right panel) created using Starlink software from an observation of 3C237 taken during the TEIFU commissioning run. The other panels show the white light image and the intervening line fitting steps of the software. The final velocity field is shown displayed in the GAIA image manipulation package, with the contours of the white light image overlayed on the velocity field. Here we build the script from several disparate Starlink packages, including KAPPA, FIGARO, and CONVERT, to carry out the velocity mapping task.

uation has arisen due to the use of the extensible N-Dimensional Data Format (NDF). This is a format for storing bulk data in the form of N-dimensional arrays of numbers. It is typically used for storing spectra, images, and similar datasets with higher dimensionality. The NDF format is based on the Hierarchical Data System (HDS) and is extensible; not only does it provide a comprehensive set of standard ancillary items to describe the data, it can also be extended indefinitely to handle additional user-defined information of any type.

While most Starlink applications were written with 2-D CCD data in mind, they were written generically to make use of the NDF format and hence a great many have the capability to handle data which has more than the anticipated two dimensions, e.g., many KAPPA and FIGARO applications are capable of being used on multi-dimensional data.

4. Prospective File Formats

The current working format for the final data product of the GEMINI (GMOS) and CIRPASS data reduction software suite, will be a multi-extension FITS (MEF) file. However, this format may be replaced by the new IRAF spectral format, which is currently in development by the IRAF group at NOAO. The MEF is similar to the standard NIRI format now used with GEMINI:

No.	Type	Name	Format	BITPI	INH
0	ifs_data.fits			16	
1	BIN TABLE	TAB	$16 \times num.\ of\ fibres$	8	
2	IMAGE	SCI	$\lambda \times num.\ of\ fibres$	-32	F
3	IMAGE	VAR	$\lambda \times num.\ of\ fibres$	-32	F
4	IMAGE	DQ	$\lambda \times num.\ of\ fibres$	16	F

Here the first extension is a binary FITS table with columns: ID, RA, DEC, and SKY. This table would hold information specific to individual lenslets/fibers like relative fibre positions on the sky (RA, DEC), whether the fibre is a sky or object spectrum (SKY), etc. The three image planes are multispec-like, each row is a separate spectrum.

However this MOS-style MEF format is not particularly natural way of handling IFS data. Indeed, under the *longslit* paradigm these files cannot be generated. A conversion program for GMOS and CIRPASS data to a more easily analysed data cube, which will involve re-binning the input spectra onto a rectangular array, is therefore desirable:

No.	Type	Name	Format	BITPI	Comment
0	ifs_data.fits				
1	IMAGE	SCI	$X \times Y \times \lambda$	-32	3-D science array
2	IMAGE	VAR	$X \times Y \times \lambda$	-32	3-D variance array
3	IMAGE	DQ	$X \times Y \times \lambda$	16	3-D data quality array

In this case, the IFU geometry information is no longer needed, but it would make sense to include the coordinates for each fibre if the user was not taking home the raw data from the telescope, presumably as a FITS binary table.

References

Allington-Smith, J. R., Content, R., Haynes, R., & Robertson, D. 2000, in ASP Conf. Ser., Vol. 195, Imaging the Universe in Three Dimensions, ed. W. van Breugel & J. Bland-Hawthorn (San Francisco: ASP), 319

Allington-Smith, J. R. & Content, R., 1998, PASP, 110, 1216

Haynes, R., et al. 1999, PASP, 111, 1451

Haynes, R., Doel, A. P., Content, R., Allington-Smith, J. R., & Lee, D. 1998, SPIE, 3355, 788

The ST5000: An Attitude Determination System with Low-Bandwidth Digital Imaging

J. W. Percival and K. H. Nordsieck
Space Astronomy Laboratory, University of Wisconsin-Madison, Madison, WI 53706

Abstract. The Space Astronomy Laboratory is building an attitude determination and digital imaging system with embedded compression. The attitude determination system uses a 30-square-degree field of view and an embedded star catalog to determine the Right Ascension and Declination of its line of sight to better than 5 arcseconds. The digital imaging subsystem uses a scheme of "progressive image transmission" in which the image is sent out over a very-low-bandwidth channel, such as a spacecraft telemetry downlink, in such a way that it can be reconstructed "on the fly" and updated as more data arrive. Large (768 × 474) useful images can be obtained over a 4-kbit downlink in as little as 10 seconds.

In addition to its use in sounding rockets and spacecraft, we are planning to use it for two ground-based applications at the Southern Africa Large Telescope (SALT). We will explore its use in generating real-time measurements of the telescope pointing, independent of the telescope control system, and we will use the low bandwidth imaging capability for public outreach.

1. Flight Test

We tested the ST5000 on a sounding rocket flight (36.172, PI K. Nordsieck) in April 1999. The ST5000 was mounted beside a Ball STRAP tracker, which was in control of the rocket, and was optically aligned with it. During the two guide star acquisitions, the ST5000 measured pitch and yaw errors, which were telemetered by the rocket ACS. During the science phase of the mission, the ST5000 generated pitch, yaw, and roll errors while progressively transmitting the acquisition field over a 19,200 baud RS-232 downlink.

Figure 1 shows the Ball STRAP pitch and yaw errors (solid line) during the first guide star acquisition. The dotted line shows the ST5000 error signals. Note that they match those of the STRAP tracker, but with much lower noise.

2. Ground Tests

We ground-tested the ST5000 while attached to the side of a small telescope (used as a convenient, pointable mount) at night from a rooftop observatory of the University of Wisconsin's Department of Astronomy. This testing had two purposes: to measure the sensor performance (noise and sensitivity) as a

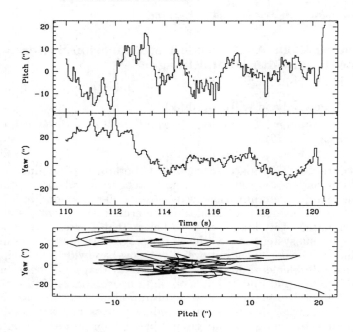

Figure 1. Pitch and Yaw errors during the test flight. Solid lines are the Ball STRAP signals; dotted lines are the ST5000 signals.

function of camera gain and exposure time, and to verify the performance and robustness of the acquisition and tracking operations on real star fields.

Figure 2 shows the Noise Equivalent Angle (NEA) single-star tracking measurements obtained with the ST5000. Also shown are two points representing (for comparison) the Ball STRAP tracker specification, and the predictions from our numerical model for both the prototype ST5000 and the ST5000 upgrade. The solid line and filled triangles represent the performance at low camera gain, set electronically by the ST5000 software. The long-dashed line and open triangles show the numerical model and data at medium gain. We tested the camera at high gain, but found that the increased pixel noise did not justify the higher sensitivity. The short-dashed line represents the model predictions for the ST5000 upgrade, with its improved lens and more sensitive CCD. We also show the measured and expected performance with an update rate of 5 Hz, allowing longer exposure time. We point out the excellent correlation between the numerical model and the measurements. The ST5000 acquired and tracked stars as faint as magnitude 6.5 with an NEA better than 7 arcseconds in the middle of the city of Madison, through more than 1 air mass, in the presence of atmospheric turbulence (seeing) larger than 1 arcsecond. The numerical model does not consider the effect of seeing, which has the effect of moving the points upward, off the model prediction. This is noticeable for the faintest acquisitions (the x's), which we would expect to be the most susceptible to turbulence in the atmosphere. The departure of the measurements from the model predictions at the bright end (filled triangles) is due to saturation of the CCD pixels during the

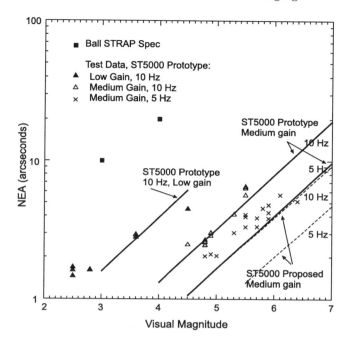

Figure 2.   ST5000 Single-star Noise Equivalent Angle.

exposures. The ST5000 upgrade will use electronic shuttering to allow shorter exposures. In any case, the ST5000 can track on any of the detected objects, easily ignoring any saturated stars.

3. Attitude Determination

The ST5000 can determine its attitude with respect to an absolute coordinate system (say, FK5 Equatorial) by analyzing the star patterns in a single frame. It can do so without any a-priori knowledge of attitude, plate scale, or image orientation. The determination is insensitive to stray objects in the field such as glints, hot pixels, airplanes, satellites, or asteroids. It does not depend on star brightnesses, colors, or any knowledge of the sensor's spectral sensitivity.

The algorithm comprises two major steps. First, the stars in the field must be identified. We use the "star triangles" technique of Valdes et al. (1995). We form all possible triangles in a given frame, and plot each one as a single point in a triangle-shape space. We compare this frame's shape space to an all-sky shape space catalog embedded inside the unit. This allows us to associate stars in the frame with specific entries in an all-sky catalog.

Second, given the association between frame and catalog stars, the attitude must be derived. We use the "q-method" of Lerner (1997) to determine the attitude in the form of a quaternion. This method is suitable for a CCD image, which produces many simultaneous vector measurements.

The principal challenges in doing this attitude determination in an embedded system are making it fast and implementing it with modest computing resources. Space-qualified processors typically lag the consumer marker in both speed and available memory, both of which are important in this application. The all-sky triangle catalog, for example, contains 43.7 million triangles with sides no longer than 6 degrees and member stars no fainter than 8th magnitude. We prune this catalog by keeping a much smaller number of "good" triangles, and we emphasize a fast catalog search during the star identification phase.

4. Progressive Image Transmission

The NASA-funded Progressive Image Transmission (PIT) System (Percival & White 1993; White & Percival 1994, NAG5-2694) offers a number of features that make it especially appropriate for supporting low-bandwidth digital imaging from telescopes or spacecraft. First, it uses a state-of-the-art wavelet transform to achieve very high compression. Second, it implements this as a fast, exactly-reversible in-place integer transform that can be easily ported to older, slower, memory-challenged flight processors. Finally, it formats and transmits the compressed data bytes in a way that allows progressive visualization: the image appears very quickly, immediately showing full-frame detail at all spatial scales and intensities, and as more bytes are received, the image keeps improving, asymptotically converging to losslessness (if time allows).

In progressive transmission, the image can be truncated at any point (say, due to fixed-length downlink windows or the unexpected arrival of a new imaging event), and the currently received bytes always allow the wavelet transform to be reversed and the image reconstructed.

In the example of a SALT-mounted aspect camera, PIT would allow intercontinental distribution of large CCD images at very low bandwidth. For a typical CCD size of 768×474 pixels sampled to 8 bits transmitted over a 19,200 baud Internet connection (many homes do not have high-speed Internet connections), the uncompressed digital image would take an unacceptable 152 seconds. With PIT, a usable version of the image could be transmitted in 3–4 seconds, and would be suitable for many outreach activities. The image would continue to improve with time, without any pre-chosen compression cutoff, allowing the scientific user to achieve the fidelity desired for more technical applications such as target identification or attitude determination.

References

Lerner, G. M. 1997, in Spacecraft Attitude and Control, ed. J. Wertz (Dordrecht: Reidel), 426

Percival, J. W. & White, R. L. 1993, in ASP Conf. Ser., Vol. 52, Astronomical Data Analysis Software and Systems II, ed. R. J. Hanisch, R. J. V. Brissenden, & J. Barnes (San Francisco: ASP), 321

Valdes, F. G., Campusano, L. E., Velasquez, J. D., & Stetson, P. B. 1995, PASP, 107, 1119

White, R. L. & Percival, J. W. 1994, Proc. SPIE 2199, 703

Astronomical Data Analysis Software and Systems X
ASP Conference Series, Vol. 238, 2001
F. R. Harnden Jr., F. A. Primini, and H. E. Payne, eds.

An Object Oriented Design for Monitoring the Chandra Science Instrument X-ray Background

J. G. Petreshock, S. J. Wolk, M. Cresitello-Dittmar, T. Isobe

Harvard-Smithsonian Center for Astrophysics, 60 Garden Street, Cambridge, MA 02138

Abstract. The Monitoring and Trends Analysis (M&TA) System for the *Chandra* X-ray Observatory consists of multiple software threads designed to monitor and visualize spacecraft behavior. The Science Instrument (SI) background monitoring is one such thread that is designed to compile a temporally and spatially ordered table of the observed flux and energy spectrum in detector coordinates. In this paper we describe the design of the tools, and applications of the data products generated, and the output product flexibility.

1. Introduction

As part of the M&TA[1] System (Wolk et al. 2000) the SI Background (SIB) monitoring thread is based on C++ tools designed with object oriented design methodologies that provide flexibility for generating SIB maps. These tools and several PERL wrapper scripts are woven together to form a pipeline (see Figure 1) used by the *Chandra*[2] X-ray Center Data Systems (CXCDS) pipeline processing system (Plummer 2001). The resultant products will provide a means for visualization and quantitative analysis of temporal variations in sky emission and instrument background.

The extremely high angular resolution and sensitivity of *Chandra* makes this a unique problem among active X-ray telescopes. But these extremes also hold the promise of producing a background map of unparalleled resolution. We will discuss and demonstrate the current suite of tools used for monitoring the SIB for a single *Chandra* Flight Instrument, the Advanced CCD Imaging Spectrometer (ACIS).

2. Goals and Importance

Understanding *Chandra's* background is an important factor in driving Science Operations (SciOps) decisions. For example, if we detect lower background levels while the spacecraft is within the magnetotail then we can schedule observations of low surface brightness targets during these times.

---

[1] http://cxc.harvard.edu/mta

[2] http://chandra.harvard.edu/

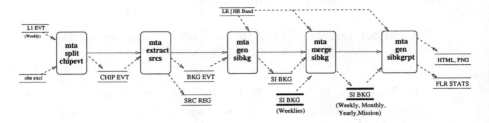

Figure 1. SIB pipeline flow chart.

3. Software Requirements

Depending upon the SI configuration and individual observation configurations, *Chandra* can collect well over 1.0GB worth of event data per week. This large data volume was another aspect of the SIB monitoring problem that had to be handled. We solved this problem by binning the data by time according to event time stamps, thus increasing data manageability.

The SIB monitor is designed to detect temporal and spatial variations in the background count rates and characterize flare properties including: rates, intensities, and durations. Flares are identified by comparing background count rates for each energy band against predefined limits. Background values that exceed these limits are flagged as flares. This is demonstrated by Figure 2, which diagrams a one dimensional view of the parameterized temporal and energy binning performed by the SIB tools. These parameters are controlled via the individual tools parameter file using the CXCDS parameter interface library. The flexibility of the parameterization provides control over the data volume and thus increased data manageability. This flexibility also allows users to create products that range from a purely instrumental background analysis to a finely binned sky map.

The SIB monitoring generates FITS compliant binary tables. The CXCDS DataModel (DM) provides the interface between the classes and the FITS file reading and generation. Each class is equipped with data members that store pointers to DM variables used to connect with the FITS file.

4. Classes

The five classes created for the SIB monitor were designed to interface using cascading levels of data encapsulation. This design provides flexibility with class handling. With only slight modifications, specifically function overloading, these classes were capable of being used within both the *mta_gen_sibkg* tool that generates the SIB monitor products and the *mta_merge_sibkg* tool that is used to merge multiple SIB monitor products together. To reduce the memory usage by the SIB tools, the classes are designed with time being the primary processing key, followed by sky coordinates (Right Ascension, and Declination) and finally energy. X-ray Events from input files are processed per time bin (see Figure 2).

An OO Design for Monitoring the Chandra Background 469

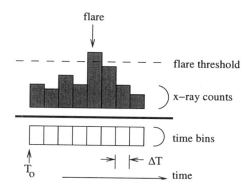

Figure 2. Single energy band Binning Diagram.

When an event from a new time bin is detected, the objects flush their data to a FITS file and the object containers are reset in preparation for receiving data from the new time bin. The SIB classes are described in the following subsections in order from base classes to higher level derived classes.

4.1. EnergyBand

This class is the basis of the M&TA SIB monitoring, the EnergyBand (EB) class stores information about energy bands such as the number of events, energy ranges, the band name, and a pointer to the output FITS column. The EB class functions to store the information for a time bin and is responsible for writing the data to FITS file.

4.2. EnergyBandContainer

The EnergyBandContainer is a container class for EB objects and stores the information for the energy spectrum as defined by the user configurable energy band definitions files. This container class provides a clean interface hiding the work of event processing from the higher level classes.

4.3. SkyLocation

The SkyLocation class is a derived class from EnergyBandContainer. A SkyLocation object describes the energy spectrum for a given parcel of the sky. For the case that the sky coordinates are undersampled (i.e., sky is resolved into one pixel) the SkyLocation object represents the SIB.

4.4. mtaSIBkgMap

The mtaSIBkgMap is a container class for SkyLocation objects and represents the full observed background map. This is the class responsible for controlling the highest level of event processing and provides a clean control interface with the lower level objects.

4.5. EnergyBandDefinitions

This class is used to read and store the EB definitions indicated by the energy band definitions parameter. The sole function for this class it to provide a template to populate new EnergyBandContainer objects. This class is not a derived class, and operates in conjunction with the mtaSkyBkgMap class to initialize the required objects for event processing.

5. Data Products

The SIB monitor products for the ACIS instrument contain a column for each energy band defined within the input energy band definitions files as well as a time column and a flare column. The flare column is defined as an N element bit array that represents each energy band. These FITS products are currently used to generate light curves, flare statistics, and temporal spectral variability plots.

6. Discussion

In any experiment, understanding the signal and background components within a data set is a crucial step, and SIB monitoring provides a useful quantitative analysis of several background properties. The analysis of SIB properties can also be used to update target scheduling, to increase *Chandra's* observational efficiency.

Our application of these simple classes allows us to handle the large volume of X-ray data from *Chandra* efficiently and effectively. The SIB monitor products are ideal for integration into the M&TA trend analysis thread and for use in multisystem correlations with data in existing M&TA databases. The rebinning of sky coordinates allows a basic means for generating the observed X-ray background map. With future versions of the M&TA SIB monitoring we plan to provide the capability to generate higher resolution background maps by applying archived aspect solutions to the background information in detector coordinates. Also, the knowledge acquired by monitoring *Chandra's* SIB can be applied toward future missions and thus potentially increase cost effectiveness.

Acknowledgments. This project is supported by the *Chandra X-ray Center* under NASA contract NAS8-39073.

References

Petreshock, J. G., Wolk, S. J., & Cresitello-Dittmar, M. 2000, in ASP Conf. Ser., Vol. 216, Astronomical Data Analysis Software and Systems IX, ed. N. Manset, C. Veillet, & D. Crabtree (San Francisco: ASP), 475

Plummer, D. 2001, this volume, 475

Wolk, S. J., et al. 2000, in ASP Conf. Ser., Vol. 216, Astronomical Data Analysis Software and Systems IX, ed. N. Manset, C. Veillet, & D. Crabtree (San Francisco: ASP), 453

Kalman Filtering in Chandra Aspect Determination

Roger Hain, Thomas L. Aldcroft, Robert A. Cameron, Mark Cresitello-Dittmar, Margarita Karovska

Harvard-Smithsonian Center for Astrophysics, 60 Garden St., Cambridge, MA 02138

Abstract. The ability of the Chandra X-ray Observatory to achieve unprecedented image resolution is due, in part, to its ability to accurately reconstruct the spacecraft attitude history. This is done with a Kalman filter and Rauch-Tung-Striebel (RTS) smoother, which are key components of the overall aspect solution software. The Kalman filter/RTS smoother work by combining data from star position measurements, which are accurate over the long term but individually noisy, and spacecraft rate information from on-board gyroscopes, which are very accurate over the short-term, but are subject to drifts in the bias rate over longer time scales. The strengths of these two measurement sources are complementary. The gyro rate data minimizes the effects of noise from the star measurements, and the long-term accuracy of the star data provides a high-fidelity estimate of the gyro bias drift. Analysis of flight data, through comparison of observed guide star position with expected position and examination of the reconstructed X-ray image point spread function, supports the conclusion that performance goals (1.0 arcsecond mean aspect error, 0.5 arcsecond aspect error spread diameter) were met.

1. Application of Kalman Filtering in Chandra

The schematic shown in Figure 1 illustrates the basic flow of the Kalman filter design for Chandra aspect determination (Gelb 1974, TRW Memorandum 1996). The main purpose of the filter is to derive an accurate estimate of the spacecraft angular position, or attitude, at any time. Together with fiducial light data for estimating spacecraft flex, this attitude estimate is used to accurately determine the source in the celestial sphere of X-rays detected by the science instruments. Details of the hardware, ground software, and flight performance are given in Aldcroft et al. (2000) and Cameron et al. (2000).

The filter estimates 3-axis error in current attitude estimate and 3-axis gyroscope drift rate, resulting in six estimated quantities referred to as the filter "state." The schematic illustrates the basic flow of the filter. Beginning with an initial state estimate, and initial statistics which characterize how well the state estimates are known (the "covariance"), the filter iterates through two steps.

First, as shown in the upper left box, gyro data are received. The state estimate of gyro bias is subtracted from the raw gyro data, and the spacecraft attitude is updated according to the angular motion derived from the corrected

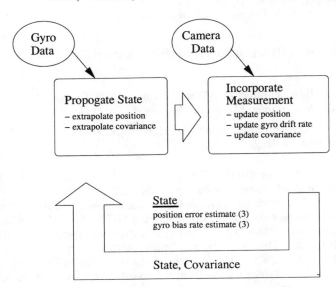

Figure 1. Chandra Aspect Kalman Filter Schematic.

gyro data. The error statistics associated with the state are also updated, since both uncertainty in gyro bias and attitude errors due to rate uncertainty grow over time. In this way, state and attitude estimates are maintained and propagated forward in time until a star camera measurement is available.

The second step is shown in the upper right box. A measurement (the location of a guide star) is received from the aspect camera and incorporated into the filter's state estimates and error statistics. The filter mathematically compares the new attitude estimate based on the current measurement with the expected attitude from the prior propagation step. Using the expected attitude error and the expected measurement error (the star camera noise), new estimates of attitude error and gyro bias drift are calculated. After incorporating these new data, the error statistics are updated to reflect more accurate attitude knowledge, and the cycle begins again.

The four charts in Figure 2 illustrate the aspect pipeline Kalman filter's behavior over time using simulated data. In the upper left graph, the solid line shows the actual pitch estimate error for a 500 second nominal dither observation. The dashed lines represent the root mean square (RMS) value of the covariance for pitch error (i.e., the filter's own estimate of the 1-σ uncertainty in pitch estimate). Note that in this plot and the yaw plot below it, the error in each axis quickly converges to a steady state value. The two charts shown on the right demonstrate the improvement in pitch and yaw estimate error with the addition of a smoothing algorithm. The use of a smoother is possible since the processing is done post-facto. The smoother allows better overall estimates and much improved performance in the very early part of the data. The smoother accomplishes this by utilizing state and measurement data from the entire time period to improve the estimate at each point in time. The filter alone is not able to use future data to produce estimates.

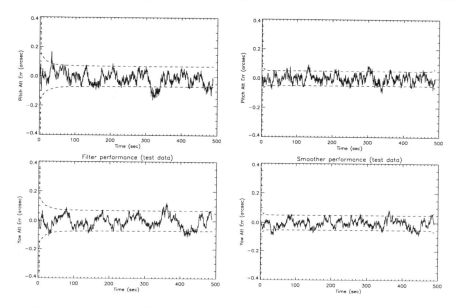

Figure 2. Sample filter and smoother results.

2. Measuring Aspect Performance

To quantify overall aspect quality a 50,000 second observation was examined. An X-ray point source was identified in this observation, and a circular region around the point source with a radius of approximately 2 arcseconds was isolated. Aspect corrected coordinates for the source photons were then examined in a coordinate system aligned with the spacecraft reference frame. This coordinate system is chosen since residual aspect errors are expected to be most visible near the dither frequency, and such errors are more clearly visible as a single frequency oscillation in these coordinates. The corrected sky coordinates did not show any visible oscillations. The power spectrum of photon positions also showed no peaks above the background noise. However, the amount of noise due to photon counting statistics is large compared to the expected aspect oscillations. To improve the SNR of any signal at the dither frequency, the data were phase modulated at the dither frequency. This was done by resetting event times to "time mod(dither period)."

The phase modulated data are shown in Figure 3. Time only goes from 0 to the dither period, and events later than an integer number of dither periods are wrapped back to start at time 0 again. In this way, any residual aspect error at the dither frequency is preserved in the data, but random noise is reduced by averaging. The top charts show phase modulated data for the sky coordinates transformed to the spacecraft X axis (left) and Y axis (right).

The two lower charts show the same data binned on 50 second intervals, with the intention of reducing noise. If no residual oscillation were present, the values of the binned data points would become smaller as the bin size becomes larger, without showing any noticeable constant offset. If, on the other hand, a clear

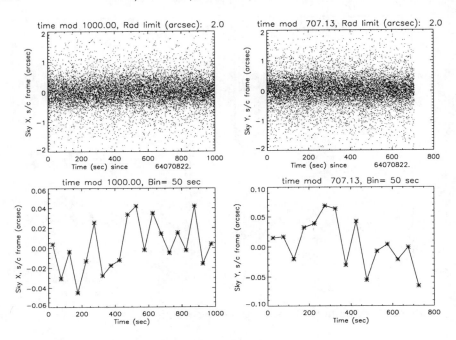

Figure 3. Aspect error in event reconstruction.

signal were present, this should become apparent in the structure of the binned data points. Although even the worst case binned data point is very small (less than 0.08 arcseconds), some slight evidence of a residual signal may be present in the Y coordinate charts. The binned data show a disproportionate number of positive data points in the first half of the dither interval, and a disproportionate number of negative data points in the second half. Note, however, that even if this error is entirely due to aspect error and not noise, the worst case still compares very favorably with the 0.5 arcsecond spread diameter requirement.

This project is supported by the Chandra X-ray Center under NASA contract NAS8-39073.

References

Gelb, A. 1974, Applied Optimal Estimation (Cambridge, MA: MIT Press)

Aldcroft, T. L., Karovska, M., Cresitello-Dittmar, M. L., Cameron, R. A., & Markevitch, M. L. 2000, Proc. SPIE, 4012, 650

Cameron, R. A., Aldcroft, T. L., Podgorski, W. A., Freeman, M. D., & Shirer, J. J. 2000, Proc. SPIE, 4012, 658

TRW Memorandum, AXAF Attitude and Aspect Determination, 52100.300.087 TRW-SE11K, January 11, 1996

The Chandra Automatic Data Processing Infrastructure

David Plummer and Sreelatha Subramanian

Harvard-Smithsonian Center for Astrophysics, 60 Garden St. MS-81, Cambridge, MA 02138

Abstract. The requirements for processing Chandra telemetry are very involved and complex. To maximize efficiency, the infrastructure for processing telemetry has been automated such that all stages of processing will be initiated without operator intervention once a telemetry file is sent to the processing input directory. To maximize flexibility, the processing infrastructure is configured via an ASCII registry. This paper discusses the major components of the Automatic Processing infrastructure including our use of the STScI OPUS system. It describes how the registry is used to control and coordinate the automatic processing.

1. Introduction

Chandra data are processed, archived, and distributed by the Chandra X-ray Center (CXC). Standard Data Processing is accomplished by dozens of "pipelines" designed to process specific instrument data and/or generate a particular data product. Pipelines are organized into levels and generally require as input the output products from earlier levels. Some pipelines process data by observation while others process according to a set time interval or other criteria. Thus, the processing requirements and pipeline data dependencies are very complex. This complexity is captured in an ASCII processing registry which contains information about every data product and pipeline. The Automatic Processing system (AP) polls its input directories for raw telemetry and ephemeris data, pre-processes the telemetry, kicks off the processing pipelines at the appropriate times, provides the required input, and archives the output data products.

2. CXC Pipelines

A CXC pipeline is defined by an ASCII profile template that contains a list of tools to run and the associated run-time parameters (e.g., input/output directory, root-names, etc.). When a pipeline is ready to run, a pipeline run-time profile is generated by the profile builder tool, pbuilder. The run-time profile is executed by the Pipeline Controller, pctr. The pipeline profiles and pctr support conditional execution of tools, branching and converging of threads, and logfile output containing the profile, list of run-time tools, arguments, exit status, parameter files, and run-time output. This process is summarized in Figure 1.

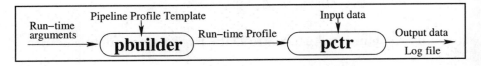

Figure 1. The CXC Pipeline Processing Mechanism.

3. Pipeline Processing Levels and Products

CXC pipeline processing is organized into different levels according to the extent of the processing. Higher levels take the output of lower levels as input. The first stage of processing is Level 0 which de-commutates telemetry and processes ancillary data. Level 0.5 processing determines the start and stop times of each observation interval and also generates data products needed for Level 1 processing. Level 1 processing includes aspect determination, science observation event processing, and calibration. Level 1.5 assigns grating data coordinates to the transmission grating data. Level 2 processing includes standard event filtering, source detection, and grating data spectral extraction. Level 3 processing generates catalogs spanning multiple observations.

4. Standard Pipeline Processing Threads

Figure 2 represents the series of pipelines that are run to process the Chandra data. Each circle represents a different pipeline (or related set of pipelines). Level 0 processing (De-commutation) will produce several data products that correspond to the different spacecraft components. Data from the various components of the spacecraft will follow different threads through the system. The arrows represent the flow of data as the output products of one pipeline are used as inputs to a pipe (or pipes) in the next level. Some pipelines are run on arbitrary time boundaries (as data are available) and others must be run on time boundaries based on observation interval start and stop times (which are determined in the level 0.5 pipe, OBI_DET).

5. Pipeline Processing Registry

The complete pipeline processing requirements for Chandra are very complex with many inter-dependencies (as can be seen in Figure 2). In order to run the pipelines efficiently in a flexible and automated fashion we configure the Automatic Processing system with a pipeline processing registry. We first register all the Chandra input and output data products. We can then capture the processing requirements and inter-dependencies by registering all the pipelines. Data products are registered with a File_ID, file name convention (using regular expressions), method for extracting start/stop times, and archive ingest keywords (detector, level, etc.). Pipelines are registered with a Pipe_ID, pipeline profile name, pbuilder arguments, kickoff criteria (detector in focal plane, gratings in/out, etc.), input and output data products (by File_ID), and method for generating the "root" part of output file names.

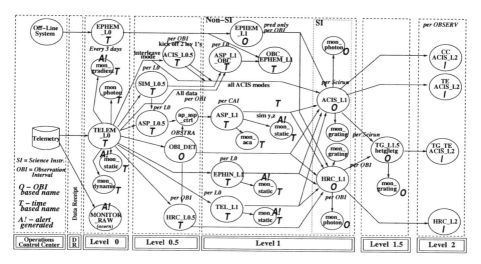

Figure 2. Standard Processing Threads.

6. Automatic Processing Components

With a processing registry, the Automatic Processing system is able to recognize data products, extract start and stop times, initiate pipeline processing, and ingest products into the archive. Figure 3 illustrates the flow of data through the AP system. Here is a brief description of each of the AP components in Figure 3:

- The OCC (Operations Control Center) sends scheduled observation and engineering requests, raw telemetry, and ancillary data (ephemeris, clock correlations, etc.) to the CXC.
- Ancillary Data Receipt (implemented via "OPUS") sends ancillary data to the Archive via "darch."
- The Data Archiver/Retriever Server (darch) and the Archive Cache Server (cache) serve as an interface to the Archive. Files sent to darch are first sent to the cache, then the Archive. Darch checks the cache before retrieving from the Archive to save time and reduce the load on the archive. Darch also sends a notification to OST for every data product cached. For more details see Subramanian (2001).
- Telemetry Data Receipt polls the input directory and picks up new raw telemetry files. It then checks counters and trims off any overlapping data sending the edited raw telemetry file to darch and DR_FlowControl.
- DR_FlowControl sends raw files to the Telemetry Processor one at a time and is used as an entry point for error recovery or reprocessing.
- The Telemetry Processor strips out telemetry into strip files by spacecraft component. It also identifies gaps in the telemetry and the start and stop of observations. The strippers run continuously on a "stream" of raw telemetry. The Extractors can then run on each strip file and decommutate the raw data to create Level 0 FITS files.

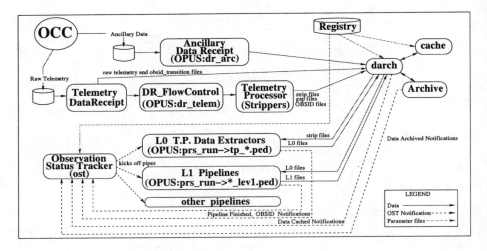

Figure 3. Automatic Processing System.

- The OST (Observation Status Tracker) knows about all data products and pipelines via the registry. It sends a message to OPUS to start pipeline processing when all inputs are available.
- The OPUS system is off-the-shelf software from the Space Telescope Science Institute. It provides distributed processing, GUIs for control and directory polling (among other useful utilities). The CXC AP system runs 3 "OPUS Pipes:" Ancillary Data Receipt, DR_FlowControl, and Pipeline Processing. Pipeline Processing consists of 6 OPUS "stages" including data retrieval, running the CXC pipeline, data archiving, notifications, and cleanup. For more OPUS information see Rose (2001).

7. Conclusion

The AP Infrastructure was designed to fulfill a complex set of Chandra processing requirements as efficiently as possible. Instead of hard-coding all the complex requirements and dependencies into software, the AP system relies upon a registry method to configure the processing. The AP infrastructure software can then remain fairly general and maintenance becomes easier as most new processing requirements, enhancements and bug fixes can be accomplished by registry updates. Also, the registry can be easily updated apart from Software Releases for special purposes such as testing, reprocessing and special processing.

Acknowledgments. This project is supported by the Chandra X-ray Center under NASA contract NAS8-39073.

References

Subramanian, S. 2001, this volume, 303
Rose, J. 2001, this volume, 325

The FITS Embedded Function Format

Arnold H. Rots, Jonathan C. McDowell, X. Helen He, Peter E. Freeman

Harvard-Smithsonian Center for Astrophysics, 60 Garden Street MS 81, Cambridge, MA 02138

Michael Wise

Massachusetts Institute of Technology, Center for Space Research

Abstract. We have developed a format convention that allows one to specify an n-dimensional function in a FITS binary table, the FITS Embedded Function (FEF). The format allows for enumerated values, constants, and analytical functions, and arithmetic combinations of those three. The parameters of the analytical functions may, again, be enumerated, constant, or function values. The concept is intended to allow the user to extract a multi-dimensional subimage from a FEF in the same way one would extract a subimage from a primary array or image extension. The format is extremely versatile and has many potential applications. Developing a generic FEF extractor is very challenging but will allow very cost-effective reuse.

1. Scope

The FITS standard provides for the definition of n-dimensional images of data points, but in certain situations it can be extremely useful to have the ability to store an image in parameterized form, as an arbitrary function of n parameters (coordinate axes). Just as one can extract a sub-image from a primary HDU or an image extension, one should be able to extract an n-dimensional sub-image of enumerated values from an extension containing such a parameterized image.

In cases where the parameterization is possible, this representation effects a huge savings in storage space. We have developed a convention that accommodates arbitrary function specifications using FITS binary tables. We have called it the *FITS Embedded Function* (FEF).

As an example, within the Chandra X-ray Center we have applied the FITS Embedded Function to the Chandra response matrix specification. As a result, a generic FEF image extractor will be able to construct response matrices using the FEF response matrix definition.

2. Definitions

Function tables are identified by:

    HDUCLASS= 'ASC   '

HDUCLAS1= 'FUNCTION'

The function is defined by:

FUNCTION= '<expression>'
FUNCNAME= '<name>'

The function expression may contain the arithmetic operators +, −, *, /, **, and five types of operands or "parameters:"
- FTYPEi: axes of the function evaluation space; these may just be specified as a range or may also appear as a table column (TTYPEj) if other parameters are to be enumerated along such an axis.
- DTYPEi: constants; the name is provided by DTYPEi, the value by DVALi, and the units by DUNITi. This is the "constant column."
- TTYPEi: parameters enumerated in table column i.
- VTYPEi: function components; the name is provided by VTYPEi, the function expression by VFUNCi. This is the "virtual function column."
- WTYPEi: arithmetic components; the name is provided by WTYPEi, the arithmetic expression by WFUNCi. This is the "virtual arithmetic column."

Except for the first, these operands can be thought of as columns in the table (real columns, constant columns, and virtual columns), and the function would thus be an arithmetic combination of these "columns." One should be warned, though, that operands may be defined in terms of other operands; hence reality is more complicated than this simplified view.

The assumption is that implementation of a FEF reader will include the following functionality.

```
typedef struct {
  char name[32] ;
  double min ;
  double max ;
  int num ;
} FefParam ;

FefParam parms[n] ;
fits_file fef ;

double* image = readFef (fef, n, parms) ;
```

readFef returns an n-dimensional image where the lengths of the axes are set by parms[i].num, the names by parms[i].name, and the minimum and maximum by parms[i].min and parms[i].max, respectively. Each parms[i].name must correspond with an FTYPE value and the minimum and maximum values must not exceed the corresponding FLMIN and FLMAX values. n should be equal to FAXIS.

Extraneous columns are allowed. However, one should take care when filtering on such a column (or any other, for that matter), that such filtering can only be guaranteed to yield a valid FEF extension if all but one of the FTYPE variables contains more than one point. Even then, it will still require that all FAXISi keywords are corrected.

3. The Coordinate Axes

The evaluation space of the function (or the image coordinates, if you like) is set by a set of "pseudo coordinate axis" keywords:
- FTYPEi:
- FAXIS: The number of coordinate axes in evaluation space
- FAXISi: The number of points along axis i (for enumerated axes only)
- FTYPEi: Name of axis i
- FLMINi: Legal minimum values for axis i (required for free-running variables)
- FLMAXi: Legal maximum values for axis i (required for free-running variables)

One should distinguish between enumerated and free-running variables (coordinates). The free-running variables only appear as parameters in function expressions while for enumerated variables the function value (or some parameter value) is explicitly given for a list of the variable's values. As an extreme example, a measured image could be represented with one row per pixel, a column for each coordinate, and one for the image value, where the coordinate position of each pixel would be explicitly given on each line: all coordinates would be enumerated. If, on the other hand, the function were a one-dimensional Gaussian, there would only be one free-running variable. Somewhere in between, one could conceive of a function in x and y, where the function value is a one-dimensional Gaussian with free-running variable x, but where the width of the Gaussian is enumerated for all values of y.

FLMIN and FLMAX are clearly needed for free-running variables to define the range over which the function is defined.

We shall assume that if function values are requested for values of enumerated variables that fall between the enumerated grip points, the values of all enumerated columns will be interpolated linearly. It is important to note that *the parameter values are interpolated, not the function values.*

If the values of an entire row are to be held constant for a range of values of an enumerated variable, one may specify bins by using a <FTYPE>_LO and <FTYPE>_HI column.

In some cases (sparse images, e.g., response matrices) it is desirable to limit the domain of a free running variable depending on the values of the enumerated variables. This may be done by incorporating two columns for FDMINi and FDMAXi.

4. FUNCTION Specification

The FUNCTION definition is an arithmetic expression in which the operands may be the *values* of any of the following: FTYPEn, DTYPEn, TTYPEn, VTYPEn, and WTYPEn, with operators $+$, $-$, $*$, $/$, $**$, and parentheses allowed, as in:

```
FUNCTION= 'Norm - Scale * (X2 + Y2)'
FUNCNAME= 'HRMA_EffArea'
BUNIT    = 'mm**2   '
```

5. VTYPE

Virtual function columns (VTYPEi) are defined through functions:

```
VFUNCi = 'G(P1, P2, P3, ...; C)'
```

where G is chosen from a defined set of function names and where P_i may be the value of any valid operand—one of the following: FTYPEn, DTYPEn, a constant number, TTYPEn, VTYPEn, or WTYPEn. C is the name of a parameter object or a coefficient object that is specific to the function G and whose attributes need to be specified as $C\_ < name >$; not all functions require a parameter object. For each parameter object attribute there has to be one TTYPE, DTYPE, VTYPE, or WTYPE that carries its name as value. VFIELDS specifies the number of VTYPEs that are defined.

Note that allowing VTYPEs and WTYPEs to be used in the definition of VTYPEs provides for the specification of nested functions. This is a powerful capability that requires particular care to prevent recursion.

6. WTYPE

In general, arithmetic virtual columns (WTYPEi) are defined as:

```
WFUNCi = 'P1 ^ P2 ^ ...'
```

Where "^" denotes an arithmetic operator $+$, $-$, $*$, $/$, $**$, and parentheses are allowed. P_i may be the value of any of the following: FTYPEn, DTYPEn, a constant number, TTYPEn, VTYPEn, or WTYPEn. WFIELDS specifies the number of WTYPEs that are defined.

Acknowledgments. This project is supported by the Chandra X-ray Center under NASA contract NAS8-39073.

New Elements of Sherpa, CIAO's Modeling and Fitting Tool

P. E. Freeman, S. Doe, A. Siemiginowska

Harvard-Smithsonian Center for Astrophysics MS-81, 60 Garden Street, Cambridge, MA 02138

Abstract. We describe enhancements made to *Sherpa* for the CIAO 2.0 release, concentrating upon those that enable a user to: (1) analyze *Chandra* X-ray Observatory grating data with wavelength- or energy-space models; (2) simultaneously fit background and source datasets; and (3) estimate and visualize confidence intervals and regions. We also list enhancements that we plan to make to *Sherpa* for future CIAO releases.

1. Introduction

Sherpa is the modeling and fitting tool of the *Chandra* Interactive Analysis of Observations (CIAO) software package (Doe et al. 1998 and references therein). We have developed it with the primary goal that a user should be able to take full advantage of *Chandra*'s unprecedented observational capabilities and be able to analyze data in up to four dimensions (energy E or wavelength λ, time t, and spatial location $[x, y]$) with a wide variety of models, optimization methods, and fit statistics. The enhancements that we have made to *Sherpa* for the CIAO 2.0 release, described below, represent major steps towards this goal.

2. Enhancements to Sherpa

2.1. Grating Analysis

Data Analysis in Wavelength and Energy Space. *Chandra* grating data are most naturally analyzed in wavelength space, while *XSPEC* line models such as `xsraymond` are defined in energy space.[1] *Sherpa* now allows one to define models in either space, while using either grating Ancillary Response Files (gARFs) or Response Matrix Files (gRMFs) or both. The `ANALYSIS` command allows one to switch between spaces.[2] One can also now apply filters defined in wavelength or energy space to single datasets, groups of datasets, or to `allsets`. See Figure 1.

---

[1] Models in the *XSPEC* v.10 library are available to users of CIAO 2.0, while the v.11 library will be available starting with CIAO 2.1.

[2] The reader will find more information about `ANALYSIS`, as well as all other *Sherpa* commands, at http://asc.harvard.edu/ciao/documents_manuals.html.

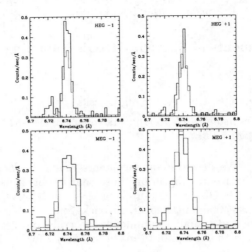

Figure 1. Best-fit of a normalized Gaussian function to an emission line observed in four first-order HEG and MEG *Chandra* grating spectra of Capella. The amplitude, full-width at half-maximum, and position values are linked between datasets. The `identify` function of *GUIDE* indicates that line is most likely due to the Si XIII 2→1 transition at 6.7403 Å.

Data Analysis with Two Background Spectra. Standard processing of *Chandra* grating data includes the extraction of background spectra, dubbed "up" and "down," from either side of the source extraction region. One can either fit both spectra simultaneously with the source spectrum (see below), or `SUBTRACT` both from the source spectrum.

The Grating User Interactive Data Extension (GUIDE). This S-lang-based extension to *Sherpa* assists the fitting of atomic lines and differential emission measure (DEM). For more information, see Doe, Noble, & Smith (2001) and http://asc.harvard.edu/ciao/download/doc/guide_doc.ps.

Saving Analysis Results. One can save and restore a *Sherpa* session using a Model Descriptor List (MDL) file, which records information about input datasets, and filter and model definitions. One example of its usefulness is in DEM fitting, where the input data are MDL-stored line fluxes and flux errors.[3]

2.2. Simultaneous Analysis of Background and Source Data

Previous versions of *Sherpa* allowed the user to input background data with the commands `BACK` or `READ BACK`, but these data could only be subtracted from

---

[3] Flux errors are easily estimated for three *Sherpa* models for which the amplitude is equal to the flux: the normalized Gaussian (`ngauss`); the delta function (`delta`); and the Lorentzian (`lorentz`).

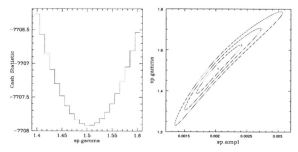

Figure 2. Examples of parameter space visualization. *Left:* A plot showing the Cash statistic as a function of power-law slope, generated using INTERVAL-PROJECTION. *Right:* contour plot showing 1, 2, and 3σ confidence regions for the power-law amplitude and slope, generated using REGION-PROJECTION. The central cross indicates the best-fit point.

the source data. *Sherpa* now allows the simultaneous analysis of one (or two) background dataset(s) for every source dataset that is read in. A background model is fit directly to the background data, and is also extrapolated to the source region, where it is added to the source model before convolution. Rescaling for different extraction region sizes is done using the values of the BACKSCAL keyword, set in the header of the PHA files containing source and background data, using the commands SETDATA and SETBACK.

2.3. Estimation of Confidence Intervals and Regions

Sherpa contains many new methods that one can use to estimate confidence intervals or visualize confidence regions for best-fit model parameters. Note that these methods are strictly valid, i.e., provide 1σ confidence intervals that actually contain 68.3% of the integrated probability, when (1) the χ^2 or $\log\mathcal{L}$ (log-likelihood) surface in parameter space is approximately shaped like a multi-dimensional paraboloid, and (2) the best-fit point is sufficiently far from parameter space boundaries.

Uncertainty. The confidence interval is determined for each parameter in turn by varying its value while holding the values of all other parameters at their best-fit values. While fast, UNCERTAINTY will underestimate a parameter's interval if it is correlated with other parameters. One can visualize spaces with INTERVAL-UNCERTAINTY and REGION-UNCERTAINTY.

Projection. The confidence interval is determined for each parameter in turn while allowing the values of all other parameters to float to new best-fit values. One can visualize spaces with INTERVAL-PROJECTION and REGION-PROJECTION (see Figure 2).

Covariance. The confidence interval for each parameter is determined using the diagonal terms of the covariance matrix. While fast, it cannot be used to visualize parameter spaces.

2.4. Other Enhancements

Below we list other enhancements to *Sherpa* made for the CIAO 2.0 release.
- We have extensively retooled the algorithms for the optimization methods POWELL, SIMPLEX, and LEVENBERG-MARQUARDT to make them more robust.
- The parameter value guessing algorithm now takes into account the exposure time and ARF if PHA spectral data are input.
- One can use ARFs and RMFs with different photon-space binning.
- One can simulate one-dimensional spectra with FAKEIT.
- One can define two-dimensional spatial models in either image coordinates or in the World Coordinate System (WCS).
- New models other than ngauss and delta include a broken power law (bpl), one- and two-dimensional constants (const and const2d), a two-dimensional delta function (delta2d), a phenomenological photoionization edge model (edge), and a line broadening model (linebroad).
- One can set preferences in a .sherparc file in the home directory.

3. Selected Future Enhancements to Sherpa

Spectral Fitting. *Sherpa* does not yet treat photon "pile-up," which can markedly affect the fitting of energy spectra of strong sources observed by either *Chandra* or *XMM*. Another *Chandra*-specific enhancement would be the ability to convolve data with analytic functions specified in Fits Embedded Function (FEF) files (Rots et al. 2001), rather than a response matrix, which could markedly decrease the time needed to analyze grating spectra.

Spatial Analysis. Currently, *Sherpa* cannot apply exposure maps in spatial analysis, nor can it calculate the fluxes in two dimensions. Also, the current *Sherpa* requirement that a one-to-one mapping exist between each background bin and source bin must be waived so that, e.g., one can define differently sized source and background regions in an image.

Statistics. Enhancements to be made include adding model comparison tests, correlation analysis and non-parametric fitting, and support for Bayesian analyses (e.g., specification of the prior and credible interval/region estimation).

Acknowledgments. This project is supported by the *Chandra* X-ray Center under NASA contract NAS8-39073.

References

Doe, S., Noble, M., & Smith, R. 2001, this volume, 310
Doe, S., Ljungberg, M., Siemiginowska, A., & Joye, W. 1998, in ASP Conf. Ser., Vol. 145, Astronomical Data Analysis Software and Systems VII, ed. R. Albrecht, R. N. Hook, & H. A. Bushouse (San Francisco: ASP), 157
Rots, A., McDowell, J., Wise, M., He, H., & Freeman, P. 2001, this volume, 479

Converting FITS into XML: Methods and Advantages

Brian Thomas,[1] Edward Shaya,[1] and Cynthia Cheung

Goddard Space Flight Center/NASA, Code 631, Greenbelt, MD 20771

Abstract. We discuss how and why FITS data should be encapsulated in XML. Our goal is not to throw away the FITS standard entirely. Rather, we seek to re-map the FITS standard into an XML-based format. The advantages of doing so are legion and include: greater interoperability, parsing by XML aware browsers and applications, hierarchical structure for improved searchability, default values for header descriptions, extensibility for specialized usage and future development, and piggybacking on industry applications.

1. Why FITS and XML are Good for Each Other

The current life cycle of astronomical data is lived through a bizarre parade of data formats. A probable path in this cycle could be: data are taken at the telescope as FITS, they are submitted to a publisher in LaTeX, the publisher converts these data into SGML, data centers convert the SGML into some form of ASCII (such as HTML), and the end-user downloads this and converts the data back into FITS (!) for use in various analysis software.

So many format translations need not occur, since XML, the eXtensible Markup Language, could provide a reasonable interchange format to stand in at every step of the cycle. XML is designed to be data-centric (unlike HTML or SGML) with the layout/presentation of the information being provided by a separate file (often referred to as a "style sheet"). This simple divorced situation allows us then, at each step of the data cycle, to re-use the same core XML data file. Another advantage of XML is its description and validation via a DTD ("Document Type Definition"). The DTD may be used both to hold information about the data (such as keyword definitions) and to provide a template for querying databases holding the XML documents. Default values for header descriptions may be obtained from the DTD when the document is parsed, allowing the document to hold only those keywords with non-default values. All XML documents may hold URLs that point back to their respective DTDs, insuring that the correct DTD may always be found. Finally, as a recommendation of the World Wide Web Consortium (W3C), XML has the weight of the IT community at large behind it. Currently, this Internet standard is receiving massive development support that easily outstrips any spending that astronomers invest in FITS. It makes sense to leverage this resource.

---

[1]Raytheon ITSS, 4500 Forbes Blvd., Lanham, MD 20706

There is no mystery as to why FITS lacks many of the favorable characteristics of XML; the FITS standard was developed in the late 1970s for a different computing environment (VAX/Fortran/tape archives). While FITS has evolved, it still contains many limitations based on its origins that need not be adhered to today, including 8 character keywords, 80 character cards, a maximum of 999 table records, etc. Yet even for its limitations, FITS still contains a good understanding of the general needs of astronomy data. Because XML is designed to be the basis of other languages, it is generic in nature and will require development effort to adequately encapsulate scientific data with it.

There is no need to re-invent the wheel here. What is really needed is a marriage between XML and FITS with the goal of re-mapping (rather than redefining) the FITS standard into an XML-based format. We refer to this new hybrid data format standard to as "FITSML" and briefly discuss important characteristics of this project below.

2. FITSML: Using XDF to XMLize FITS

The Astronomical Data Center (ADC) at Goddard Space Flight Center is in the process of developing a science data interchange language in XML. This new language, XDF (eXtensible Data Format) is formulated to contain only the most basic needs of encapsulating scientific (not just astronomical) data.

We have found the translation of basic FITS data types into XDF-based FITSML is possible without content loss or redefining important FITS definitions (keywords). This is possible because the XDF data model is fairly sophisticated and allows for many types of ASCII and binary data, unevenly distributed data cubes, grouping of data with mixed data formats, and vector spaces, among other things. Furthermore, the XDF kernel is designed to allow discipline specific keywords to be included within XDF-based data formats. This property allows FITSML to look familiar to old hands. XDF also allows for two other important properties of FITS: the addition of user-defined keywords and extension of FITSML into sub-field/mission-specific varieties of FITS.

XDF provides more than a means to XMLize FITS however. It brings in new (and we feel needed) properties that FITS currently lacks. Some of the most important of these features include the following:

Inheritablity: While discipline-specific keywords may be layered on top of XDF to create new discipline specific languages these new languages still share common base conventions which are understood by the others. This property in the object-oriented world of programming is known as inheritance and provides that software designed to read FITSML will also be able to read any other XDF file, regardless of whether the data is astronomical in nature or not. Properly formulated, the structure of FITSML may be designed such that the interchange works (in a limited fashion) in reverse, e.g., any XDF software may read in FITSML.

Infinite hierarchical nature: XDF allows for both infinitely regressing hierarchical structure/array and keyword associations. This means that FITSML files may contain more sophisticated data structures than FITS which in turn provides greater freedom to application developers and archivists to seek appropriate software solutions for storing data. As for the keywords, this kind of flexible keyword hierarchy, if properly implemented, may provide improved searchability of the data by allowing for a more "natural-language" understanding of the indexing and easier formulation of queries by a human to the database. An example of the kind of keyword hierarchies that may be held in FITSML is shown below.

```
<!-- some FITS keywords, but in hierarchy -->
   <observation>
      <telescope>VLA</telescope>
      <observer>Syke</observer>
      <imageType>object</imageType>
      <datesAndTimes>
          <observationDate>27/10/1982</observationDate>
      </datesAndTimes>
      <positions>
          <astroObject>3C405</astroObject>
      </positions>
   </observation>
   ... continues ...
</XDF>
```

Machine understandable scientific units: One of the central problems in comparing science data is the determination of compatibility from data units. All science units break down into eight principle SI units (meter, gram, second, radian, Ampere, degree Kelvin, candela) and "number". These basic units are defined in the XDF DTD and are used to build up other common scientific units. For example, the unit "Newton" can be defined as an entity:

```
<!ENTITY newton
'<unitGroup name="N">
    <apply><times/>
         <meter />
         <kilogram />
         <apply><power/><second /><cn>-2</cn></apply>
    </apply>
</unitGroup>' >
```

Unit entities may also be used to create other units, such as for the unit "Pascal" in the example below which uses "&newton;" entity in its definition:

```
<!ENTITY pascal
'<unitGroup name="Pa">
    <apply><divide/>
         &newton;
         <apply><power/><meter /><cn>2</cn></apply>
    </apply>
</unitGroup>' >
```

With units expressed as entities the FITSML parser may easily examine and decompose unit definitions into the constituent nine basic units.

3. Summary, Future Progress, and Resources

3.1. The FITSML Project and its Future Direction

XDF and FITSML are works in progress. Although we have examined the encapsulation of standard forms of astronomical data, such as images, spectra, tables, and sky atlases, we continue to examine other types of data in order to give XDF the generality we desire, and to work with the FITS community to further develop the FITSML DTD in order to meet their needs. Towards this end, we look forward to developing a prescription for FITSML to wrap legacy FITS files and to test translation of more advanced FITS data formats into FITSML. Other plans include investigating various combinations of style sheets for viewing (and perhaps editing FITSML) within web browsers, and releasing a beta software package for XDF and an alpha software package for FITSML (both releases have software written in Java/Perl) in Spring 2001. We hope to provide a beta software package and simple translation tools between FITS and FITSML before the end of 2001.

3.2. Resources

Space limitations prevent our fully describing FITSML or disclosing the FITSML DTD or samples within this brief article. Please refer to the following web pages for more information:

- XDF Homepage[2] (including links to FITSML/XDF DTDs)
- Software Download[3] (including samples of FITSML/XDF):
- ADC Homepage[4]

---

[2] http://xml.gsfc.nasa.gov/XDF/

[3] http://xml.gsfc.nasa.gov/ADCSoftwareDownload.html

[4] http://adc.gsfc.nasa.gov/

SVDFIT: An IRAF Task for Eigenvector Sky Subtraction

Douglas J. Mink and Michael J. Kurtz

Harvard-Smithsonian Center for Astrophysics

Abstract. A new method has been developed for estimating and removing the spectrum of the sky from deep spectroscopic observations. This method does not rely on simultaneous measurement of the sky spectrum with the object spectrum. The technique is based on the iterative subtraction of continuum estimates and eigenvector sky models derived from singular value decompositions of sky spectra and sky spectra residuals. IRAF tasks have been developed to implement this technique. Using simulated data derived from small-telescope observations, we demonstrate that the method is effective for faint objects on large telescopes.

1. Introduction

Multi-fiber spectrograph apertures are too small to return simultaneous sky spectra for each fiber, so the sky contribution which is removed from the observed object+sky is not exactly what was observed by that fiber. The current solution is to aim some fibers at blank sky and subtract them from the all of the object+sky spectra, using the closest sky or a mean image sky and fitting it in some way, such as by matching the OI line at 5577 Angstroms.

In a previous paper (Kurtz & Mink 2000), we described a way to model the night sky by decomposing a set of observed night sky spectra into orthogonal eigenvectors using singular value decomposition. A subset of these eigenvectors is then fit to the object+sky spectra and the fit removed, leaving residual object spectra. This is made easier by the fact that for redshift work, any continuum signal is removed anyway, as only the positions and shapes of absorption and emission lines are needed.

This paper describes the software package we developed to carry out the work. Because software written as part of the RVSAO package (Kurtz & Mink 1998) already did much of the work of dealing with spectra, we adapted that and produced a new IRAF package, SVDFIT. The code was written in SPP to take advantage of many existing useful subroutines, though we experimented with various numbers of iterations using CL scripts and simpler SPP tasks.

2. New IRAF Tasks

svdprep is an SPP task, based on RVSAO's **sumspec** task, which rebins spectra to a common dispersion, with additional processing at the same time. Specific portions of the spectra, such as the brightest night sky lines, may be re-

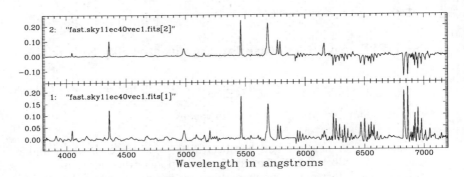

Figure 1. First and second eigenvectors of the night sky after major lines and continuum have been removed.

moved, as may a continuum fit using the same continuum removal subroutines as in RVSAO. Spectra may be normalized to their mean values so they have a common range of values, and stacked into a single two-dimensional file for further processing by **svdvec** or **svdres**. The spectra may also be redshifted to a common radial velocity, though this feature isn't used for sky subtraction.

svdvec is a Fortran program, called by IRAF as a foreign task, which decomposes an array of spectra, such as that created by **svdprep**, into eigenvectors using Singular Value Decomposition. An array of eigenvector spectra with the same dispersion is returned.

svdres is an SPP task which fits a set of eigenvectors to a spectrum, stacked spectra, or a list of spectra, returning the residuals as spectra in the same format as input. The processing features of **svdprep** are included in **svdres**, so it can be run as a single task on unrebinned spectra.

svdres2 is a modification of svdres to fit two sets of eigenvectors to a spectrum, removing and optional continuum in between, returning the final residuals, which may be sky-subtracted spectra. This avoids a lot of I/O and more easily parameterizes the entire two-pass process.

3. Making Eigenvectors

Sky spectra from long exposures on moonless nights were rebinned using the **svdprep** task to make an array of spectra with identical dispersion. Before rebinning each spectrum, the brightest night sky lines (OI at 5577, 6300, and 6363 Angstroms and Na at 5890 Angstroms) were replaced by interpolated continuum, and the entire continuum signal was removed. We assume that any spectral features on these bright lines would have been overwhelmed by the Poisson noise in a long-exposure sky spectrum.

Eigenvectors for these spectra were then computed using **svdvec**. Any singular valued decomposition program could be used. Figure 1 shows the first two eigenvectors.

SVDFIT: An IRAF Task for Eigenvector Sky Subtraction

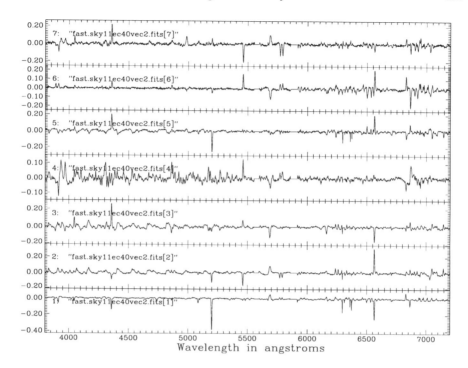

Figure 2. Next seven eigenvectors of the night sky after first two eigenvectors and second continuum have been removed.

Major sky spectrum features are then removed by fitting the first two eigenvectors—3 or 4 might be OK, too—and removing the fit from the array of spectra using **svdres**.

Using **svdprep**, the continuum is removed from the residual spectra. It turns out that many of the eigenvectors in the first pass account for night to night sky continuum variation on a scale smaller by the sky spectrum signal. Removing this spectrum by spectrum leaves only sky and object spectrum features.

We again compute eigenvectors for the array of residual spectra using **svdvec**. Fitting the first 7–10 of these eigenvectors using **svdres** removes the rest of the sky features from our object+sky spectrum. Figure 2 shows the first seven eigenvectors.

4. Removing Sky Contributions from Spectra

The **svdres2** task fits and removes night sky eigenspectra from observed object+sky spectra. Figure 3 shows the process from bottom to top:

1. Original spectrum.
2. Remove continuum and brightest night sky lines from spectrum and rebin spectrum to match dispersion of eigenspectra.

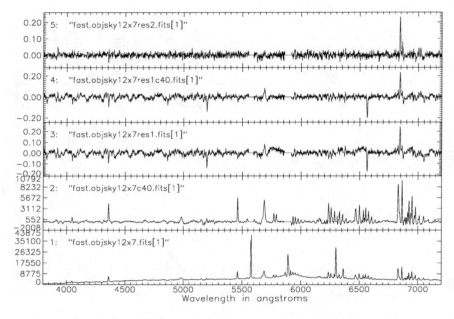

Figure 3. Removal of the night sky contribution to an object spectrum using **svdres2**.

3. Remove major sky spectrum features by fitting first two eigenvectors and remove the fit sky from the spectrum.
4. Remove the continuum from the residual spectrum.
5. Fit the first seven second pass eigenspectra and remove the sky spectrum thus fit to leave a spectrum of the observed object, which may then be run through RVSAO to get a redshift.

5. Obtaining this Software

For the current status of this package, access to source code, additional documentation, and examples, see http://tdc-www.harvard.edu/iraf/svdfit/.

References

Kurtz, Michael J. & Mink, Douglas J. 1998, PASP, 110, 934
Kurtz, Michael J. & Mink, Douglas J. 2000, ApJ, 533, L183

Astronomical Data Analysis Software and Systems X
ASP Conference Series, Vol. 238, 2001
F. R. Harnden Jr., F. A. Primini, and H. E. Payne, eds.

BIMA Xfiles: Empowering the Observer with Tcl/Tk Applications

Theodore Yu

Hat Creek Radio Observatory, Hat Creek, CA 96040

Abstract. Described is a Tcl/Tk application suite that helps expedite operations at Hat Creek Radio Observatory (HCRO). This is the site of a ten element millimeter interferometer operated by the Berkeley-Illinois-Maryland Association (BIMA). Xscribe provides an integrated environment for maintaining the schedule and summary of observations. Xplore is a sky chart based calibrator selector. Xplotwatch and Xfixwatch are a telemetry monitor and critical alarm interface respectively. Xaudio is an audio reminder selector.

1. Introduction

The predecessors of the BIMA Xfiles[1] were Unix C-shell (csh) scripts with command line interfaces. Tcl/Tk scripting allowed for rapid development of graphical user interfaces (GUIs). By leveraging existing programs rather than rebuilding everything from scratch, Tcl/Tk also allowed for rapid deployment. Observers familiar with the predecessor scripts quickly grasped the new functionality and appreciated the ease of use. By bundling together many previously separate tasks under a common graphical interface, these applications have made for a much more efficient and pleasant experience for the BIMA observer.

2. Xscribe

Before Xscribe, the BIMA schedule was composed simply by using a text editor. Time was allocated in an ASCII graph format and Scribe then parsed this into an observing program that drove the array. Scribe was Xscribe's predecessor, a Unix csh script which relied heavily on the awk and sed pattern matching utilities. This scheme was deployed for three years until it could not cope with new scheduling pressure.

The availability of 1 mm receivers meant there was a need for dynamic scheduling. Successful 1 mm observing requires lower atmospheric opacity and higher atmospheric stability than for 3 mm observing. Because such conditions are unpredictable in the long term, 1 mm observing cannot be scheduled *a priori*. Thus, when the atmospheric window opens and closes, there must be a scheme to efficiently substitute appropriate projects.

---

[1] http://bima2.astro.uiuc.edu/tcltk/xfiles.html

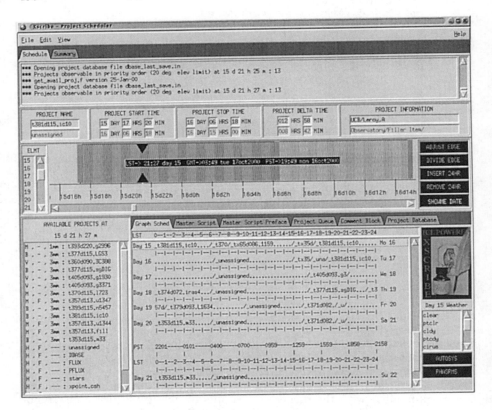

Figure 1. Xscribe GUI: the schedule maintenance page.

2.1. A New Paradigm

Xscribe addresses this need by rendering the schedule in a graphical editor. The Xscribe GUI is modeled as a tabbed notebook, with two primary pages for maintaining the schedule and summary of observations. Figure 1 shows the schedule page, which includes the graphical editor. The editor is a scrollable viewport with vertical (green) bars representing project time boundaries. Using click and drag mouse operations:

- All available projects in priority order and their LST ranges may be viewed at any given time.
- Existing slot boundaries may be adjusted and new slots may be created.
- Projects may easily be substituted.
- The schedule may be displaced at any point by arbitrary 24 hr LST increments.

The schedule page also features a secondary tabbed notebook. There are pages here for exporting the Xscribe native graphic format into ASCII list, ASCII graph, and master script formats. The ASCII formats are automatically made

BIMA Xfiles: Empowering the Observer with Tcl/Tk

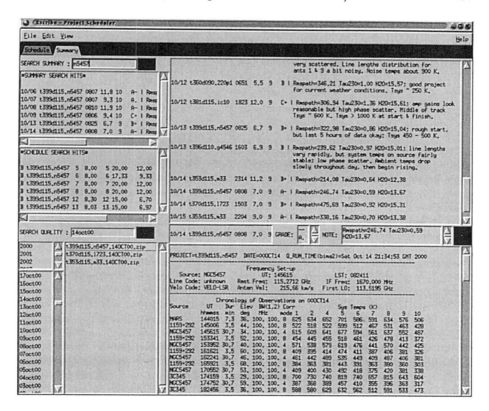

Figure 2. Xscribe GUI: the summary maintenance page.

available on the WWW. There is also a page for editing the project database, which lists priority and coordinate information.

The master script drives the array and launches an automated quality analysis after each project is done. Each archived quality report includes diagnostic plots in GIF format. The entire quality archive is accessible from Xscribe's summary page, shown in Figure 2. Xscribe calls on John Bradley's well known Xv image viewer (not a Tcl/Tk application) to display the GIF plots, and provides a convenient template for summarizing an observation. An advantage of scripting is integration of external components and applications, such as the Xv image viewer.

2.2. On-line and Off-line Usage

The site observer by definition uses Xscribe on-line. At the start of an observing season, the BIMA scheduler sends two initial lists to the observatory: the schedule and project database. These templates are generated as part of the automated proposal pipeline at the University of Illinois. They are imported into Xscribe, after which the schedule and database evolve as the observing season proceeds.

Xscribe is flexible enough to set up for off-line users. For example, the program is installed for the BIMA scheduler at the University of Illinois. It has HTTP transfer capability so that off-line users may download the updated ASCII lists from the observatory and import them into the local Xscribe session.

3. Xplore

Interferometric observations usually require a point source calibrator to be observed together with the science source. The goal is to remove time dependent atmospheric phase fluctuations. To reduce any direction dependent errors, the angular separation between the two targets should be minimal. At the same time, the calibrator should be sufficiently strong to satisfy signal to noise requirements.

To help make a decision, the BIMA calibrator database is plotted in the Xplore GUI, with both symbol color and size indicating flux strength. The scrollable viewport has an adjustable field of view. Clicking on a calibrator shows information such as specific flux and angular separation from the viewport center (science target position). A graph above the star chart shows how much the two targets overlap each other in terms of LST range.

If the Tcl/Tk extension BLT is available, then clicking on a calibrator will also show its flux history on a line graph. If the BIMA data analysis package (MIRIAD) is available, then planetary positions are calculated as well. These optional features again take advantage of external components and applications. Like Xscribe, Xplore is flexible enough to set up for off-line users. It has HTTP transfer capability so that the off-line user may download an updated calibrator database from the observatory.

4. Xplotwatch, Xfixwatch, and Xaudio

Sensors continually monitor the array and the archived data are a valuable resource for diagnosing problems. Xplotwatch is a data mining tool. The Xplotwatch GUI allows the user to easily choose arbitrary combinations of antennas, sensors, and dates. The data are rendered as either antenna based or sensor based plots. The time (abscissa) axis may be a fraction of a day or several concatenated days, for studying short or long scale trends respectively.

Several critical sensors have an associated alarm. An alarm will trigger if the monitored data falls out of tolerance range. Often the problem cannot be repaired immediately. The Xfixwatch GUI may be used to selectively disable the alarm. The GUI allows any combination of sensors and antennas to be easily selected. It is also trivial to disable a given sensor for all antennas, or disable all sensors for a given antenna.

The most critical time of an observation is during the beginning. The Xaudio GUI offers a choice of entertaining sound effects which will be played when a project starts.

Acknowledgments. I am grateful to Rick Forster, Marc Pound, and Peter Teuben for enlightening discussions and encouragement. Thanks to the users of the BIMA Xfiles for their useful feedback.

Astronomical Data Analysis Software and Systems X
ASP Conference Series, Vol. 238, 2001
F. R. Harnden Jr., F. A. Primini, and H. E. Payne, eds.

Immersive 4-D Interactive Visualization of Large-Scale Simulations

Peter Teuben

Astronomy Department, University of Maryland, College Park, MD

Piet Hut

Institute for Advanced Study, Princeton, NJ

Stuart Levy

National Center for Supercomputing Applications, University of Illinois Urbana-Champaign, Urbana, IL

Jun Makino

Department of Astronomy, The University of Tokyo, Bunkyo-ku, Tokyo 113-0033, JAPAN

Steve McMillan

Department of Physics and Atmospheric Science Drexel University, Philadelphia, PA

Simon Portegies Zwart[1]

Massachusetts Institute of Technology, Cambridge, MA

Mike Shara, Carter Emmart

American Museum of Natural History, New York, NY

Abstract. In dense clusters a bewildering variety of interactions between stars can be observed, ranging from simple encounters to collisions and other mass-transfer encounters. With faster and special-purpose computers like GRAPE, the amount of data per simulation is now exceeding 1 TB. Visualization of such data has now become a complex 4-D data-mining problem, combining space and time, and finding interesting events in these large datasets. We have recently starting using the virtual reality simulator, installed in the Hayden Planetarium in the American Museum for Natural History, to tackle some of these problem. This work[2] reports on our first "observations," modifications needed for our specific experiments, and perhaps field ideas for other fields in science which can

---

[1]Hubble Fellow

[2]http://www.astro.umd.edu/nemo/amnh/

benefit from such immersion. We also discuss how our normal analysis programs can be interfaced with this kind of visualization.

1. NEMO, Starlab and GRAPE

NEMO[3] (Teuben 1994) and Starlab[4] are traditional programming environments with which N-body simulations can be setup, run, and analyzed. NEMO also has a number of tools to import and export data in tables, CCD type images, FITS files, and a large number of other N-body formats. NEMO is geared more towards collisionless stellar dynamics, while Starlab has more sophisticated programs to deal with close encounters, and can now also incorporate stellar evolution through the SEBA package (Portegies Zwart et al. 2001). NEMO and Starlab present themselves to a user as a large set of programs, often glued together using pipes in shell scripts to set up and run complex simulations. For the programmer, a large set of classes and functions are available to construct new integrators and analysis programs. For example, in the following Starlab example an anisotropic King model with 2048 particles has been evolved with 50% binaries (i.e., 3096 actual stars) and stellar evolution:

```
mk_aniso_king -i -n 2048 -u -w 4 -F 3                    |\
   mkmass -i -u 100 -l 0.1 -f 3                          |\
   mksecondary -f 0.5 -l 0.1                             |\
   addstar -Q 0.5 -R 2.5                                 |\
   scale -M 1 -E -0.25 -Q 0.5                            |\
   mkbinary -f 2 -l 1 -u 1000000 -o 2                    |\
   kira -a 0.1 -d 1 -D 25 -n 25 -t 4000 -Q -G 2 -u -B -z 1 > run001
```

The GRAPE special purpose hardware (Hut & Makino 1999), now running at 100 TerraFlops speed, has been successfully interfaced with Starlab, and now is starting to produce massive datasets. Analysis and visualization techniques of those dataset are becoming increasingly challenging.

2. AMNH, Virtual Director, and Partiview

The American Museum for Natural History (AMNH) in New York City has recently renovated its planetarium, and converted it into a state-of-the-art digital planetarium with capabilities for scientific visualization. Their computer system consists of an Onyx2, with 28 CPUs, 14 GB of memory, 2 TB diskspace and 7 graphics pipes. Each graphics pipe controls one of 7 projectors which illuminate the dome in a dodecahedral pattern. The software that drives most visualization is an NCSA product called Virtual Director[5] (virdir), that we have now been

---

[3] http://www.astro.umd.edu/nemo/

[4] http://www.manybody.org

[5] http://virdir.ncsa.uiuc.edu/virdir/virdir.html

Figure 1.   Partiview in action: after loading a 4-D dataset, the mouse in the window controls motion and spatial orientation, the jog-wheel at the top right controls animation, either manually or via the CD-like control buttons to the right. Partiview also has a command language, commands are entered in the Cmd window in the middle, right below the scrolling logfile window.

using during a number of night sessions in the dome, much like optical observers (during daytime the planetarium is of course used for public viewing). It allows us to "fly" through the data, in space and time. By adding complete orbital information for a select number of stars we have started fully interactive data mining of our 4-D spacetime histories of these star cluster simulations runs. In order for us to test new visualization techniques, algorithms and interfaces with the Starlab environment, we used an existing program `partiview`, which had been derived from `virdir`, and which can be run on a normal workstation or laptop. It uses the FLTK and MESA/OpenGL libraries for its user interface and fast graphics. A screenshot of `partiview` in action can be seen in Figure 1. We have modified `partiview` to understand our Starlab simulation data, and added interfaces that allow this workstation version to animate and move in time and space. `partiview` comes with a small but powerful set of commands with which dataselections and viewing can be made, and we hope to expand this into a more mature scripting language. It is also fairly straightforward for other packages to benefit from using `partiview`.

3. Future Plans

In the spirit of the federated model of archiving observational data, recently proposed by the National Virtual Observatory (NVO) initiative, we will develop a Starlab-based archive. A simulation of a globular cluster with a million stars stars for ten billion years will generate 100 TB of raw data, of which we would like to store at least 1 TB, and preferably more, for 4-D visualization of the full history of the evolution of a star cluster. Although our main goal will be to enable rapid and intelligent access to our simulation output files, we will simultaneously develop a flexible and transparent interface with the NVO database and protocols. Our Starlab policy will be to make all simulation results freely and publicly available to 'guest observers.'

Acknowledgments. Part of this paper was written while we were visiting the American Museum of Natural History. We acknowledge the hospitality of their astrophysics department and visualization group. We thank the Alfred P. Sloan Foundation for a grant to Hut for observing astrophysical computer simulations in the Hayden Planetarium at the Museum. NCSA's Virtual Director group comprises of Donna Cox, Robert Patterson, and Stuart Levy.

References

Hut, P. & Makino, J. 1999, Science, 283, 64

Portegies Zwart, S. F., McMillan, S. L. W., Hut, P., & Makino, J. 2000, MNRAS, 321, 199

Teuben, P. J. 1994, in ASP Conf. Ser., Vol. 77, Astronomical Data Analysis Software and Systems IV, ed. R. A. Shaw, H. E. Payne, & J. J. E. Hayes (San Francisco: ASP), 398

An Integrated Procedure for Tree N-body Simulations: FLY and AstroMD

U. Becciani, V. Antonuccio-Delogu

Osservatorio Astrofisico di Catania, Catania - Italy

F. Buonomo, C. Gheller

Cineca, Casalecchio di Reno (BO) - Italy

Abstract. We present a new code for evolving three-dimensional self-gravitating collisionless systems with a large number of particles $N \geq 10^7$. **FLY** (Fast Level-based N-bodY code) is a fully parallel code based on a tree algorithm. It adopts periodic boundary conditions implemented by means of the Ewald summation technique. FLY is based on the one-side communication paradigm for sharing data among the processors that access remote private data, avoiding any kind of synchronization. The code was originally developed on a CRAY T3E system using the *SHMEM* library and it was ported to SGI ORIGIN 2000 and IBM SP (on the latter making use of the *LAPI* library). FLY[1] version 1.1 is open source, *freely available code.*

FLY output data can be analysed with AstroMD, an analysis and visualization tool specifically designed for astrophysical data. AstroMD can manage different physical quantities. It can find structures without well defined shape or symmetries, and perform quantitative calculations on selected regions. AstroMD[2] is *freely available.*

1. Introduction

FLY is the N-body tree code we designed and developed to run very big Large Scale Structure simulations of the universe using MPP and SMP parallel systems. FLY uses the Leapfrog numerical integration scheme for performance reasons, and incorporates fully periodic boundary conditions using the Ewald method. The I/O data format is integrated with the AstroMD package.

AstroMD is an analysis and visualization tool specifically designed for astrophysical data. AstroMD can find structures not having a well defined shape or symmetries, and perform quantitative calculations on a selected region or structure. AstroMD makes use of Virtual Reality techniques, which are particularly effective for understanding the three dimensional distribution of the

---

[1] http://www.ct.astro.it/fly/

[2] http://www.cineca.it/astromd

fields: their geometry, topology, and specific patterns. The data display gives the illusion of a surrounding medium into which the user is immersed. The result is that the user has the impression of traveling through a computer-based multi-dimensional model which can be manipulated by hand.

2. FLY Code

The FLY code, written in Fortran 90 and C languages, uses the one-side communication paradigm: it has been developed on the CRAY T3E using the SHMEM library. It adopts a simple domain decomposition, a grouping strategy, and a data buffering that allows us to minimize data communication.

2.1. Domain Decomposition

FLY does not split the domain with orthogonal planes. Instead, domain decomposition is done by assigning an equal number of particles to each processor. The input data are a sorted file containing the position and velocity fields, so that particles with nearby tag number are also close in the physical space. The arrays containing the tree properties are distributed using a fine grain data distribution.

2.2. Grouping

During the tree walk procedure, FLY builds a single interaction list (IL) to be applied to all particles inside a grouping cell (C_{group}). This reduces the number of tree accesses required to build the IL. We consider a hypothetical particle we call *Virtual Body* (VB) placed in the center of mass of the C_{group}: the VB interaction list IL_{VB} is formed by two parts:

$$IL_{VB} = IL_{far} + IL_{near} \qquad (1)$$

where IL_{far} includes the elements more distant than a threshold parameter from VB, and IL_{near} includes the elements near VB. Using the two lists, it is possible to compute the force F_p of each particle p in C_{group} as the sum of two components:

$$F_p = F_{far} + F_{near} \qquad (2)$$

The component F_{far} is computed only once for VB, and it is applied to all the particles, while the F_{near} component is computed separately for each particle. The size of the C_{group}, and the tree-level where it can be considered, is constrained by the maximum allowed value of the overall error of this method. In this sense the performance of FLY is a level-based code.

2.3. Data Buffering

The data buffer is managed as a simulated cache in local RAM. Every time the PE has to access a remote element, at first it looks in the local simulated cache and, if the element is not found, the PE executes GET calls to download the remote element, and stores it in the buffer. In a simulation with 16-million-particles clustered, with 32 PEs and 256 MB of local memory, without the use of the simulated cache, the PEs execute about 2.1×10^{10} remote GETs. Using the

Procedure for Tree N-body Simulations: FLY and AstroMD 505

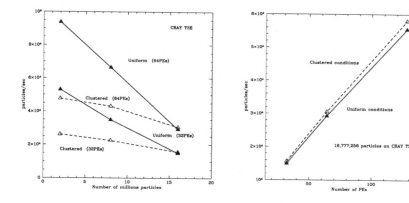

Figure 1. CRAY T3E/1200e. FLY particles/second using 32 and 64 PEs in uniform and clustered conditions, and scalability from 32 PEs to 128 PEs in a 16-million-particle simulation.

data buffering, this value decreases at 1.6×10^8 remote GETs, with an enormous advantage in terms of scalability and performance.

Figure 1 shows the code performance on CRAY T3E system, obtained by running simulations with 32 and 64 PEs, and FLY scalability, considering the case of 16,777,216 particles, where a speed-up factor of 118 is reached using 128 PEs. The highest performance obtained in a clustered configuration is a positive effect of the grouping characteristic. These results show that FLY has very good scalability and very high performance, and can be used to run very big simulations.

3. AstroMD Package

AstroMD is developed using the Visualization Toolkit (VTK) by Kitware, a freely available software portable to several platforms which range from the PC to the most powerful visualization systems, with a good scalability. Data are visualized with respect to a box which can describe the whole computational mesh or a sub-mesh. AstroMD can find structures not having well defined shape or symmetries, and performs quantitative calculations on a selected region or structure. The user can choose the sample by the loaded particle type (i.e., stars, dark matter, and gas particles), the size of the visualization box, and the starting time from which he wants to show the evolution of the simulation. The *Density* entries control the visualization of the iso-surfaces. These are calculated on a grid whose resolution can be selected by the user. To allow a more accurate investigation of a subset of the visualized system, it is possible to use a cubic sampler. If the sampler is selected (*Show/Hide* menu), visualized in the scene and enabled (*Sampler* menu), all the computations are performed only inside the region of the sampler (Figure 2). AstroMD can also show the evolution in time of the simulated system over the entire interval of time for which data are available, performing interpolations at intermediate frames. During the evolu-

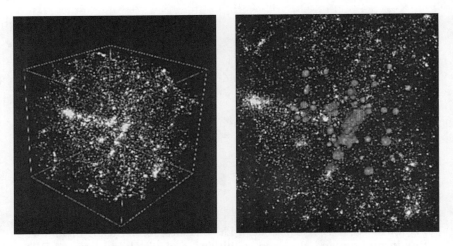

Figure 2. Typical structure of voids and filaments and Iso-surfaces, in a sub-region at the end of a simulation.

tion, the updated time of evolution is displayed in the *Time* entry of the *Cloud* section. Finally, snapshots of displayed images can be created using the button *Take it* in the *Screenshot* menu. AstroMD is developed by the VISIT (Visual Information Technology) laboratory at CINECA (Casalecchio di Reno - Bologna) in collaboration with the Astrophysical Observatory of Catania.

ACE: Astronomical Cataloging Environment

Francisco G. Valdes

IRAF Group, NOAO[1], PO Box 26732, Tucson, AZ 85726

Abstract. A new IRAF package under development for detecting, evaluating and cataloging objects in astronomical images is summarized.

Astronomers are producing ever more and deeper images of the sky. While an image contains all the information about that view of the sky within some passband and at some resolution, for many astronomical investigations what is needed is a catalog of the sources in the image. Reducing the image to the significant information about the sources – position, brightness, classification, etc. – makes working with very large areas of the sky more tractable.

The Astronomical Cataloging Environment (ACE) is a new IRAF package being developed for cataloging sources in images and manipulating the catalogs. The reasons for developing ACE, given a number of other similar programs, are to implement new or improved features and algorithms and to provide integration with IRAF. While ACE will include a variety of tools for cataloging and analysis, this paper outlines the features and algorithms of the basic detection and cataloging tool, excluding object classification.

The output of the detection task consists of a combination of object regions, a catalog, and log information. The object regions are recorded in a pixel mask with values combining the object number and various flags. The pixel mask uses a run-length encoding that is compact and efficient to use in the algorithms. These masks can be used for a variety of purposes and algorithms outside of ACE. The catalog may be in a variety of standard formats (e.g., binary table or text file). The contents of the catalog are specified by the user in a catalog definition file. This can select any of the measured values and define expressions based upon them.

All algorithms provide for application of user specified pixel masks and exposure maps. Maps include a constant value, a low resolution image to interpolate or a full image. The pixel masks select pixels to be ignored. An exposure map is used to adjust estimates of the standard deviation of a pixel by the square root of the map value. This is typically used when multiple, dithered exposures are combined by *averaging* the overlapping pixels where the number of overlapping pixels at each point in the final image may vary. When all images have the same exposure time the exposure map would be the number of images that contributed to each pixel.

---

[1]National Optical Astronomy Observatories, operated by the Association of Universities for Research in Astronomy, Inc. (AURA) under cooperative agreement with the National Science Foundation.

1. Sky Determination

Detection of sources in images begins with estimating the probability distribution of the background sky signal *at every image pixel*. In practice only the mode μ and the standard deviation σ are estimated, and a normal distribution is assmumed for estimating the significance of deviations from the mode. Sky information may be speicified by the user, separately for both the mode and σ in the form of maps, or determined by ACE from the image data. When ACE measures the sky from a science image, it uses one or two passes through the data and then updates the estimates when detecting the objects. The first pass fits 1-D functions to a subsample of lines using sigma clipping iteration to eliminate objects. Unrejected pixels are used to fit 2-D surfaces for μ and σ, and coefficients are recorded in the image header.

A second pass divides the image into large, equal-sized blocks and estimates μ and σ in each block using histograms of the deviations $\delta = (I - \mu)/\sigma$, where I is the pixel value and μ and σ are from the first pass sky. The deviation histogram is truncated at some number (e.g., 2.5) to reject object pixels. When the truncated histogram contains too few pixels, the block and its neighbors are rejected to eliminate effects from large galaxies and bright stars. The μ and σ of the deviation histograms are evaluated and scaled back to data values using the means $\langle\mu\rangle$ and $\langle\sigma\rangle$ from the computation of δ. Blocks are averaged into larger blocks, and any missing blocks are interpolated. The block values are stored as reduced-resolution sky and sky sigma maps that are interpolated to get sky values for every pixel in the science image.

2. Object Detection

Object detection consists of finding pixels with deviations δ outside specified limits. Object pixels have positive deviation, and "negative" objects (i.e., negative pixel deviations) may either be detected or ignored. Deviations within the limits are used to update the sky by accumulating and evaluating histograms as before.

Adjacent object pixels (user-selected either along the four pixel edges or also along the diagonals) are assigned to the same object region. A candidate object is detected when no further pixels are added to an object region.

The depth of detection is set by pixel noise that may be reduced by convolving the image data at the expense of reducing resolution. Matching the PSF size produces optimal convolution, but any function will improve faint object detection. ACE permits use of any convolution function and recognizes symmetries in the kernel to optimize computation. When image data are convolved, the values of σ used in computing δ are reduced by the root sum of the kernel weights.

Typical detection thresholds δ are 3–5, which ensures that some pixels outside the thresholds will not be part of real objects. To avoid too many spurious detections ACE provides two detection-stage criteria: that the number of pixels in a candidate object must exceed a certain value, and that at least one of the mean and maximum of δ must exceed a specified statistical significance.

ACE: Astronomical Cataloging Environment

3. Splitting Merged Objects

Regions of connected pixels that are just above the background sky will sometimes include more than one astronomical object due to the overlap of their light. While all such overlapping objects can't be separated, many can be by setting a higher threshold. Conceptually this means finding saddles in the light distribution between the objects.

ACE finds such thresholds by taking the detected regions and repeatedly raising the threshold by a small amount to see whehter or not the pixels separate into two or more subregions. When a separation occurs the original region is marked as a "multiple" object and the subregions are added to the list of detected objects. The threshold is then raised further, with the split regions included as separate objects, to look for further splitting. This continues until there are fewer pixels in a region than the criteria for defining a new object.

The amount of time taken to split merged objects is related to the number of thresholds used. To limit the number of thresholds while still considering separations at all thresholds from near sky to the peaks of the brightest objects requires using non-linearly spaced thresholds. Near sky background, where most objects merge and faint objects quickly disappear, a linear spacing is used with steps that are a fraction of the sky standard deviation. When the threshold reaches a certain value the spacings become non-linear.

Rather than read through the image for each threshold, possibly convolving the data, a special in-memory splitting image of the positive deviations is created during the detection stage. Pixels that are not in objects, i.e., those below the positive detection threshold, are set to zero in the splitting image. The other pixels are converted into bin numbers corresponding to the chosen thresholds to produce a discrete number of short integer values. The splitting image is compressed using run-length encoding of the non-zero values. This format, which excludes pixels not in objects, can be efficiently searched for objects by running through the run-length segments. As the splitting goes through the thresholds, lower thresholds are removed from the run-length segments of the splitting image.

4. Growing the Object Regions

Threshold detection always misses the wings of the light distribution in the detected objects. Also the method of increasing thresholds shrinks the regions identified with the split objects. To improve this situation, once the object regions have been identified there is an option to "grow" the regions, i.e., the borders of the object regions are extended by adding pixels that touch the borders. Objects are permitted to grow on the other side, even after they've touched a neighbor on one side. When adding a pixel in one pass that touches more than one object region, there is a contention for that pixel. This is resolved by assigning the pixel to the object with the greater flux defined as the sum of sky-subtracted pixel values. This operation can be done efficiently using the run-length mask encoding of the object regions.

5. Cataloging

The preceding steps all deal with defining the regions occupied by the objects in the image. The next step is to compute catalog parameters and statistical errors over the object regions, efficiently accessed through the object region mask. The variances of object pixels are represented by $\sigma^2_{object} = (\sigma^2 + (I-\mu)/g)/t$ where g is a gain factor scaling the pixel values to a Poisson variance and t is the relative exposure from the exposure map. All the terms are functions of position as specified by maps. If no exposure map is given, a default of unity is used; if no gain map is given, the Poisson term is omitted. The μ and σ may be values initially specified or sky values derived earlier, or new values may be specified. By not giving a gain map and using an externally derived σ, any noise behavior for the objects can be defined.

Rather than giving a list all possible quantities, a categorized summary follows. Position information includes the sky-subtracted intensity centroid over the object region, the position of the maximum pixel and the limits of the object region along the image axes. The positions may be measured in image pixel coordinates and in celestial coordinates. Celestial coordinates require a coordinate system in the image header.

Photometric quantities include fluxes within the object region and within circular apertures of specified radii, a Kron flux and an apportioned flux for split objects. The apportioned flux takes the original flux of a merged object and divides it among the split objects in the same ratio as the fluxes of the split objects measured at the higher isophote where the objects separated. The fluxes may be expressed in magnitudes if desired. The circular aperture fluxes include a simple partial pixel correction at the boundaries.

Shape information includes the second moments of the light distribution from which an ellipticity and position angle are derived. Miscellaneous information includes catalog entry identifications, the number of pixels in the object and flags of various kinds.

6. Multi-image and Difference-image Cataloging

For these two common types of astronomical imaging programs, ACE uses the celestial coordinate systems to align the images. ACE provides the ability to use one of two or more image (or a combination thereof) to define object regions and then determine the sky and evaluate objects in individual images using the common regions.

ACE also allows specifying two images (one of which could be a stack of all exposures) that are registered and differenced internally to detect sources in the difference. Distinguishing real differences from artificial ones (e.g., caused by point spread function variations and registration errors) can be difficult. Such effects will produce significant positive and negative differences centered on bright and narrow sources such as bright stars. Such false flux differences are minimized by requiring that a flux ratio threshold be exceeded in a comparison between the two images.

Automated Photometric Calibration Software For IRAF

Lindsey E. Davis

National Optical Astronomy Observatories, Tucson, AZ 85719

Abstract. New IRAF tools for the photometric calibration of optical/IR images are presented. The new tools are standard star catalog and world coordinate driven, requiring both accurate standard star catalog coordinate and accurate image coordinate data. Network access to standard star catalog servers is supported. The new tools are suitable for embedded use in pipeline reduction software.

1. Introduction

The new tools described here were written to support the photometric calibration of NOAO Deep Wide Survey data and are sufficiently general that, with minor modifications, they can be used with any standard star catalog server and applied to any optical/IR image data.

Sections 2 and 3 describe the current standard star catalog and image data requirements. Sections 4 and 5 provide an overview of the software and a summary of its principle features. The current status of and future plans for the software are summarized in section 6.

2. The Standard Star Catalog Requirements

The standard star catalogs must satisfy the following requirements.

- The standard star catalog must be a text file or a catalog server either local or remote. The catalog server must support simple region extraction queries and text output. The standard star text files and catalog servers must provide ids, celestial coordinates, and one or more magnitudes for all the extracted standard stars.
- The standard star catalog coordinates must be defined in a standard celestial coordinate system. The coordinates must be accurate enough to permit unambiguous location of the standard stars in the input image data.
- The standard star catalog magnitudes must be defined in a standard photometric system. Extraction of catalog magnitude errors is desirable but not required.

3. The Image Data Requirements

The input image data must satisfy the following requirements.

- The image coordinate system must be accurate enough to permit unambiguous location of the standard stars in the image. The image coordinate system need not be the same as the standard star catalog coordinate system. The software will transform from one coordinate system to the other.
- The effective exposure time, the effective airmass, the filter name, the effective time of observation, the effective gain, and the effective readout noise must be stored in the input image headers before beginning photometric calibrations.
- Basic CCD reductions including overscan, zero level, dark current, flat field, and fringe corrections must be completed before beginning photometric calibrations.
- Mosaic specific CCD reductions such as crosstalk and interchip gain corrections must be completed before beginning photometric calibrations.
- The photometric zero point must be constant across the images. Variable image scale effects must be removed before beginning photometric calibrations.

4. Scientific Functionality

Given a standard star catalog and an input image list which satisfy the criteria defined in sections 2 and 3, the photometric calibration software performs the following functions. An outline of the calibration software is shown in Table 1.

- Selects a standard star catalog from the list of supported catalogs.
- Selects the images containing standard stars from the list of input images using the standard star coordinates and the image coordinate system.
- Groups the standard star images by filter and time of observation using information in the image headers.
- Creates standard star pixel position and magnitude tables for each standard star image using the standard star coordinates and the image coordinate system.
- Computes aperture photometry for the extracted standard stars using parameter settings appropriate for the instrument and exposure times in the image headers.
- Creates a standard star observations files by combining data for the same field taken through different filters and at different effective airmasses.
- Calibrates the photometry by solving the system of transformation equations defined by the user.

5. Software Features

The photometric calibration software supports the following features.

- The standard star catalog may be either a file or a local or remote catalog server which supports region extraction. The software is independent of the type and format of the standard star catalog and does not need to be modified to support new standard star catalogs.
- Standard star list filtering can be performed by the catalog server as part of the extraction or by the photometric calibration software after extrac-

Table 1. The Photometric Calibration Package

		Main Entry Point Tasks
stdsets	-	Group standard star images by filter
stdphot	-	Create a standard star observations file
phparams	-	Solve the transformation equations
lan4mphot	-	Create a Landolt star 4m mosaic observations file
lan4mparams	-	Solve the transformation equations using Landolt stars
		Individual Tasks
phlist	-	List the photometric processing information
phimfind	-	Select standard star images and / or
phimfind	-	Extract standard star lists
phfiltpars	-	Edit the catalog file filtering parameters
phimsets	-	Group standard star images by filter
phshifts	-	Compute the relative shifts of images by group
phphot	-	Do aperture photometry on standard star images
dpars	-	Edit the data dependent parameters
cpars	-	Edit the centering parameters
spars	-	Edit the sky fitting parameters
ppars	-	Edit the photometry parameters
phobsfile	-	Prepare a standard star observations file
		Individual Tasks Optimized for NOAO 4m Mosaic Data
ph4mphot	-	Do aperture photometry on combined 4m mosaic images
d4mpars	-	Edit the 4m mosaic data dependent parameters
c4mpars	-	Edit the 4m mosaic centering parameters
s4mpars	-	Edit the 4m mosaic sky fitting parameters
p4mpars	-	Edit the 4m mosaic aperture photometry parameters

tion. Common filtering operations include changing coordinate systems, imposing magnitude limits, and sorting by magnitude.
- All the standard celestial coordinate systems are supported. The photometric calibration software will automatically convert the standard star coordinates from the standard star coordinate system to the image coordinate system.
- Standard star image selection, standard star list extraction, standard star identification, and standard star measurement are world coordinate system driven and fully automated.
- The image grouping and final calibration steps still require some user interaction mostly in the interests of quality assessment.

6. Current Status and Future Plans

The current photometric calibration software uses a set of catalog access tasks and catalog access API developed for use with astrometric catalogs but not yet part of the standard IRAF distribution (Davis 2000). The code which does the actual aperture photometry, matches the observations, and computes the photometric transformations is an adaptation of the existing IRAF APPHOT and PHOTCAL packages (Davis and Gigoux 1993).

Although most of the software is automated and runs without intervention by the user, some user input is still required in the areas of grouping the standard star images and interacting with the fitting process to get an optimal fit. More work is required in these areas in order to make the code fully automated in a pipeline environment.

The author would like to thank Daniel Durand for help in the early stages of developing the catalog interface, the NOAO Deep Wide Survey team for providing motivation for the project.

References

Davis, L. E. 2000, in ASP Conf. Ser., Vol. 216, Astronomical Data Analysis Software and Systems IX, ed. N. Manset, C. Veillet, & D. Crabtree (San Francisco: ASP), 667

Davis, L. E. & Gigoux, P. 2000, in ASP Conf. Ser., Vol. 52, Astronomical Data Analysis Software and Systems II, ed. R. J. Hanisch, R. J. V. Brissenden, & J. Barnes (San Francisco: ASP), 479

CCD Charge Shuffling Improvements for ICE

Robert L. Seaman

IRAF Group[1], NOAO[2], PO Box 26732, Tucson, AZ 85726

Abstract. NOAO has been using IRAF at its telescopes since Unix workstations were first placed in the domes. At the Kitt Peak National Observatory, this has included data acquisition using the ICE (*IRAF Control Environment*) package that was developed in coordination with Skip Schaller at Steward Observatory. ICE continues to be used both inside KPNO and Steward and at other observatories.

Improvements to ICE are described that support a dual exposure mode implemented via charge shuffling techniques. Charge shuffling involves repeatedly shifting the charge back-and-forth from side-to-side of a CCD while nodding the telescope alternately from an object to a blank sky position. The CCD is optically masked such that the sky pixels are kept dark while the object pixels are exposed and vice versa. The nodding and shuffling and opening and closing of the camera shutter occurs on a short enough time scale that the sky brightness variations are frozen.

The output of this process is a dual exposure of contemporaneous object and sky spectra accumulated through the exact same optical path. This mode is beneficial, for instance, for multi-slitlet observations such that the width of each slitlet can be minimized to allow many more slits per exposure. New parameters added to ICE include the number of nods and the number of pixels to shift for each exposure. A variety of different nodding patterns are supported, such as a simple ABAB object/sky pattern and a bracketed pattern that begins and ends with a half-length sky subexposure. The on-object and on-sky exposure times may be specified separately.

1. Discussion

The nod-and-shuffle spectroscopic observing mode was first brought to the attention of NOAO personnel by Karl Glazebrook who helped implement this mode on the LDSS++ spectrograph at the Anglo Australian Observatory.

---

[1] Image Reduction and Analysis Facility, distributed by the National Optical Astronomy Observatories

[2] National Optical Astronomy Observatories, operated by the Association of Universities for Research in Astronomy, Inc. (AURA) under cooperative agreement with the National Science Foundation

The acquisition of charge shuffled images requires the tight coupling of several software systems typically maintained by different personnel at a given observatory. The fundamental method relies on using familiar CCD charge shifting techniques in new ways. At the same time, the telescope is required to nod back-and-forth between object and sky guide stars with as little overhead as possible. The data acquisition package (ICE at KPNO) must synchronize and manage both of these systems.

The microcode for the standard KPNO CCD 2901 controllers has always supported the feature of being able to shift charge both toward and away from the readout amplifier. No changes were necessary to this software (maintained by the CCD laboratory).

The mountain programming group (MPG) was responsible for implementing the new telescope offsetting feature. Several possible designs were discussed between the MPG and IRAF programmers that positioned the functional divide either closer to or farther away from the telescope control or data acquisition sides of the equation. The final choice was to fully implement a high level TCS multi-target offset function.

The IRAF group was responsible for modifications to the ICE CCD data acquisition software. The nodding/charge shuffling exposure mode requires that only the middle third of the chip be unmasked. After each object or sky subexposure is taken, the shutter must be closed, the charge shifted to one side or the other and the telescope nodded to the other offset position. This is repeated through many exposures, back-and-forth.

Given the requirement to repeat this on time scales of a minute or quicker for an hour or more at a time for each exposure, any successful implementation of charge shuffling must parameterize the most useful exposure parameters while hiding the complexity of the multiple offset exposures from the observers and the telescope operator.

The familiar IRAF parameter mechanism has provided this service for several other unique observing modes in the past (e.g., the 4m scan table and the Fabry-Perot). New parameters added to ICE include the number of nods to include in each exposure and the number of pixels to shift for each exposure. It is possible to turn the mode on and off at will and to easily change the parameters with each new exposure.

In addition, since preferred NOAO observing recommendations for charge shuffling are still being evaluated, access was provided to a variety of different nodding patterns. These include a simple ABAB object/sky pattern versus a bracketed pattern that begins and ends with a half-length sky subexposure. The idea of this feature is to ensure that the mid-UT of the sky exposure coincides with the mid-UT of the object exposure. The trade-off is that the sky spans a slightly longer open shutter duration. A second evaluation feature allows the on-object and on-sky exposure times to differ. Whether these options are preserved as user parameters, or are rather implemented as hardwired defaults, will depend on the scientific evaluation of their utility.

The added complexity of the shuffling mode required revamping various abort procedures and other infrastructure and the addition of several new keywords to the image headers. We are still finalizing the additions to the keyword dictionary.

A fully supported implementation of charge shuffling as a KPNO facility class instrument will require further discussions and effort on both observing run preparation (to better generate slit masks, for instance) and on the data reduction facilities for these new image types.

2. Status

One full engineering run has been completed and another is in progress contemporaneously with the ADASS X conference. The current engineering run is also testing a new auto-guider and a new engineering grade CCD. Results are encouraging but several areas of instrument level concern have been identified:

1. During the first engineering run we found that the slit masks were likely to move around during these more demanding exposures.
2. The physical nodding of the telescope takes more time than desired.
3. The heuristics used by the active support system for the primary were never intended for such a purpose. This has the result of adjusting all the actuators with each separate offset.

The corresponding overhead of such effects was quite significant during the first engineering run, amounting to about 900 seconds for a typical one hour (grand total) exposure of 30 60-second on target plus 30 60-second off target exposures. (Or about 15 seconds per offset.) We are trying a variety of strategies to mitigate this overhead. For instance, it is possible to turn off the active support during the offsets while leaving the normal tracking updates functioning.

Of course, these difficulties can also be considered positive factors since the nodding mode will drive improvements for normal operations as well.

References

Schaller, S. 1992, "IRAF Data Acquisition Software", in ASP Conf. Ser., Vol. 25, Astronomical Data Analysis Software and Systems I, ed. D. M. Worrall, C. Biemesderfer, & J. Barnes (San Francisco: ASP), 482

Seaman, R. 1997, "Asteroseismology - Observing for a SONG", in ASP Conf. Ser., Vol. 125, Astronomical Data Analysis Software and Systems VI, ed. G. Hunt & H. E. Payne (San Francisco: ASP), 190

Glazebrook, K., et. al. 1998, AAO Newsletter 84, 9

Glazebrook, K. 1998, AAO Newsletter 87, 11

Dey, A., et. al. 2000, internal NOAO report

Enhancements of MKRMF

X. Helen He

Harvard-Smithsonian Center for Astrophysics, 60 Garden Street, MA 02138

Michael Wise

MIT Center for Space Research, Cambridge, MA 02139

Kenny J. Glotfelty

Harvard-Smithsonian Center for Astrophysics, 60 Garden Street, MA 02138

Abstract. MKRMF, a data analysis tool of the Chandra X-ray Science Center (CXC), has evolved to more effectively create response matrix files (RMF). It provides new and enhanced features: uniform binning syntax, all-inclusive FITS embedded function (FEF) file extraction, and non-linear EBOUNDS calculation. This paper describes the algorithm, application interfaces and highlights of the future development.

1. Introduction

MKRMF generates a response matrix for an arbitrary redistribution function over a 2-dimensional grid plane. Since last reported by He (1999) it has evolved to include:

- uniform binning syntax and rebinned RMF output,
- application of generic FEF file, and
- non-linear EBOUNDS calculations to closer reflect real gains.

The tool's parameter file (mkrmf.par) has been accordingly updated, as listed in Table 1. The table highlights the most important parameters.

2. Uniform Binning Syntax and Rebinned Output

In conjunction with CXC Data Model (DM) filtering, MKRMF adopts the DM grid binning syntax for the "axis1" and "axis2" parameters. The binning syntax is consistent with other CIAO [1] tools. As previously defined, the binning syntax is grouped into two categories: command line and ASCII/FITS table file. The command line input allows the user to specify discrete grids in lower bound, upper bound, and binning type, separated by ":". The grid type can be specified

---

[1] http://asc.harvard.edu/ciao/

Table 1. Input Parameters for MKRMF

Name	Default Value(if any)	Description
infile		name of FEF input file
outfile		name of RMF output file
axis1		axis-1(name=lo:hi:btype)
axis2		axis-2(name=lo:hi:btype)
logfile		name of log file
thresh	1e-05	low threshold of energy cut-off in keV
outfmt	legacy	RMF output format (legacy—cxc)
clobber	yes	overwrite existing output file (yes—no)?
verbose	0	verbosity level (0 = no display)

as linear or logarithmic binning for either the binning step or the total number of bins. The file input allows the user to tabulate an arbitrary grid. The ASCII file format contains two columns: lower and upper bounds. Two columns extracted from the FITS file define the bounds by following DM filtering syntax. Table 2 summarizes the binning syntax.

Table 2. Binning Syntax

Syntax	Description	Example
`<axis-name>=<min>:<max>:<bin-num>`	linear bin in bin number	`pi=1:1024:#1024`
`<axis-name>=<min>:<max>:<bin-step>`	linear bin in bin step	`energy=0.1:10.0:0.05`
`<axis-name>=<min>:<max>:<bin-step>L`	logarithmic bin in bin step	`energy=0.1:10.0:0.05L`
`<axis-name>=<min>:<max>:<bin-num>L`	logarithmic bin in bin number	`energy=0.1:10.0:#1500L`
`<axis-name>=<file>`	grids tabulated in ASCII file	`energy=grid(eng.txt)`
`<axis-name>=<file>`	grids tabulated in FITS format	`energy=grid(rmf.fits[MATRIX][cols ENERG_LO, ENERG_HI])`

MKRMF always computes the matrix (and EBOUNDS) at full resolution at the specified `<min>:<max>` range. The matrix can be binned by specifying bin type and step size. MKRMF will use this information to scale down the output to the requested bin size.

3. New FEF File

A FEF FITS binary table is a CXC format convention to allow specification of an n-dimensional image in the form of an analytic function of n variables in a FITS binary table HDU. This file format has been developed to be extremely generic and to allow very cost effective reuse (Rots 2001).

MKRMF takes the analytic redistribution functions which are conventionally expressed by columns of independent function variables. These function variable columns are stored in the FEF format and used by MKRMF to create standard response matrices. Prior to the release of CIAO 2.0, a FEF file existed for each individual spatial region for which (different) redistribution functions were defined. Because of this, those who wished to perform analyses over large spatial regions or over multiple chips potentially had to keep track of a large number of files. In CIAO 2.0, these files have been merged into a single file, an all-inclusive and spatial varying FEF file. In addition to columns containing redistribution function parameter values, the new FEF file contains new columns CCD_ID, CHIPX, and CHIPY, which provide the location and bounds for each spatial region. Another new column, REGNUM, contains an integer value that identifies each spatial region uniquely. The new FEF format also encapsulates the gain relationship of the CCD for energy in a specific region. This relationship is defined by a new column, CHANNEL (or PHA), mapped to ENERGY in that region. By introducing the (ENERGY, CHANNEL) pair columns, the generic FEF extractor can effectively result in data block consisting of one region for each energy when a DM filter is applied to the FEF file. Therefore, the generic FEF file is effectively reusable and backward compatible.

The MKRMF user must now use DM syntax to retrieve information for one spatial region, as shown in this example:

```
mkrmf infile=fef.fits[ccd_id=7, chipx=(1:100),chipy=(1:32)]
```

When mkrmf executes the program with an input FEF file named "fef.fits", it extracts the data containing (ENERGY, CHANNEL) gain and redistribution function variables internally sorted for CCD_ID of 7 with chip pixel range of 1 to 100 and 1 to 32 along X- and Y-Axis, respectively. Figure 1 illustrates two MATRIX components generated on a FEF file with the filter above for a combination of ten embedded Gaussian functions on linear binned PHA-energy grids.

4. Non-linear EBOUNDS Interpolation

In previous releases, the EBOUNDS array was calculated through a linear analytic expression of two constant parameters (keywords): SCALE and OFFSETS. MKRMF now employs a new algorithm for the gain calculation from data tabulated in the new FEF file. The original linear interpolation still exists for backwards compatibility.

The CHANNEL and ENERGY pairs in the new FEF file encapsulate the piecewise linear gain information within a given (CHIPX, CHIPY) region for each specific energy value. Our new scheme to calculate the EBOUNDS array is to linearly interpolate the energy for a given PHA value from the pairs of channel and corresponding energies that bound the PHA value. The EBOUNDS energy

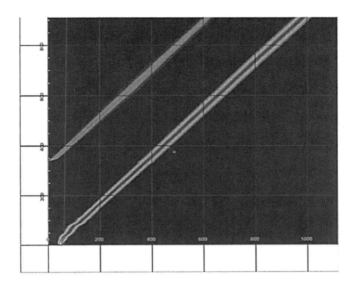

Figure 1.   RMF MATRIX Output.

derived in this way represents the variation of the CCD gain in that region, or an effectively non-linear and spatial varying relationship.

5. Future Developments

Future focus will be on multi-region RMF calculations. The current tool provides RMF output limited to one region characterized by the same response function within the given (ENERGY, CHANNEL) range. However, those parameters are also functions of spatial variables. This limitation will be removed by taking a weighting factor into consideration for each region, so a multi-region RMF output can be achieved by applying the weighting to redistribution functions of the interesting regions. As such, MKRMF data I/O is expected also to be updated in the future. The user interface to MKRMF will also be evaluated to make it easier for users.

Acknowledgments. We are grateful for many fruitful discussions with various CXC members. This project is supported by the Chandra X-ray Science Center as part of NASA contract NAS8-39073.

References

He, H. Wise, M., & Ljungberg, M. 1999, in ASP Conf. Ser., Vol. 216, Astronomical Data Analysis Software and Systems IX, ed. N. Manset, C. Veillet, & D. Crabtree (San Francisco: ASP), 636

Rots, A. H. 2001, this volume, 479

Astronomical Data Analysis Software and Systems X
ASP Conference Series, Vol. 238, 2001
F. R. Harnden Jr., F. A. Primini, and H. E. Payne, eds.

Development of Radio Astronomical Data Reduction Software NEWSTAR

Miho Ikeda, Kota Nishiyama

Japan Science and Technology Corporation, Honchou, Kawaguchi, Saitama 332-0012, Japan

Masatoshi Ohishi

National Astronomical Observatory, Osawa, Mitaka, Tokyo 181-8588, Japan

Ken'ichi Tatematsu

Nobeyama Radio Observatory, Nobeyama, Minamimaki, Minamisaku, Nagano, 384-1305, Japan

Abstract. NEWSTAR was developed for processing data from the 45-m radio telescope of the Nobeyama Radio Observatory (NRO). Based on AIPS, and developed under Solaris and IRIX, NEWSTAR has been expanded to include LINUX and HP-UX versions that have considerably improved the reduction performance of NRO observational data.

1. Introduction

NEWSTAR is a radio astronomical data reduction package for the 45-m radio telescope and the Nobeyama Millimeter Array (NMA) of the Nobeyama Radio Observatory (NRO), Japan. Since NEWSTAR was originally developed for data reduction within NRO on Solaris, Sun and IRIX platforms, users wanting off-site access to NEWSTAR formerly had to obtain computers similar to those at NRO or access NRO machines remotely. In order to improve this situation, LINUX and HP-UX versions of NEWSTAR were created, giving users access to NEWSTAR virtually anywhere.

2. Structure of NEWSTAR

NEWSTAR consists of an AIPS core plus the original application. The AIPS core provides file management, mapping processes for interferometer data, and so on. The original application was designed to reduce spectral data obtained by the 45-m telescope and make data cubes from the spectral data. Users can also reduce UV-data of the NMA using NEWSTAR before further processing with AIPS.

NEWSTAR also adds graphical user interfaces (GUIs). Users can input parameters for each task from the GUI, and the parameters are saved, obviating the need for learning native AIPS commands and permitting NEWSTAR use

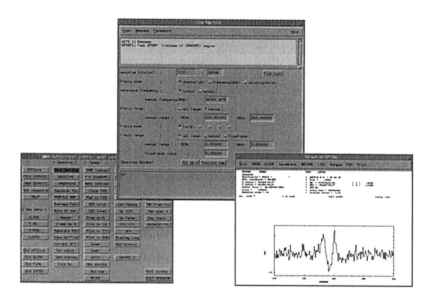

Figure 1. NEWSTAR has several window panels, e.g., the Task-Selection panel (bottom left), a parameter-input window (top center) and the Tektronics window (bottom right).

easily without references. (Users can use still invoke native AIPS commands from the AIPS window.)

3. UIs

NEWSTAR has several window panels, samples of which are shown in Figure 1. The Task-Selection panel lists tasks included in NEWSTAR, and users can select tasks from this panel. Each task has a parameter-input window. Users can see mapping points and a list of files from this panel. The Tektronics window displays spectra and/or maps.

NEWSTAR also has an AIPS window permitting users to execute native AIPS commands. From the "File List" window, users can list, delete, rename and backup data files.

4. Treatment of data

The flow of the data reduction of NEWSTAR is shown in Figure 2. Raw data obtained by the 45-m telescope are saved through the archival system. The saved data are under AIPS control. Users can craete spectra by integrating the raw data, make map data (cube data) from the spectra, and handle them again by using several AIPS tasks.

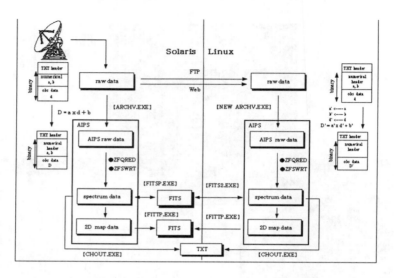

Figure 2. NEWSTAR data reduction flow under Solaris (left side) and LINUX (right side).

At NRO, Solaris 2.7 is adopted as the operating system of the workstations for data reduction. When users continue data reduction using NEWSTAR on other operating systems, difference of data endian should be considered. For example, in the case of LINUX, it is necessary to convert from big to little endians. The flow of the data reduction in the LINUX version is shown in the right side of Figure 2. The raw data transfered from the NRO archival system are transformed by task "NEW ARCHV.EXE" for LINUX. The data endian is transformed at that time. The spectral data and the map (cube) data made in the Solaris NEWSTAR are transformed to FITS files, and the FITS files are read using the LINUX version of NEWSTAR. Conversely, FITS file made in the LINUX version can be reduced under Solaris NEWSTAR.

For other operating systems, the flow of the data reduction is similar to that above.

5. Summary

The NEWSTAR radio astronomical data reduction software was originally developed for data reduction within NRO. By porting it to LINUX, HP-UX, and IRIX, users now have access to its data reduction capabilities in their home institutions, and the performance of NRO data reduction software has be considerably improved.

Acknowledgments. The authors would like to thank Jun Maekawa and Kiyohiko Yanagisawa for giving very useful advice. This research is supported by "Research and Development Applying Advanced Computational Science and Technology" of Japan Science and Technology Corporation.

ESO C Library for an Image Processing Software Environment (eclipse)

Nicolas Devillard

European Southern Observatory

Abstract. Written in ANSI C, eclipse is a library offering numerous services related to astronomical image processing: FITS data access, various image and cube loading methods, binary image handling and filtering (including convolution and morphological filters), 2-D cross-correlation, connected components, cube and image arithmetic, dead pixel detection and correction, object detection, data extraction, flat-fielding with robust fit, image generation, statistics, photometry, image-space resampling, image combination, and cube stacking. It also contains support for mathematical tools like random number generation, FFT, curve fitting, matrices, fast median computation, and point-pattern matching. The main feature of this library is its ability to handle large amounts of input data (up to 2 GB in the current version) regardless of the amount of memory and swap available on the local machine. Another feature is the very high speed allowed by optimized C, making it an ideal base tool for programming efficient number-crunching applications, e.g., on parallel (Beowulf) systems.

Running on all Unix-like platforms, eclipse is portable. A high-level interface to Python is foreseen that would allow programmers to prototype their applications much faster than through C programs.

1. Introduction

The eclipse library is intended as a low-level software facility to offer image-processing and data analysis functionality through calls to a C library. End-users can also access most functions through Unix commands, but the main idea is to program stand-alone pipeline commands running without user interactions.

Begun in 1995 to provide Adonis users with a usable pipeline environment to reduce their data in real-time in the telescope control-room, eclipse has been further extended to support more instruments at ESO (ISAAC, CONICA, WFI, possibly VIRMOS) and also in other institutes (Subaru, CFH).

2. Base requirements

From the eclipse developer's manual, its base requirements include:

- It shall be freely distributed through the Net.
- It shall be easily compiled on most workstations.

- It shall respect POSIX and ANSI standards whenever possible.
- It shall be stand-alone, i.e., self consistent. Users should not be required to download any other piece of software to use it. Additional software download may be required to extend functionality, but that should remain optional.
- As a data processing engine, eclipse is to run without user interaction or displays of any sort. These are left to external software packages. Stubs might be provided, e.g., to display an image or a plot, but they should never be mandatory in a command.
- Image processing or data analysis functionality is offered in a C library, accessible through a standard header (.h file) and a compiled library. Access is given to end-users through Unix commands.
- All Unix commands shall be documented with manual pages.
- A complete User's Manual and cookbook shall be available from the Web, and possibly with the software distribution.
- Only standard and widespread formats should be used for inputs and outputs.
- If eclipse replicates a functionality already present in another data reduction software facility, it must be justified either by speed gain, algorithm enhancement, user control over the algorithm, or ease of use.

3. Portability

As a portable library, eclipse should not only be able to be compiled on various platforms of today, but it also should be compilable on future platforms. Adhering to programming standards whatever they are (POSIX, ANSI, or simply coding conventions) ensures a reasonable survival of the library over time.

More generally speaking, writing portable software increases the quality of the code. By compiling the source with various compilers and running it on different platforms, weaknesses and defects are put under stress and are therefore revealed, whereas they might hide indefinitely or be very hard to find with a single development environment. It is a good idea to respect global programming conventions whenever possible.

Specializing a piece of software for a given platform is usually done for reasons of speed, to access the dedicated hardware or software present on the local system. Specializing software also means reducing its life expectancy, because underlying hardware or software may not be vendor-supported in the future.

4. File formats

The following file formats are supported by eclipse:
- FITS through an internal lightweight library
- PAF files: parameter files, internal to ESO
- ini files: windows-like parameter description
- PGM/PNM/PBM: to allow universal pixel translations
- various ASCII formats.

5. Coding standards

Coding style for eclipse conforms to that specified in the document "Recommended C Style and Coding Standards." Any eclipse developer should have most of the rules described there in mind when coding.

Following coding rules is often seen as a matter of taste. After all, if a programmer writes thousands of lines without indentation, it will also compile, and if it runs in the end, who cares about the style!

The philosophy behind coding rules and conventions is that a piece of software is defined by more than just the executable file that is produced after compilation. Pages of code are very often modified to correct bugs, enhance functionality, include into other (unforeseen) contexts, etc. It is vital that any developer in the team be able to modify somebody else's code with minimal overhead.

6. Modules

Most eclipse modules can be detached from the library to be reused as code snippets in other applications, thereby offering versatile functionality:

- File format handling: FITS images, cubes, tables, ini file handling, PNM output, various ASCII parsers
- Image processing primitives: Efficient bulk data loading, binary, integer, floating-point image handling, image filtering, both convolution and morphologic filters, 2-D cross-correlation, connected components (segmentation), cube and image arithmetic operations, dead pixel detection and correction, object detection, data extraction, flat-fielding with robust fit, image generation, statistics, photometry, image space resampling (warping), image combination, cube stacking with 3-D filtering
- Math primitives: Random number generation according to a given distribution law, FFT, curve fitting, matrices, fast median computation, point-pattern matching
- Unix handling: Portable endian detection and byte-swapping, message printing and logging, time measurement, compressed file reading/writing, generic dictionary handling, software configuration through environment variables, file locking, file system information query, command-line option handling, extended memory handling, memory-mapped file support through a generic interface, token-based ASCII file parsing, socket support, time printing in ISO8601 format, user identification
- Infrared specific: Database of 832 infrared standard stars included in the code, ISAAC-specific data reduction procedures.

7. Memory handling

Memory handling in eclipse has been carefully studied to avoid the usual limitations met in image processing software systems. The upper limit for allocatable memory is not fixed by the amount of RAM and swap found on the system, but by the amount of disk space accessible to the current user. The price paid is degraded performance as soon as the system runs out of memory, but the gain is a uniform interface to the programmer who does not have to care at all about memory handling. The same source code will process a gigabyte of data no matter what the locally available memory.

8. Algorithms

Several ESO pipelines are now based on eclipse for number-crunching developments. This has produced a number of high-level algorithms specialized for data processing, for one instrument or another. Some of these algorithms are adaptable enough to be used for other instruments not initially foreseen (e.g., the high-level data reduction tasks written for CONICA, a VLT instrument, also perform very well on CFHT Adaptive Optics data).

9. Parallel processing

Since eclipse offers its functionality through a Unix command-line call, it is possible to parallelize it through coarse-grain parallelism. A master machine can send a list of Unix commands to run to a group of slave nodes and get the results back. Since eclipse is C-based, it is easy to add potential communication procedures (based e.g., on MPI or PVM) to make the algorithms fully parallel.

Wide-Field Imager (WFI) data from the 2.2-m telescope at La Silla is currently processed on the ESO Beowulf system using eclipse.

10. Python/eclipse

Like any other C library, eclipse can easily be interfaced to appear as a Python extension. This would allow image processing programs to be object-oriented, and to benefit from all the advantages of using Python.

The eclipse source code is currently under modification to make it easier to interface to Python. The wrapper code will be generated by SWIG and a Python-compatible procedure will be distributed to offer eclipse as a Python extension.

Astronomical Data Analysis Software and Systems X
ASP Conference Series, Vol. 238, 2001
F. R. Harnden Jr., F. A. Primini, and H. E. Payne, eds.

Developing a Wavelet CLEAN Algorithm for Radio-Interferometer Imaging

S. Horiuchi

National Astronomical Observatory, Mitaka, Tokyo 181-8588, Japan

S. Kameno

National Astronomical Observatory, Mitaka, Tokyo 181-8588, Japan

M. Ohishi

National Astronomical Observatory, Mitaka, Tokyo 181-8588, Japan

Abstract. A new CLEAN-type algorithm searches for model components not in the dirty image but in the multi-resolution wavelet transform of the dirty image based on the dirty beam. A single-resolution prototype algorithm that searches for CLEAN components after convolution with the dirty beam has been tested and shown to be advantageous for the case of sparsely sampled and noisy u-v data, without specifying regions to be CLEANed.

1. Introduction

The image formed by simple Fourier transformation of the visibilities observed by radio interferometers has defects due to limited sampling of the u-v plane. Nonlinear deconvolution is required to correct these defects, CLEAN being the most popular algorithm (e.g., Cornwell, Braun, & Briggs 1999). However, it sometimes requires use of *a priori* knowledge of source structure, to avoid creating CLEAN components where the source 'should not' exist. This is the case partly because the standard CLEAN algorithm just takes the peak of the dirty image as the position of the next CLEAN component. For complex sources the peak brightness feature may be just a region of overlapping sidelobes plus noise; therefore choosing the area to be cleaned is crucial. In this paper an alternative CLEAN-type algorithm is proposed that uses wavelets to overcome the subjectivity of setting CLEAN boxes, so that reliable images can automatically be produced without *a priori* knowledge of source structure.

2. A prototype Wavelet CLEAN

If one can use not only the peak in the dirty map but also knowledge of the sidelobes to decide the position of CLEAN components, one may be able to establish an alternative CLEAN algorithm without CLEAN boxes, which leads to

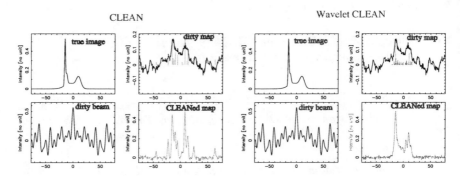

Figure 1.   CLEAN and Wavelet CLEAN for 1-D model data.

a more efficient automatic imaging program not dependent on *a priori* knowledge of sources. Multi-dimensional wavelet expansion of the dirty image may provide the possibility of realizing such an algorithm, if one can construct a wavelet function based on the dirty beam.

Although such an ideal function for fast wavelet transformation is not known to date, as a first trial to demonstrate the power of the wavelet CLEAN, test have been conducted of a zero-order prototype that searches for CLEAN components not in the dirty image $I_d(x,y)$ or the residual image $I_r(x,y)$, but after convolution of those with the dirty beam $B(x,y)$. Convolving the dirty image with dirty beam $I_d(x,y) * *B(x,y)$ corresponds to a wavelet expansion of the dirty image at single resolution.

3. Sample images

Example 1-D deconvolution using CLEAN and the prototype Wavelet CLEAN for model data are shown in Figure 1. In both cases, the dirty map $I_d(x)$ is a convolution of the true image $I_{true}(x)$ with the dirty beam $B(x)$, with 10% gaussian noise added. For Wavelet CLEAN, the normal CLEAN procedure was performed not on $I_d(x)$ and $I_r(x)$, but on $I_d(x) * B(x)$ and $I_r(x) * B(x)$. Two hundred iterations were performed with a loop gain of 0.1 and no CLEAN boxes specified. The results show that the chance of picking wrong CLEAN components is significantly reduced for the Wavelet CLEAN. The two methods provided equally good results for a noise-free model image.

An example of two-dimensional Wavelet CLEAN for 5 GHz VLBA data is shown in Figure 2, which also illustrates the order of computations for the procedure. The observed source is the radio quasar 0108+388, which consists of only a few major bright components at 5 GHz at VLBA resolution. One hundred iterations were taken with a gain of 0.3 and no CLEAN boxes specified. The positions of CLEAN components at each iteration for both CLEAN and Wavelet CLEAN are shown in Figure 3. The plot suggests that the positions found by standard CLEAN are fluctuating during the iteration while those of Wavelet CLEAN are rather stable around the peak positions in the original dirty image.

Wavelet CLEAN for Radio-Interferometer Imaging 531

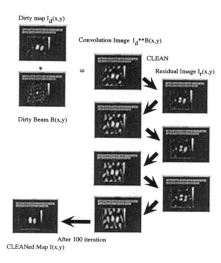

Figure 2. 2-D experiment of Wavelet CLEAN with real data.

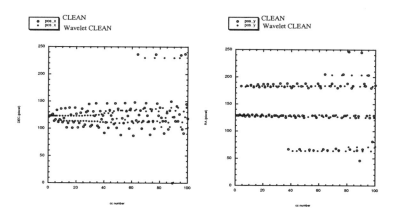

Figure 3. Positions of CLEAN components at each iteration of Figure 2.

4. Discussion

For the case of sparsely sampled and noisy u-v data, and without specifying regions to be CLEANed, the Wavelet method is shown to be advantageous. In fact taking the convolution of the dirty image with the dirty beam was discussed briefly in the original proposal of the CLEAN algorithm by Högbom (1974). In the original CLEAN algorithm the convolution was not performed because in Fourier space it differs from the original dirty image by only a multiplicative factor, and the author felt it was unnecessary. Nevertheless, a difference between CLEAN in two different image domains can be observed. The improvement of this prototype Wavelet CLEAN, which does convolve with the dirty beam, can be attributed to a suppression of noise and aliasing in the dirty image by use of information in the sidelobes of the whole image domain. So far this method has the advantage for compact sources with several components. Although the possible extension of the algorithm for multi-dimensional imaging of extended radio sources, introducing wavelet functions based on the dirty beam, remains to be studied, the current analysis of the prototype provides the basic concept.

Acknowledgments. This research is supported by "Research and Development Applying Advanced Computational Science and Technology" of Japan Science and Technology Corporation (JST).

References

Högbom, J. A. 1974, ApJ, 15, 417
Cornwell, T. J., Braun, R., & Briggs, D. S 1999, in ASP Conf. Ser., Vol. 180, Synthesis Imaging in Radio Astronomy II, ed. by Taylor, G. B., Carilli, C. L., & Perley R. A. (San Francisco: ASP), 151

Author Index

Abrams, D. C., 287
Accomazzi, A., 321
Adams-Wolk, N. R., **419**
Adamson, A. J., 137, 314
Albrecht, R., 447
Aldcroft, T. L., 439, 471
Alexov, A., **178**
Ali, B., 170
Allan, A., **459**
Allington-Smith, J., 459
Amico, P., 148
Antonuccio-Delogu, V., 503
Aoki, K., 70
Archibald, E. N., 299
Arnaud, K. A., 67, 415
Arnouts, S., 283
Arviset, C., 90

Baas, F., 93
Backus, C., 369
Ballet, J., **381**
Bambery, R. J., **369**
Barbati, M., 404
Barg, M. I., **241**, 349
Barth, A. J., **385**
Baruffolo, A., 160
Becciani, U., **503**
Behnke, J., 133
Beikman, S. J., 435
Benacchio, L., **119**, 160
Benfante, L., **160**
Benoist, C., 283
Benvenuti, P., 141
Berriman, G. B., **36**, 52
Berry, D. S., **129**
Bertero, M., 404
Bienaymé, O., 74
Binegar, S., 90
Boccacci, P., 404
Boër, M., **103**, **111**, 388
Bonnarel, F., 56, **74**
Borra, E., 97
Bottini, D., **455**
Brandt, M., 90
Bridger, A. B., 137, 314
Brissenden, R. J., **22**
Bristow, P., 178

Brosch, N., 125
Buchholtz, G., 111
Buonomo, F., 503
Bushouse, H., **427**
Butler, R. F., **431**

Calderwood, T., **443**
Cameron, R. A., 471
Carbillet, M., **249**, 253, 404
Carbognani, F., 199
Cavanagh, B., 314
Chan, S. J., **365**
Cheung, C., 217, 487
Chiu, N.-M., 36, 52
Christian, C. A., **107**
Christian, D., 174
Claeskens, J. F., 97
Clark, G., 115
Cornwell, T., 408
Correia, S., **404**
Cresitello-Dittmar, M., **439**, 467, 471
Crutcher, R., 279
Currie, M. J., 137

da Costa, L., 283
Dabrowski, Y., 93
Daly, P. N., **265**, **333**
Davis, L. E., **511**
De Cuyper, J.-P., **125**
de la Fuente, A. J., 152, 156
De La Peña, M. D., **59**
Delaney, M., 170
Delplancke, F., 249
Demleitner, M., **321**
Devillard, N., **525**
Dobrzycki, A., 443
Dodd, R. J., 306
Doe, S., **310**, 483
Dolensky, M., **90**
Donahue, M., 174
Dorman, B., **415**
Dubois, P., 56
Durand, D., 63

Ebisuzaki, T., 103, 111
Economou, F., 137, 299, **314**
Egret, D., 56, 74
Eichhorn, G., 321

Ellison, B., 93
Elst, E., 125
Elvis, M. S., 435
Emmart, C., 499
Esposito, S., 249

Fang, F., 329, **337**
Faye, S., 103
Femenía, B., 249
Fernique, P., 56, **63**, 74
Ferro, A. J., 241
Filippi, G., **199**
Fini, L., 249, **253**, 404
Fishman, M., 44
Fitzpatrick, M., **261**
Fixsen, D. J., 257, 396
Flanagan, J. M., 435
Folger, M., 137
Fosbury, R. A. E., 447
Foschini, L., 144
Freeman, P. E., 479, **483**
Freudling, W., 447
Frisée, L., 213

Gabriel, C., **400**
Gaetz, T., 435
Gaffney, N., 90
Gan, N., 67
García, J., 156
Garilli, B., 451, 455
Gastaud, R., 170
Genova, F., **56**, 74
Gheller, C., 503
Gianotti, F., **245**
Glotfelty, K. J., 435, 518
Golap, K. , **408**
Golden, A., 431
González-Riestra, R., 152, **156**
Good, J., 36, **52**
Gosset, E, 373
Grant, C. S., 321
Graybeal, J., **189**
Green, W. B., **32**
Greenfield, P., 59
Greimel, R., **287**
Grella, V., 44
Grosvenor, S., 48
Guest, S., 170
Guillaume, D. R., **221**, 279

Gunn, J. E., 269

Hain, R., **471**
Hambly, N. C., 182
Handley, T., 36, 52, 229
Hanisch, R. J., 257
Harris, D. E., 443
He, X. H., 479, **518**
Heikkila, C. W., **67**
Hensberge, H., 125
Hernandez, J., 90
Hills, R., 93
Hirst, P., 314
Hobbs, R., 82
Holland, W. S., 299
Honsa, J., 423
Hook, R. N., **283**, 447
Hopkins, E., 291, 295
Horiuchi, S., **529**
Hsu, E., 115
Hubin, N., 249
Hudec, R., 125
Hut, P., 499
Hutchinson, M. G., 365

Ikeda, M., **522**
Imhoff, C., 174
Irwin, M. J., 287, 317
Isaac, M., 103, 111
Ishihara, Y., 237
Isobe, T., 467
Ivezić, Z., 269

Jean, C., 97
Jenness, T., **299**, 314
Jerius, D., 435
Jessop, H., 443
Jessop, N., 299
Johnson, A., 36, 52
Johnson, R., 459
Jones, J., **44**, 48

Kackley, R. D., 137
Kadowaki, N., 115
Kalpakis, K., **133**
Kameno, S., 529
Karovska, M., **435**, 471
Katz, D. S., 257
Kemball, A., 408
Kent, S., 269

Kepner, J., **209**
Kerbel, U., 44
Kerber, F., 178
Kerr, T. H., 314
Kimball, T., 174
Klotz, A., 111
Knapp, G. R., 269
Kolhatkar, S., **329**
Kong, M., 36, 52
Koratkar, A., 48
Kosugi, G., 237
Kroll, P., 125
Krzaczek, R., 189
Kunszt, P. Z., 40
Kurtz, M. J., 321, 491

Lampens, P., 125
Lamy, P., 377
Le Fèvre, O., 451
Lee, W.-P., 52
Leeks, S. J., 365
Lesteven, S., 56, **78**
Levay, K., 174
Levy, S., 499
Lewis, J. R., 287, 317
Li, C., 48
Li, J., 337
Lin, G., **392**
Llebaria, A., **377**
Lord, S., 365
Lupton, R. H., **269**

Ma, J., 52
Makino, J., 499
Maks, L., 44
Mandel, E., **225**
Mandel, J., 233
Manfroid, J., 97, **373**
Martínez, J., 156
Mather, J. C., 257, 396
Matusow, D., 48
McDowell, J. C., 435, 479
McGlynn, T. A., 67
McMahon, R. G., 317
McMillan, S., 499
Mehringer, D., 279
Melchior, A.-L., 103, 111
Micol, A., 141, **148**
Mignani, R., 283

Milburn, J., 189
Miller III, W. W., **325**
Miller, B., 459
Mink, D. J., **491**
Misra, D., 44
Miura, A., 70
Miville-Deschênes, M.-A., 170
Mizumoto, Y., 237
Molinari, S., 365
Monkewitz, S., 36
Montesinos, B., 156
Montfort, F., 97
Moreau, O., 97
Morita, Y., 237
Morris, D., 439
Moshir, M. M., 329, 337
Murray, S. S., **13**, 225, 321
Murtagh, F., 78

Nakamoto, H., 237
Nakos, Th., 97
Narron, R., 337
Nieto-Santisteban, M. A., 257, 396
Nishiyama, K., 522
Noble, M., 310
Nordsieck, K. H., 463
Norton, S. W., 36

O'Neil, E. J., 241
Ochsenbein, F., 56
Offenberg, J. D., 257, **396**
Ogasawara, R., 237
Ohishi, M., 522, 529
Okumura, K., 170
Olivera, F., 93
Ortiz, P. F., 56
Ott, S., **170**

Pack, H., 103
Padovani, P., **174**
Pasad, M., 133
Pasquali, A., 447
Pauwels, T., 125
Peachey, J., 67
Pell, V., 44
Pence, W. D., 67
Pennypaker, C., 103, 111
Percival, J. W., **463**
Petreshock, J. G., **467**

Pickup, D. A., 137
Pierfederici, F., **141**
Pirenne, B., 90, 141
Pirzkal, N., **447**
Plante, R., 221, **279**
Plummer, D. A., 303, **475**
Poels, J., **97**
Poinçot, P., 78
Ponz, J. D., 152, 156
Portegies Zwart, S., 499
Postman, M., 174
Pound, M. W., **82**
Purves, M., 137
Puttagunta, V., 133

Rauw, G., 373
Read, M. A., **182**
Rees, N. P., 93, 137
Reid, M. L., **306**
Riccardi, A., 249, 253
Richer, J., 93
Richichi, A., 404
Riggs, M., 133
Rité, C., 283
Robson, E. I., 299
Rodríguez, F., 152, 156
Roll, J. B., 225, **233**
Rosa, M., 178
Rose, J. F., 325
Rots, A. H., 435, **479**
Royer, P., 373

Sabbey, C. N., **317**
Sabol, E., 67
Sanz, L., 156
Sauvage, M., 170
Schirmer, M., 283
Scholl, I., **86**
Scodeggio, M., **451**
Scollick, K., 107
Scott, S., 82
Seaman, R. L., **515**
Sengupta, R., **257**, 396
Shara, M., 499
Shaw, R. A., **349**
Shaya, E., **217**, 487
Shearer, A., 431
Shopbell, P. L., **115**
Shortridge, K., **343**

Sidher, S. D., 365
Siemiginowska, A., 483
Sivera, P., 199
Skillen, I., 152, 156
Škoda, P, **423**
Šlechta, M., 423
Slijkhuis, R., 283
Smirnov, O., **411**
Smith, H., 93
Smith, I. A., **93**
Smith, M., 174
Smith, R., 310
Solano, E., **152**, 156
Springer, P. L., 257
Stephen, J. B., **144**
Stevens, J. A., 299
Stobie, E. B., 241, 349
Stockman, H. S., 257, 396
Subramanian, S., **303**, 475
Sullivan, D. J., 306
Surdej, J., 97
Swade, D. A., 291, **295**
Swam, M. S., **291**, 295
Swings, J. P., 97
Swinyard, B. M., 365
Szalay, A. S., **3**, 40

Takata, T., 237
Talavera, A., 152, 156
Tan, M., 137
Tatematsu, K., 522
Terlouw, J. P., 358
Teuben, P. J., **499**
Thakar, A. R., **40**
Thernisien, A., 377
Thiébaut, C., 103, 111, **388**
Thomas, B., 217, **487**
Thompson, R., 174
Tilanus, R. P. J., 299
Tody, D., 261
Tresse, L., 455
Trifoglio, M., 245
Tsutsumi, T., 164
Tsvetkov, M., 125
Tuffs, R. J., 400
Turner, J., 459

Unger, S. J., 365
Uno, S., 70

Valdes, F. G., 459, **507**
Vallenari, A., 404
Vandame, B., 283
Vangeyte, B., 97
van Dessel, E., 97, 125
Vettolani, G., 451
Viard, É., 249
Vogelaar, M. G. R., **358**
Volpato, A., 160

Walsh, J. R., 447
Walton, N. A., 287
Wamsteker, W., 152, 156
Watanabe, M., **70**
Watanabe, N., 237
Waterson, C., 337
Wenger, M., 56, **213**
White, R. L., 59
Wicenec, A., 141, 283
Wilson, P., 67
Wise, M., 479, 518
Wolf, K. R., **48**
Wolk, S. J., 467
Wright, G. S., **137**, 314

Yagi, M., **237**
Yasuda, N., 70
Yoshida, M., 237
Young, W., 408
Yu, T., **495**

Zanichelli, A., 451
Zhang, A., 36, **229**
Zhao, J.-H., **164**
Zimmerman, L., **168**

Subject Index

ACS, 447
adaptive optics, 249
AIPS++, 279, 408, 411
algorithms, 269
    cosmic ray removal, 396
    deconvolution, 431, 529
    drizzling, 431
    field-matching, 388
    fitting, 483
    genetic, 388
    matrix, 518
    modeling, 483
    self-organizing-map (SOM), 78
analysis
    interactive, 170, 400
    software, 439, 459
    wavelet, 388
animation, 369
AOT P32, 400
applet, 321
applications software
    AIPS++, 279, 408, 411
    FTOOLS, 67
    GIPSY, 358
    IDL, 133, 249, 253, 369, 385, 404
    IRAF, 59, 261, 423, 491, 507, 511, 515
    NEMO, 499
    SPEFO, 423
    Starlab, 499
    Starlink, 129, 459
    STSDAS, 178
    XSPEC, 415
archives, 36, 52, 74, 107, 125, 141, 152, 156, 164, 178, 299, 303
    HST, 90
    interoperability, 148
    miscellaneous, 70
    multimission, 174
aspect, 471
association
    spectrum, 90
AST library, 129
astronomy
    archives, 40, 182
    data analysis, 170, 310
    databases, 125

education, 119
extragalactic, 329
gamma-ray, 144, 245
HST, 178
infrared, 36, 189, 369, 400
JCMT, 93
network, 63
optical, 249, 269, 404
outreach, 119
parallel computing, 503
radio, 82, 279, 408, 411, 495, 522
solar, 86, 377
source detection, 443
spectroscopy, 459
star trackers, 463
submillimeter, 164
X-ray, 13, 22, 245, 381, 467, 518
atlases, 70
attitude determination, 463, 471
AXAF, 471, 475

bibliography, 78
BIMA, 279, 495

C++, 467
calibration, 178, 245, 299, 365, 373
    HST, 427
    ionosphere, 411
    on-the-fly (OTFC), 291
    photometry, 511
catalogs, 70, 107, 160, 182, 511
    stellar, 133
CCD, 515
    pile-up, 381
CDS, 213
Chandra, 13, 22, 381, 435, 439, 467, 471, 475, 483, 518
Chandra X-ray Center (CXC), 22, 475
    Data Systems (CXCDS), 419, 467
charge shuffling, 515
CIAO, 310, 483
CL, 59
CLEAN, 529
client/server systems, 233
clustering, 78
computing

Subject Index 539

distributed, 189, 237, 279
    parallel, 209, 279, 408
coordinate conversion, 229
CORBA, 82, 160, 189, 213, 237, 325
    ILU, 189
    TAO, 189
cosmic ray removal, 396

data
    acquisition, 137, 515
    analysis, 67, 144, 170, 237, 303, 404, 483, 503, 518
    archives, 52, 74, 107, 125, 152, 156, 178, 299, 303
    event-driven, 358
    extraction, 321
    flow, 148, 164
    formats, 217
    mining, 36, 97, 148, 152, 156
    previewing, 90
    processing, 295, 377
    reduction, 137, 237, 314, 423, 451, 525
    structures, 129
    warehouse, 148
databases, 40, 56, 97, 241, 392
    distributed, 3, 86, 152, 156
    object-relational, 160
    spatial, 133
    window searches, 133
Datablade, 229
declarative knowledge, 221
deconvolution, 431, 529
digitisation, 125
distributed, 40
    computing, 189, 237, 279
    databases, 86
DOM, 221
drizzling, 431
dynamics, 499

education, 103, 107, 111, 115, 119
Electrical Ground Support Equipment (EGSE), 245
European Southern Observatory (ESO), 199, 283
    ESO Imaging Survey (EIS), 283
extragalactic astronomy, 329

Extreme Ultraviolet Explorer (*EUVE*), 174

Far Ultraviolet Spectroscopic Explorer (*FUSE*), 174
Fast Light Tool Kit (FLTK), 499
fault tolerance, 257
field-matching, 388
FITS data format, 225, 459, 487
    convention, 479
    function representation, 479
fitting, 483
flat field, 373
FOS, 178
FTOOLS, 67

genetic algorithm, 388
GIPSY, 358
GRAPE, 499
graphics, 59
gravitational lens, 97, 306
grid, 3, 279
Ground Support Equipment (GSE), 245
Guide Star Catalog (GSC)-II, 160

Hera, 67
holography, 93
HP-UX, 522
Huairou Solar Observing Station (HSOS), 392
Hubble Space Telescope (*HST*), 174, 291, 295, 431, 447

IDL, 133, 249, 253, 369, 385, 404
ILU, 189
image
    analysis, 321
    compression, 463
    display, 261, 358
    parameterized, 479
    processing, 74, 209, 269, 317, 377, 404, 435, 525
    sequences, 377
information
    retrieval, 56, 78
    services, 56
infrared
    astronomy, 36, 369, 400
    data reduction, 317, 365

Infrared Space Observatory (*ISO*), 170, 400
infrastructure, 475
integral field units, 459
interactive analysis, 170, 400
interferometry, 82, 249, 369, 404
International Ultraviolet Explorer (*IUE*), 152, 156, 174
Internet, 119
interoperability, 63, 213
ionosphere, 411
IRAF, 59, 261, 423, 507, 511, 515
    packages, 491
ISOCAM, 170
ISOPHOT, 400

Java, 44, 52, 82, 86, 90, 160, 221, 241, 321, 325, 369

kalman filtering, 471

LabVIEW, 333
languages
    C++, 467
    IDL, 133, 249, 253, 369, 385, 404
    Java, 44, 52, 82, 86, 90, 160, 221, 241, 321, 325, 369
    Perl, 241, 299, 314
    PHP, 392
    Python, 59, 241, 447
    S–Lang, 310
    scripting, 59, 310
    Tcl/Tk, 495
laser guide stars, 249
libraries
    AST, 129
    NDF, 129
Linux, 152, 156, 265, 287, 333, 522
liquid mirror, 97

mapping, 400
MAST, 174
matrix algorithm, 518
metadata, 221
minimal buy-in, 225
modeling, 483
monitoring, 467
multiresolution, 388
MySQL, 392

N-body, 499
NDF, 459
    library, 129
NEMO, 499
neural networks, 78
NICMOS, 427

objects, 182, 213
observing, 515
    coordinated, 44
    planning, 168
    preparation, 137
    remote, 115
    scheduling, 168
    tools, 137, 495
on-the-fly calibration (OTFC), 291
on-the-fly reprocessing (OTFR), 291, 295
Ondrejov 2m telescope, 423
OpenGL, 59, 499
operations, 22
optical astronomy, 404
OPUS, 291, 295, 325, 475
ORAC, 299, 314
outreach, 103, 111, 119

parallel computing, 40, 209, 279, 287, 408
Parametrized Ionospheric Model (PIM), 411
pattern recognition, 97
Perl, 241, 299, 314
photographic plates, 125
photometry, 306, 373, 511
    crowded fields, 431
PHP, 392
pipelines, 178, 237, 279, 287, 291, 295, 299, 314, 325, 329
    data reduction, 337
plotting, 90
plug-in, 261
previewing, 90
problem reporting, 199
processing, 475
production environments, 325
proposal tools, 507
PSF, 435
Python, 59, 241, 447

R-trees, 133, 160

Subject Index

radio astronomy, 279, 495, 522
real-time, 333
remote observing, 115
reprocessing
    on-the-fly (OTFR), 291

S–Lang, 310
scheduling, 44, 495
Schmidt, 182
scripting languages, 59, 310
SCUBA, 299
SEA, 48
self-organizing-map (SOM) algorithm, 78
simulations, 48, 249, 253, 329, 404
singular value decomposition, 491
SIRTF, 32, 329
sky surveys, 283
sociology, 269
software, 427, 467
    algorithm, 529
    architecture, 3, 32, 36, 225
    calibration, 511
    data analysis, 306
    development, 170, 253, 419
    embedded query, 229
    engineering, 199
    fault tolerance, 257
    history, 343, 349
    methodology, 487
    package, 365
    standards, 199, 209
    systems, 295
    testing, 168
solar astronomy, 86, 377
space missions
    Chandra, 13, 22, 381, 435, 439, 467, 471, 475, 483, 518
    EUVE, 174
    FUSE, 174
    HST, 174, 291, 447
    ISO, 400
    IUE, 152, 156, 174
    SIRTF, 32, 329
space science, 107
Space Telescope Science Institute (STScI), 291
spectroscopy, 314, 451, 491, 507
    computing, 455

spectrum archive, 423
SPEFO, 423
Spike, 168
standards, 487
Starlab, 499
Starlink, 129, 459
STSDAS, 178
Subaru, 237
Submillimeter Array (SMA), 164
submillimeter astronomy, 164
surface measurement, 93
surveys, 70, 74
    sky, 283

TAO, 189
Tcl/Tk, 495
techniques
    image processing, 144
    photometry, 306
telescopes
    automatic, 103, 111
    control, 423
    liquid mirror, 97
    Ondrejov 2m, 423
    SMA, 164
    Subaru, 237
    VLT, 199
testing, 199, 329
time delay integration (TDI), 97
tools, 369, 427

UKIRT, 137
user interfaces, 52, 107

Virtual Observatory, 3, 36, 40, 48, 74, 141, 148
VisiBroker, 189
visualization, 44, 52, 82, 385, 499, 503
VLT, 199

wavelet analysis, 388
WCS, 129
web-based tools, 67, 392
WFPC2, 431
wide-field imaging, 125, 408

X-ray astronomy, 13, 381, 467, 518
XDF, 217, 487
XImtool, 261

XML, 86, 217, 221, 241, 279, 487
XMM-Newton, 381
XSLT, 221
XSPEC, 415

Zope, 241

ASTRONOMICAL SOCIETY OF THE PACIFIC
CONFERENCE SERIES

and

INTERNATIONAL ASTRONOMICAL UNION
VOLUMES

Published
by

The Astronomical Society of the Pacific
(ASP)

ASP CONFERENCE SERIES VOLUMES
Published by the Astronomical Society of the Pacific

PUBLISHED: 1988 (* asterisk means OUT OF STOCK)

Vol. CS-1   PROGRESS AND OPPORTUNITIES IN SOUTHERN HEMISPHERE
              OPTICAL ASTRONOMY: CTIO 25TH Anniversary Symposium
              eds. V. M. Blanco and M. M. Phillips
              ISBN 0-937707-18-X

Vol. CS-2   PROCEEDINGS OF A WORKSHOP ON OPTICAL SURVEYS FOR QUASARS
              eds. Patrick S. Osmer, Alain C. Porter, Richard F. Green, and Craig B. Foltz
              ISBN 0-937707-19-8

Vol. CS-3   FIBER OPTICS IN ASTRONOMY
              ed. Samuel C. Barden
              ISBN 0-937707-20-1

Vol. CS-4   THE EXTRAGALACTIC DISTANCE SCALE:
              Proceedings of the ASP 100th Anniversary Symposium
              eds. Sidney van den Bergh and Christopher J. Pritchet
              ISBN 0-937707-21-X

Vol. CS-5   THE MINNESOTA LECTURES ON CLUSTERS OF GALAXIES
              AND LARGE-SCALE STRUCTURE
              ed. John M. Dickey
              ISBN 0-937707-22-8

PUBLISHED: 1989

Vol. CS-6   SYNTHESIS IMAGING IN RADIO ASTRONOMY: A Collection of Lectures
              from the Third NRAO Synthesis Imaging Summer School
              eds. Richard A. Perley, Frederic R. Schwab, and Alan H. Bridle
              ISBN 0-937707-23-6

PUBLISHED: 1990

Vol. CS-7   PROPERTIES OF HOT LUMINOUS STARS: Boulder-Munich Workshop
              ed. Catharine D. Garmany
              ISBN 0-937707-24-4

Vol. CS-8*   CCDs IN ASTRONOMY
              ed. George H. Jacoby
              ISBN 0-937707-25-2

Vol. CS-9   COOL STARS, STELLAR SYSTEMS, AND THE SUN: Sixth Cambridge Workshop
              ed. George Wallerstein
              ISBN 0-937707-27-9

Vol. CS-10*   EVOLUTION OF THE UNIVERSE OF GALAXIES:
              Edwin Hubble Centennial Symposium
              ed. Richard G. Kron
              ISBN 0-937707-28-7

Vol. CS-11   CONFRONTATION BETWEEN STELLAR PULSATION AND EVOLUTION
              eds. Carla Cacciari and Gisella Clementini
              ISBN 0-937707-30-9

Vol. CS-12   THE EVOLUTION OF THE INTERSTELLAR MEDIUM
              ed. Leo Blitz
              ISBN 0-937707-31-7

PUBLISHED: 1991

Vol. CS-13   THE FORMATION AND EVOLUTION OF STAR CLUSTERS
              ed. Kenneth Janes
              ISBN 0-937707-32-5

ASP CONFERENCE SERIES VOLUMES
Published by the Astronomical Society of the Pacific

PUBLISHED: 1991 (* asterisk means OUT OF STOCK)

Vol. CS-14 ASTROPHYSICS WITH INFRARED ARRAYS
ed. Richard Elston
ISBN 0-937707-33-3

Vol. CS-15 LARGE-SCALE STRUCTURES AND PECULIAR MOTIONS IN THE UNIVERSE
eds. David W. Latham and L. A. Nicolaci da Costa
ISBN 0-937707-34-1

Vol. CS-16 Proceedings of the 3rd Haystack Observatory Conference on ATOMS, IONS, AND MOLECULES: NEW RESULTS IN SPECTRAL LINE ASTROPHYSICS
eds. Aubrey D. Haschick and Paul T. P. Ho
ISBN 0-937707-35-X

Vol. CS-17 LIGHT POLLUTION, RADIO INTERFERENCE, AND SPACE DEBRIS
ed. David L. Crawford
ISBN 0-937707-36-8

Vol. CS-18 THE INTERPRETATION OF MODERN SYNTHESIS OBSERVATIONS OF SPIRAL GALAXIES
eds. Nebojsa Duric and Patrick C. Crane
ISBN 0-937707-37-6

Vol. CS-19 RADIO INTERFEROMETRY: THEORY, TECHNIQUES, AND APPLICATIONS, IAU Colloquium 131
eds. T. J. Cornwell and R. A. Perley
ISBN 0-937707-38-4

Vol. CS-20 FRONTIERS OF STELLAR EVOLUTION:
50th Anniversary McDonald Observatory (1939-1989)
ed. David L. Lambert
ISBN 0-937707-39-2

Vol. CS-21 THE SPACE DISTRIBUTION OF QUASARS
ed . David Crampton
ISBN 0-937707-40-6

PUBLISHED: 1992

Vol. CS-22 NONISOTROPIC AND VARIABLE OUTFLOWS FROM STARS
eds. Laurent Drissen, Claus Leitherer, and Antonella Nota
ISBN 0-937707-41-4

Vol CS-23 ASTRONOMICAL CCD OBSERVING AND REDUCTION TECHNIQUES
ed. Steve B. Howell
ISBN 0-937707-42-4

Vol. CS-24 COSMOLOGY AND LARGE-SCALE STRUCTURE IN THE UNIVERSE
ed. Reinaldo R. de Carvalho
ISBN 0-937707-43-0

Vol. CS-25 ASTRONOMICAL DATA ANALYSIS, SOFTWARE AND SYSTEMS I - (ADASS I)
eds. Diana M. Worrall, Chris Biemesderfer, and Jeannette Barnes
ISBN 0-937707-44-9

Vol. CS-26 COOL STARS, STELLAR SYSTEMS, AND THE SUN:
Seventh Cambridge Workshop
eds. Mark S. Giampapa and Jay A. Bookbinder
ISBN 0-937707-45-7

Vol. CS-27 THE SOLAR CYCLE: Proceedings of the
National Solar Observatory/Sacramento Peak 12th Summer Workshop
ed. Karen L. Harvey
ISBN 0-937707-46-5

ASP CONFERENCE SERIES VOLUMES
Published by the Astronomical Society of the Pacific

PUBLISHED: 1992 (asterisk means OUT OF STOCK)

Vol. CS-28　AUTOMATED TELESCOPES FOR PHOTOMETRY AND IMAGING
eds. Saul J. Adelman, Robert J. Dukes, Jr., and Carol J. Adelman
ISBN 0-937707-47-3

Vol. CS-29　Viña del Mar Workshop on CATACLYSMIC VARIABLE STARS
ed. Nikolaus Vogt
ISBN 0-937707-48-1

Vol. CS-30　VARIABLE STARS AND GALAXIES
ed. Brian Warner
ISBN 0-937707-49-X

Vol. CS-31　RELATIONSHIPS BETWEEN ACTIVE GALACTIC NUCLEI
AND STARBURST GALAXIES
ed. Alexei V. Filippenko
ISBN 0-937707-50-3

Vol. CS-32　COMPLEMENTARY APPROACHES TO DOUBLE
AND MULTIPLE STAR RESEARCH, IAU Colloquium 135
eds. Harold A. McAlister and William I. Hartkopf
ISBN 0-937707-51-1

Vol. CS-33　RESEARCH AMATEUR ASTRONOMY
ed. Stephen J. Edberg
ISBN 0-937707-52-X

Vol. CS-34　ROBOTIC TELESCOPES IN THE 1990's
ed. Alexei V. Filippenko
ISBN 0-937707-53-8

PUBLISHED: 1993

Vol. CS-35*　MASSIVE STARS: THEIR LIVES IN THE INTERSTELLAR MEDIUM
eds. Joseph P. Cassinelli and Edward B. Churchwell
ISBN 0-937707-54-6

Vol. CS-36　PLANETS AROUND PULSARS
ed. J. A. Phillips, S. E. Thorsett, and S. R. Kulkarni
ISBN 0-937707-55-4

Vol. CS-37　FIBER OPTICS IN ASTRONOMY II
ed. Peter M. Gray
ISBN 0-937707-56-2

Vol. CS-38　NEW FRONTIERS IN BINARY STAR RESEARCH: Pacific Rim Colloquium
eds. K. C. Leung and I.-S. Nha
ISBN 0-937707-57-0

Vol. CS-39　THE MINNESOTA LECTURES ON THE STRUCTURE
AND DYNAMICS OF THE MILKY WAY
ed. Roberta M. Humphreys
ISBN 0-937707-58-9

Vol. CS-40　INSIDE THE STARS, IAU Colloquium 137
eds. Werner W. Weiss and Annie Baglin
ISBN 0-937707-59-7

Vol. CS-41　ASTRONOMICAL INFRARED SPECTROSCOPY:
FUTURE OBSERVATIONAL DIRECTIONS
ed. Sun Kwok
ISBN 0-937707-60-0

ASP CONFERENCE SERIES VOLUMES
Published by the Astronomical Society of the Pacific

PUBLISHED: 1993 (* asterisk means OUT OF STOCK)

Vol. CS-42	GONG 1992: SEISMIC INVESTIGATION OF THE SUN AND STARS ed. Timothy M. Brown ISBN 0-937707-61-9
Vol. CS-43	SKY SURVEYS: PROTOSTARS TO PROTOGALAXIES ed. B. T. Soifer ISBN 0-937707-62-7
Vol. CS-44	PECULIAR VERSUS NORMAL PHENOMENA IN A-TYPE AND RELATED STARS, IAU Colloquium 138 eds. M. M. Dworetsky, F. Castelli, and R. Faraggiana ISBN 0-937707-63-5
Vol. CS-45	LUMINOUS HIGH-LATITUDE STARS ed. Dimitar D. Sasselov ISBN 0-937707-64-3
Vol. CS-46	THE MAGNETIC AND VELOCITY FIELDS OF SOLAR ACTIVE REGIONS, IAU Colloquium 141 eds. Harold Zirin, Guoxiang Ai, and Haimin Wang ISBN 0-937707-65-1
Vol. CS-47	THIRD DECENNIAL US-USSR CONFERENCE ON SETI -- Santa Cruz, California, USA ed. G. Seth Shostak ISBN 0-937707-66-X
Vol. CS-48	THE GLOBULAR CLUSTER-GALAXY CONNECTION eds. Graeme H. Smith and Jean P. Brodie ISBN 0-937707-67-8
Vol. CS-49	GALAXY EVOLUTION: THE MILKY WAY PERSPECTIVE ed. Steven R. Majewski ISBN 0-937707-68-6
Vol. CS-50	STRUCTURE AND DYNAMICS OF GLOBULAR CLUSTERS eds. S. G. Djorgovski and G. Meylan ISBN 0-937707-69-4
Vol. CS-51	OBSERVATIONAL COSMOLOGY eds. Guido Chincarini, Angela Iovino, Tommaso Maccacaro, and Dario Maccagni ISBN 0-937707-70-8
Vol. CS-52	ASTRONOMICAL DATA ANALYSIS SOFTWARE AND SYSTEMS II - (ADASS II) eds. R. J. Hanisch, R. J. V. Brissenden, and Jeannette Barnes ISBN 0-937707-71-6
Vol. CS-53	BLUE STRAGGLERS ed. Rex A. Saffer ISBN 0-937707-72-4

PUBLISHED: 1994

Vol. CS-54*	THE FIRST STROMLO SYMPOSIUM: THE PHYSICS OF ACTIVE GALAXIES eds. Geoffrey V. Bicknell, Michael A. Dopita, and Peter J. Quinn ISBN 0-937707-73-2
Vol. CS-55	OPTICAL ASTRONOMY FROM THE EARTH AND MOON eds. Diane M. Pyper and Ronald J. Angione ISBN 0-937707-74-0

ASP CONFERENCE SERIES VOLUMES
Published by the Astronomical Society of the Pacific

PUBLISHED: 1994 (* asterisk means OUT OF STOCK)

Vol. CS-56	INTERACTING BINARY STARS ed. Allen W. Shafter ISBN 0-937707-75-9
Vol. CS-57	STELLAR AND CIRCUMSTELLAR ASTROPHYSICS eds. George Wallerstein and Alberto Noriega-Crespo ISBN 0-937707-76-7
Vol. CS-58*	THE FIRST SYMPOSIUM ON THE INFRARED CIRRUS AND DIFFUSE INTERSTELLAR CLOUDS eds. Roc M. Cutri and William B. Latter ISBN 0-937707-77-5
Vol. CS-59	ASTRONOMY WITH MILLIMETER AND SUBMILLIMETER WAVE INTERFEROMETRY, IAU Colloquium 140 eds. M. Ishiguro and Wm. J. Welch ISBN 0-937707-78-3
Vol. CS-60	THE MK PROCESS AT 50 YEARS: A POWERFUL TOOL FOR ASTROPHYSICAL INSIGHT, A Workshop of the Vatican Observatory --Tucson, Arizona, USA eds. C. J. Corbally, R. O. Gray, and R. F. Garrison ISBN 0-937707-79-1
Vol. CS-61	ASTRONOMICAL DATA ANALYSIS SOFTWARE AND SYSTEMS III - (ADASS III) eds. Dennis R. Crabtree, R. J. Hanisch, and Jeannette Barnes ISBN 0-937707-80-5
Vol. CS-62	THE NATURE AND EVOLUTIONARY STATUS OF HERBIG Ae/Be STARS eds. Pik Sin Thé, Mario R. Pérez, and Ed P. J. van den Heuvel ISBN 0-9837707-81-3
Vol. CS-63	SEVENTY-FIVE YEARS OF HIRAYAMA ASTEROID FAMILIES: THE ROLE OF COLLISIONS IN THE SOLAR SYSTEM HISTORY eds. Yoshihide Kozai, Richard P. Binzel, and Tomohiro Hirayama ISBN 0-937707-82-1
Vol. CS-64*	COOL STARS, STELLAR SYSTEMS, AND THE SUN: Eighth Cambridge Workshop ed. Jean-Pierre Caillault ISBN 0-937707-83-X
Vol. CS-65*	CLOUDS, CORES, AND LOW MASS STARS: The Fourth Haystack Observatory Conference eds. Dan P. Clemens and Richard Barvainis ISBN 0-937707-84-8
Vol. CS-66*	PHYSICS OF THE GASEOUS AND STELLAR DISKS OF THE GALAXY ed. Ivan R. King ISBN 0-937707-85-6
Vol. CS-67	UNVEILING LARGE-SCALE STRUCTURES BEHIND THE MILKY WAY eds. C. Balkowski and R. C. Kraan-Korteweg ISBN 0-937707-86-4
Vol. CS-68*	SOLAR ACTIVE REGION EVOLUTION: COMPARING MODELS WITH OBSERVATIONS eds. K. S. Balasubramaniam and George W. Simon ISBN 0-937707-87-2
Vol. CS-69	REVERBERATION MAPPING OF THE BROAD-LINE REGION IN ACTIVE GALACTIC NUCLEI eds. P. M. Gondhalekar, K. Horne, and B. M. Peterson ISBN 0-937707-88-0

ASP CONFERENCE SERIES VOLUMES
Published by the Astronomical Society of the Pacific

PUBLISHED: 1995 (* asterisk means OUT OF STOCK)

Vol. CS-70*	GROUPS OF GALAXIES eds. Otto-G. Richter and Kirk Borne ISBN 0-937707-89-9
Vol. CS-71	TRIDIMENSIONAL OPTICAL SPECTROSCOPIC METHODS IN ASTROPHYSICS, IAU Colloquium 149 eds. Georges Comte and Michel Marcelin ISBN 0-937707-90-2
Vol. CS-72	MILLISECOND PULSARS: A DECADE OF SURPRISE eds. A. S Fruchter, M. Tavani, and D. C. Backer ISBN 0-937707-91-0
Vol. CS-73	AIRBORNE ASTRONOMY SYMPOSIUM ON THE GALACTIC ECOSYSTEM: FROM GAS TO STARS TO DUST eds. Michael R. Haas, Jacqueline A. Davidson, and Edwin F. Erickson ISBN 0-937707-92-9
Vol. CS-74	PROGRESS IN THE SEARCH FOR EXTRATERRESTRIAL LIFE: 1993 Bioastronomy Symposium ed. G. Seth Shostak ISBN 0-937707-93-7
Vol. CS-75	MULTI-FEED SYSTEMS FOR RADIO TELESCOPES eds. Darrel T. Emerson and John M. Payne ISBN 0-937707-94-5
Vol. CS-76	GONG '94: HELIO- AND ASTERO-SEISMOLOGY FROM THE EARTH AND SPACE eds. Roger K. Ulrich, Edward J. Rhodes, Jr., and Werner Däppen ISBN 0-937707-95-3
Vol. CS-77	ASTRONOMICAL DATA ANALYSIS SOFTWARE AND SYSTEMS IV - (ADASS IV) eds. R. A. Shaw, H. E. Payne, and J. J. E. Hayes ISBN 0-937707-96-1
Vol. CS-78	ASTROPHYSICAL APPLICATIONS OF POWERFUL NEW DATABASES: Joint Discussion No. 16 of the 22nd General Assembly of the IAU eds. S. J. Adelman and W. L. Wiese ISBN 0-937707-97-X
Vol. CS-79*	ROBOTIC TELESCOPES: CURRENT CAPABILITIES, PRESENT DEVELOPMENTS, AND FUTURE PROSPECTS FOR AUTOMATED ASTRONOMY eds. Gregory W. Henry and Joel A. Eaton ISBN 0-937707-98-8
Vol. CS-80*	THE PHYSICS OF THE INTERSTELLAR MEDIUM AND INTERGALACTIC MEDIUM eds. A. Ferrara, C. F. McKee, C. Heiles, and P. R. Shapiro ISBN 0-937707-99-6
Vol. CS-81	LABORATORY AND ASTRONOMICAL HIGH RESOLUTION SPECTRA eds. A. J. Sauval, R. Blomme, and N. Grevesse ISBN 1-886733-01-5
Vol. CS-82*	VERY LONG BASELINE INTERFEROMETRY AND THE VLBA eds. J. A. Zensus, P. J. Diamond, and P. J. Napier ISBN 1-886733-02-3
Vol. CS-83*	ASTROPHYSICAL APPLICATIONS OF STELLAR PULSATION, IAU Colloquium 155 eds. R. S. Stobie and P. A. Whitelock ISBN 1-886733-03-1

ASP CONFERENCE SERIES VOLUMES
Published by the Astronomical Society of the Pacific

PUBLISHED: 1995 (* asterisk means OUT OF STOCK)

ATLAS   INFRARED ATLAS OF THE ARCTURUS SPECTRUM, 0.9 - 5.3 μm
eds. Kenneth Hinkle, Lloyd Wallace, and William Livingston
ISBN: 1-886733-04-X

Vol. CS-84   THE FUTURE UTILIZATION OF SCHMIDT TELESCOPES, IAU Colloquium 148
eds. Jessica Chapman, Russell Cannon, Sandra Harrison, and Bambang Hidayat
ISBN 1-886733-05-8

Vol. CS-85*   CAPE WORKSHOP ON MAGNETIC CATACLYSMIC VARIABLES
eds. D. A. H. Buckley and B. Warner
ISBN 1-886733-06-6

Vol. CS-86   FRESH VIEWS OF ELLIPTICAL GALAXIES
eds. Alberto Buzzoni, Alvio Renzini, and Alfonso Serrano
ISBN 1-886733-07-4

PUBLISHED: 1996

Vol. CS-87   NEW OBSERVING MODES FOR THE NEXT CENTURY
eds. Todd Boroson, John Davies, and Ian Robson
ISBN 1-886733-08-2

Vol. CS-88*   CLUSTERS, LENSING, AND THE FUTURE OF THE UNIVERSE
eds. Virginia Trimble and Andreas Reisenegger
ISBN 1-886733-09-0

Vol. CS-89   ASTRONOMY EDUCATION: CURRENT DEVELOPMENTS, FUTURE COORDINATION
ed. John R. Percy
ISBN 1-886733-10-4

Vol. CS-90   THE ORIGINS, EVOLUTION, AND DESTINIES OF BINARY STARS IN CLUSTERS
eds. E. F. Milone and J. -C. Mermilliod
ISBN 1-886733-11-2

Vol. CS-91   BARRED GALAXIES, IAU Colloquium 157
eds. R. Buta, D. A. Crocker, and B. G. Elmegreen
ISBN 1-886733-12-0

Vol. CS-92*   FORMATION OF THE GALACTIC HALO .... INSIDE AND OUT
eds. Heather L. Morrison and Ata Sarajedini
ISBN 1-886733-13-9

Vol. CS-93   RADIO EMISSION FROM THE STARS AND THE SUN
eds. A. R. Taylor and J. M. Paredes
ISBN 1-886733-14-7

Vol. CS-94   MAPPING, MEASURING, AND MODELING THE UNIVERSE
eds. Peter Coles, Vicent J. Martinez, and Maria-Jesus Pons-Borderia
ISBN 1-886733-15-5

Vol. CS-95   SOLAR DRIVERS OF INTERPLANETARY AND TERRESTRIAL DISTURBANCES:
Proceedings of 16th International Workshop National Solar Observatory/Sacramento Peak
eds. K. S. Balasubramaniam, Stephen L. Keil, and Raymond N. Smartt
ISBN 1-886733-16-3

Vol. CS-96   HYDROGEN-DEFICIENT STARS
eds. C. S. Jeffery and U. Heber
ISBN 1-886733-17-1

ASP CONFERENCE SERIES VOLUMES
Published by the Astronomical Society of the Pacific

PUBLISHED: 1996 (* asterisk means OUT OF STOCK)

Vol. CS-97	POLARIMETRY OF THE INTERSTELLAR MEDIUM eds. W. G. Roberge and D. C. B. Whittet ISBN 1-886733-18-X
Vol. CS-98	FROM STARS TO GALAXIES: THE IMPACT OF STELLAR PHYSICS ON GALAXY EVOLUTION eds. Claus Leitherer, Uta Fritze-von Alvensleben, and John Huchra ISBN 1-886733-19-8
Vol. CS-99	COSMIC ABUNDANCES: Proceedings of the 6th Annual October Astrophysics Conference eds. Stephen S. Holt and George Sonneborn ISBN 1-886733-20-1
Vol. CS-100	ENERGY TRANSPORT IN RADIO GALAXIES AND QUASARS eds. P. E. Hardee, A. H. Bridle, and J. A. Zensus ISBN 1-886733-21-X
Vol. CS-101	ASTRONOMICAL DATA ANALYSIS SOFTWARE AND SYSTEMS V – (ADASS V) eds. George H. Jacoby and Jeannette Barnes ISBN 1080-7926
Vol. CS-102	THE GALACTIC CENTER, 4th ESO/CTIO Workshop ed. Roland Gredel ISBN 1-886733-22-8
Vol. CS-103	THE PHYSICS OF LINERS IN VIEW OF RECENT OBSERVATIONS eds. M. Eracleous, A. Koratkar, C. Leitherer, and L. Ho ISBN 1-886733-23-6
Vol. CS-104	PHYSICS, CHEMISTRY, AND DYNAMICS OF INTERPLANETARY DUST, IAU Colloquium 150 eds. Bo Å. S. Gustafson and Martha S. Hanner ISBN 1-886733-24-4
Vol. CS-105	PULSARS: PROBLEMS AND PROGRESS, IAU Colloquium 160 ed. S. Johnston, M. A. Walker, and M. Bailes ISBN 1-886733-25-2
Vol. CS-106	THE MINNESOTA LECTURES ON EXTRAGALACTIC NEUTRAL HYDROGEN ed. Evan D. Skillman ISBN 1-886733-26-0
Vol. CS-107	COMPLETING THE INVENTORY OF THE SOLAR SYSTEM: A Symposium held in conjunction with the 106th Annual Meeting of the ASP eds. Terrence W. Rettig and Joseph M. Hahn ISBN 1-886733-27-9
Vol. CS-108	M.A.S.S. -- MODEL ATMOSPHERES AND SPECTRUM SYNTHESIS: 5th Vienna - Workshop eds. Saul J. Adelman, Friedrich Kupka, and Werner W. Weiss ISBN 1-886733-28-7
Vol. CS-109	COOL STARS, STELLAR SYSTEMS, AND THE SUN: Ninth Cambridge Workshop eds. Roberto Pallavicini and Andrea K. Dupree ISBN 1-886733-29-5
Vol. CS-110	BLAZAR CONTINUUM VARIABILITY eds. H. R. Miller, J. R. Webb, and J. C. Noble ISBN 1-886733-30-9

ASP CONFERENCE SERIES VOLUMES
Published by the Astronomical Society of the Pacific

PUBLISHED: 1996 (* asterisk means OUT OF STOCK)

Vol. CS-111 MAGNETIC RECONNECTION IN THE SOLAR ATMOSPHERE:
Proceedings of a Yohkoh Conference
eds. R. D. Bentley and J. T. Mariska
ISBN 1-886733-31-7

Vol. CS-112 THE HISTORY OF THE MILKY WAY AND ITS SATELLITE SYSTEM
eds. Andreas Burkert, Dieter H. Hartmann, and Steven R. Majewski
ISBN 1-886733-32-5

PUBLISHED: 1997

Vol. CS-113 EMISSION LINES IN ACTIVE GALAXIES: NEW METHODS AND TECHNIQUES, IAU Colloquium 159
eds. B. M. Peterson, F.-Z. Cheng, and A. S. Wilson
ISBN 1-886733-33-3

Vol. CS-114 YOUNG GALAXIES AND QSO ABSORPTION-LINE SYSTEMS
eds. Sueli M. Viegas, Ruth Gruenwald, and Reinaldo R. de Carvalho
ISBN 1-886733-34-1

Vol. CS-115 GALACTIC CLUSTER COOLING FLOWS
ed. Noam Soker
ISBN 1-886733-35-X

Vol. CS-116 THE SECOND STROMLO SYMPOSIUM:
THE NATURE OF ELLIPTICAL GALAXIES
eds. M. Arnaboldi, G. S. Da Costa, and P. Saha
ISBN 1-886733-36-8

Vol. CS-117 DARK AND VISIBLE MATTER IN GALAXIES
eds. Massimo Persic and Paolo Salucci
ISBN-1-886733-37-6

Vol. CS-118 FIRST ADVANCES IN SOLAR PHYSICS EUROCONFERENCE:
ADVANCES IN THE PHYSICS OF SUNSPOTS
eds. B. Schmieder. J. C. del Toro Iniesta, and M. Vázquez
ISBN 1-886733-38-4

Vol. CS-119 PLANETS BEYOND THE SOLAR SYSTEM
AND THE NEXT GENERATION OF SPACE MISSIONS
ed. David R. Soderblom
ISBN 1-886733-39-2

Vol. CS-120 LUMINOUS BLUE VARIABLES: MASSIVE STARS IN TRANSITION
eds. Antonella Nota and Henny J. G. L. M. Lamers
ISBN 1-886733-40-6

Vol. CS-121 ACCRETION PHENOMENA AND RELATED OUTFLOWS, IAU Colloquium 163
eds. D. T. Wickramasinghe, G. V. Bicknell, and L. Ferrario
ISBN 1-886733-41-4

Vol. CS-122 FROM STARDUST TO PLANETESIMALS:
Symposium held as part of the 108th Annual Meeting of the ASP
eds. Yvonne J. Pendleton and A. G. G. M. Tielens
ISBN 1-886733-42-2

Vol. CS-123 THE 12th 'KINGSTON MEETING': COMPUTATIONAL ASTROPHYSICS
eds. David A. Clarke and Michael J. West
ISBN 1-886733-43-0

Vol. CS-124 DIFFUSE INFRARED RADIATION AND THE IRTS
eds. Haruyuki Okuda, Toshio Matsumoto, and Thomas Roellig
ISBN 1-886733-44-9

ASP CONFERENCE SERIES VOLUMES
Published by the Astronomical Society of the Pacific

PUBLISHED: 1997 (* asterisk means OUT OF STOCK)

Vol. CS-125   ASTRONOMICAL DATA ANALYSIS SOFTWARE AND SYSTEMS VI – (ADASS VI)
eds. Gareth Hunt and H. E. Payne
ISBN 1-886733-45-7

Vol. CS-126   FROM QUANTUM FLUCTUATIONS TO COSMOLOGICAL STRUCTURES
eds. David Valls-Gabaud, Martin A. Hendry, Paolo Molaro, and Khalil Chamcham
ISBN 1-886733-46-5

Vol. CS-127   PROPER MOTIONS AND GALACTIC ASTRONOMY
ed. Roberta M. Humphreys
ISBN 1-886733-47-3

Vol. CS-128   MASS EJECTION FROM AGN (Active Galactic Nuclei)
eds. N. Arav, I. Shlosman, and R. J. Weymann
ISBN 1-886733-48-1

Vol. CS-129   THE GEORGE GAMOW SYMPOSIUM
eds. E. Harper, W. C. Parke, and G. D. Anderson
ISBN 1-886733-49-X

Vol. CS-130   THE THIRD PACIFIC RIM CONFERENCE ON
RECENT DEVELOPMENT ON BINARY STAR RESEARCH
eds. Kam-Ching Leung
ISBN 1-886733-50-3

PUBLISHED: 1998

Vol. CS-131   BOULDER-MUNICH II: PROPERTIES OF HOT, LUMINOUS STARS
ed. Ian D. Howarth
ISBN 1-886733-51-1

Vol. CS-132   STAR FORMATION WITH THE INFRARED SPACE OBSERVATORY (ISO)
eds. João L. Yun and René Liseau
ISBN 1-886733-52-X

Vol. CS-133   SCIENCE WITH THE NGST (Next Generation Space Telescope)
eds. Eric P. Smith and Anuradha Koratkar
ISBN 1-886733-53-8

Vol. CS-134   BROWN DWARFS AND EXTRASOLAR PLANETS
eds. Rafael Rebolo, Eduardo L. Martin, and Maria Rosa Zapatero Osorio
ISBN 1-886733-54-6

Vol. CS-135   A HALF CENTURY OF STELLAR PULSATION INTERPRETATIONS:
A TRIBUTE TO ARTHUR N. COX
eds. P. A. Bradley and J. A. Guzik
ISBN 1-886733-55-4

Vol. CS-136   GALACTIC HALOS: A UC SANTA CRUZ WORKSHOP
ed. Dennis Zaritsky
ISBN 1-886733-56-2

Vol. CS-137   WILD STARS IN THE OLD WEST: PROCEEDINGS OF THE 13th NORTH
AMERICAN WORKSHOP ON CATACLYSMIC VARIABLES
AND RELATED OBJECTS
eds. S. Howell, E. Kuulkers, and C. Woodward
ISBN 1-886733-57-0

Vol. CS-138   1997 PACIFIC RIM CONFERENCE ON STELLAR ASTROPHYSICS
eds. Kwing Lam Chan, K. S. Cheng, and H. P. Singh
ISBN 1-886733-58-9

ASP CONFERENCE SERIES VOLUMES
Published by the Astronomical Society of the Pacific

PUBLISHED: 1998 (* asterisk means OUT OF STOCK)

Vol. CS-139   PRESERVING THE ASTRONOMICAL WINDOWS:
Proceedings of Joint Discussion No. 5 of the 23rd General Assembly of the IAU
eds. Syuzo Isobe and Tomohiro Hirayama
ISBN 1-886733-59-7

Vol. CS-140   SYNOPTIC SOLAR PHYSICS --18th NSO/Sacramento Peak Summer Workshop
eds. K. S. Balasubramaniam, J. W. Harvey, and D. M. Rabin
ISBN 1-886733-60-0

Vol. CS-141   ASTROPHYSICS FROM ANTARCTICA:
A Symposium held as a part of the 109th Annual Meeting of the ASP
eds. Giles Novak and Randall H. Landsberg
ISBN 1-886733-61-9

Vol. CS-142   THE STELLAR INITIAL MASS FUNCTION: 38th Herstmonceux Conference
eds. Gerry Gilmore and Debbie Howell
ISBN 1-886733-62-7

Vol. CS-143*   THE SCIENTIFIC IMPACT OF THE GODDARD HIGH RESOLUTION SPECTROGRAPH (GHRS)
eds. John C. Brandt, Thomas B. Ake III, and Carolyn Collins Petersen
ISBN 1-886733-63-5

Vol. CS-144   RADIO EMISSION FROM GALACTIC AND EXTRAGALACTIC COMPACT SOURCES, IAU Colloquium 164
eds. J. Anton Zensus, G. B. Taylor, and J. M. Wrobel
ISBN 1-886733-64-3

Vol. CS-145   ASTRONOMICAL DATA ANALYSIS SOFTWARE AND SYSTEMS VII – (ADASS VII)
eds. Rudolf Albrecht, Richard N. Hook, and Howard A. Bushouse
ISBN 1-886733-65-1

Vol. CS-146   THE YOUNG UNIVERSE GALAXY FORMATION AND EVOLUTION AT INTERMEDIATE AND HIGH REDSHIFT
eds. S. D'Odorico, A. Fontana, and E. Giallongo
ISBN 1-886733-66-X

Vol. CS-147   ABUNDANCE PROFILES: DIAGNOSTIC TOOLS FOR GALAXY HISTORY
eds. Daniel Friedli, Mike Edmunds, Carmelle Robert, and Laurent Drissen
ISBN 1-886733-67-8

Vol. CS-148   ORIGINS
eds. Charles E. Woodward, J. Michael Shull, and Harley A. Thronson, Jr.
ISBN 1-886733-68-6

Vol. CS-149   SOLAR SYSTEM FORMATION AND EVOLUTION
eds. D. Lazzaro, R. Vieira Martins, S. Ferraz-Mello, J. Fernández, and C. Beaugé
ISBN 1-886733-69-4

Vol. CS-150   NEW PERSPECTIVES ON SOLAR PROMINENCES, IAU Colloquium 167
eds. David Webb, David Rust, and Brigitte Schmieder
ISBN 1-886733-70-8

Vol. CS-151   COSMIC MICROWAVE BACKGROUND AND LARGE SCALE STRUCTURES OF THE UNIVERSE
eds. Yong-Ik Byun and Kin-Wang Ng
ISBN 1-886733-71-6

Vol. CS-152   FIBER OPTICS IN ASTRONOMY III
eds. S. Arribas, E. Mediavilla, and F. Watson
ISBN 1-886733-72-4

ASP CONFERENCE SERIES VOLUMES
Published by the Astronomical Society of the Pacific

PUBLISHED: 1998 (* asterisk means OUT OF STOCK)

Vol. CS-153 LIBRARY AND INFORMATION SERVICES IN ASTRONOMY III -- (LISA III)
eds. Uta Grothkopf, Heinz Andernach, Sarah Stevens-Rayburn,
and Monique Gomez
ISBN 1-886733-73-2

Vol. CS-154 COOL STARS, STELLAR SYSTEMS AND THE SUN: Tenth Cambridge Workshop
eds. Robert A. Donahue and Jay A. Bookbinder
ISBN 1-886733-74-0

Vol. CS-155 SECOND ADVANCES IN SOLAR PHYSICS EUROCONFERENCE:
THREE-DIMENSIONAL STRUCTURE OF SOLAR ACTIVE REGIONS
eds. Costas E. Alissandrakis and Brigitte Schmieder
ISBN 1-886733-75-9

PUBLISHED: 1999

Vol. CS-156 HIGHLY REDSHIFTED RADIO LINES
eds. C. L. Carilli, S. J. E. Radford, K. M. Menten, and G. I. Langston
ISBN 1-886733-76-7

Vol. CS-157 ANNAPOLIS WORKSHOP ON MAGNETIC CATACLYSMIC VARIABLES
eds. Coel Hellier and Koji Mukai
ISBN 1-886733-77-5

Vol. CS-158 SOLAR AND STELLAR ACTIVITY: SIMILARITIES AND DIFFERENCES
eds. C. J. Butler and J. G. Doyle
ISBN 1-886733-78-3

Vol. CS-159 BL LAC PHENOMENON
eds. Leo O. Takalo and Aimo Sillanpää
ISBN 1-886733-79-1

Vol. CS-160 ASTROPHYSICAL DISCS: An EC Summer School
eds. J. A. Sellwood and Jeremy Goodman
ISBN 1-886733-80-5

Vol. CS-161 HIGH ENERGY PROCESSES IN ACCRETING BLACK HOLES
eds. Juri Poutanen and Roland Svensson
ISBN 1-886733-81-3

Vol. CS-162 QUASARS AND COSMOLOGY
eds. Gary Ferland and Jack Baldwin
ISBN 1-886733-83-X

Vol. CS-163 STAR FORMATION IN EARLY-TYPE GALAXIES
eds. Jordi Cepa and Patricia Carral
ISBN 1-886733-84-8

Vol. CS-164 ULTRAVIOLET–OPTICAL SPACE ASTRONOMY BEYOND HST
eds. Jon A. Morse, J. Michael Shull, and Anne L. Kinney
ISBN 1-886733-85-6

Vol. CS-165 THE THIRD STROMLO SYMPOSIUM: THE GALACTIC HALO
eds. Brad K. Gibson, Tim S. Axelrod, and Mary E. Putman
ISBN 1-886733-86-4

Vol. CS-166 STROMLO WORKSHOP ON HIGH-VELOCITY CLOUDS
eds. Brad K. Gibson and Mary E. Putman
ISBN 1-886733-87-2

Vol. CS-167 HARMONIZING COSMIC DISTANCE SCALES IN A POST-HIPPARCOS ERA
eds. Daniel Egret and André Heck
ISBN 1-886733-88-0

ASP CONFERENCE SERIES VOLUMES
Published by the Astronomical Society of the Pacific

PUBLISHED: 1999 (* asterisk means OUT OF STOCK)

Vol. CS-168   NEW PERSPECTIVES ON THE INTERSTELLAR MEDIUM
eds. A. R. Taylor, T. L. Landecker, and G. Joncas
ISBN 1-886733-89-9

Vol. CS-169   11th EUROPEAN WORKSHOP ON WHITE DWARFS
eds. J.-E. Solheim and E. G. Meištas
ISBN 1-886733-91-0

Vol. CS-170   THE LOW SURFACE BRIGHTNESS UNIVERSE, IAU Colloquium 171
eds. J. I. Davies, C. Impey, and S. Phillipps
ISBN 1-886733-92-9

Vol. CS-171   LiBeB, COSMIC RAYS, AND RELATED X- AND GAMMA-RAYS
eds. Reuven Ramaty, Elisabeth Vangioni-Flam, Michel Cassé, and Keith Olive
ISBN 1-886733-93-7

Vol. CS-172   ASTRONOMICAL DATA ANALYSIS SOFTWARE AND SYSTEMS VIII – (ADASS VIII)
eds. David M. Mehringer, Raymond L. Plante, and Douglas A. Roberts
ISBN 1-886733-94-5

Vol. CS-173   THEORY AND TESTS OF CONVECTION IN STELLAR STRUCTURE:
First Granada Workshop
ed. Álvaro Giménez, Edward F. Guinan, and Benjamín Montesinos
ISBN 1-886733-95-3

Vol. CS-174   CATCHING THE PERFECT WAVE: ADAPTIVE OPTICS AND
INTERFEROMETRY IN THE 21st CENTURY,
A Symposium held as a part of the 110th Annual Meeting of the ASP
eds. Sergio R. Restaino, William Junor, and Nebojsa Duric
ISBN 1-886733-96-1

Vol. CS-175   STRUCTURE AND KINEMATICS OF QUASAR BROAD LINE REGIONS
eds. C. M. Gaskell, W. N. Brandt, M. Dietrich, D. Dultzin-Hacyan,
and M. Eracleous
ISBN 1-886733-97-X

Vol. CS-176   OBSERVATIONAL COSMOLOGY: THE DEVELOPMENT OF GALAXY SYSTEMS
eds. Giuliano Giuricin, Marino Mezzetti, and Paolo Salucci
ISBN 1-58381-000-5

Vol. CS-177   ASTROPHYSICS WITH INFRARED SURVEYS: A Prelude to SIRTF
eds. Michael D. Bicay, Chas A. Beichman, Roc M. Cutri, and Barry F. Madore
ISBN 1-58381-001-3

Vol. CS-178   STELLAR DYNAMOS: NONLINEARITY AND CHAOTIC FLOWS
eds. Manuel Núñez and Antonio Ferriz-Mas
ISBN 1-58381-002-1

Vol. CS-179   ETA CARINAE AT THE MILLENNIUM
eds. Jon A. Morse, Roberta M. Humphreys, and Augusto Damineli
ISBN 1-58381-003-X

Vol. CS-180   SYNTHESIS IMAGING IN RADIO ASTRONOMY II
eds. G. B. Taylor, C. L. Carilli, and R. A. Perley
ISBN 1-58381-005-6

Vol. CS-181   MICROWAVE FOREGROUNDS
eds. Angelica de Oliveira-Costa and Max Tegmark
ISBN 1-58381-006-4

Vol. CS-182   GALAXY DYNAMICS: A Rutgers Symposium
eds. David Merritt, J. A. Sellwood, and Monica Valluri
ISBN 1-58381-007-2

ASP CONFERENCE SERIES VOLUMES
Published by the Astronomical Society of the Pacific

PUBLISHED: 1999 (* asterisk means OUT OF STOCK)

Vol. CS-183   HIGH RESOLUTION SOLAR PHYSICS: THEORY, OBSERVATIONS, AND TECHNIQUES
eds. T. R. Rimmele, K. S. Balasubramaniam, and R. R. Radick
ISBN 1-58381-009-9

Vol. CS-184   THIRD ADVANCES IN SOLAR PHYSICS EUROCONFERENCE: MAGNETIC FIELDS AND OSCILLATIONS
eds. B. Schmieder, A. Hofmann, and J. Staude
ISBN 1-58381-010-2

Vol. CS-185   PRECISE STELLAR RADIAL VELOCITIES, IAU Colloquium 170
eds. J. B. Hearnshaw and C. D. Scarfe
ISBN 1-58381-011-0

Vol. CS-186   THE CENTRAL PARSECS OF THE GALAXY
eds. Heino Falcke, Angela Cotera, Wolfgang J. Duschl, Fulvio Melia, and Marcia J. Rieke
ISBN 1-58381-012-9

Vol. CS-187   THE EVOLUTION OF GALAXIES ON COSMOLOGICAL TIMESCALES
eds. J. E. Beckman and T. J. Mahoney
ISBN 1-58381-013-7

Vol. CS-188   OPTICAL AND INFRARED SPECTROSCOPY OF CIRCUMSTELLAR MATTER
eds. Eike W. Guenther, Bringfried Stecklum, and Sylvio Klose
ISBN 1-58381-014-5

Vol. CS-189   CCD PRECISION PHOTOMETRY WORKSHOP
eds. Eric R. Craine, Roy A. Tucker, and Jeannette Barnes
ISBN 1-58381-015-3

Vol. CS-190   GAMMA-RAY BURSTS: THE FIRST THREE MINUTES
eds. Juri Poutanen and Roland Svensson
ISBN 1-58381-016-1

Vol. CS-191   PHOTOMETRIC REDSHIFTS AND HIGH REDSHIFT GALAXIES
eds. Ray J. Weymann, Lisa J. Storrie-Lombardi, Marcin Sawicki, and Robert J. Brunner
ISBN 1-58381-017-X

Vol. CS-192   SPECTROPHOTOMETRIC DATING OF STARS AND GALAXIES
ed. I. Hubeny, S. R. Heap, and R. H. Cornett
ISBN 1-58381-018-8

Vol. CS-193   THE HY-REDSHIFT UNIVERSE:
GALAXY FORMATION AND EVOLUTION AT HIGH REDSHIFT
eds. Andrew J. Bunker and Wil J. M. van Breugel
ISBN 1-58381-019-6

Vol. CS-194   WORKING ON THE FRINGE:
OPTICAL AND IR INTERFEROMETRY FROM GROUND AND SPACE
eds. Stephen Unwin and Robert Stachnik
ISBN 1-58381-020-X

PUBLISHED: 2000

Vol. CS-195   IMAGING THE UNIVERSE IN THREE DIMENSIONS:
Astrophysics with Advanced Multi-Wavelength Imaging Devices
eds. W. van Breugel and J. Bland-Hawthorn
ISBN 1-58381-022-6

ASP CONFERENCE SERIES VOLUMES
Published by the Astronomical Society of the Pacific

PUBLISHED: 2000 (* asterisk means OUT OF STOCK)

Vol. CS-196	THERMAL EMISSION SPECTROSCOPY AND ANALYSIS OF DUST, DISKS, AND REGOLITHS eds. Michael L. Sitko, Ann L. Sprague, and David K. Lynch ISBN: 1-58381-023-4
Vol. CS-197	XVth IAP MEETING DYNAMICS OF GALAXIES: FROM THE EARLY UNIVERSE TO THE PRESENT eds. F. Combes, G. A. Mamon, and V. Charmandaris ISBN: 1-58381-24-2
Vol. CS-198	EUROCONFERENCE ON "STELLAR CLUSTERS AND ASSOCIATIONS: CONVECTION, ROTATION, AND DYNAMOS" eds. R. Pallavicini, G. Micela, and S. Sciortino ISBN: 1-58381-25-0
Vol. CS-199	ASYMMETRICAL PLANETARY NEBULAE II: FROM ORIGINS TO MICROSTRUCTURES eds. J. H. Kastner, N. Soker, and S. Rappaport ISBN: 1-58381-026-9
Vol. CS-200	CLUSTERING AT HIGH REDSHIFT eds. A. Mazure, O. Le Fèvre, and V. Le Brun ISBN: 1-58381-027-7
Vol. CS-201	COSMIC FLOWS 1999: TOWARDS AN UNDERSTANDING OF LARGE-SCALE STRUCTURES eds. Stéphane Courteau, Michael A. Strauss, and Jeffrey A. Willick ISBN: 1-58381-028-5
Vol. CS-202	PULSAR ASTRONOMY – 2000 AND BEYOND, IAU Colloquium 177 eds. M. Kramer, N. Wex, and R. Wielebinski ISBN: 1-58381-029-3
Vol. CS-203	THE IMPACT OF LARGE-SCALE SURVEYS ON PULSATING STAR RESEARCH, IAU Colloquium 176 eds. L. Szabados and D. W. Kurtz ISBN: 1-58381-030-7
Vol. CS-204	THERMAL AND IONIZATION ASPECTS OF FLOWS FROM HOT STARS: OBSERVATIONS AND THEORY eds. Henny J. G. L. M. Lamers and Arved Sapar ISBN: 1-58381-031-5
Vol. CS-205	THE LAST TOTAL SOLAR ECLIPSE OF THE MILLENNIUM IN TURKEY eds. W. C. Livingston and A. Özgüç ISBN: 1-58381-032-3
Vol. CS-206	HIGH ENERGY SOLAR PHYSICS – *ANTICIPATING HESSI* eds. Reuven Ramaty and Natalie Mandzhavidze ISBN: 1-58381-033-1
Vol. CS-207	NGST SCIENCE AND TECHNOLOGY EXPOSITION eds. Eric P. Smith and Knox S. Long ISBN: 1-58381-036-6
ATLAS	VISIBLE AND NEAR INFRARED ATLAS OF THE ARCTURUS SPECTRUM 3727-9300 Å eds. Kenneth Hinkle, Lloyd Wallace, Jeff Valenti, and Dianne Harmer ISBN: 1-58381-037-4
Vol. CS-208	POLAR MOTION: HISTORICAL AND SCIENTIFIC PROBLEMS, IAU Colloquium 178 eds. Steven Dick, Dennis McCarthy, and Brian Luzum ISBN: 1-58381-039-0

ASP CONFERENCE SERIES VOLUMES
Published by the Astronomical Society of the Pacific

PUBLISHED: 2000 (* asterisk means OUT OF STOCK)

Vol. CS-209   SMALL GALAXY GROUPS, IAU Colloquium 174
              eds. Mauri J. Valtonen and Chris Flynn
              ISBN: 1-58381-040-4

Vol. CS-210   DELTA SCUTI AND RELATED STARS: Reference Handbook
              and Proceedings of the 6th Vienna Workshop in Astrophysics
              eds. Michel Breger and Michael Houston Montgomery
              ISBN: 1-58381-043-9

Vol. CS-211   MASSIVE STELLAR CLUSTERS
              eds. Ariane Lançon and Christian M. Boily
              ISBN: 1-58381-042-0

Vol. CS-212   FROM GIANT PLANETS TO COOL STARS
              eds. Caitlin A. Griffith and Mark S. Marley
              ISBN: 1-58381-041-2

Vol. CS-213   BIOASTRONOMY '99: A NEW ERA IN BIOASTRONOMY
              eds. Guillermo A. Lemarchand and Karen J. Meech
              ISBN: 1-58381-044-7

Vol. CS-214   THE Be PHENOMENON IN EARLY-TYPE STARS, IAU Colloquium 175
              eds. Myron A. Smith, Huib F. Henrichs and Juan Fabregat
              ISBN: 1-58381-045-5

Vol. CS-215   COSMIC EVOLUTION AND GALAXY FORMATION:
              STRUCTURE, INTERACTIONS AND FEEDBACK
              The 3rd Guillermo Haro Astrophysics Conference
              eds. José Franco, Elena Terlevich, Omar López-Cruz, and Itziar Aretxaga
              ISBN: 1-58381-046-3

Vol. CS-216   ASTRONOMICAL DATA ANALYSIS SOFTWARE AND SYSTEMS IX -- (ADASS IX)
              eds. Nadine Manset, Christian Veillet, and Dennis Crabtree
              ISBN: 1-58381-047-1   ISSN: 1080-7926

Vol. CS-217   IMAGING AT RADIO THROUGH SUBMILLIMETER WAVELENGTHS
              eds. Jeffrey G. Mangum and Simon J. E. Radford
              ISBN: 1-58381-049-8

Vol. CS-218   MAPPING THE HIDDEN UNIVERSE: THE UNIVERSE BEHIND THE MILKYWAY
              THE UNIVERSE IN HI
              eds. Renée C. Kraan-Korteweg, Patricia A. Henning, and Heinz Andernach
              ISBN: 1-58381-050-1

Vol. CS-219   DISKS, PLANETESIMALS, AND PLANETS
              eds. F. Garzón, C. Eiroa, D. de Winter, and T. J. Mahoney
              ISBN: 1-58381-051-X

Vol. CS-220   AMATEUR - PROFESSIONAL PARTNERSHIPS IN ASTRONOMY:
              The 111th Annual Meeting of the ASP
              eds. John R. Percy and Joseph B. Wilson
              ISBN: 1-58381-052-8

Vol. CS-221   STARS, GAS AND DUST IN GALAXIES: EXPLORING THE LINKS
              eds. Danielle Alloin, Knut Olsen, and Gaspar Galaz
              ISBN: 1-58381-053-6

PUBLISHED: 2001

Vol. CS-222   THE PHYSICS OF GALAXY FORMATION
              eds. M. Umemura and H. Susa
              ISBN: 1-58381-054-4

ASP CONFERENCE SERIES VOLUMES
Published by the Astronomical Society of the Pacific

PUBLISHED: 2001 (* asterisk means OUT OF STOCK)

Vol. CS-223	COOL STARS, STELLAR SYSTEMS AND THE SUN: Eleventh Cambridge Workshop eds. Ramón J. García López, Rafael Rebolo, and María Zapatero Osorio ISBN: 1-58381-056-0
Vol. CS-224	PROBING THE PHYSICS OF ACTIVE GALACTIC NUCLEI BY MULTIWAVELENGTH MONITORING eds. Bradley M. Peterson, Ronald S. Polidan, and Richard W. Pogge ISBN: 1-58381-055-2
Vol. CS-225	VIRTUAL OBSERVATORIES OF THE FUTURE eds. Robert J. Brunner, S. George Djorgovski, and Alex S. Szalay ISBN: 1-58381-057-9
Vol. CS-226	12th EUROPEAN CONFERENCE ON WHITE DWARFS eds. J. L. Provencal, H. L. Shipman, J. MacDonald, and S. Goodchild ISBN: 1-58381-058-7
Vol. CS-227	BLAZAR DEMOGRAPHICS AND PHYSICS eds. Paolo Padovani and C. Megan Urry ISBN: 1-58381-059-5
Vol. CS-228	DYNAMICS OF STAR CLUSTERS AND THEY MILKY WAY eds. S. Deiters, B. Fuchs, A. Just, R. Spurzem, and R. Wielen ISBN: 1-58381-060-9
Vol. CS-229	EVOLUTION OF BINARY AND MULTIPLE STAR SYSTEMS A Meeting in Celebration of Peter Eggleton's 60th Birthday eds. Ph. Podsiadlowski, S. Rappaport, A. R. King, F. D'Antona, and L. Burderi IBSN: 1-58381-061-7
Vol. CS-230	GALAXY DISKS AND DISK GALAXIES eds. Jose G. Funes, S. J. and Enrico Maria Corsini ISBN: 1-58381-063-3
Vol. CS-231	TETONS 4: GALACTIC STRUCTURE, STARS, AND THE INTERSTELLAR MEDIUM eds. Charles E. Woodward, Michael D. Bicay, and J. Michael Shull ISBN: 1-58381-064-1
Vol. CS-232	THE NEW ERA OF WIDE FIELD ASTRONOMY eds. Roger Clowes, Andrew Adamson, and Gordon Bromage ISBN: 1-58381-065-X
Vol. CS-233	P CYGNI 2000: 400 YEARS OF PROGRESS eds. Mart de Groot and Christiaan Sterken ISBN: 1-58381-070-6
Vol. CS-234	X-RAY ASTRONOMY 2000 eds. R. Giacconi, S. Serio, and L. Stella ISBN: 1-58381-071-4
Vol. CS-235	SCIENCE WITH THE ATACAMA LARGE MILLIMETER ARRAY ed. Alwyn Wootten ISBN: 1-58381-072-2
Vol. CS-236	ADVANCED SOLAR POLARIMETRY –THEORY, OBSERVATION, AND INSTRUMENTATION, The 20th Sacramento Peak Summer Workshop ed. M. Sigwarth ISBN: 1-58381-073-0

ASP CONFERENCE SERIES VOLUMES
Published by the Astronomical Society of the Pacific

PUBLISHED: 2001 (* asterisk means OUT OF STOCK)

Vol. CS-237 GRAVITATIONAL LENSING: RECENT PROGRESS AND FUTURE GOALS
eds. Tereasa G. Brainerd and Christopher S. Kochanek
ISBN: 1-58381-074-9

Vol. CS-238 ASTRONOMICAL DATA ANALYSIS SOFTWARE AND SYSTEMS X
eds. F. R. Harnden, Jr., Francis A. Primini, and Harry E. Payne
ISBN: 1-58381-075-7

Vol. CS-239 MICROLENSING 2000: A NEW ERA OF MICROLENSING ASTROPHYSICS
ed. J. W. Menzies and P. D. Sackett
ISBN: 1-58381-076-5

All book orders or inquiries concerning ASP or IAU volumes listed should be directed to the:

The Astronomical Society of the Pacific Conference Series
390 Ashton Avenue
San Francisco   CA 94112-1722 USA

Phone: 415-337-2126
Fax: 415-337-5205
E-mail: catalog@aspsky.org
Web Site: http://www.aspsky.org

IAU VOLUMES ON NEXT PAGE

INTERNATIONAL ASTRONOMICAL UNION (IAU) VOLUMES
Published by the Astronomical Society of the Pacific

PUBLISHED: 1999

Vol. No. 190 NEW VIEWS OF THE MAGELLANIC CLOUDS
eds. You-Hua Chu, Nicholas B. Suntzeff, James E. Hesser,
and David A. Bohlender
ISBN: 1-58381-021-8

Vol. No. 191 ASYMPTOTIC GIANT BRANCH STARS
eds. T. Le Bertre, A. Lèbre, and C. Waelkens
ISBN: 1-886733-90-2

Vol. No. 192 THE STELLAR CONTENT OF LOCAL GROUP GALAXIES
eds. Patricia Whitelock and Russell Cannon
ISBN: 1-886733-82-1

Vol. No. 193 WOLF-RAYET PHENOMENA IN MASSIVE STARS AND STARBURST GALAXIES
eds. Karel A. van der Hucht, Gloria Koenigsberger, and Philippe R. J. Eenens
ISBN: 1-58381-004-8

Vol. No. 194 ACTIVE GALACTIC NUCLEI AND RELATED PHENOMENA
eds. Yervant Terzian, Daniel Weedman, and Edward Khachikian
ISBN: 1-58381-008-0

PUBLISHED: 2000

Vol. XXIVA TRANSACTIONS OF THE INTERNATIONAL ASTRONOMICAL UNION
REPORTS ON ASTRONOMY 1996-1999
ed. Johannes Andersen
ISBN: 1-58381-035-8

Vol. No. 195 HIGHLY ENERGETIC PHYSICAL PROCESSES AND MECHANISMS FOR
EMISSION FROM ASTROPHYSICAL PLASMAS
eds. P. C. H. Martens, S. Tsuruta, and M. A. Weber
ISBN: 1-58381-038-2

Vol. No. 197 ASTROCHEMISTRY: FROM MOLECULAR CLOUDS TO PLANETARY SYSTEMS
eds. Y. C. Minh and E. F. van Dishoeck
ISBN: 1-58381-034-X

Vol. No. 198 THE LIGHT ELEMENTS AND THEIR EVOLUTION
eds. L. da Silva, M. Spite, and J. R. de Medeiros
ISBN: 1-58381-048-X

PUBLISHED: 2001

IAU SPS ASTRONOMY FOR DEVELOPING COUNTRIES
Special Session of the XXIV General Assembly of the IAU
ed. Alan H. Batten
ISBN: 1-58381-067-6

Vol. No. 196 PRESERVING THE ASTRONOMICAL SKY
eds. R. J. Cohen and W. T. Sullivan, III
ISBN: 1-58381-078-1

Vol. No. 200 THE FORMATION OF BINARY STARS
eds. Hans Zinnecker and Robert D. Mathieu
ISBN: 1-58381-068-4

Vol. No. 203 RECENT INSIGHTS INTO THE PHYSICS OF THE SUN AND HELIOSPHERE:
HIGHLIGHTS FROM SOHO AND OTHER SPACE MISSIONS
eds. Pål Brekke, Bernhard Fleck, and Joseph B. Gurman
ISBN: 1-58381-069-2

INTERNATIONAL ASTRONOMICAL UNION (IAU) VOLUMES
Published by the Astronomical Society of the Pacific

PUBLISHED: 2001

Vol. No. 204   THE EXTRAGALACTIC INFRARED BACKGROUND AND ITS COSMOLOGICAL IMPLICATIONSe
eds. Martin Harwit and Michael G. Hauser
ISBN: 1-58381-062-5

Vol. No. 205   GALAXIES AND THEIR CONSTITUENTS AT THE HIGHEST ANGULAR RESOLUTIONS
eds. Richard T. Schilizzi, Stuart N. Vogel, Francesco Paresce, and Martin S. Elvis
ISBN: 1-58381-066-8

Complete lists of proceedings of past IAU Meetings are maintained at the
IAU Web site at the URL: http://www.iau.org/publicat.html

Volumes 32 - 189 in the IAU Symposia Series may be ordered from
Kluwer Academic Publishers
P. O. Box 117
NL 3300 AA Dordrecht
The Netherlands